ESO ASTROPHYSICS SYMPOSIA
European Southern Observatory

Series Editor: Bruno Leibundgut

B. Aschenbach   V. Burwitz   G. Hasinger
B. Leibundgut

Relativistic Astrophysics and Cosmology – Einstein's Legacy

Proceedings of the MPE/USM/MPA/ESO
Joint Astronomy Conference
Held in Munich, Germany, 7-11 November 2005

 Springer

Volume Editors

Bernd Aschenbach
Max-Planck-Institut für
extraterrestrische Physik
Giessenbachstraße
85748 Garching
Germany
Email: bra@mpe.mpg.de

Günther Hasinger
Max-Planck-Institut für
extraterrestrische Physik
Giessenbachstraße
85748 Garching
Germany
Email: ghasinger@mpe.mpg.de

Vadim Burwitz
Max-Planck-Institut für
extraterrestrische Physik
Giessenbachstraße
85748 Garching
Germany
Email: burwitz@mpe.mpg.de

Bruno Leibundgut
European Southern Observatory
Karl-Schwarzschild-Straße 2
85748 Garching
Germany
Email: bleibund@eso.org

Series Editor

Bruno Leibundgut
European Southern Observatory
Karl-Schwarzschild-Straße 2
85748 Garching
Germany

Library of Congress Control Number: 2007936612

ISBN   978-3-540-74712-3 Springer Berlin Heidelberg New York

Springer is a part of Springer Science+Business Media
springer.com
© Springer-Verlag Berlin Heidelberg 2007

Cover design: WMXDesign, Heidelberg
Typesetting: by the authors
Production: Integra Software Services Pvt. Ltd., Puducherry, India

Printed on acid-free paper   55/3180/Integra   5 4 3 2 1 0

Preface

The year 2005 featured the 100th anniversary of the 'annus mirabilis', the year in which Albert Einstein published three of his outstanding scientific papers. The Max-Planck Society with their institutes for extraterrestrische Physik (MPE) and Astrophysik (MPA) together with the Technische Universität München (TUM), the Ludwig-Maximilians-Universität München (LMU) and the European Southern Observatory (ESO) as well as the Deutsche Akademie der Naturforscher Leopoldina, the Bayerische Akademie der Wissenschaften and the Berlin-Brandenburgische Akademie der Wissenschaften considered this an excellent opportunity for celebrating Einstein and his work by an International Conference on Relativistic Astrophysics and Cosmology – Einstein's Legacy. The conference took place, on the premises of the TUM downtown Munich from November 7 to 11, 2005 with about 200 participants from 30 countries from all over the world. The scientific sessions were organised in 10 invited talks, 48 contributed talks and almost 90 posters on display in the Audimax of the TUM. The topics presented and discussed were focusing on cosmology, active galactic nuclei, clusters of galaxies, X-ray binaries and jets, gamma-ray bursts and gravity.

A scientific highlight was the well-attended public evening lecture on 'Der relativistische Kosmos – und was Astrophysiker aus Einsteins Ideen gemacht haben' held by Jürgen Ehlers in the Plenarsaal of the Bayerische Akademie der Wissenschaften on November 8. Another highlight, although more of the social kind, was the Conference Dinner at the MPE in Garching on November 10. Despite a brief shortage of beer, which broke out around 10 p.m., the spirits were high and still improved considerably after an emergency order of further beer, such that quite a number of attendees tried to delay the busses leave to the hotels in downtown Munich well after midnight.

From all of what has been brought to my attention I may conclude that the conference as seen by the participants was quite a success both scientifically and socially, and the weather was fine as well with lots of sunshine for the entire week. The conference organisation went smooth and efficient, thanks to the excellent support by the TUM administration, the teams of the Congress & Seminar Management (CSM) and of the Milde-Marketing company, the MPE personnel and, last but not least, because of the tireless engagement and commitment of all members of the Local Organizing Committee. I would like to express my sincere thanks to all who made this conference an enjoyable event.

Garching, March 2007   *Bernd Aschenbach*

Contents

Part III   Black Holes

List of Participants

Gopakumar Achamveedu
Friedrich-Schiller-Universität
Theoretisch-Physikalisches Institut
Max-Wien-Platz 1
D-07743 Jena, Germany
gopu@tpi.uni-jena.de

Lorenzo Amati
IANF, IASF BO
via P. Gobetti 101
I-40129 Bologna, Italy
amati@bo.iasf.cnr.it

Sepehr Arbabi
IPM Tehran, Physics
P.O.Box 19395-5531
Tehran, Iran
arbabi@ipm.ir

Patricia Arevalo
Max-Planck-Institut für extraterr. Physik
Giessenbachstrasse
D-85748 Garching, Germany
parevalo@mpe.mpg.de

Tigran Arshakian
Max-Planck-Institut für Radioastronomie
Auf dem Hügel 69
D-53121 Bonn, Germany
tigar@mpifr-bonn.mpg.de

Bernd Aschenbach
Max-Planck-Institut für extraterr. Physik
Giessenbachstrasse
D-85748 Garching, Germany
bra@mpe.mpg.de

Carsten Aulbert
Max Planck Institute for
Gravitational Physics
(Albert Einstein Institute)
Am Mühlenberg 1
D-14476 Potsdam, Germany
carsten.aulbert
@aei-potsdam.mpg.de

Marco Baldi
Max-Planck-Institut für Astrophysik
K. Schwarzschild-Str. 1
D-85748 Garching, Germany
mbaldi@mpa-garching.mpg.de

Italo Balestra
Max-Planck-Institut für
extraterrestrische Physik
Giessenbachstrasse
D-85748 Garching, Germany
balestra@mpe.mpg.de

Werner Becker
Max-Planck-Institut für extraterr. Physik
Giessenbachstrasse
D-85748 Garching, Germany
web@mpe.mpg.de

Tomaso Belloni
INAF
Osservatorio Astronomico di Brera
Via E. Bianchi 46
I-23807 Merate, Italy
belloni@merate.mi.astro.it

Ralf Bender
Max-Planck-Institut für extraterr. Physik
Giessenbachstrasse
D-85748 Garching, Germany
bender@mpe.mpg.de

Erika Benitez
IAUNAM DAEC
Circuito de la Investigacion
04510 Mexico DF, Mexico
erika@astroscu.unam.mx

Stefano Bianchi
ESAC
P.O.Box 50727
E-28080 Madrid, Spain
Stefano.Bianchi@sciops.esa.int

Carlo Luciano Bianco
University "La Sapienza"
and ICRAnet, Physics
Piazzale Aldo Moro 5
I-00185 Roma, Italy
maria.bernadini@icra.it

Roger Blandford
Kavli Institute for Particle
Astrophysics and Cosmology
Stanford, CA, USA
rdb@slac.stanford.edu

Hans Boehringer
Max-Planck-Institut für
extraterrestrische Physik
Giessenbachstrasse
D-85748 Garching, Germany
hxb@mpe.mpg.de

Thomas Boller
Max-Planck-Institut für
extraterrestrische Physik
Giessenbachstrasse
D-85748 Garching, Germany
bol@mpe.mpg.de

Valentin Bosch-Ramon
Universidad de Barcelona
Departamento d'Astron.i Meteorologia
Diagonal Av. 647 7th fl.
E-08028 Barcelona, Spain
vbosch@am.ub.es

William Brandt
Penn State University
Astronomy and Astrophysics
525 Davey Lab
University Park, Pennsyl. USA
niel@astro.psu.edu

Wolfgang Brinkmann
Max-Planck-Institut für
extraterrestrische Physik
Giessenbachstrasse
D-85748 Garching, Germany
wpb@mpe.mpg.de

Hermann Brunner
Max-Planck-Institut für
extraterrestrische Physik
Giessenbachstrasse
D-85748 Garching, Germany
hbrunner@mpe.mpg.de

Vadim Burwitz
Max-Planck-Institut für
extraterrestrische Physik
Giessenbachstrasse
D-85748 Garching, Germany
burwitz@mpe.mpg.de

Andrej Cadez
University in Ljubljana
FMF, Department of Physcis
Jadranska 19
1000 Ljubljana, Slovenia
andrej.cadez@fmf.uni-lj.si

Max Camenzind
University of Heidelberg, ZAH,
Königstuhl 12
D-69117 Heidelberg, Germany
m.camenzind@lsw.uni-
heidelberg.de

Nico Cappelluti
Max-Planck-Institut für extraterr. Physik
Giessenbachstrasse
D-85748 Garching, Germany
cappelluti@mpe.mpg.de

Carmelita Carbone
SISSA/ISAS
via Beirut 2
I-34014 Trieste, Italy
carbone@sissa.it

Andrea Cattaneo
Institut d'Astrophysique de Paris
98 bis Bd. Arago
F-75014 Paris, France
cattaneo@iap.fr

Alfonso Cavaliere
Dip. Fisica, Universitá Tor Vergata
Department of Physics
Via Ricerca Scientifica 1
I-00130 Roma, Italy
alfonso.cavaliere@roma2.infn.it

Indranil Chattopadhyay
Chungnam National University
Dept. of Astronomy & Space Science
Kung-Dong 220,Yusung-gu
305-764 Taejon, South Korea
indra@canopus.cnu.ac.kr

Vahram Chavushyan
INAOE
and IA UNAM
Luis Enrique Erro No.1
72840 Tonantzintla, Mexico
vahram@inaoep.mx

Eugene Churazov
Max-Planck-Institut für Astrophysik
Karl-Schwarzschild-Str. 1
D-85748 Garching, Germany
churazov@mpa-garching.mpg.de

Monica Colpi
Dipartimento di Fisica G. Occhialini
Università di Milano Bicocca
Piazza della Scienza 3
I-20126 Milano, Italy
colpi@mib.infn.it

Andrea Comastri
INAF-Osservatorio
Astronomico Bologna
via Ranzani 1
I-40127 Bologna, Italy
andrea.comastri@bo.astro.it

Robinson Cortés-Huerto
Universidad Nacional de Colombia
Physics Department
Ciudad
Universitaria
1 Bogotá D.C., Columbia
robicor@gmail.com

Jamie Crummy
Institute of Astronomy
Madingley Road
Cambridge CB3 0HA, UK
jc@ast.cam.ac.uk

Jorge Cuadra
Max-Planck-Institut für Astrophysik
Karl-Schwarzschild-Str. 1
D-85748 Garching, Germany
jcuadra@MPA-Garching.MPG.DE

Ujjal Debnath
Bengal Engineering & Science University
Dept. of Mathematics
Shibpur
711103 Howrah, India
ujjaldebnath@yahoo.com

Paul Dempsey
University College Dublin
Mathematical Sciences
Belfield
Dublin 4, Ireland
paul.dempsey@ucd.ie

Wilfried Domainko
University of Innsbruck
Institute of Astrophysics
Technikerstrasse 25
A-6020 Innsbruck, Austria
wilfried.domainko@uibk.ac.at

Anton Dorodnitsyn
Space Research Insitute
Space Plasma Physics
Profsoyuznaya st. 84/32
117997 Moscow, Russia
dora@mx.iki.rssi.ru

Massimo Dotti
Università degli Studi dell'Insubria
Fisica e Matematica
Via Valleggio 11
22100 Como, Italy
dotti@mib.infn.it

Robert Dunn
Institute of Astronomy
Madingley Road
Cambridge CB3 0HA, UK
rjhd2@ast.cam.ac.uk

Wolfgang Duschl
Institut für Theoretische Astrophysik
der Universität Heidelberg
Tiergartenstraße 15
69121 Heidelberg, Germany
wjd@ita.uni-heidelberg.de

Ioana Dutan
Max-Planck-Institut für Radioastronomie
Auf dem Hügel 69
53121 Bonn, Germany
idutan@mpifr-bonn.mpg.de

Sergey Dyadechkin
Physical Institute of St. Petersburg
Earth Physics
1, Uiyanouskaya
St. Petersburg, Russia
ego@geo.phys.spbu.ru

Andreas Eckart
I.Physikalisches Institut
Universität zu Köln
Zülpicher Str.77
D-50937 Köln, Germany
eckart@ph1.uni-koeln.de

Jürgen Ehlers
Max Planck Institute for Gravitational
Physics
(Albert Einstein Institute)
D-14476 Potsdam, Germany
juergen.ehlers@
aei-potsdam.mpg.de

Mary Clare Erlund
Institute of Astronomy
Madingley Road
Cambridge CB3 0HA, UK
mce@ast.cam.ac.uk

Oscar Esquivel
Max-Planck-Institut für Kernphysik
Tiergartenstr. 15
D-69117 Heidelberg, Germany
oscar.esquivel@mpi-hd.mpg.de

Andrew Fabian
Institute of Astronomy
Madingley Road, Cambridge
CB3 0HA, UK
acf@ast.cam.ac.uk

Heino Falcke
ASTRON
LOFAR
Oude Hoogeveensedijk 4
NL-7991 PD Dwingeloo
The Netherlands
falcke@astron.nl

Rene Fassbender
Max-Planck-Institut für
extraterrestrische Physik
Giessenbachstrasse
D-85748 Garching, Germany
rfassben@mpe.mpg.de

Georg Feulner
Universitäts-Sternwarte München
Scheinerstrasse 1
D-81679, München, Germany
feulner@usm.lmu.de

William Forman
Harvard-Smithsonian Center
for Astrophysics, HEA
60 Garden Street
Cambridge, MA 02138, USA
wrf@cfa.harvard.edu

Walmar Freitas Porto
Science on the Web, Consultant
Rua Manoel Medeiros Guedes, 9
58038-360 Joao Pessoa, Brazil
walmarporto@uol.com.br

Sándor Frey
FÖMI
Satellite Geodetic Observatory
P.O.Box 585
1592 Budapest, Hungary
frey@sgo.fomi.hu

Jorge Manuel García Añó
Grasmaierstr. 11
D-80805 München,Germany
jormaga@gmail.com

Merse Elod Gaspar
Eotvos University
Atomic Physics
Pazmany Peters s., 1/A
1592 Budapest, Hungary
merse@complex.elte.hu

Neil Gehrels
NASA-GSFC
Exploration of the Universe Division
8800 Greenbelt Rd.
20771 Greenbelt, USA
gehrels@milkyway.gsfc.nasa.gov

Reinhard Genzel
Max-Planck-Institut für
extraterrestrische Physik
Giessenbachstrasse
D-85748 Garching, Germany
genzel@mpe.mpg.de

Dimitrios Giannios
Max-Planck-Institut für Astrophysik
Karl-Schwarzschild-Str. 1
D-85748 Garching, Germany
giannios@mpa-garching.mpg.de

Andreja Gomboc
University Ljubljana
FMF, Department of Physcis
Jadranska 19
1000 Ljubljana, Slovenia
andreja.gomboc@fmf.uni-lj.si

James Graham
Institute of Astronomy
Madingley Road
Cambridge CB3 0HA, UK
jg307@ast.cam.ac.uk

Rainer Gruber
Max-Planck-Institut für
extraterrestrische Physik
Giessenbachstrasse
D-85748 Garching, Germany
gru@mpe.mpg.de

Matteo Guainazzi
ESAC
P.O.Box 50727
28080 Madrid, Spain
matteo.guainazzi@sciops.esa.int

Cristiano Guidorzi
Liverpool John Moores University
Astrophysics Research Institute
Twelve Quays House, Egerton
Birkenhead CH41 1LD, UK
crg@astro.livjm.ac.uk

Alvaro Hacar
UCM
Avda. Complutense s/n
28040 Madrid, Spain
alvaro.hacar@googlemail.com

Günther Hasinger
Max-Planck-Institut für
extraterrestrische Physik
Giessenbachstrasse
D-85748 Garching, Germany
ghasinger@mpe.mpg.de

Florian Hildebrand
Bayerischer Rundfunk
Redaktion Wissenschaft
Rundfunkplatz 1
D-80300 München, Germany
Florian.Hildebrand@t-online.de

Wolfgang Hillebrand
Max-Planck-Institut für Astrophysik
Karl-Schwarzschild-Str. 1
D-85748 Garching, Germany
wfh@mpa-garching.mpg.de

Hannes Horst
European Southern Observatory
Casilla 19001
Santiago 19, Chile
hhorst@eso.org

Hsiu-Hui Huang
Max-Planck-Institut für
extraterrestrische Physik
Giessenbachstrasse
D-85748 Garching, Germany
rhuang@mpe.mpg.de

Anatoli Iyudin
Moscow State University, SINP
Space Research Department
Vorobevy Gory
119992 Moscow, Russia
aiyudin@srd.sinp.msu.ru

Vijayasarathi Janardhanam
Munich University of Technology
Electrical Engineering
Enzianstr. 1, Zi 0103
D-85748 Garching b. München, Germany
vijayasarathi.j@mytum.de

Claus H. Jaroschek
Univertätssternwarte der LMU
Scheinerstr. 1
D-81679 München, Germany
cjarosch@usm.uni-muenchen.de

Magd Elias Kahil
The American University in Cairo
Dept. Mathematics
113 Kasr El Aini
11511 Cairo, Egypt
kahil@aucegypt.edu

Wolfgang Kapferer
University of Innsbruck
Institute of Astrophysics
Technikerstrasse 25
6020 Innsbruck, Austria
wolfgang.e.kapferer@uibk.ac.at

Shigeyuki Karino
SISSA/ISAS
via Beirut 2
I-34014 Trieste, Italy
karino@sissa.it

Sergey Karpov
Special Astrophysical Observatory
of the RAS
Relativistic Astrophysics Group
Bukovo, 369167 Nizhniy Arkhyz
Zelenchukskaya, Russia
karpov@sao.ru

Wolfgang Kausch
University of Innsbruck
Institute of Astrophysics
Technikerstrasse 25
6020 Innsbruck, Austria
wolfgang.kausch@uibk.ac.at

Irek Khamitov
TÜBITAK National Observatory of
Turkey
Akdeniz Universitesi Yerleskisi
07058 Antalya, Turkey
irekk@tug.tug.tubitak.gov.tr

John Kirk
Max-Planck-Institut für Kernphysik
Tiergartenstr. 15
D-69117 Heidelberg, Germany
john.kirk@mpi-hd.mpg.de

Kenta Kiuchi
Waseda University, Dept. of Physics
Okubo 3-4-1
169-0072 Shinjuku-ku, Japan
kiuchi@gravity.phys.
waseda.ac.jp

Rüdiger Kneissl
University of California
Department of Physics
366 Le Conte Hall
Berkeley 94720-7300, USA
rkneissl@berkeley.edu

Daniel Kobras
Universität Tübingen
Auf der Morgenstelle 10
72076 Tübingen, Germany
kobras@
tat.physik.uni-tuebingen.de

Bence Kocsis
Eotvos University
Atomic Physics
Pazmany P.s.1/a
1592 Budapest, Hungary
bkocsis@complex.elte.hu

Shinji Koide
Toyama University
Faculty of Engineering
3190 Gofuku
930-8555 Toyama, Japan

Stefanie Komossa
Max-Planck-Institut für
extraterrestrische Physik
Giessenbachstrasse
D-85748 Garching, Germany
skomossa@mpe.mpg.de

Uros Kostic
University Ljubljana
FMF, Department of Physcis
Jadranska 19
1000 Ljubljana, Slovenia
uros.kostic@fmf.uni-lj.si

Michael Kramer
Jodrell Bank Observatory
University of Manchester
Jodrell Bank
Macclesfield, UK
mkramer@jb.man.ac.uk

Robert Laing
European Southern Observatory
Karl-Schwarzschild-Str. 2
D-85748 Garching, Germany
rlaing@eso.org

Tomasz Lanczewski
Institute of Nuclear Physics PAS
Astrophysics
Radzikowskiego 152
31-342 Krakow, Poland
tomasz.lanczewski@ifj.edu.pl

Tatiana Larchenkova
Lebedev Physical Institute
Astro Space Center
Profsoyuznaya str. 84/32
117997 Moscow, Russia
tanya@lukash.asc.rssi.ru

Bruno Leibundgut
European Southern Observatory
Karl-Schwarzschild-Str. 2
D-85748 Garching, Germany
bleibund@eso.org

Andrei Lobanov
Max-Planck-Institut für Radioastronomie
Auf dem Hügel 69
D-53121 Bonn, Germany
alobanov@mpifr-bonn.mpg.de

Dirk Lorenzen
Wildgansstr. 32c
D-22145 Hamburg-Meiendorf
Germany

Vincenzo Mainieri
Max-Planck-Institut für
extraterrestrische Physik
Giessenbachstrasse
D-85748 Garching, Germany
vmainieri@mpe.mpg.de

Magdalena Mair
University of Innsbruck
Institute of Astrophysics
Technikerstrasse 25
6020 Innsbruck, Austria
magdalena.mair@uibk.ac.at

Subhabrata Majumdar
CITA
60 St. George St.
Toronto, ON M5S3H8, Canada
subha@cita.utoronto.ca

Karl Mannheim
Universität Würzburg
ITPA
Am Hubland
D-97074 Würzburg, Germany
mannheim@astro.uni-wuerzburg.de

Laura Maraschi
INAF-Osservatorio Astronomico di Brera
Via Brera 28
Milano, Italy
maraschi@brera.mi.astro.it

Herman Marshall
Center for Space Research, MIT
77 Mass. Ave.
Cambridge, MA 02139, USA
hermanm@space.mit.edu

Carlos Martins
Universidade do Porto
CFP
Rua do Campo Alegre, 687
4169-007 Porto, Portugal
cmartins@fc.up.pt

Sabino Matarrese
University of Padova
Physics Dept. "G. Galilei"
v. Marzolo 8
35131 Padova, Italy
matarrese@pd.iufn.it

Lucio Mayer
ETH Zürich
Institut für Astronomie
Hoengerrberg Campus
8057 Zürich, Switzerland
lucio@phys.ethz.ch

Klaus Meisenheimer
Max-Planck-Institut für Astronomie
Königstuhl 17
D-69117 Heidelberg, Germany
meise@mpia.de

Paulo Mendes
Max-Planck-Institut für
extraterrestrische Physik
Giessenbachstrasse
D-85748 Garching, Germany
p.mendes@mpe.mpg.de

Andrea Merloni
Max-Planck-Institut für Astrophysik
Karl-Schwarzschild-Str. 1
D-85748 Garching, Germany
am@mpa-garching.mpg.de

Hinrich Meyer
DESY
Notkestrasse 85
22607 Hamburg, Germany
Hinrich.Meyer@desy.de

Leonhard Meyer
Universitäet Köln
1. Phys. Institut
Zülpicher Str. 77
50937 Köln, Germany
leo@ph1.uni-koeln.de

Alain Milsztajn
CEA-Saclay
Dapnia-SPP
91191 Gif-Sur-Yvette, France

Felix Mirabel
European Southern Observatory
Casilla 19001
Santiago 19, Chile
fmirabel@eso.org

Yosuke Mizuno
National Space Science and Technology
Gamma-ray Astrophysic
320 Sparkman Drive, XD-12
Huntsville, AL35802, USA
Yosuke.Mizuno@msfc.nasa.gov

Shuntaro Mizuno
Waseda University
Department of Physics
3-4-1 Okubo Shinjuku
169-8050 Tokyo, Japan
shuntaro@gravity.phys.
wased.ac.jp

Akira Mizuta
Max-Planck-Institut für Astrophysik
Karl-Schwarzschild-Str. 1
D-85748 Garching, Germany
mizuta@mpa-garching.mpg.de

Rainer Moll
University of Innsbruck
Institute of Astrophysics
Technikerstrasse 25
6020 Innsbruck, Austria
rainer.moll@uibk.ac.at

James Moran
Harvard-Smithsonian
Center for Astrophysics
60 Garden Street
Cambridge MA 02138, USA
moran@cfa.harvard.edu

Masahiro Morikawa
Ochanomizu University
Department of Physics
2-1-1 Otsuka Bunkyo-ku
112 Tokyo, Japan
hiro@phys.ocha.ac.jp

Andreas Müller
Max-Planck-Institut für
extraterrestrische Physik
Giessenbachstrasse
D-85748 Garching, Germany
amueller@mpe.mpg.de

Akika Nakamichi
Gunma Astronomical Observatory
6860-86 Nakayama Takayama-n
377-0702 Gunma, Japan
akika@astron.pref.gunma.jp

Kenichiro Nakazato
Waseda University
Department of Physics
Okubo 3-4-1
169-0072 Shinjuku-ku, Japan
nakazato@heap.phys.waseda.ac.jp

Hagai Netzer
Tel Aviv University
School of Physics and Astronmy
University St.
69978 Tel Aviv, Israel
netzer@wise.tau.ac.il

Gernot Neugebauer
Friedrich-Schiller-Universität
Theoretisch-Physikalisches Institut
Max-Wien-Platz 1
D-07743 Jena, Germany
g.neugebauer@tpi.uni-jena.de

Ken-Ichi Nishikawa
National Space Science and Technology
Gamma-ray Astrophysic
320 Sparkman Drive, XD-12
Huntsville, AL35802, USA
ken-ichi.nishikawa@msfc.
nasa.gov

Hyerim Noh
Korea Astronomy & Space Science
61-1, Hwaam-dong,Yusung-gu
305-348 Taejon, South Korea
hr@kasi.re.kr

Nina Nowak
Max-Planck-Institut für
extraterrestrische Physik
Giessenbachstrasse
D-85748 Garching, Germany
nnowak@mpe.mpg.de

Lyman Page
Princeton University
Department of Physics
Jadwin Hall
Princeton, 08544, USA
page@princeton.edu

Jochen Peitz
Universität Tübingen
Theoretical Physics
Auf der Morgenstelle 10
72076 Tübingen, Germany
peitz@tat.physik.uni-
tuebingen.de

Manuel Perucho Pla
Max-Planck-Institut für Radioastronomie
Auf dem Hügel 69
D-53121 Bonn, Germany
perucho@mpifr-bonn.mpg.de

Jerome Petri
Max-Planck-Institut für Kernphysik
Saupfercheck 1
D-69117 Heidelberg, Germany
jerome.petri@mpi-hd.mpg.de

Sterl Phinney
Theoretical Astrophysics
California Institute of Technology
Pasadena, CA 91125, USA
esp@tapir.caltech.edu

Christopher Pilot
Maine Maritime Acadamy
Dismuke Hall
Castine, ME 04420
USA
cpilot@mma.edu

Paul Plucinsky
Harvard-Smithsonian
Center for Astrophysics
High Energy Astrophysics Division
60 Garden St., MS-70
Cambridge
MA 02138, USA
plucinsky@cfa.harvard.edu

Gabriele Ponti
University of Cambridge
Institute of Astronomy
Madingley Road
Cambridge CB3 0HA, Great Britain
ponti@ast.cam.ac.uk

Alexei Pozanenko
Space Research Institute
Gamma-ray spectroscopy
Profsoyuznaya
11997 Moscow, Russia
apozanen@iki.rssi.ru

Almudena Prieto
Max-Planck-Institut für Astronomie
Königstuhl 17
D-69117 Heidelberg, Germany
prieto@mpia.de

Georg Raffelt
Max-Planck-Institut für Physik
Föhringer Ring 6
D-80805 München, Germany
raffelt@mppmu.mpg.de

Paola Rebusco
Max-Planck-Institut für Astrophysik
Karl-Schwarzschild-Str. 1
D-85748 Garching, Germany
pao@mpa-garching.mpg.de

Thomas Reiprich
Max-Planck-Institut für Radioastronomie
Auf dem Hügel 71
D-53121 Bonn, Germany
thomas@reiprich.net

Frank M. Rieger
University College Dublin
Mathematical Sciences
Belfield
Dublin 4, Ireland
frank.rieger@ucd.ie

Hans-Walter Rix
Max-Planck-Institut für Astronomie
Königstuhl 17
D-69117 Heidelberg, Germany
rix@mpia.de

Gustavo E. Romero
Instituto Argentino de Radioastronomia
C.C.No.5
1894 Villa Elisa, Argentina
romero@fcaglp.unlp.edu.ar

Bronislaw Rudak
Nicolaus Copernicus Astronomical
Center
Rabianska 8
87100 Torun, Poland
bronek@ncac.torun.pl

Wilton Sanders
NASA-Headquarters
Science Mission Directorate
300 E Street, SW
20546 Greenbelt, USA
wsanders@hq.nasa.gov

Hidetomo Sawai
Waseda University
Department of Physics
Okubo 3-4-1
169-0072 Shinjuku-ku, Japan
hsawai@heap.phys.waseda.ac.jp

Sergey Sazonov
Max-Planck-Institut für Astrophysik
Karl-Schwarzschild-Str. 1
D-85748 Garching, Germany
sazonov@mpa-garching.mpg.de

Norbert Schartel
ESAC
XMM-Newton SOC
P.O.Box 50727
28080 Madrid, Spain
Norbert.Schartel@sciops.esa.int

Marc Schartmann
Max-Planck-Institut für Astronomie
Königstuhl 17
D-69117 Heidelberg, Germany
schartmann@mpia.de

Sabine Schindler
University of Innsbruck
Institute of Astrophysics
Technikerstrasse 25
A-6020 Innsbruck, Austria
sabine.schindler@uibk.ac.at

Maarten Schmidt
Astronomy
California Institute of Technology
Pasadena, CA 91125, USA
mxs@astro.caltech.edu

Peter Schneider
Max-Planck-Institut für Radioastronomie
Auf dem Hügel 71
D-53121 Bonn, Germany
peter@astro.uni-bonn.de

Bernard Schutz
Max Planck Institute for Gravitational
Physics
(Albert Einstein Institute)
D-14476 Potsdam, Germany
schutz@aei-potsdam.mpg.de

Stella Seitz
Universitätssternwarte
Scheinerstr. 1
D-81679 München, Germany
stella@usm.uni-muenchen.de

Alberto Sesana
Dipartimento di Fisica e Matematica
Università dell'Insubria
via Valleggio 11
I-22100 Como, Italy
Alberto.Sesana@mib.infn.it

Prajval Shastri
Indian Institute of Astrophysics
Sanjapur Road
560034 Bangalore, India
pshastri@iiap.res.in

Peter Shaver
European Southern Observatory
Karl-Schwarzschild-Str. 2
D-85748 Garching, Germany
pshaver@eso.org

Debora Sijacki
Max-Planck-Institut für Astrophysik
Karl-Schwarzschild-Str. 1
D-85748 Garching, Germany
deboras@mpa-garching.mpg.de

John Silverman
Max-Planck-Institut für
extraterrestrische Physik
Giessenbachstrasse
D-85748 Garching, Germany
jsilverman@mpe.mpg.de

Hajime Sotani
Waseda University
Department of Physics
55N-307B,Okubo 3-4-1
169-8555 Tokyo, Japan
sotani@gravity.phys.waseda.ac.jp

Henk Spruit
Max-Planck-Institut für Astrophysik
Karl-Schwarzschild-Str. 1
D-85748 Garching, Germany
henk@mpa-garching.mpg.de

Rashid Sunyaev
Max-Planck-Institut für Astrophysik
Karl-Schwarzschild-Str. 1
D-85748 Garching, Germany
sunyaev@mpa-garching.mpg.de

Yasuo Tanaka
Max-Planck-Institut für
extraterrestrische Physik
Giessenbachstrasse
D-85748 Garching, Germany
ytanaka@mpe.mpg.de

Takayuki Tatekawa
Ochanomizu University
Department of Physics
2-1-1 Otsuka Bunkyo-ku
112 Tokyo, Japan
tatekawa@cosmos.phys.ocha.ac.jp

Fabrizio Tavecchio
INAF, Osservatorio Astronomico di Brera
Via Brera
I-20121 Milan, Italy
tavecchio@merate.mi.astro.it

Joachim Truemper
Max-Planck-Institut für
extraterrestrische Physik
Giessenbachstrasse
D-85748 Garching, Germany
jtrumper@mpe.mpg.de

Alexander Unzicker
Pestalozzi-Gymnasium München
Eduard-Schmid-Str. 1
D-81541 München, Germany
alexander.unzicker@
lrz.uni-muenchen.de

Marta Volonteri
University of Cambridge
Institute of Astronomy, Madingley Road,
CB3 0HA Cambridge, UK
marta@ast.cam.ac.uk

Ulf von Rauchhaupt
Frankfurter Allgemeine Zeitung
Redaktion Sonntagszeitung
D-60267 Frankfurt am Main, Germany

Jian-Min Wang
Laboratory for High Energy Astrophysics
Institute of High Energy Physics
Chinese Academy of Sciences
Beijing 100039, P.R. China
wangjm@mail.ihep.ac.cn

Edward Wright
UCLA Astronomy
P.O.Box 951547
Los Angeles, CA 90095-15, USA
wright@astro.ucla.edu

Gerard Williger
University Louisville/JHU
Physics and Astronomy
102 Natural Sciences Bldg
Louisville, KY 40205, USA
williger@pha.jhu.edu

Fei Xiang
Max-Planck-Institut für Astrophysik
K. Schwarzschild-Str. 1
D-85748 Garching, Germany
arthawks@mpa-garching.mpg.de

Dawei Xu
National Astronomical Observatories
Chinese Academy of Sciences
Datun Road 20A, Chaoyang Dist.
Beijing 100012, P.R. China
dwxu@bao.ac.cn

Chengmin Zhang
National Astronomical Observatories
Chinese Academy of Sciences
Datun Road 20A, Chaoyang Dist.
Beijing 100012, P.R. China
zhangcm@bao.ac.cn

Shuang-Nan Zhang
Tsingua University
Physics Department
Beijing 100084, P.R. China
zhangcm@bao.ac.cn

Xiao-Ling Zhang
Max-Planck-Institut für
extraterrestrische Physik
Giessenbachstrasse
D-85748 Garching, Germany
zhangx@mpe.mpg.de

Relativistic Astrophysics and Cosmology
— Einstein's Legacy —

Scientific Advisory Comittee

R. Blandford, J. Ehlers, R. Genzel, G. Hasinger (chair), B. Leibundgut, G. Neugebauer, M. Rees, H.-W. Rix, P. Schneider, B. Schutz, R. Sunyaev, J. Trümper

Local Organizing Comittee

B. Aschenbach (chair), V. Burwitz, W. Frankenhuizen, M. Freyberg, I. Jacobs, M. Lindner (TUM)

Invited Speakers

Roger Blandford, Jürgen Ehlers, Neil Gehrels, Reinhard Genzel, Riccardo Giacconi, Piero Madau, Felix Mirabel, Hagai Netzer, Lyman Page, Sterl Phinney, Edward L. Wright

Part I

Cosmology

Structure Formation in a Variable Dark Energy Model and Observational Constraints

S. Arbabi-Bidgoli[1] and M.S. Movahed[1,2,3]

[1] Institute for Studies in Theoretical Physics and Mathematics Iran, P.O.Box 19395-5531, Tehran, Iran, arbabi@ipm.ir
[2] Department of Physics, Sharif University of Technology, P.O.Box 11365–9161, Tehran, Iran
[3] Iran Space Agency, P.O.Box 199799-4313, Tehran, Iran

Summary. The interpretation of a vast number of cosmological observations in the framework of FRW models suggests that the major part of the energy density of the universe is in form of dark energy with still unknown physical nature. In some models for dark energy, which are motivated by particle physics theory, the equation of state and the contribution of dark energy to the energy density of the universe can be variable. Here we study structure formation in a parameterized dark energy model, and compare its predictions with recent observational data, from the Supernova Ia gold sample and the parameters of large scale structure determined by the 2-degree Field Galaxy Redshift Survey (2dFGRS), and put some constraints on the free parameters of this model.

1 Introduction

Since the FRW models have been the framework of cosmology, the determination of the density parameter has always been an important task of cosmological observations. In the last 10-15 years cosmology has entered an era with an increasing amount of precise data, coming from different and independent observations. The interpretation of these data in the FRW framework, suggests that the density parameter is close to unity with good precision [5]. Also the contribution of the different constituents has been determined to be about 70% for dark energy, 26% for cold dark matter and 4% for baryonic matter. This result is mainly obtained from 3 different kinds of observations. First, there is the measurement of the anisotropies in the cosmic microwave background radiation, specially those from the WMAP experiment. The first peak of spectrum of the oscillations is sensitive to the total density of the universe and the measurements show with high precision that we live in a universe close to flat. The second type of observation are the light curves of distant supernovae Ia (SNIa), which have been discovered in the past 10 years at high redshifts. If we suppose that the samples of distant supernovae Ia are comparable to the close ones, we obtain the result that the universe has entered a phase of acceleration around redshift 0.5. This can be interpreted as the presence of a dark energy term, if we describe the expansion of the universe according to the Friedmann equation. The third type of observation is the amount of structure which has formed in the universe up to the present time and is often expressed by the abundance of galaxy clusters. This quantity is sensitive to the difference of matter and dark energy density. The combination of these three types of data suggests that we live in a flat ΛCDM universe dominated by dark energy Λ and cold dark matter CDM. In the now standard ΛCDM scenario there are two fundamental problems with the Λ term. The first problem arises if we try to interpret Λ as the vacuum energy. There is a huge discrepancy of more than 120 orders of magnitude with the expected scales of particle physics theories. The second problem arises if we assume a constant value for the dark energy density of the universe, as suggested by the simple cosmological constant term in Einstein's equation. It becomes necessary to fine tune the amount of dark energy with very high precision at early times to the present value, to obtain its domination at redshift of $z = 0.5$, because the energy densities of matter and radiation change with the power 3 and 4 of redshift. There is a class of models which try to tackle these two problems by assuming the dark energy density to be variable with redshift. For instance variable dark energy could be due to a decaying scalar field. In this work, we study variable dark energy without specifying a physical model. For this purpose we use a simple parametrization scheme given by [7], that expresses the properties of the model in three parameters. These are the present density of dark energy and its present state parameter, Λ and w_0, and a parameter b, which determines how much the dark energy equation of state of changes with redshift:

$$w(z; b, w_0) = \frac{w_0}{[1 + b\ln(1 + z)]^2}. \tag{1}$$

Parameter b is the most important parameter for our present study. Our aim is to study the formation of structure in this model and find observational constraints for its free parameters, particularly b. To study structure formation we use the linear perturbation theory. For obtaining the observational constraints, we use the data of distant supernova Ia of the gold sample [4], and the large scale structure data of the 2dF Galaxy Redshift Survey [2]. In this analysis we assume for the present state parameter $w_0 = -1$ and $\Omega_{tot} = \Omega_m + \Omega_\lambda = 1$ in order to be equal to the usual ΛCDM model at the present time. In a later paper we will relax this assumption and use current data to find also the best fitting value of w_0.

2 Structure Formation in the Variable Dark Energy Model

In the linear perturbation theory, the equation governing the growth of the density contrast $\delta = (\rho - \rho_0)/\rho_0$ is:

$$\ddot{\delta} + 2\frac{\dot{a}}{a}\dot{\delta} - 4\pi G\rho_0\delta = 0 \tag{2}$$

This non-relativistic equation is valid, if we study perturbations that are larger than the Jeans length, but smaller than the horizon scale. At shorter wavelengths, there is an additional term which would cause the perturbation to oscillate and for perturbations larger than the horizon it is necessary to use relativistic theory. We solve this equation taking into account the dynamics of the universe, by introducing Eq. 1 as the equation of state of dark energy into the Friedmann equation. Here we also assume that the sound horizon of dark energy is much larger than the wavelength of the perturbations, so we do not need to consider the clustering of dark energy. We only take into account the effects of changed dynamics of the universe caused by the presence of dark energy and its influence on structure formation. The growth of the density contrast as the numerical solution of Eq. 1 depending on parameter b is shown in Fig. 1, left panel.

As discussed in our preliminary paper [1], changing the equation of state of dark energy towards a more dust like behavior may result in an earlier domination of this component and there will be less growth of structure in such a universe. We use this effect to test predictions of this generic model of variable dark energy against current observations. For this purpose we use the growth index of structure f, which is defined as $f = d\ln(\delta)/d\ln(a)$. The result of our linear perturbation calculations is shown in Fig. 1, right panel. The behavior of the present value of the growth index for different b can be approximated by the following fitting formula:

$$f(\Omega_m, b, z = 0) \approx \Omega_m{}^{0.6} + \left(0.03 - 0.01b^{0.75}\right)\left(1 - \Omega_m\right)^{1.33-0.12b} - \frac{b^{0.16}}{100} \quad (3)$$

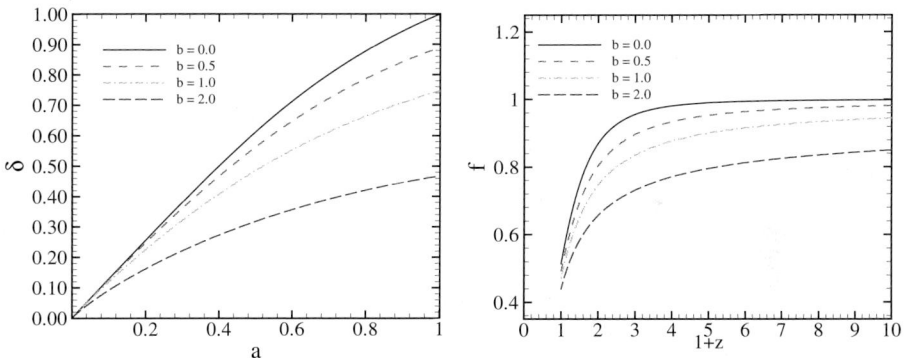

Fig. 1. Left Panel: Evolution of density contrast as a function of scale factor for different bending parameters in the flat universe with $\Omega_m = 0.3$, $\Omega_\lambda = 0.7$ and $w_0 = -1.0$. Right Panel: The growth index f at the present time as a function of the bending parameter b. We assume a flat universe with $\Omega_m = 0.3$ and $\Omega_\lambda = 0.7$.

3 Observational Constraints on the Dark Energy Model

To find the allowed regions for the free parameters of the model we make use of the method of likelihood analysis. To compare predictions of the model with the luminosity distance of the observed supernova Ia gold sample we define χ^2 as:

$$\chi^2 = \sum_{i=1}^{N} \frac{\left[\mu_{obs}(z_i) - \mu_{th}(z_i)\right]^2}{\sigma_i{}^2} \quad (4)$$

The difference between observations and predictions is minimized and the allowed region is determined under marginalization of the rest parameters. The results of this part are given in Tab. 1, which shows that the free parameters are not satisfactorily determined by this data alone. To make narrower limitations of the allowed regions and have a more precise estimate of the parameters we add the observational constraints obtained from the large scale structure data from the 2dF galaxy redshift survey, [2], containing about 220000 galaxy redshifts. In [3] a redshift distortion parameter β is given, which is defined as the relation of the growth index f to the bias factor B: $\beta := f/B$. β was reported at the characteristic redshift of the survey at $z = 0.15$: $\beta = 0.49 \pm 0.09$. The bias was measured by [6] as $B = 1.04 \pm 0.11$. Using these values, we obtain the growth index at this redshift, $f = 0.51 \pm 0.1$. Adding this comparison with the SNIa data, there is an additional term in the likelihood analysis above, $\chi^2 = \chi_{SNIa}^2 + \chi_{LSS}^2$, with $\chi_{LSS}^2 = (f_{obs} - f_{th})^2 / \sigma_f^2$.

As reported in Table 1 combining the observational constraints from the supernovae Ia and the large scale structure the allowed region of the parameters becomes more limited and also the estimated values becomes more reliable. In Fig. 2 the allowed regions for the free parameters of the model are shown using both observational constraints from the supernovae Ia and the large scale structure measurements of the 2dFGRS.

Table 1. The observational constraints for the free parameters of the model, (mean and 68.3% C.L. in brackets) for the density of cold dark matter, the parameter b, the Hubble constant and the age of the universe

Observation	CDM	b	H_0 (km/s/hMpc)	T_0 (Gyr)
SNIa	0.01 [0.0 - 0.15]	2.08 [1.1 - 2.48]	65.0 [64.3 - 65.7]	14.23
SNIa + LSS	0.02 [0.15 - 0.27]	0.08 [0.2 - 1.25]	64.5 [64.0 - 64.8]	14.55

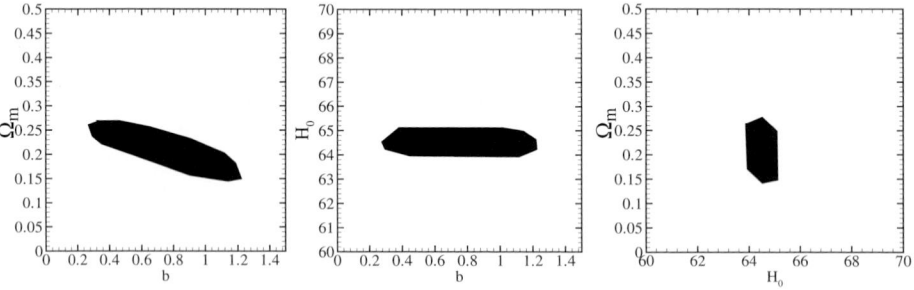

Fig. 2. 1-σ confidence contours in the plane of (b, Ω_m), left panel, (b, H_0), middle panel and (H_0, Ω_m), right panel.

4 Conclusions

In this paper, the formation of structure in the variable dark energy model of [7] has been analyzed and its predictions have been compared to observational data of supernovae Ia and the large scale structure. The calculation of the formation of structure in linear approximation shows that increasing the variability in the equation of state of dark energy reduces the growth of structure. Comparing the predictions of this model for the luminosity distance of remote supernovae Ia shows that this data do not put enough constraints on the free parameters of this model. Adding the data of galaxy redshift surveys reduces the allowed ranges and makes the results of the analysis more reliable. A more detailed analysis of the observational constraints on this model by relaxing some of the assumptions made here is under preparation by the authors.

References

1. Arbabi-Bidgoli, S., Movahed, M.S., Rahvar, S., astro-ph/0508323
2. Colless M.M. et al., astro-ph/0306581 (2003)
3. Demianski, M., Piedipalumbo, E., Rubano, C., Tortora, C., Astronomy and Astrophysics, **431**, 27 (2005)
4. Riess, A., G., et al. Astrophys. J. **607**, 665 (2004)
5. Spergel D. N. *et al.* [WMAP Collaboration], Astrophys. J. Suppl. **148** (2003) 175
6. Verde L., Kamionkowski M., Mohr J. J., Benson A.J., MNRAS, **321**, L7 (2001)
7. Wetterich, C., Physics Lett, B **594**, 17 (2004)

Effect of the Rotation of the Universe on the Energy Levels of Hydrogen Atoms

R. Cortés-Huerto[1] and J.M. Tejeiro[2]

[1] Physics Department, Universidad Nacional de Colombia `rcortesh@unal.edu.co`
[2] National Astronomical Observatory, Universidad Nacional de Colombia

Summary. We use two models in order to deal with a rotating universe. The first one is a thin rotating spherical shell. When we introduce in this shell an Hydrogen atom we found that the gravitomagnetic field of this universe can split the energy levels of the atom in a way analogous to the Zeeman effect.

The second model is the Gödel universe. There we use the solution of the Dirac equation on an arbitrary spacetime to find the shifts on energy levels of Hydrogen atoms caused by the rotation of the universe.

In both cases the interaction energy is very small, so we have to study the effect of cosmic rotation on Hydrogen atoms in a rotating expanding universe.

1 Shell Model and Spin-Spin Interaction

The spacetime outside a rotating object could be described by the Kerr metric [1]. In the weak field slow motion limit and when we solve the parallel transport equation for the spin, we obtain that the spin of a test particle processes with angular frequency given by

$$\boldsymbol{\Omega} = -\frac{3}{2}\boldsymbol{v} \times \boldsymbol{\nabla}\phi - \frac{1}{2}\boldsymbol{\nabla} \times \boldsymbol{h}\,, \tag{1}$$

where $\boldsymbol{v}$ is the velocity of the particle, ϕ is the Newtonian potential and $\boldsymbol{h}$ is a vectorial potential analogous to the magnetic dipolar potential. The second term on the right hand side of (1), called the Lense-Thirring term, takes the form

$$\boldsymbol{\Omega}_{LT} = \frac{2}{r^3}\left(3(\boldsymbol{J}\cdot\hat{r})\hat{r} - \boldsymbol{J}\right)\,, \tag{2}$$

where $\boldsymbol{J}$ is the angular momentum of the source.

The Kerr metric could describe the exterior spacetime of a thin rotating spherical shell [2]. In the weak field slow motion limit, a gyroscope in the interior of such shell processes with an angular frequency given by

$$\Omega_{IN} = \frac{4}{3}\frac{GM}{c^2 R}\boldsymbol{\omega}\,,\tag{3}$$

where M, R and $\boldsymbol{\omega}$ are the mass, radius and angular frequency of the shell. In this approximation, the gravitational fields of such a shell are completely analogous to the electromagnetic fields of a charged thin rotating spherical shell. This fact, among others, let us to identify the gravitational fields with the electromagnetic fields in a scheme called gravitomagnetism.

A hydrogen atoms in presence of an uniform magnetic field, shows a splitting in its fundamental energy levels of a magnitude that depends on the intensity of the applied magnetic field.

$$H_{int} = -\boldsymbol{\mu}\cdot\boldsymbol{B}\,,\tag{4}$$

where $\boldsymbol{\mu} = e\,\boldsymbol{s}/2mc$ is the magnetic moment of the electron, $\boldsymbol{s}$ is the orbital angular momentum and $\boldsymbol{B}$ is the external magnetic field.

In the gravitational scenario, the gravitomagnetic field interacts with the spin of an electron by [3]

$$H_{int} = \boldsymbol{\sigma}\cdot\boldsymbol{\Omega}_{IN}\,,\tag{5}$$

where $\boldsymbol{\sigma} = \boldsymbol{s}\hbar$.

If we assume that the universe could be modelled like a rotating shell, and that we live in its interior, a hydrogen atom must show a splitting of its fundamental energy levels produced by $\boldsymbol{\Omega}_{IN}$. In order to get the magnitude of the splitting, we must evaluate $\hbar\,\Omega_{IN}$. Considering that the current upper limit for the rate of rotation of the universe is about 10^{-24} s^{-1} [4], we obtain that the interaction energy between spins is about 1.5×10^{-39}eV.

2 Gödel Universe and Lamb Shift of Hydrogen

The metric of a stationary and spatially homogeneous Gödel universe has the form

$$ds^2 = -dt^2 - 2\sqrt{2}U(x)dt\,dy + dx^2 - U^2(x)dy^2 + dz^2\,,\tag{6}$$

where $U(x) = e^{\sqrt{2}\omega x}$.When a Fermi coordinate system comoving with the cosmological fluid is introduced, we obtain from (6) [4]

$$^{F}R_{0101} = {}^{F}R_{0202} = {}^{F}R_{1212} = \omega^2\,.\tag{7}$$

Where the upper index ^{F} refers to a Fermi coordinate system. With these components of the Riemann tensor we obtain the components of Ricci tensor in this coordinates, particularly

$$^{F}R_{00} \approx 2\omega^2 \, . \tag{8}$$

In 1980 Parker found, by solving the Dirac equation for an arbitrary space time, shifts produced by curvature on the lowest energy levels of Hydrogen like atoms [5].

In terms of R and R_{00} we have for such levels

$$E^{(1)}(1S_{1/2}) = \frac{1}{2}\zeta^{-2}m^{-1}R_{00} + \frac{1}{8}m^{-1}(R - 3R_{00}) + \cdots , \tag{9}$$

where the degeneracy of this level is not removed to first order in the curvature.

The energy of the $2S_{1/2}$ states is separated from that of the $2P_{1/2}$ states (Lamb shift) by

$$E^{(1)}(2S_{1/2}) = 7\zeta^{-2}m^{-1}R_{00} \, . \tag{10}$$

$$E^{(1)}(2P_{1/2}) = 5\zeta^{-2}m^{-1}R_{00} \, . \tag{11}$$

Where $\zeta = Ze$ and m is the reduced mass of the system.

Considering the current upper limit for the rate of rotation of the universe $10^{-24}\mathrm{s}^{-1}$, we have that the energy gap between $E^{(1)}(2S_{1/2})$ and $E^{(1)}(2P_{1/2})$ levels is about $10^{-58}\mathrm{eV}$.

3 Discussion

The rotating shell model for the universe, in spite of its lack of realism, give us a nice view of the Zeeman effect produced by the gravitomagnetic field of a rotating source on Hydrogen atoms.

Although for both models, shell and Gödel universe, the magnitude orders for splitting and shifts respectively are very small, we must consider a rotating expanding universe because in such a model the rate of rotation could be greater for early cosmological epochs. Particularly the window $(30 \leq z \leq 70)$ is important from the point of view of 21 cm tomography. This point will be treated in a future paper.

References

1. R. P. Kerr: Phys. Rev. Lett. **11** (1963)
2. V. De La Cruz, W. Israel: Phys. Rev. **170** (1968)
3. B. Mashhoon: Class. Quant. Grav. 17, 2399 (2000)
4. I. Ciufolini et al: Phys. Lett. A 308, 101 (2003)
5. L. Parker: Phys. Rev. Lett. **44**, 23 (1980)

Modified Chaplygin Gas and Accelerated Universe

U. Debnath

Department of Mathematics, Bengal Engineering and Science University, Shibpur, Howrah-711 103, India, ujjaldebnath@yahoo.com

Summary. In this paper we have considered a model of modified Chaplygin gas and its role in accelerating phase of the universe. We have assumed that the equation of state of this modified model is valid from the radiation era to ΛCDM model. We have used recently developed statefinder parameters in characterizing different phase of the universe diagrammatically.

1 Introduction

Recent observations of the luminosity of type Ia supernovae indicate [1, 2] an accelerated expansion of the universe and lead to the search for a new type of matter which violates the strong energy condition $\rho + 3p < 0$. The matter consent responsible for such a condition to be satisfied at a certain stage of evolution of the universe is referred to as *dark energy*. There are different candidates to play the role of the dark energy. The type of dark energy represented by a scalar field is often called *quintessence*. The transition from a universe filled with matter to an exponentially expanding universe does not necessarily require the presence of the scalar field as the only alternative. In particular one can try another alternative by using an exotic type of fluid - the so-called Chaplygin gas which obeys an equation of state like $p = -B/\rho$ [3] where p and ρ are respectively the pressure and energy density and B is a positive constant. Subsequently the above equation was modified to the form $p = -B/\rho^{\alpha}$ with $0 \leq \alpha \leq 1$. This model gives the cosmological evolution from an initial dust like matter to an asymptotic cosmological constant with an epoch that can be seen as a mixture of a cosmological constant and a fluid obeying an equation of state $p = \alpha\rho$. This generalized model has been studied previously [4, 5].

In the paper [4], the flat Friedmann model filled with Chaplygin fluid has been analyzed in terms of the recently proposed "*statefinder*" parameters [6]. In fact trajectories in the $\{s, r\}$ plane corresponding to different cosmological models demonstrate qualitatively different behaviour. The statefinder diagnostic along with future SNAP observations may perhaps be used to discriminate between different dark energy models. The above statefinder diagnostic pair are constructed from the scale factor $a(t)$ and its derivatives upto the third order as follows:

$$r = \frac{\dddot{a}}{aH^3} \quad \text{and} \quad s = \frac{r-1}{3\left(q - \frac{1}{2}\right)} \tag{1}$$

where H and q $\left(= -\frac{a\ddot{a}}{\dot{a}^2}\right)$ are the Hubble parameter and the deceleration parameter respectively. These parameters are dimensionless and allow us to characterize the properties of dark energy in a model independent manner. The parameter r forms the next step in the hierarchy of geometrical cosmological parameters after H and q.

In section 2, all the discussions are valid in general for $k = 0, \pm 1$, but in section 3 we have specifically considered the simple case of a spatially flat universe ($k = 0$), which naturally follows from the simplest version of the inflationary scenario and is confirmed by recent CMB experiments [7]. In our present work we consider a more general modified Chaplygin gas obeying an equation of state [8]

$$p = A\rho - \frac{B}{\rho^\alpha} \quad \text{with} \quad 0 \leq \alpha \leq 1 \tag{2}$$

This equation of state shows radiation era (when $A = 1/3$) at one extreme (when the scale factor $a(t)$ is vanishingly small) while a ΛCDM model at the other extreme (when the scale factor $a(t)$ is infinitely large). At all stages it shows a mixture. Also in between there is also one stage when the pressure vanishes and the matter content is equivalent to a pure dust.

2 Modified Chaplygin Gas in FRW Model

The metric of a homogeneous and isotropic universe in FRW model is

$$ds^2 = dt^2 - a^2(t)\left[\frac{dr^2}{1 - kr^2} + r^2(d\theta^2 + sin^2\theta d\phi^2)\right] \tag{3}$$

where $a(t)$ is the scale factor and k $(= 0, \pm 1)$ is the curvature scalar.

The Einstein field equations are

$$\frac{\dot{a}^2}{a^2} + \frac{k}{a^2} = \frac{1}{3}\rho \tag{4}$$

and

$$\frac{\ddot{a}}{a} = -\frac{1}{6}(\rho + 3p) \tag{5}$$

where ρ and p are energy density and isotropic pressure respectively (choosing $8\pi G = c = 1$).

The energy conservation equation is

$$\dot{\rho} + 3\frac{\dot{a}}{a}(\rho + p) = 0 \tag{6}$$

Using equation (2) we have the solution of ρ as

$$\rho = \left[\frac{B}{1+A} + \frac{C}{a^{3(1+A)(1+\alpha)}} \right]^{\frac{1}{1+\alpha}} \tag{7}$$

where C is an arbitrary integration constant.

Now for small value of scale factor $a(t)$, we have

$$\rho \simeq \frac{C^{\frac{1}{1+\alpha}}}{a^{3(1+A)}} \,, \tag{8}$$

which is very large and corresponds to the universe dominated by an equation of state $p = A\rho$.

Also for large value of scale factor $a(t)$,

$$\rho \simeq \left(\frac{B}{1+A} \right)^{\frac{1}{1+\alpha}} \quad \text{and} \quad p \simeq - \left(\frac{B}{1+A} \right)^{\frac{1}{1+\alpha}} = -\rho \tag{9}$$

which correspond to an otherwise empty universe with a cosmological constant $\left(\frac{B}{1+A} \right)^{\frac{1}{1+\alpha}}$.

For accelerating universe q must be negative i.e., $\ddot{a} > 0$ i.e.,

$$a^{3(1+\alpha)(1+A)} > \frac{C(1+A)(1+3A)}{2B} \tag{10}$$

This expression shows that for small value of scale factor we have a decelerating universe while for large values of scale factor we have an accelerating universe and the transition occurs when the scale factor has the expression $a = \left[\frac{C(1+A)(1+3A)}{2B} \right]^{\frac{1}{3(1+\alpha)(1+A)}}$.

3 The Role of StateFinder Parameters in FRW Universe

The statefinder parameters have been defined in equation (1). Trajectories in the $\{r, s\}$ plane corresponding to different cosmological models, for example ΛCDM model diagrams correspond to the fixed point $s = 0$, $r = 1$.

For Friedmann model with flat universe (i.e., $k = 0$),

$$H^2 = \frac{\dot{a}^2}{a^2} = \frac{1}{3}\rho \quad \text{and} \quad q = -\frac{\ddot{a}}{aH^2} = \frac{1}{2} + \frac{3}{2}\frac{p}{\rho} \tag{11}$$

So from equation (1) we get

$$r = 1 + \frac{9}{2}\left(1 + \frac{p}{\rho}\right)\frac{\partial p}{\partial \rho}, \quad s = \left(1 + \frac{\rho}{p}\right)\frac{\partial p}{\partial \rho} \tag{12}$$

Thus we get the ratio between p and ρ:

$$\frac{p}{\rho} = \frac{2(r-1)}{9s} \tag{13}$$

For modified Chaplygin gas, velocity of sound can be written as

$$v_s^2 = \frac{\partial p}{\partial \rho} = A(1+\alpha) - \frac{\alpha p}{\rho} \tag{14}$$

From equations (12) and (13) we get the relation between r and s:

$$18(r-1)s^2 + 18\alpha s(r-1) + 4\alpha(r-1)^2 = 9sA(1+\alpha)(2r+9s-2) \tag{15}$$

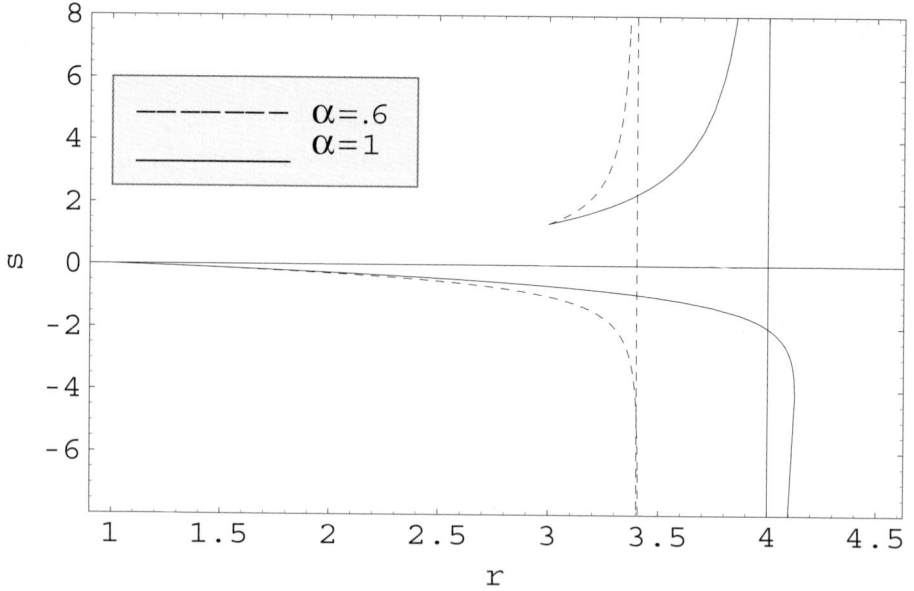

Fig. 1. shows the variation of s against r for different values of α (= 0.6, 1) and for $A = 1/3$.

Figure 1 shows the variation of s with the variation of r for $A = 1/3$ and for $\alpha = 0.6, 1$. The portion of the curve on the positive side of s, which is physically admissible, is only the values of r greater than $\left\{1 + \frac{9}{2}A(1+A)\right\}$. The part of the curve between $r = 1$ and $r = 1 + \frac{9}{2}A(1+A)$ with positive value of s is not admissible (we have not shown that part in the figure 1) because for the Chaplygin gas under consideration we face a situation, where the magnitude of the constant B becomes negative.

Thus the curve in the positive side of s starts from radiation era and goes asymptotically to the dust model. But the portion in the negative side of s represents the evolution from dust state ($s = -\infty$) to the ΛCDM model ($s = 0$). Thus the total curve represents the evolution of the universe starting from the radiation era to the ΛCDM model.

4 Discussions

In this work, we have presented a model for modified Chaplygin gas. In this model we are able to describe the universe from the radiation era ($A = 1/3$ and ρ is very large) to ΛCDM model (ρ is small constant). So compared with the Chaplygin gas model, the present model describe universe to a large extent. Also if we put $A = 0$ with $\alpha = 1$, then we can recover the results of the Chaplygin gas model. In figure 1, for $\{r, s\}$ diagram the portion of the curve for $s > 0$ between $r = 1$ to $r = 1 + \frac{9}{2}A(1 + A)$ is not describable by the modified Chaplygin gas under consideration. For example, if we choose $r = 1.03$, $A = 1/3$ then from the curve $s = 0.01$ which corresponds to $q = 3/2$ and hence we have from the equation of state, $B < 0$ which is not valid for the specific Chaplygin gas model considered here. At the large value of the scale factor we must have some stage where the pressure becomes negative and hence B has to be chosen positive. It follows therefore that a portion of the curve as mentioned above should not remain valid.

References

1. N. A. Bahcall, J. P. Ostriker, S. Perlmutter and P. J. Steinhardt, *Science* **284** 1481 (1999)
2. S. J. Perlmutter et al, *Astrophys. J.* **517** 565 (1999)
3. A. Kamenshchik, U. Moschella and V. Pasquier, *Phys. Lett. B* **511** 265 (2001); V. Gorini, A. Kamenshchik, U. Moschella and V. Pasquier, *gr-qc*/0403062 (2004)
4. V. Gorini, A. Kamenshchik and U. Moschella, *Phys. Rev.D* **67** 063509 (2003); U. Alam, V. Sahni , T. D. Saini and A.A. Starobinsky, *Mon. Not. Roy. Astron. Soc.* **344**, 1057 (2003)
5. M. C. Bento, O. Bertolami and A. A. Sen, *Phys. Rev. D* 66 043507 (2002)
6. V. Sahni, T. D. Saini, A. A. Starobinsky and U. Alam, *JETP Lett.* **77** 201 (2003)
7. A. Benoit, P. Ade, A. Amblard et al, *Astron. Astrophys.* **399** L25 (2003)
8. H. B. Benaoum, *hep-th*/0205140 (2002)

Spherically Symmetric, Static Spacetimes in a Tensor-Vector-Scalar Theory

D. Giannios

Max Planck Institute for Astrophysics, Box 1317, D-85741 Garching, Germany,
giannios@mpa-garching.mpg.de

Summary. Recently, a relativistic gravitation theory has been proposed by Bekenstein [1] that gives the Modified Newtonian Dynamics (or MOND) in the weak acceleration regime. The theory is based on three dynamic gravitational fields and succeeds in explaining a large part of extragalactic and gravitational lensing phenomenology without invoking dark matter. Here, I consider the strong gravity regime of Tensor-Vector-Scalar (TeVeS) and study spherically symmetric, static and vacuum spacetimes. Two branches of solutions are identified: in the first the vector field is aligned with the time direction while in the second the vector field has a non-vanishing radial component. The Parameterized Post-Newtonian (PPN) coefficients are calculated for these cases. For the first branch of solutions, I derive exact analytic expressions for the physical metric and apply them to the case of black holes. Implications are discussed.

1 Introduction

On cosmological scales, Newtonian gravitational theory under-predicts the acceleration of stars and gas. Furthermore, galaxies and clusters of galaxies show anomalously large gravitational lensing when only baryonic matter is taken into account. A natural "cure" for these discrepancies is to assume the existence of dark matter, which dominates over visible matter (for review see [2]). Such dark matter might solve the "missing mass" problem within the standard theory of gravity (i.e., general relativity or GR).

A second approach to the acceleration discrepancy and lensing anomaly is to look for alternative theories to GR. Among many other attempts, the MOND paradigm has been proposed [3, 4]. It is characterized by an acceleration scale a_0 so that $\tilde{\mu}(|a|/a_0)a = -\nabla \Phi_N$, where Φ_N is the Newtonian potential, $\tilde{\mu}(x) = x$ for $x \ll 1$, while $\tilde{\mu}(x) = 1$ for $x \gg 1$.

This empirical law has been very successful in explaining the rotation curves in a large number of galaxies using the observed distributions of gas and stars as input [5] and the observed correlation between the infrared luminosity of a disk galaxy L_K and the asymptotic rotational velocity v_a [6].

However, MOND is merely a prescription for gravity and not a self consistent theory. A theory of gravity is needed that has the MOND characteristics in the weak acceleration limit but also of full predictive power. In this context, a new relativistic theory of gravity has been introduced by Bekenstein [1]. It consists of three dynamical gravitation fields: a tensor field ($g_{\alpha\beta}$), a vector field (U_α) and a scalar field (ϕ) leading to the acronym TeVeS. The theory involves a free function F, a length scale ℓ and two positive dimensionless constants κ, K.

TeVeS gives MOND in the weak acceleration limit (inheriting the successes of MOND on large scale), makes similar prediction on gravitational lensing as GR (with dark matter) and provides a formulation for constructing cosmological models [7]. The strong gravity regime of TeVeS has been studied in Ref. [8]. The basic findings of that study are summarized here.

2 The Basic Equations of TeVeS

TeVeS is based on three dynamical gravitational fields: a tensor field (the Einstein metric $g_{\alpha\beta}$), a 4-vector field U_α and a scalar field ϕ. The physical metric $\tilde{g}_{\alpha\beta}$ in TeVeS is connected to these fields through the expression

$$\tilde{g}_{\alpha\beta} \equiv e^{-2\phi} g_{\alpha\beta} - 2U_\alpha U_\beta \sinh(2\phi). \tag{1}$$

The total action in TeVeS is the sum of 4 terms S_g, S_s, S_U and S_m (see Ref. [1]), where S_g is identical to the Hilbert-Einstein action and is the part that corresponds to the tensor field, while S_s, S_U, S_m are the actions of the scalar fields, the vector field and the matter respectively. The basic equations of TeVeS are derived by varying the total action S with respect to the fields.

Doing so for $g_{\alpha\beta}$, one arrives to the metric equations

$$G_{\alpha\beta} = 8\pi G[\tilde{T}_{\alpha\beta} + (1 - e^{4\phi})U^\mu \tilde{T}_{\mu(\alpha} U_{\beta)} + \tau_{\alpha\beta}] + \Theta_{\alpha\beta}, \tag{2}$$

where a pair of indices surrounded by parenthesis stands for symmetrization, i.e. $A_{(\alpha} B_{\beta)} = A_\alpha B_\beta + A_\beta B_\alpha$, the $G_{\alpha\beta}$ denotes the Einstein tensor for $g_{\alpha\beta}$, $\tilde{T}_{\alpha\beta}$ is the energy momentum tensor and $\tau_{\alpha\beta}$, $\Theta_{\alpha\beta}$ stand for the energy momentum contributed by the scalar and the vector fields respectively.

Similarly one derives a scalar equation which depends on a free function F of the theory. The results presented here are essentially independent of the exact choice of F and quite general. The vector equation is derived through variation of S with respect to U_α and includes a Lagrange multiplier λ that enforces the normalization condition $g^{\alpha\beta} U_\alpha U_\beta = -1$.

3 Spherical Symmetric, Static Spacetimes

I focus on the strong gravity limit of TeVeS and explore the spacetime in the vicinity of a spherically symmetric mass. The isotropic form of a spherical symmetric, static metric is

$$g_{\alpha\beta} dx^\alpha dx^\beta = -e^\nu dt^2 + e^\zeta (dr^2 + r^2 d\theta^2 + r^2 \sin^2 \theta d\varphi^2), \tag{3}$$

where both ν, ζ are only functions of r. For a static system the vector field has two non-vanishing components $U^\alpha = (U^t, U^r, 0, 0)$, functions of r.

The r component of the vector equation can be brought into the form

$$U^r \left(\lambda + \frac{\kappa (Gm_s)^2}{2\pi} \frac{e^{-(\nu+4\zeta)}}{r^4 [e^{-\zeta} - (U^r)^2]^2} \right) = 0. \tag{4}$$

This equation shows that there are two cases: either U^r vanishes or one has a constraint on the Lagrange multiplier λ. In the former case the vector field is aligned with the time direction, while in the latter there is a non-vanishing radial component of the vector field.

When U^r vanishes, the vector field is determined its normalization expression which yields $U^\alpha = (e^{-\nu/2}, 0, 0, 0)$, while the differential equations resulting from the tt and rr components of the metric equation (2) are

$$\zeta'' + \frac{(\zeta')^2}{4} + \frac{2\zeta'}{r} = -\frac{\kappa (Gm_s)^2}{4\pi} \frac{e^{-(\zeta+\nu)}}{r^4} - K \left(\frac{(\nu')^2}{8} + \frac{\nu''}{2} + \frac{\nu'\zeta'}{4} + \frac{\nu'}{r} \right) \tag{5}$$

and

$$\frac{(\zeta')^2}{4} + \frac{\zeta'\nu'}{2} + \frac{\zeta'+\nu'}{r} = \frac{\kappa (Gm_s)^2}{4\pi} \frac{e^{-(\zeta+\nu)}}{r^4} - K \frac{(\nu')^2}{8}. \tag{6}$$

The study of the properties of these equations for the appropriate boundary conditions constitutes most of the rest of this work.

In the non-aligned case, $U^r \neq 0$, the Lagrange multiplier λ is given by setting the term inside the parenthesis of Eq. (4) to zero. The components of the vector field are connected to the functions ν and ζ through the normalization condition for the vector that gives $e^\nu (U^t)^2 - e^\zeta (U^r)^2 = 1$ and the t component of the vector equation. The last two expressions can be combined with the tt and rr components of the metric equation (2) to arrive to a closed system of four differential equations with four unknown functions.

3.1 The PPN Coefficients

Far from the source (but not too far, so that the MOND corrections can be safely neglected), the metric can be taken to be asymptotically flat. Expanding the e^ζ, e^ν to powers of $1/r$ we have

$$e^\nu = 1 - r_g/r + a_2(r_g/r)^2 + \cdots \quad \text{and} \quad e^\zeta = 1 + b_1 r_g/r + b_2(r_g/r)^2 + \cdots. \tag{7}$$

If the vector field is aligned with the time direction one can substitute the expansions (7) into the metric equations (5) and (6), match coefficients of like powers if $1/r$ and solve for the coefficients a_i, b_i. The detailed calculation shows that the physical metric is *identical* to GR up to post-Newtonian order [1, 8]. No constraints can be set to the parameters of TeVeS κ and K from measurements of the standard Post-Newtonian coefficients, if the radial component of the vector field vanishes.

In the non-aligned case one needs to consider the asymptotic expansion of the vector field components. For $r_g/r \ll 1$, the vector field relaxes to its cosmological value, i.e.,

$U^t \to 1$ and $U^r \to 0$ (since there is no preferred spatial direction). Assuming $K \ll 1$, $\kappa \ll 1$ and furthermore that $\phi_c \ll 1$, I calculate the standard β, γ PPN coefficients, as predicted by TeVeS for $U^r \neq 0$. While the γ coefficient coincides with the GR prediction, the β differs from unity. The best observational determination of β yields $\beta - 1 < 10^{-4}$. For the choice of the function F made in Ref. [1], κ is constrained to be $\simeq 0.03$ which results in $\beta - 1 > 2.5 \times 10^{-3}$. This is in conflict with observations.

4 Analytic Solution when U^r Vanishes

Until now, I have kept the study of spherical symmetric spacetimes in TeVeS quite general. From this point on, I focus on the branch of solutions for which $U^r = 0$ where exact analytic solutions are possible. In the simplest case where both $\kappa = 0$ and $K = 0$, the metric equations in TeVeS coincide with these in GR and their right hand side is zero (i.e. no source terms appear). In this limit the integration of Eqs. (5), (6) leads to the familiar GR solution.

In the general case both κ and K are non-zero. and the *exact* spherical symmetric, vacuum solutions for the physical metric have the form [8]

$$\tilde{g}_{tt} = -\left(\frac{r - r_c}{r + r_c}\right)^a \quad \text{and} \quad \tilde{g}_{rr} = \frac{(r^2 - r_c^2)^2}{r^4}\left(\frac{r - r_c}{r + r_c}\right)^{-a}, \tag{8}$$

where $a \equiv \frac{r_g}{2r_c} + \frac{\kappa G m_s}{4\pi r_c}$ and $r_c \equiv \frac{r_g}{4}\sqrt{1 + \frac{\kappa}{\pi}\left(\frac{G m_s}{r_g}\right)^2 - \frac{K}{2}}$.

One can check that, for the derived solution, the energy density contributed by the vector field is negative. The fact that $\Theta_{tt} < 0$ can have important consequences for the theory since it may lead to instability of the vacuum from the quantum point of view.

The scalar field is given by the expression

$$\phi(r) = \phi_c + \frac{\kappa G m_s}{8\pi r_c}\ln\left(\frac{r - r_c}{r + r_c}\right), \tag{9}$$

where ϕ_c stands for the cosmological value of the scalar field at a specific epoch and m_s is a (non-negative) particular integral over the proper energy density and pressure of the star's matter. Unless $\kappa m_s = 0$, the scalar field diverges logarithmically at $r = r_c$ and there is always a radius $r_1 > r_c$ where $\phi(r_1) = 0$ and becomes negative further in.

One can apply this solution to the case of a non-rotating black hole. A characteristic radius of the physical metric (8) is r_c where the tt component of the metric vanishes. This radius corresponds to the event horizon if the surface area and the curvature invariants are finite there. Combined, these two constraints imply that the value $a = 2$ describes a black hole. For $a = 2$ the physical metric has a form identical to that predicted by GR.

One can check that m_s is non-zero for this solution, which means that there is always a region close to r_c where the scalar field turns negative. Bekenstein in Ref. [1], on the other hand, has shown that TeVeS becomes acausal (i.e. it suffers from superluminal propagation of field disturbances) when $\phi < 0$. As a result, the theory appears to behave in an unphysical way in the vicinity of a black hole (for $U^r = 0$).

5 Conclusions

Bekenstein's relativistic gravitational theory (TeVeS) that leads to MOND in the relevant limit, has been proposed as a modification to GR. In Ref. [8], I have looked at TeVeS in the strong gravity limit. Two branches of solutions are identified: the first is characterized by the vector field being aligned with the time direction while in the (not previously explored) second branch the vector field has a non-vanishing radial component. It is shown that the β and γ PPN coefficients in TeVeS are identical to these of GR in the first branch of solutions while the β PPN coefficient differs in the two theories in the second, where TeVeS is in conflict with recent observational determinations of it (for the choice of the free function F made in Ref. [1]).

For the first branch of solutions, I derive exact analytic expression for the physical metric that depends on the two dimensionless parameters κ, K of TeVeS [8]. The energy density contributed by the vector field is negative possibly turning the vacuum unstable from the quantum point of view. In the case of a black hole, the solution for the metric is identical to that of GR but is shown to be acausal in the vicinity of the black hole.

References

1. J. D. Bekenstein: Phys. Rev. D **70**, 083509 (2004)
2. V. Trimble: Astron. Astrophys. **25**, 425 (1987)
3. M. Milgrom: Astrophys. J. **270**, 365 (1983)
4. M. Milgrom: Astrophys. J. **270**, 371 (1983)
5. K. G. Begeman, A. H. Broeils, R. H. Sanders: Month. Not. Roy. Astron. Soc. **249**, 523 (1991)
6. R. P. Tully, J. R. Fisher: Astron. Astrophys. **54**, 661 (1977)
7. C. Skordis, D. F. Mota, P. G. Ferreira, C. Boehm: Phys. Rev. Lett. **96**, 011301 (2006)
8. D. Giannios: Phys. Rev. D **71**, 103511 (2005)

Type Ia Supernovae and Cosmology

W. Hillebrandt and F.K. Röpke

Max-Planck-Institut für Astrophysik, Karl-Schwarzschild-Str. 1, D-85748 Garching, Germany, (wfh,fritz)@mpa-garching.mpg.de

Summary. Recent progress in modeling type Ia supernovae by means of 3-dimensional hydrodynamic simulations as well as several of the still open questions are addressed in this article. It will be shown that the new models have considerable predictive power which allows us to study observable properties such as light curves and spectra without adjustable non-physical parameters. This is a necessary requisite to improve our understanding of the explosion mechanism and to settle the question of the applicability of SNe Ia as distance indicators for cosmology. We explore the capabilities of the models by comparison with observations and show in a preliminary approach, how such a model can be applied to study the origin of the diversity of SNe Ia which could be a source of considerable systematic errors in their distances.

1 Introduction

During the past couple of years type Ia supernovae (SNe Ia) have become a major tool in observational cosmology through their role as distance indicators. The now generally accepted view that the Universe is presently in a phase of accelerated expansion, caused by either a cosmological constant or an unknown form of 'dark energy' with negative pressure, rests mainly on the interpretation of supernova luminosities. While, however, the number of supernovae observed at cosmological redshifts is steadily increasing [1, 2, 3, 4], thus reducing the statistical errors of the cosmological parameters derived from them, a sound theoretical understanding of these objects – justifying in particular the calibration techniques applied in distance measurements – is still lacking.

The most recent analysis of supernova data by the CFHT-SNLS team [3] gives for a flat ΛCDM cosmology $\Omega_M = 0.264 \pm 0.042(stat) \pm 0.032(sys)$. Combining their data for 71 SNe and an additional sample of 45 nearby SNe from the literature with the analysis of baryon-acoustic oscillations in SDSS [5] they arrive at $\Omega_M = 0.271 \pm 0.021(stat) \pm 0.007(sys)$ and for the equation-of-state parameter $w = -1.02 \pm 0.09(stat) \pm 0.054(sys)$, in good agreement with results of other groups [1, 2]. It is obvious that one can reduce the total errors further only if the systematic errors, including those coming from the SNe themselves and their light curve calibration, are under control.

2 Modeling type Ia supernovae

Early attempts to model SNe Ia were based on one-dimensional numerical simulations. Such models gave valuable insight into the basic mechanism of the explosions. However, their predictive power was limited due to the fact that underlying physical processes enter the models in a parametrized way. This is overcome by three-dimensional modeling of SNe Ia [6, 7]. The until now most successful model assumes that a white dwarf (WD) consisting of carbon and oxygen accretes matter from a non-degenerate binary companion until its mass approaches the Chandrasekhar limit (for a review see [8]). Due to the rapid increase of the central density nuclear reactions set in giving rise to a stage of convective carbon burning. This stage lasts for several hundred years. Subsequently, a thermonuclear runaway of small convective temperature fluctuations ignites a thermonuclear flame. In principle, two distinct modes of flame propagation are allowed for. In a sub-sonic deflagration burning is mediated by thermal conduction while a super-sonic detonation is driven by a shock wave. On the basis of one-dimensional simulations a prompt detonation is ruled out for SNe Ia since it drastically under-produces the intermediate intermediate mass elements observed in their spectra [9].

As a laminar deflagration, however, the flame propagates too slowly to explain the energy release necessary to explode the WD. Thus, any valid SN Ia model needs to provide means of flame acceleration. Two mechanisms are conceivable. Firstly, the flame propagation continues in the deflagration mode but is significantly accelerated by the interaction with turbulence. A one-dimensional (1D) version of this mechanism, the model W7 of [10], has demonstrated that such a model in principle is capable of reproducing the main observational features of SNe Ia. An alternative way to speed up the flame is to assume a deflagration-to-detonation transition (DDT) at later stages of the explosion. The weak point of these delayed detonation models (e.g. [11]) is that a physical mechanism providing a DDT in SNe Ia could not be identified yet [12, 13, 14, 15].

To reach the goal outlined in the introduction it is necessary to have a SN Ia explosion model that contains no tunable parameters and, thus, can be used to extract the impact of physical parameters on the light curves from a comparison with observations. Therefore we set aside the possibility of a delayed detonation and focus on the turbulent (Chandrasekhar-mass) deflagration model. Turbulence is induced here by generic hydrodynamic instabilities. The main challenge in numerically implementing this model is the vast range of relevant length scales involved. Not only is the width of a thermonuclear flame in the degenerate carbon/oxygen material at the onset of the explosion 9 orders of magnitude below the radius of the WD. The turbulent cascade extends to even smaller scales and interacts with the flame down to the Gibson length at which the laminar flame speed equals the turbulent velocity fluctuations (10^4 cm and decreasing in the explosion process). This problem can be tackled in a Large Eddy Simulation (LES) approach. Here, only the largest scales of the problem are directly resolved and turbulence effects on unresolved scales are included via a subgrid-scale model [16, 17]. The thermonuclear flame is modeled as a sharp discontinuity separating the burnt from the unburnt material. Its evolution is followed utilizing the level-set method [18]. Since the structure of the flame is not resolved, the flame propagation velocity must be provided externally. This, however, does not introduce undetermined parameters to the model since the theory of turbulent combustion [19] predicts that for most stages of the SN Ia explosion the flame propagation proceeds in the flamelet regime where it completely decouples from the microphysics of the burning and is determined by the turbulent velocity fluctuations that can be derived from the subgrid-scale model.

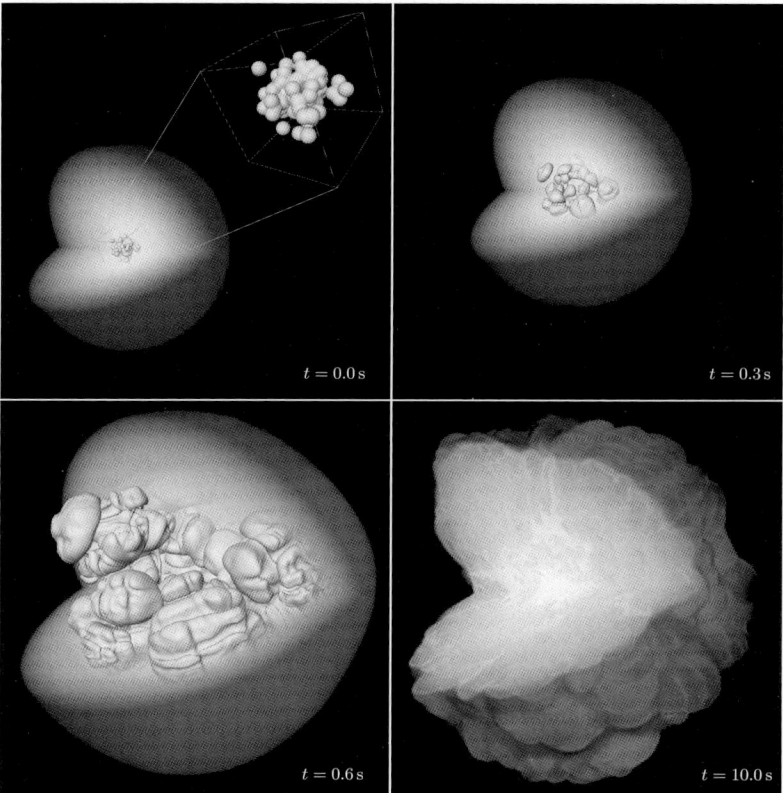

Fig. 1. Snapshots from a full-star SN Ia simulation starting from a multi-spot ignition scenario. The logarithm of the density is volume rendered indicating the extent of the WD star and the iso-surface corresponds to the thermonuclear flame. The last snapshot marks the end of the simulation and is not to scale with the earlier snapshots.

3 Some simulation results

A typical evolution of a SN Ia explosion modeled as described above is shown in Fig. 1. Starting from an ignition in multiple sparks the flame propagates outward. At $t = 0.3$ s, the mushroom-shaped features due to the buoyancy instability are clearly visible. Subsequently, the flame becomes increasingly corrugated and is accelerated by interaction with turbulence. It therefore burns through a large fraction of the WD material. The snapshot at $t = 0.6$ s shows the flame evolution around the peak of energy production due to nuclear burning. Up to this point, the burning terminated in nuclear statistical equilibrium (NSE) and the carbon/oxygen material was primarily converted to iron group elements. The expansion of the WD decreases the fuel density steadily and once it falls below 5×10^7 g cm^{-3} nuclear burning becomes incomplete and produces mainly intermediate mass elements. About 2 s after ignition, expansion quenches the burning and the following evolution is characterized by the relaxation to homologous expansion of the ejecta, which is reached to a reasonable accuracy $\sim$10 s after ignition [21]. The density structure of the ejecta at this stage is shown in Fig. 1, where the traces of turbulent flame propagation are clearly visible.

We have also 'post-processed' several of our models in order to see whether or not also reasonable isotopic abundances are obtained [20]. With a few exceptions, the isotopic abundances are within the expected range and are not too different from the 1D model *W7* that gives good fits of observed spectra of 'normal' SNe Ia. Exceptions include the high abundance of (unburned) C and O, and the overproduction of ^{48,50}Ti, ^{54}Fe, and ^{58}Ni, a general feature of our present models. We think that the latter reflects a deficiency of the models which burn C and O at densities too high and temperatures too low to have almost all the NSE material in form of ^{56}Ni. The high mass fraction unburned C and O results in part from the fact that in most of our simulations we stopped the burning at a density of 10^7 g/cm^3.

4 Predictions for observable quantities

Apart from the initial conditions simulations as described above contain no free parameters. Therefore the question arises whether such models are capable of reproducing observations without any fitting. The explosion energies achievable in the outlined scenario reach up to $\sim$8 $\times$ 10^{50} erg and the models produce $\sim$0.4 $M_\odot$ of ^{56}Ni. This falls into the range of observational expectations, although on the side of the weaker SN Ia explosions [22].

Nonetheless, the synthetic lightcurves derived from models of the class described here fit the observations in the B and V bands around maximum luminosity rather well [23, 24] (see also Fig. 2). A much harder constraint on the explosion model is posed by spectral observations, since spectra are particularly sensitive to the chemical composition of the ejecta. [25] pointed out a potential problem of deflagration SN Ia models. In late time "nebular" spectra, unburnt material (transported towards the center in downdrafts due to the large-scale buoyancy-unstable flame pattern) gives rise to a strong oxygen line of low-velocity material which is in conflict with observations. However, the synthetic spectrum of [25] was derived from a simplistic centrally ignited model. Recently, [26] showed that multi-spot ignition models may succeed to burn out the central parts of the WD, reducing the amount of oxygen at low velocities.

Detailed spectral observations allow to determine the chemical composition of the ejecta in velocity space [27]. The mixed composition of the ejecta observed there points to a deflagration phase being at least a significant contribution to the SN Ia explosion process. The central parts are found to be clearly dominated by iron group elements, which are mixed out to velocities of about 12000 km s^{-1}. Intermediate mass elements

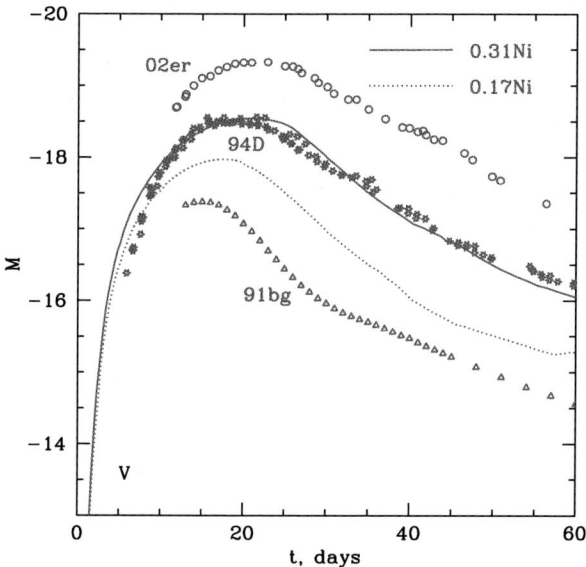

Fig. 2. Theoretical predictions (solid and dashed lines) of two 3D models with different Ni masses in comparison with observed V-band lightcurves. SN 1994D was a 'normal' SN Ia, while SN 2003er and SN 1991bg were bright and faint, respectively [23]. Note that the highest Ni-masses we could get until now are around 0.4 $M_\odot$.

are distributed over a wide range in radii and no unburnt material is found at velocities $\lesssim 5000\,\mathrm{km\,s^{-1}}$. A recent high-resolution full-star deflagration SN Ia simulation [28] demonstrated that such models can get close to these observational constraints.

Although it cannot be ruled out that the pure deflagration model of SNe Ia is incomplete, the results indicate that it may be at least a dominant part of the mechanism. Therefore it is justified to ask how such models are affected by initial parameters of the exploding WD. This may give a hint to the origin of the diversity of SNe Ia. To moderate the computational expenses, simplified setups may be used to study the effects of physical parameters on the explosion models. Such an approach was recently taken by [29] and resulted in the first systematic study of progenitor parameters in three-dimensional models. The basis of this study was a single-octant setup with moderate (yet numerically converged) resolution. However, the lack in resolution did not allow for a reasonable multi-spot ignition scenario and thus only weak explosions can be expected. It is therefore not possible to set the absolute scale of effects in this approach, but trends can clearly be identified. The parameters chosen for the study were the WD's carbon-to-oxygen ratio, its central density at ignition and its ^{22}Ne mass fraction resulting from the metallicity of the progenitor. All parameters were varied independently to study the individual effects on the explosion process. The results of this survey are given in Tab. 1.

Table 1. Variation of initial parameters in SN Ia explosion models

Parameter	Range of variation	Effect on ^{56}Ni production	Effect on total energy
$X(^{12}C)$	[0.30, 0.62]	$\leq 2\%$	$\sim 14\%$
ρ_c [10^9 g/cm^3]	[1.0, 2.6]	$\sim 6\%$	$\sim 17\%$
Z [$Z_\odot$]	[0.5, 3.0]	$\sim 20\%$	none

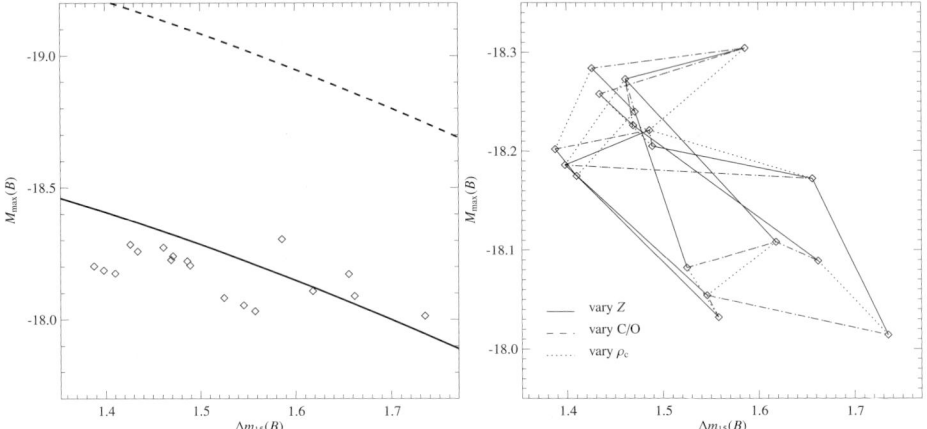

Fig. 3. Peak luminosity vs. decline rate of the light curve in the B band (diamonds correspond to SN Ia explosion models; the dashed curve in the left panel indicates the original relation by [30] and a shifted relation is marked as a solid curve). The right panel of shows that the slope is dominated by the variation of the progenitor's metallicity.

To determine the effects of these variations on observables, synthetic light curves were derived from all models [24]. From these, the peak luminosities and decline rates (in magnitudes 15 days after maximum; Δm_{15}) were determined.

A comparison with the relation given by [30] (forming the basis of the calibration of cosmological distance measurements) is provided in the left panel of Fig. 3. Obviously, the absolute magnitude of the *Phillips relation* is not met by our set of models. Moreover, the range of scatter in Δm_{15} is much narrower than that of the set of observations used by [30]. But there is a trend of our models consistent with the slope of the Phillips relation. The right panel of Fig. 3 shows that this slope is dominated by the variation of the progenitor's metallicity.

5 Summary and conclusions

In this article, we have presented a new class of deflagration models for SNe Ia, based on parameter-free 3D simulations of turbulent thermonuclear combustion. All models we have computed, differing in the ignition conditions, the grid resolution, and the properties of the Chandrasekhar-mass WD, explode. The explosion energy and the

Ni-masses are only moderately dependent on the way the nuclear flame is ignited making the explosions robust. However, since ignition is a stochastic process, the differences we find can explain part of the spread in observed SNe Ia, though not all, and the same holds for variations of the WD's properties.

Based on our models we can predict light curves, spectra, and abundances, and the first results look promising. The light curves seem to be in good agreement with observations although, on average the peak luminosities and the expansion velocities are a bit low. Also the nuclear abundances of elements and their isotopes are found to be in the expected range. A major outcome of our study is that the prime suspect for (cosmological) evolution of SN Ia properties, the C/O ratio of the WD, has little effect on the peak luminosity while changing the explosion energy. This is a first indication from modeling that a one-parameter calibration of the light curves might be insufficient. We also confirmed the results of analytic studies that the WD's metallicity has a systematic, although small, effect on the SN luminosity, which could help to reduce the scatter in the SN Ia Hubble diagram further.

However it appears to be necessary to consider modifications of the 'standard model' in order to explain the full spread of observed SN Ia properties. These modifications could include other progenitor channels, such as WD mergers or sub-Chandrasekhar mass models, or deflagration-detonation-transition models, with yet unknown consequences for their use in cosmology.

References

1. J. L. Tonry et al.: ApJ **594**, 1 (2003)
2. A. G. Riess et al.: ApJ **607**, 665 (2004)
3. Astier, P. et al.: A&A **447**, 31 (2006)
4. S. Blondin et al.: ApJ **131**, 1648 (2006)
5. D. J. Eisenstein et al.: ApJ **633**, 560 (2005)
6. M. Reinecke, et al.: A&A **391**, 1167 (2002)
7. V. N. Gamezo et al.: Science **29**, 77 (2003)
8. W. Hillebrandt, and J. C. Niemeyer: Ann. Rev. Astron. Astrophys. **38**, 191 (2000)
9. W. D. Arnett: Astrophys.Space Sci. **5**, 180 (1969)
10. K. Nomoto, F.-K. Thielemann, and K. Yokoi: ApJ **286**, 644 (1984)
11. V. N. Gamezo, A. M. Khokhlov, and E. S. Oran: PRL **92**, 211102 (2004)
12. J. C. Niemeyer: ApJ **523**, L57 (1999)
13. A. M. Lisewski, W. Hillebrandt, and S. E. Woosley: ApJ **538**, 831 (2000)
14. F. K. Röpke, W. Hillebrandt, and J. C. Niemeyer: A&A **420**, 411 (2004)
15. F. K. Röpke, W. Hillebrandt, and J. C. Niemeyer: A&A **421**, 783 (2004)
16. J. C. Niemeyer, and W. Hillebrandt: ApJ **452**, 769 (1995)
17. W. Schmidt: A&A **450**, 283-294 (2006)
18. S. Osher, and J. A. Sethian: J. Comp. Phys. **79**, 12 (1988)
19. N. Peters: *Turbulent Combustion*, Cambridge University Press, Cambridge (2000).
20. C. Travaglio: A&A **425**, 1029 (2004)
21. F. K. Röpke: A&A **432**, 969 (2005)
22. M. Stritzinger:
23. E. Sorokina, and S. Blinnikov, in *From Twilight to Highlight: The Physics of Supernovae*, edited by W. Hillebrandt, and B. Leibundgut, ESO Astrophysics Symposia, Springer-Verlag, 268–275 (2003)

24. S. I. Blinnikov et al.: A&A **453**, 229-240 (2006)
25. C. Kozma et al.: A&A **437**, 983 (2005)
26. F. K. Röpke, et al.: A&A **448**, 1-14 (2006)
27. M. Stehle, et al.: MNRAS **360**, 1231 (2005)
28. F. K. Röpke, and et al. (2006), in preparation.
29. F. K. Röpke et al.: A&A **453**, 203-217 (2006)
30. M. M. Phillips et al.: Astron. J. **118**, 1766 (1999)

Path and Path Deviation Equations in Kaluza-Klein Type Theories: A Brief Introduction*

M.E. Kahil[1,2]

[1] Mathematics Department, Modern Sciences and Arts University, Giza, EGYPT
[2] Mathematics Department, The American University in Cairo, Cairo, EGYPT,
`kahil@aucegypt.edu`

Summary. The problem of motion in higher dimensions is an intriguing problem. This can be seen by studing path and path deviation equations for charged, spinning and spinning charged objects in higher dimensions. The significance of motion in higher dimensions may yield some physical implications on the effect of the extra dimension on the current motion. In this approach, we apply the Bazanski method to derive path and path deviation equations for different objects in five dimensions.

1 The Bazanski Approach in 5D

In an attempt to study the motion in higher dimensions we derive path and path deviation equations for charged, spinning, and charged spinning particles in 5-dimensions using the Bazanski Lagrangian [1] in 5D:

$$L = g_{AB} U^A \frac{D\Psi^B}{DS} \tag{1}$$

where $(A = 1, 2, 3, 4, 5)$. By taking the variation with respect to the deviation vector Ψ^C and the tangent vector U^C, we obtain the well known geodesic and geodesic deviation equations respectively.

In Compact Spaces The process to unify electromagnetism (gauge fields) and gravity depends on extra component(s) of the metric using the cylinder condition A charged particle whose behavior is described by the Lorentz equation in 4D behaves as a test particle moving on a geodesic in 5D as well as its deviation equation becomes like the well known geodesic deviation equation [2]. This result is obtained from the usual Basanski method in 5D.

But, in Non-Compact Spaces the path equation has two main defects:
(i) it is not gauge invariant,
(ii) the additional extra force from an extra dimension is parallel to the four vector velocity i.e. $f_\mu U^\mu \neq 0$.

---

* An extended version of this article may be found at: gr-qc/0511023

Ponce de Leon [3] has introduced some types of transformations in order overcome the above mentioned problems to express the path equation in five dimensions for a test particle moving in non-compactified extra dimension like the ususal geodesic equation i.e.:

$$\frac{d^2\xi^A}{dS^2} + \left\{ \begin{array}{c} A \\ BC \end{array} \right\} \frac{d\xi^B}{dS} \frac{d\xi^C}{dS} = 0, \tag{2}$$

where ξ^A is the projected 5D velocity. This allows us to introduce its corresponding Bazanski Lagrangian:

$$L = g_{AB} \frac{d\xi^A}{dS} \frac{D\Psi^B}{DS} \tag{3}$$

which gives its geodesic deviation equation as:

$$\frac{D^2\Psi^A}{DS^2} = 0. \tag{4}$$

2 Rotation in 5D

The concept of rotation in higher dimensions is related to obtaining the governing equation of the current spinning object [6]. For a spinning gyroscope it is well known that the fifth equation is testing the rate of precession [7]. Some authors believe that the study of two nearby free-falling gyroscopes could be used to examine the question of the existence of gravitational waves [8].

We may apply the Bazanski approach on the following Lagrangian:

$$L = g_{AB} U^A \frac{D\Psi^B}{DS} + \frac{1}{2m} R_{ABCD} S^{CD} U^B \Psi^A \tag{5}$$

to derive the path equation of a spinning object [8] in 5D:

$$\frac{dU^C}{dS} + \left\{ \begin{array}{c} C \\ AB \end{array} \right\} U^A U^B = \frac{1}{2m} R^C_{.ABD} S^{BD} U^A \tag{6}$$

The above equation describes spinning objects in compact spaces which satisfy the cylinder condition i.e $g_{AB,5} = 0$. This is identical to the Dixon equation [9] if we project it onto four dimensions. The fifth coordinate will contribute the electromagnetic tensor, which has already appeared in the Dixon equation.

Moreovere, if we suggest the following Lagrangian:

$$L = g_{AB}(mU^A + U_C \frac{DS^{AC}}{Ds}) \frac{D\Psi^B}{Ds} + R_{ABDE} S^{DE} U^B \Psi^A, \tag{7}$$

we obtain the original Papapetrou equation [10] in 5D :

$$\frac{D}{DS}(mU^A + U_E \frac{DS^{AE}}{DS}) = \frac{1}{2} R^A_{.BCD} S^{CD} U^B \tag{8}$$

and its corresponding deviation equation will take the following form:

$$\frac{D^2\Psi^A}{DS^2} = R^A_{.BCD}U^B(mU^C + U_E\frac{DS^{CE}}{DS})\Psi^D + g^{AC}g_{BE}(mU^E + U_o\frac{DS^{EO}}{DS})_{;C}\frac{D\Psi^B}{DS}$$

$$+ R^A_{.BCD}S^{CD}\frac{D\Psi^B}{DS} + R^A_{BCE}S^{CE}_{.;D}U^B\Psi^D + R^A_{BCE;D}S^{CE}U^B\Psi^D. \qquad (9)$$

These equations could be used to study the behavior of spinning charged objects that exhibit precession, e.g. neutron stars, compact objects, etc.

In non-compact spaces with $R_{ABCD} = 0$, it is found that a spinning particle is moving on a geodesic in 5D [10] rather than the Papapetrou equation [11]. This leads us to suggest that in non-compact spaces, satisfying the Campbell-Magaard theorem, spinning particles and spinning charged particles as well as test particles are moving along geodesics in 5D, and the path deviation equation for each of the above mentioned object becomes like the usual geodesic deviation equation. But, if we consider the case of the ususal Papapetrou equation, it becomes:

$$\frac{D}{Ds}(mV^A + V_B\frac{DS^{AB}}{DS}) = 0. \qquad (10)$$

On the contrary, its corresponding deviation equation is identical to (4).

Acknowledgments. The author would like to thank Professors M.I. Wanas , M. Abdel-Megied , G. De Young for their remarks and comments in Cairo. Also, thanks to Professor K. Buchner for his kind hospitality and invaluable comments during my stay in Munich in June 2004 and in November 2005 . A word of thankfullness to Professor G. Hasinger and the LOC for their support and hospitality during the conference.

References

1. Bazanski, S.I.: J. Math. Phys., **30**, 1018 (1989)
2. Kerner, R. Martin, J. Mignemi, S. and van Holten, J-W.: Phys Rev D, **63**, 027502 (2003)
3. Ponce de Leon, J.: Grav. Cosmol. **8**, 272 (2002)
4. Wanas, M.I. and Kahil, M.E.: Gen. Rel. Grav.,**31**, 1921 (1999)
5. Wanas, M.I., Melek, M. and Kahil, M.E.: Astrophys. Space Sci.,**228**, 273 (1995)
6. Kalinowski, M.W.: *Non-Symmetric Field Theory and its Applications*, Institute of Theoretical Physics, Warsow University (1989)
7. Seahra, S.: Phys. Rev. **D65**, 124004 (2002)
8. Liu, H. and Mashhoon, B.: Phys. Lett. A, **272**,26 (2000)
9. Nieto, J.A., Saucedo, J. and Villanueva, V.M.: Phys. Lett. **A312**, 175 (2003)
10. Dixon, W.G.: Proc. Roy. Soc. Lond A**314**,499 (1970)
11. Liu,H. and Wesson, P.S : Class Quan. Grav., **11**, 1341 (1996)
12. Papapetrou, A.: Proc. Roy.Soc. Lond A**208**, 248 (1951)

Studying Dark Energy with Galaxy Clusters

S. Majumdar

Canadian Institute for Theoretical Astrophysics, 60 St George St, Toronto, ON, M5B2M3, Canada, subha@cita.utoronto.ca

Summary. Large yield cluster surveys, having well understood the cluster redshift distribution up to high redshifts, are promising probes of precision cosmology. However, unbiased and precise constraints require understanding the nature, evolution and scatter of the 'cluster mass–observable' relation. This requirement can be met with the so–called 'self–calibration' techniques. With 'self–calibration' future surveys (for example, the SPT survey) having tens of thousands of clusters, can deliver strong constraints on the equation of state parameter w of dark energy and its time evolution. That 'self-calibration' works in practice is shown by applying it to a much smaller sample of ~ 1000 optical clusters detected by the first Red–Sequence Cluster Survey.

1 Introduction

The main probes to study the nature of dark energy include distance measurements using SNe Ia, weak gravitational lensing studies of the matter distribution, galaxy power spectrum studies, and galaxy cluster surveys. Of these, cluster survey technique provides leverage on the dark energy through the dark energy sensitivity of the expansion history of the universe and the growth rate of structures in our expanding universe There are a number of large yield cluster surveys underway or planned capable of detecting thousands of clusters to $z \sim 1$. Examples of ongoing surveys are the Red Sequence Cluster Survey (RCS-2[1]) in optical (see section 4 for preliminary results from RCS-1) and the Massive Abell Cluster survey (MACS) based on the ROSAT all sky survey in X–ray. Examples of upcoming surveys include the South Pole Telescope (SPT[2]) using Sunyaev-Zel'dovich (SZ) detected clusters or the very ambitious Large Synoptic Survey Telescope (LSST[3]) covering half the sky. The above examples are chosen so as to give an idea of the largest surveys in terms of numbers of clusters and sky coverage. So called 'self-calibration' technique (see next section) can be effectively used to constrain w only for such surveys.

---

[1] http://rcs2.org/
[2] http://spt.uchicago.edu
[3] www.lsst.org

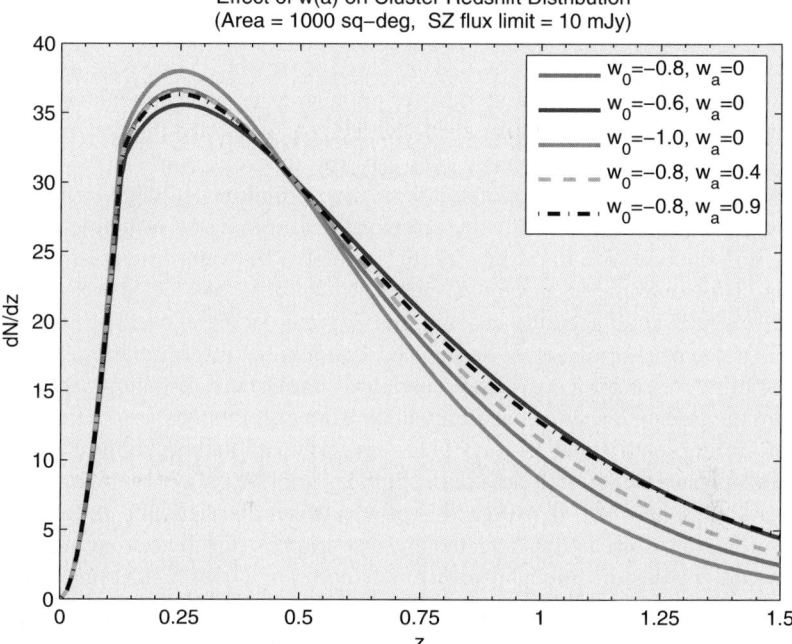

Fig. 1. The expected redshift distribution, $\frac{dN}{dz}$ for models where only the dark energy equation of state parameter $w(a) = w_0 + w_a(1 - a)$ is varied. The dark energy properties affect both the numbers as well as the shape of the cluster counts. The solid lines are for constant equation of state, with a more -ve w_0 giving lesser clusters at higher redshifts. Notice the degeneracy between w_0 and w_a.

2 Self Calibration in Cluster Surveys

The cluster redshift distribution [1] is the product of the volume surveyed and the number density of clusters which is obtained by integrating over the cluster mass function [2] above a mass limit. A selection function connects the mass limit to the observable limit of the survey. Other than a strong theoretical understanding of the space densities of dark matter haloes leading to cosmology dependent redshift distribution of clusters (see Figure 1) and a good understanding of the selection function of a survey, the most important issue concerning cluster surveys is knowing the 'right' cluster mass–observable relations, the so called 'cluster scaling relations'. Typically, targeted observations of a carefully selected sample of clusters are used to have an *a priori* handle on these scaling relations. Uncertainties in any one of the above requirements would lead to degradation of constraints and biased estimation of cosmological parameters from cluster redshift distribution.

Fortunately, other than cluster redshift counts, for wide area large yield cluster surveys there are additional observables. These include the clustering power spectrum $P_{cl}(k)$ of the observed clusters and the flux-function of the clusters at each redshift.

These functions are connected to theory of cluster formation through cluster bias [3] and the mass function [2]. One can also try to combine *rms* fluctuations with number counts, especially for SZ surveys to break degeneracies. Moreover, a direct mass measurement of a sub-sample of the cluster population can provide a handle on cluster physics degeneracies. It has been demonstrated [4, 5, 6, 7, 8] that a combination of cluster $\frac{dN}{dz}$ with the additional observables can lead to strong and unbiased estimation of cosmology, in spite of mass uncertainties. This technique of including mass uncertainties in cosmological constraints by effectively combining all available cluster data is called 'self–calibration'. In the next section, we give an example of 'self-calibration' forecasting for one of the upcoming surveys.

Finally, note that an accurate accounting of scatter in the cluster mass–observable relation is a crucial ingredient in doing 'self-calibration' in cluster surveys [9]. Any biased assumption (or prior) of scatter can significantly bias cosmological parameters away from the true underlying cosmology [10]. Both observations and simulations suggest tight scaling relations in X–Rays [11]. Similarly simulations suggest little scatter in SZ mass measurements [12]. Scatter in optically selected galaxy correlation/numbers versus mass is higher [13]. However, it is now realized that optically selected clusters differ significantly from X–Ray selected clusters and this difference may be the source of excess scatter when the optical properties are correlated with X–Ray derived masses.

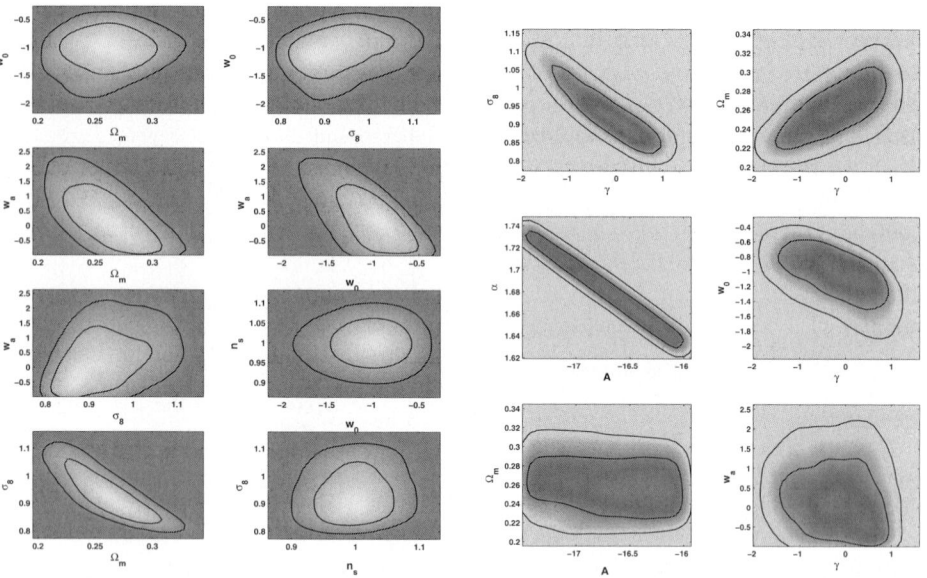

Fig. 2. Self-calibration using only $\frac{dN}{dz}$ showing parameter degeneracies for SPT. The left set of 8 panels shows degeneracies between cosmological parameters. The right set of 6 panels shows degeneracies involving cluster physics. The solid lines in all cases show 1,2-σ regions in parameter space marginalized over 8 other parameters. Note that cluster parameters A and α are degenerate by construction.

3 Dark Energy Constraints from Upcoming Surveys: An Example

In this section we take a look at what a large cluster survey would be able to deliver in terms of cosmological constraints. As an example, we chose the South Pole Survey, which is a survey covering 4000 deg^2. The SZ flux limit for the survey is taken to be a more conservative $10mJy$ at 150 GHz. This gives ~ 20000 objects for a survey up to $z \sim 1$. The SZ-flux is connected to the mass of a cluster through $f_{SZ}d_A^2 = f(\nu)f_{gas}A_{SZ}M_{cl}^\alpha E(z)^{2/3}(1 + z)^\gamma$, where $f(\nu)$ carries the frequency dependence of the SZ effect, f_{gas} is the fraction of the cluster mass in gas, d_A is the angular diameter distance to the cluster, $E(z)$ depends on the expansion history of the Universe and A_{SZ}, α & γ describes the cluster structure and its evolution. Additionally, one would like to include scatter as well in the analysis.

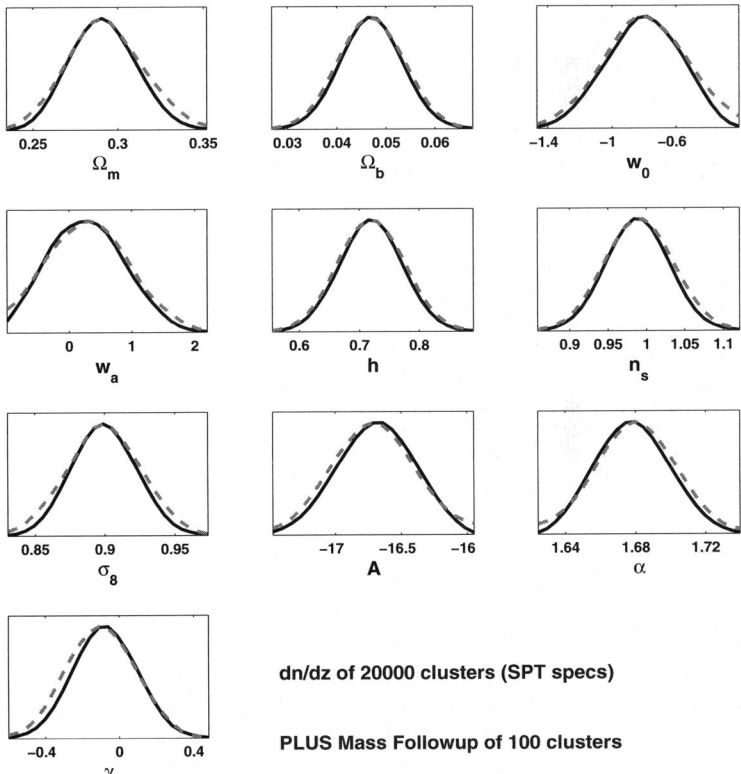

Fig. 3. Solid lines show the marginalized 1-D distribution when 100 cluster followup mass measurements are added to the SPT $\frac{dN}{dz}$. The dashed lines are for maximum Likelihood values for the global fit.

The forecasts were done using Markov Chain Monte Carlo methods on mock cluster redshift distributions between $0 < z < 1$ in bins of width $\Delta z = 0.1$. WMAP priors have been put on Ω_B, H_0 & n_s. To describe the nature of dark energy we take a variable

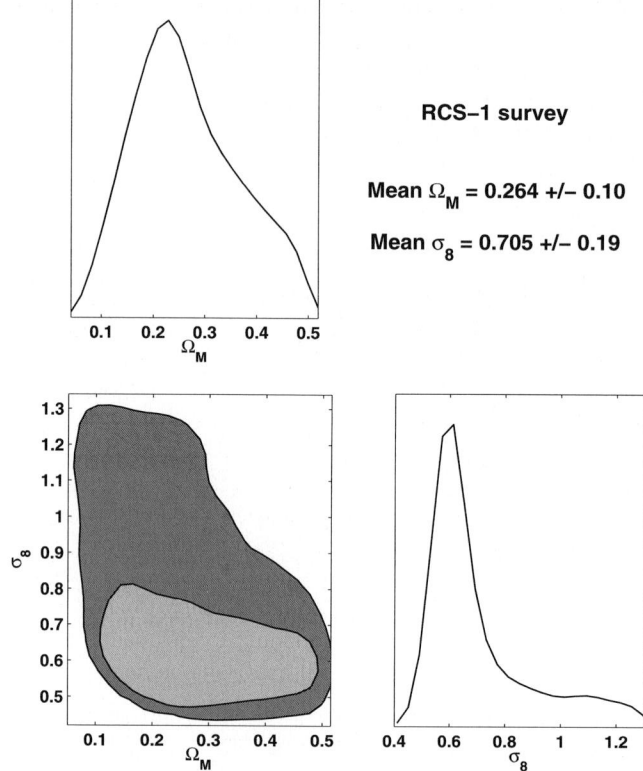

Fig. 4. RCS-1 marginalized constraints on $\Omega_M - \sigma_8$ from redshift distribution of ~ 1000 clusters between $0.35 < z < 0.95$. The high σ_8 tail is a direct consequence of unknown scatter in mass

equation of state $w(a) = w_0 + w_a(1 - a)$ such that $w_a = 0 \ \& \ w_0 = -1$ corresponds to a cosmological constant. It is clear from Figure 2 that there exists a huge amount of degeneracies between all the cosmological as well a cluster parameters. However, note that in spite of the degeneracies, most of the parameters are well constrained (see table 1). Next, we add independent followup mass measurements of 100 clusters between $0.3 < z < 0.8$ uniformly in redshift and masses $10^{14} < M_\odot < 10^{15}$ having mass uncertainty of 30%. This improves 'self–calibration' and significantly tightens

Table 1. 10-parameter MCMC forecasts for cluster survey

Parameters	Error Forecast ($\frac{dN}{dz}$)	Error Forecast ($\frac{dN}{dz}$ + Followup)
$\Delta\Omega_M$	0.025	0.018
$\Delta\sigma_8$	0.071	0.039
Δw_0	0.352	0.170
Δw_a	0.768	0.552

the constraints. The final marginalized probability distributions are shown in Figure 3 and the cosmological forecasts are tabulated in Table 1. Note that follow-up improves the constraints on σ_8 & $w(a)$ by a factor of 2. The main reason for improvement of constraints for dark energy is the fact that extra mass measurements help constrain the unknown evolution γ in the cluster scaling relation.

4 Final Comments

We have shown that 'self–calibration' in cluster surveys does hold the promise to constrain dark energy when multiple cluster observables are taken together. This makes cluster surveys competitive and complementary to other probes of dark energy. However, the question that we are left asking is whether there is any 'proof–of–the–principle'. The initial results from the RCS-1 survey shows [14] that even with only ~ 1000 clusters, 'self–calibration' is possible and we get reasonable and unbiased constraints on cosmological parameters. This is shown in Figure 4 where we show constraints marginalized over 4 other cluster parameters, including unknown scatter. With 1000 clusters, the constraints are still weak. However, this proves beyond doubt that future large yield surveys will be able to 'self–calibrate' and give tight constraints on dark energy along with other cosmological parameters.

Acknowledgments. I wish to thank Joe Mohr for many discussions on cluster surveys and Graham Cox for working on Markov Chains. I also wish to thank the RCS team for allowing me to show initial results.

References

1. Z. Haiman, J.J. Mohr and G. Holder: ApJ **553**, 545 (2001)
2. A. Jenkins, C.S. Frenk, S.D.M. White, J.M. Colberg, S. Cole, A.E. Evrard, H.M.P. Couchman and N. Yoshida: MNRAS **321**, 372 (2001)
3. H.J. Mo and S.D.M. White: MNRAS **282**, 347 (1996)
4. J.M. Diego and S. Majumdar: MNRAS **352**, 993 (2004)
5. W. Hu: Physical Review D **67**, 081304 (2003)
6. M. Lima and W. Hu: Physical review D **70**, 043504 (2004)
7. S. Majumdar and J.J. Mohr: ApJ **585**, 603 (2003)
8. S. Majumdar and J.J. Mohr: ApJ **613**, 41 (2004)
9. M. Lima and W. Hu: Physical review D **72**, 043006 (2005)
10. S. Majumdar and G. Cox: to be submitted (2007)
11. R. Stanek, A.E. Evrard, H. Böhringer, P. Schuecker and B. Nord: ApJ **648**, 956 (2006)
12. P.M. Motl, E.J. Hallman, J.O. Burns and M.L. Norman: ApJ **623**, L23 (2005)
13. H.K.C. Yee and E. Ellingson: ApJ **585**, 215 (2003)
14. M.D. Gladders et al.: ApJ **655**, 128-134 (2007)

Astrophysical Tests of Fundamental Physics

C.J.A.P. Martins[1,2]

[1] CFP, Universidade do Porto, R. do Campo Alegre 687, 4169-007 Porto, Portugal
[2] DAMTP, University of Cambridge, Wilberforce Rd., Cambridge CB30WA, U.K.
   C.J.A.P.Martins@damtp.cam.ac.uk

Summary. I describe the theoretical motivation for and possible roles of cosmological scalar fields, focusing on the spacetime variation of fundamental couplings. I then present our recent measurements of the fine-structure constant using quasar spectroscopy data from ESO's VLT telescope. Finally, I briefly discuss some ongoing and planned work and prospects for future improvements.

The most fundamental question of modern physics is whether or not there are fundamental scalar fields in nature. They are a key ingredient in the standard model of particle physics (cf. the Higgs particle, which is supposed to give mass to all other particles and makes the theory gauge-invariant), but there is so far no evidence that nature has any use for them. Indeed the observed particles are described by Fermi spinors, while the gauge forces are represented by boson vector fields. Moreover, Einstein gravity uses only a 2-tensor (the metric). Yet in recent years we have come to realize that the early universe is an ideal place to search for scalar fields, if they exist at all, and there have been some possible hints for them in various contexts. Two of these, that will be the focus of this contribution, are dark energy and varying fundamental couplings.

Observations suggest that the recent universe is dominated by an energy component whose gravitational behaviour is quite similar to that of a cosmological constant (as first introduced by Einstein). This could of course be the right answer, but the observationally required value is so much smaller than what would be expected from particle physics that a dynamical scalar field is arguably a more likely explanation. Theoretical motivation for such a field is not hard to find. In string theory, for example, dimensionful parameters are expressed in terms of the string mass scale and a scalar field vacuum expectation value. Now, the slow-roll of this field (which is mandatory so as to yield negative pressure) and the fact that it is presently dominating the universe imply (if the minimum of the potential vanishes) that the field vacuum expectation value today must be of order m_{Pl}, and that its excitations are very light, with $m \sim H_0 \sim 10^{-33}$ eV [1]. But a further consequence of this is seldom emphasized: couplings of this field lead to observable long-range forces and time-dependence of the constant of nature (with corresponding violations of the Einstein Equivalence Principle).

Provided no symmetry cancels it, there will be a term in the effective Lagrangian coupling baryons to the scalar field. The cosmological evolution of the field will therefore lead to spacetime variation of the coupling constants of baryonic matter. In GUT quintessence scenarios, varying couplings can also arise from couplings of the scalar field to the kinetic term of the gauge fields. Bounds on varying couplings therefore restrict the evolution of the scalar field and provide constraints on dark energy that are complementary (and, in a sense to be described later on, more powerful than) those obtained by traditional means. As an aside, note that a scalar field that is responsible for the dark energy will necessarily yield sizeable varying couplings, but the opposite need not be true. An example are the so-called runaway dilaton scenarios in string theory. A spacetime varying scalar field coupling to matter mediates a new interaction. If the recent evidence for a varying fine-structure constant [2] is explained by a dynamical scalar field, this automatically implies the existence of a new force. A series of space missions (ACES, μSCOPE, STEP) will improve on current bounds on the Einstein Equivalence Principle by as many as 6 orders of magnitude. These must find violations if the current data is correct [3]. Note that since the scalar field is effectively massless on solar system scales, the new force will be similar to gravity in many respects. However, if the coupling is not precisely proportional to to mass, the new force will be composition-dependent, manifesting itself as a violation of the Weak Equivalence Principle (otherwise known as the Universality of Free Fall).

Another point that is often neglected is that in theories where a dynamical scalar field is responsible for varying α, the other gauge and Yukawa couplings are also expected to vary. Specifically, in GUTs there is a relation between the variation of α and that of the QCD scale, Λ_{QCD}, implying that the nucleon mass will vary when measured in units of an energy scale that is independent of QCD, such as the electron mass. We therefore expect variations of the proton-to-electron mass ratio, $\mu = m_p/m_e$. Some specific models are discussed in [4]. As a typical example, for the MSSM embedded on a GUT one has $\dot{\mu}/\mu \sim \dot{\Lambda}_{QCD}/\Lambda_{QCD} \sim Q\dot{c}\dot{\alpha}/\alpha$. In particular, the parameter Q is very model-dependent, not only in magnitude but even in sign. It seems to be the case that if the variation of the fine-structure constant is due to a variation of the respective unified coupling, then Q will be positive, whereas if it is due to a variation of the unification scale, Λ_{QCD}, then Q will be negative. Hence in the former case the variations of α and μ will have the same sign, whereas in the latter will be opposite. However, this should be taken with some care, since the range of models that has been explored is quite restriced, and it is therefore somewhat unclear if these results are generic. Be that as it may, however, one thing is clear. The wide range of α-μ relations implies that simultaneous measurements of both quantities are a powerful discriminating tool between competing models. Indeed, by measuring both as a function of redshift we can in principle test GUT scenarios without ever needing to detect any GUT model particles, say at accelerators.

Since evidence for such a scalar field is likely to be hard to find in the solar system, one should look for it in the early universe. The most direct method is to measure the value of various (dimensionless) coupling constants at high redshift. Several such techniques are well known, in particular using absorption systems along the line of sight of quasars. The method always consists of finding two (or more) transitions which depend differently on a given coupling (or combination of couplings). A joint observation of these transitions then allows the value of the coupling to be determined at the relevant redshift. For example, the fine-structure constant, α, can be measured by comparing various fine-structure atomic doublets, while the proton-to-electron mass ratio, μ, can be measured by comparing rotational and vibrational modes of molecules (typically H_2). Other methods (including radio observations) allow the measurement of products of α, μ and g_p.

The state of the art of absorption measurements of α is [2]. As first shown in [5], the proton-to-electron mass ratio can be measured through H_2 absorption lines, exploiting the fact that the vibro-rotational energy levels depend on the reduced mass of the molecule, and the dependence is different for different transitions. Recent results using this method also hint at a variation [6]. It is interesting to contrast measurements of α and μ. Measurements of α (including radio measurements) compare line shifts from different atoms, ionizations, or excitations. These may conceivably come from different components and have different velocities. On the other hand, molecular hydrogen has many lines with different shifts from the same lower state, ensuring they all come from the same gas. These measurements are also immune to contamination by other isotopic species. Finally, since the shift is different for each line, errors in redshift determinations can't mimic a varriation of μ. On the negative side, there are by now hundreds of absorption systems where one can measure α, but only 9 DLAs are currently known with H_2 absorption.

Impreved measurements of both α and μ are extremely useful. It is known that $\alpha(z)$ can be used to reconstruct the equation of state of dark energy [7, 8]. But if one also has $\mu(z)$, this reconstruction will be asier, not to mention less model-dependent. This not only complements results anticipated by hypothetical future experiments (such as JDEM and DUNE), but given reasonable expectations for forthcoming improvements in spectroscopic measurements, is easily competitive with the standard methods for dark energy equation of state reconstruction. Note that simply measuring $w(z)$ does not fully unravel the mystery of dark energy (as is sometimes naively put): in some sense it will only tell us what the answer isn't. In other words, it is not the end of the quest, though it is perhaps the end of the beginning.

More to the point, reconstructing the dark energy equation of state using varying couplings has several fundamental advantages over the standard methiods. Firstly, one has the advantage of a much larger lever arm in terms of redshift, since such measurements can be made up to redshifts of $z \sim 4$. This may not not seem a big advantage, since dark energy is only dynamically relevant at relatively low redshift, but in fact it should be a key one, since the additional redshift coverage, in addition to directly probing the underlying scalar field dynamics, means that one can reduce (and possibly elliminate) the model-dependence that is unavoidable in the standard methods (where parametrisations like $w = w_0 + w_1 z$ are dangerously naive). This is also aided by the fact that one has two different observables (the fine-structure constant and the proton-to-electron mass ratio), so one can in fact check the self-consistence of the reconstruction. Last but not least, it provides direct evidence distinguishing dynamical dark energy from a cosmological constant, which given the current data may be very challenging for the standard cosmological tests (for example, think of a scalr field with a near-constant equation of state $w = -0.99$).

There is an alternative way of measuring α, which uses emission rather than absorption lines. Specifically, this uses the $[OIII]$ emission lines at the wavelengths of 5007 and 4959Å. The wavelength separation is set by the LS-coupling, and so depends on α^4. Such a direct measurement would require extremely good absolute calibration and hence it is not really feasible. However, the wavelengths themselves depend on α^2, so all one has to do is to measure $R = (\lambda_1 - \lambda_2)/(\lambda_1 + \lambda_2) \propto \alpha^2$. Recently people took advantage of the improving quantity and quality of large-scale structure survey data [9]. These measurements are about one order of magnitude less sensitive that the absorption ones, and also at very low redshift. Nevertheless the method has its advantages. It is quite simple and straightforward, and the lines originate from the same excited level and are very strong (so can be observed with high S/N). Moreover the transitions are strongly forbidden (so there are no optical depth effects) and the separation is small (so differential reddening is unimportant). Finally, the method is insensitive to multiple components, variable excitation, and variable isotopic mixtures, which are of concern in absorption measurements.

This has prompted us to try to apply it at high redshift, say $z \sim 3$, which is the most interesting region. Such high-redshift measurements have never been attempted before. At this redshift the $[OIII]$ lines are in the near-IR, which is a clear disadvantage: the lines are fainter because of the distance, and the measurements are much harder because of the bright sky and strong skylines. On the other hand, quasars are intrinsically brighter at high redshift, and the many skylines (specifically, the OH lines from roto-vibrational transitions in the atmosphere) can be used to obtain an accurate wavelength callibration. With this in mind we have obtained 10 hours of observation time with the VLT's ISAAC spectrograph, in order to carry out a pilot study for this method. Radio galaxies and quasars were observed between redshifts 2 and 3, and the data reduction was done with a special purpose pipeline (not the VLT one). Our results [10] are consistent with zero change, but show consistently high values. The significance of this is currently being investigated in more detail. As things stand, we have shown that emission lime measurements can be made at high redshift, even though the sensitivity of the method is not yet competitive with absorption measurements. The method has few systematic uncertainties in the physics, and is therefore well suited for evolution studies. It is also worth pointing that our wavelength calibration is the best ever for ISAAC data, and currently good enough to detect $\Delta\alpha/\alpha \sim few \times 10^{-5}$. There are three main concerns regarding systematics in our method. Firstly, the OH line wavelenghts could be systematically offset (this is very unlikely, but not inconceivable). Secondly, the $[OIII]$ lines could be affected by other emission lines (typically iron lines). Finally, H_β emission could affect the 4959Å line (though not the peak position). Clearly, improvements will require larger samples, and we will be applying to becom an ESO Large Programme in the future.

The prospects for further, more accurate measurements of fundamental constants are definitely bright. The methods described above and other completely new ones that may be devised thus offer the real prospect of an accurate mapping of the cosmological evolution of the fine-structure constant, $\alpha = \alpha(z)$, and the proton to electron mass ratio, $\mu = \mu(z)$..

This may well prove to be the most exciting area of research in the coming years. Indeed the worse that can happen to cosmology is the scenario where a number of cosmological parameters are fixed by WMAP, then nothing new happens until Planck comes along and merely adds one (or in some cases maybe two) digits to the precision of each already-known parameter. After that cosmology may well be dead: there will be little incentive to pushing research further to figure out what the next digit is. However, if in the meantime violations of the Equivalence Principle and/or varying fundamental constants are confirmed, then one will (finally) have evidence for the existence of new physics—most likely in the form of scalar fields—in nature (which one may legitimately hope that Planck is able to probe) and an entirely new era begins.

Acknowledgments. This work was funded by grant POCTI/CTE-AST/60808/2004 from FCT (Portugal), in the framework of the POCI2010 program and FEDER.

References

1. Carroll, S. M.: Phys. Rev. Lett., 81, 3067 (1998)
2. Murphy, M. T., Webb, J. K., & Flambaum, V. V.: M.N.R.A.S., 345, 609 (2003)
3. Damour, T.: Astrophys. Space Sci., 283, 445 (2003)
4. Calmet, X. & Fritzsch, H.: Phys. Lett., B540, 173 (2002)
5. Thompson, R. I.: Astrophys. Lett., 16, 3 (1975)
6. Reinhold, E. et al.: Phys. Rev. Lett. 96, 151101 (2006)
7. Parkinson, D., Bassett, B. A., & Barrow, J. D.: Phys. Lett., B578, 235 (2004)
8. Nunes, N. J. & Lidsey, J. E.: Phys. Rev., D69, 123511 (2004)
9. Bahcall, J. N., Steinhardt, C. L., & Schlegel, D.: Astrophys. J., 600, 520 (2004)
10. Brinchmann, J. et al.: AIP Conf. Proc., 736, 117 (2005)

Slow-roll Corrections to Inflation Fluctuations on a Brane

S. Mizuno[1], K. Koyama[2], and D. Wands[2]

[1] Department of Physics, Waseda University, Okubo 3-4-1, Shinjuku, Tokyo 169-8555, Japan,
s.mizuno@aoni.waseda.jp
[2] Institute of Cosmology & Gravitation, University of Portsmouth, Portsmouth PO1 2EG,
United Kingdom, (kazuya.koyama,david.wands)@port.ac.uk,

1 Introduction and Motivation

The slow-roll approximation [1] is a useful tool to study the fluctuations generated during inflation. If we can neglect the coupling to metric perturbations and the effective mass of the field then the perturbations are described by the fluctuations of a free scalar field in de Sitter spacetime. This gives the familiar result that the power spectrum of scalar field perturbations at horizon-crossing is given by $(H/2\pi)^2$. One can then calculate the comoving curvature perturbation which is conserved on super-horizon scales for adiabatic perturbations.

However inflaton perturbations will be coupled to gravity (metric perturbations) at first-order in the slow-roll parameters. In four-dimensional general relativity it is known how to consistently include linear metric perturbations by working in terms of the gauge invariant combination of scalar field and curvature perturbations, the so-called Mukhanov-Sasaki variable, which obeys a simple wave equation [2]. Exact solutions are known for the special case of power-law inflation in general relativity which generalise the de Sitter result and have been used to calculate first-order slow-roll corrections in more general inflation models [3].

In this presentation, we consider inflation in the brane world model. New ideas in the string theory suggest that our observable universe is a 4-dimensional hypersurface, or brane, in a higher dimensional bulk spacetime [4]. The simplest example of this model is the Randall-Sundrum model where there is a brane embedded in a 5-dimensional anti-de Sitter (AdS) spacetime [5]. An AdS spacetime has a characteristic curvature scale μ associated with the negative cosmological constant in the bulk. The spacetime shrinks exponentially away from the brane and this geometry effectively compactifies the 5-dimensional spacetime with the effective size μ^{-1}. On large length scales $L > \mu^{-1}$, 4-dimensional Einstein gravity is recovered, while on small scales, the gravity becomes 5-dimensional [6]. In the early universe when the Hubble horizon is smaller than μ^{-1}, we expect significant effects from higher dimensional bulk spacetime. Indeed, the Friedmann equation is modified from the conventional 4-dimensional theory for $H\mu \gg 1$ [7].

In Ref.[8], the amplitude of the curvature perturbation is calculated by taking into account the modification of the Friedmann equation. But there the effect of coupling to five-dimensional gravity has been neglected and in particular it is assumed that the power spectrum of inflaton perturbations at horizon crossing is given by $(H/2\pi)^2$. This assumption is only valid to zeroth order in slow-roll parameters. At first order the inflaton perturbations will be coupled to metric perturbations. In the brane world, metric perturbations live in the 5-dimensional spacetime, and thus we must check if 5-dimensional effects change the result of conventional 4-dimensional theory. Especially, at small scales/high energies, the 5-dimensional effects could be large.

The first attempt to study the backreaction due to metric perturbations was made in Ref [9]. There, perturbations are solved perturbatively in slow-roll parameters. We should emphasize that this is the only possible way to perform the calculations analytically. In this paper, we extend earlier studies and investigate the backreaction due to higher-dimensional perturbations using a slow-roll expansion.

2 Summary and Discussions

In this paper we have studied the effect of metric perturbations upon inflaton fluctuations during inflation, at first-order in slow-roll parameters ϵ and η, which describe the dimensionless slope and curvature of the potential. If we neglect the slope and curvature of the inflaton potential then we obtain the familiar results for free field fluctuations in de Sitter spacetime, with a scale invariant power spectrum on large (super-horizon) scales. We take this as our zeroth-order result in a slow-roll expansion.

On a four-dimensional brane-world, embedded in a five-dimensional bulk, there are no exact solutions for cosmological perturbations (for a vacuum bulk described by Einstein gravity) except for the case of an exactly de Sitter brane. Thus the only way to calculate slow-roll corrections is perturbatively in a slow-roll expansion. We have calculated the leading order bulk metric perturbations sourced by the zeroth-order inflaton fluctuations on the brane. We find that inflaton fluctuations support an infinite tower of discrete bulk perturbations, with negative effective mass-squared.

Including the effect of the metric perturbations as an inhomogeneous source term in the wave equation for the first-order inflaton fluctuations we find that the effect of bulk metric perturbations becomes small on large scales, and we recover the usual result that the comoving curvature perturbation becomes constant outside the horizon.

However at small scales (or early times for a given comoving wavelength) the effect of bulk metric perturbations cannot be neglected. We are able to give an approximate solution for inflaton fluctuations at high energies and on sub-horizon scales using a truncated tower of bulk modes. This shows that the bulk metric perturbations change the amplitude of inflaton field fluctuations on the brane. By including a large number of bulk modes we can model this effect for many oscillations, but ultimately this change of amplitude becomes a large effect leading to a breakdown of our perturbative analysis.

It is not surprising in some ways that we see a large effect at small scales as these are high momentum modes which are expected to be strongly coupled to the bulk. Nonetheless this invalidates the usual assumption that gravitational effects are small far inside the cosmological horizon. It seems necessary to consistently solve for the coupled evolution of brane and bulk modes. We numerically tried to solve this problem and verified the validity of our perturbative approach as long as perturbations remain good. But it was also found that the change of the amplitude depends on the initial conditions for bulk metric perturbations. In order to give definite predictions for the amplitude of scalar perturbations in high energy inflation, we must specify the quantum vacuum state for coupled inflaton fluctuations and metric perturbations consistently and determine initial conditions. For this purpose, it would be useful to study the quantum theory of the toy model for a coupled bulk-brane oscillators in more details where we can consistently quantise a coupled system [10].

Our result implies the possibility that the assumption that the power spectrum of inflaton perturbations at horizon crossing on a brane is given by $(H/2\pi)^2$ could be invalid and we may have significant effects on the amplitude of perturbations from the backreaction due to the bulk metric perturbations. For the detailed analysis and calculation of this presentation, please see Ref. [11].

References

1. P. J. Steinhardt and M. S. Turner, Phys. Rev. **D29**, 2162 (1984).
2. M. Sasaki, Prog. Theor. Phys. **76**, 1036 (1986); V. F. Mukhanov, Sov. Phys. JETP **67**, 1297 (1988) [Zh. Eksp. Teor. Fiz. **94N7**, 1 (1988)]. V. F. Mukhanov, H. A. Feldman and R. H. Brandenberger, Phys. Rept. **215**, 203 (1992).
3. E. D. Stewart and D. H. Lyth, Phys. Lett. **B 302**, 171 (1993)
4. R. Maartens, Living Rev. Rel. **7**, 7 (2004).
5. L. Randall and R. Sundrum, Phys. Rev. Lett. **83**, 4690 (1999).
6. J. Garriga and T. Tanaka, Phys. Rev. Lett. **84**, (2000) 2778.
7. P. Binetruy, C. Deffayet, U. Ellwanger and D. Langlois, Phys. Lett. B **477**, 285 (2000).
8. R. Maartens, D. Wands, B. A. Bassett and I. Heard, Phys. Rev. D **62**, 041301 (2000).
9. K. Koyama, D. Langlois, R. Maartens and D. Wands, JCAP **0411** (2004) 002.
10. A. George, arXiv:hep-th/0412067 (2005); K. Koyama, A. Mennim and D. Wands, arXiv:hep-th/0504201 (2005).
11. K. Koyama, S. Mizuno and D. Wands, JCAP **0508** (2005) 009.

Statistical Mechanics of the SDSS Galaxy Distribution

A. Nakamichi[1] and M. Morikawa[2]

[1] Gunma Astronomical Observatory, 6860-86 Nakayama, Takayama, Gunma 377-0702, Japan
akika@astron.pref.gunma.jp
[2] Ochanomizu University, 2-1-1 Otsuka, Bunkyo-ku, Tokyo 112-8610, Japan,
hiro@phys.ocha.ac.jp

1 Introduction

Large-scale distributions of clusters and voids are recently observed in depth and in width. Such a large structure is considered to be formed mainly by gravity. Since gravity is a long-range unshielded force, self-gravitating systems often show (a) non-extensivity, and (b) the long-tail in various distribution functions. Therefore it seems apparent that ordinary Boltzmann statistical mechanics cannot be applied in naïve form. We will have to consider generalization or new statistical mechanics in order to explain the large scale structure. In this report, we utilize the brand new large size of survey data: Sloan Digital Sky Survey DR4 spectroscopic catalog.

2 Theoretical Models

We apply the following four theoretical models of statistical mechanics classified by (non)-extensivity of the model and the long/short tails of the associated distribution function.

-**Boltzmann statistical mechanics** : extensive, short tail

We consider the galaxy distribution is supposed to obey the Boltzmann statistical mechanics with grand canonical ensemble [1]. Virial parameter b, which measures the deviation from the dynamical-equilibrium: $pV = NT(1 - b)$, is introduced. The distribution function of voids is defined to be the probability of finding no galaxy in any part of the volume V : $f(0) = e^{-\bar{N}(1-b)}$ [2] where $\bar{N} = nV$. This is the generating function of the general function $f(N)$, which is defined to be the probability of finding N galaxies in the fixed volume V : [3],[1]. $f(N) = \frac{(-n)^N}{N!} \frac{\partial^N}{\partial n^N} f(0)$.

-**Boltzmann with Fractal Matter** : *non*- extensive, short tail In the fractal model, the void probability is given by $f(0) = e^{-\bar{N}} = \exp[-n \frac{\pi^{\frac{\alpha}{2}} r^\alpha}{\Gamma(\frac{\alpha}{2}+1)}]$.

-**Renyi Statistical Mechanics** : extensive, *long* tail Renyi statistical mechanics are a generalization of the ordinary Boltzmann statistical mechanics by introducing a new form of entropy. $S[p] = \frac{1}{1-q} \ln \left(\sum_i p_i^q \right)$, which is extensive: $S_{N\,galaxy} = sN$. The distribution function:

$$p_{N,E} = \left\{ 1 - \frac{1-q}{T} \left(E - \bar{E} - \mu \left(N - \bar{N} \right) \right) \right\}^{\frac{1}{1-q}}$$

has a *long-tail* with power-law shape, which maximizes the above Renyi entropy. The void probability is given by $f(0) = \{1 + (1-q)\,Ns\}^{-1}$.

-**Tsallis Statistical Mechanics** : *non-* extensive, *long* tail

Tsallis proposed a non-extensive entropy [4]: $S[P] = \frac{1}{1-q} \left((\sum_i P_i^{\frac{1}{q}})^{-q} - 1 \right)$.

This is non-extensive: $S_{N\,galaxy} = \frac{\{1+(1-q)\,s\}^N - 1}{1-q}$. The distribution function $p_{N,E} =$

$\frac{1}{\Xi_q} \left\{ 1 - \frac{1-q}{\bar{T}} \left(E - \bar{E} - \mu \left(N - \bar{N} \right) \right) \right\}^{\frac{1}{1-q}}$ maximizes the Tsallis entropy. Using this non-extensivity, we obtain the void probability as

$$f(0) = \{1 + (1-q)\,s\}^{\frac{-N}{1-q}} \left[1 + N \ln \{1 + (1-q)\,s\} \right]^{\frac{q}{1-q}} .$$

3 Comparison with Observation

We choose the observational data of the SDSS DR4 spectroscopic catalog [5]. This large and deep new survey includes 565,715 galaxies in a 4783 square degree spectroscopic area. In order to apply the count-in-cell method, we select the date of RA 150 - 210 deg., DEC 48 – 67 deg. From Figure 1, we further select the date of the absolute magnitude brighter than -20.53. The K-correction is also included.

Volume Limited Sample and appropriate theoretical models

We make four volume limited samples in order to investigate the distance dependence:

-*Sample A*: From z=0.065 to z=0.1, 11440 galaxies.
-*Sample B*: From z=0.101 to z=0.155, 17593 galaxies.
-*Sample C*: From z=0.156 to z=0.24, 8163 galaxies.
-*Sample D*: From z=0.241 to z=0.37, 3821 galaxies.

First we fix the parameters of the void probability function $f(0)$, then we calculate higher order distribution functions $f(N)$ for each theoretical model. For all volume limited samples from A to D, we find that Renyi and also Tsallis statistics better describe the observational data.

From these results, we conclude that the *long-tail in the distribution function is the essential property for describing distribution of galaxies.*

Acknowledgments. Work supported by Japan Grants-In-Aid for Scientific Research 15740169 (Wakate B).

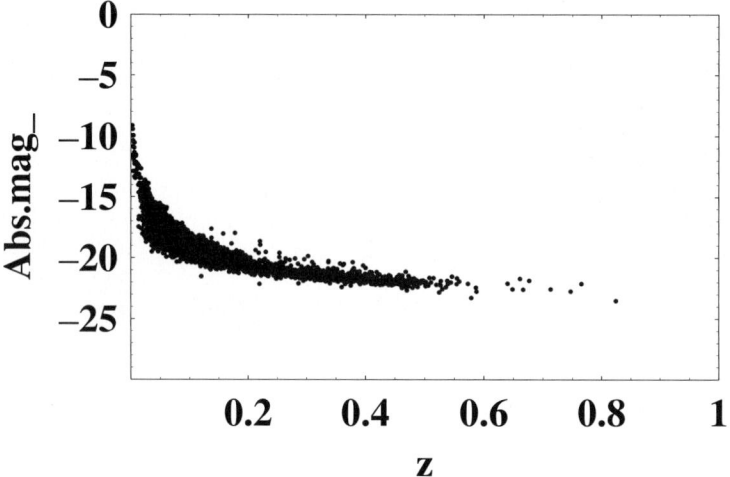

Fig. 1. Figure 1. Absolute magnitude v.s. redshift of SDSS DR4 spectroscopic data of galaxies.

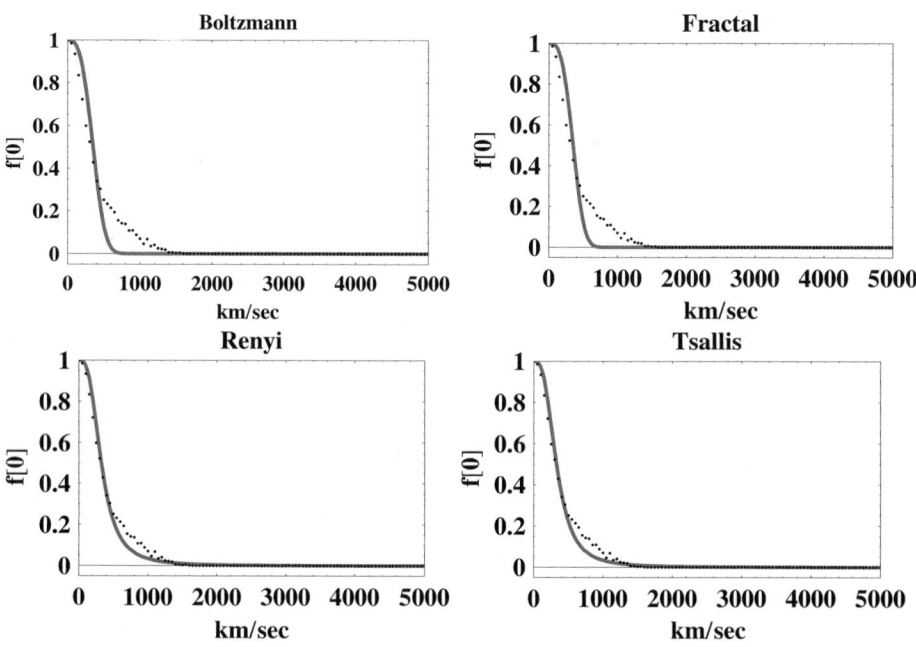

Fig. 2. Case of Sample B. Void distribution function for each of the theoretical models. Dots indicate observation and thick lines theoretical calculations.

References

1. A. Nakamichi, M. Morikawa, "Is Galaxy Distribution Non-extensive and Non-Gaussian?": Physica A, **341** 215-233 (2004).
2. W. C. Saslaw, A. J. S. Hamilton: ApJ **276** 13 (1984).
3. S. D. M. White: MNRAS **186** 145 (1979).
4. C. Thallis: J. Stat. Phys **52** 479 (1988).
5. http://www.sdss.org/dr4

The Second-Order Cosmological Perturbation and the Large Scale Structure Formation

H. Noh[1] and J.-C. Hwang[2]

[1] Korea Astronomy and Space Science Institute, Taejon, Korea, `hr@kasi.re.kr`
[2] Kyungpook National Univ., Taegu, Korea, `jchan@bh.knu.ac.kr`

1 Introduction

The evolution of linear structures in the expanding universe was first studied in Einstein's gravity. In the zero-pressure medium like the current universe Newton's gravity can be used, and the Newtonian results coincide with the ones in Einstein's theory. In this work, for the first time, we found such a relativistic-Newtonian correspondence even to the weakly nonlinear order. We show that even to the weakly nonlinear order the Newtonian equations can be used in all cosmological scales including the super-horizon scale.

The known equations are

$$\frac{\dot{a}^2}{a^2} = \frac{8\pi G}{3}\varrho - \frac{\text{const.}}{a^2} + \frac{\Lambda c^2}{3} \tag{1}$$

with $\varrho \propto a^{-3}$ for the background, and

$$\ddot{\delta} + 2\frac{\dot{a}}{a}\dot{\delta} - 4\pi G\varrho\delta = 0, \tag{2}$$

for the linear perturbation. The variable $a(t)$ is the scale factor, and $\delta \equiv \delta\varrho/\varrho$ with ϱ and $\delta\varrho$ the background and perturbed parts of the density field, respectively. The "const." part is interpreted as the spatial curvature in Einstein's gravity and the total energy in the Newton's gravity. We consider the flat background universe. Equation (2) is valid even in the presence of the cosmological constant Λ, so the above two equations describe very well our current universe and its large-scale structures. However, in the small scale, the structures are in nonlinear stage, and even in the large scale weakly nonlinear study is important. So far, a weakly nonlinear evolution has been studied using Newton's gravity only.

We show that, except for the gravitational wave contribution, the relativistic second-order perturbation is described by the same equations known in the Newtonian system. In the relativistic perturbations, due to the covariance of field equation we have freedom to fix the spacetime coordinate by choosing some of the metric or energy-momentum variables. By properly arranging the equations using various coordinate-invariant variables we can obtain the Newtonian correspondence.

2 Equations

We start from the completely nonlinear and covariant equations. The momentum conservation gives vanishing acceleration vector $\tilde{a}_a$ to all orders. The energy conservation and the Raychaudhury equation give

$$\dot{\tilde{\mu}} + \tilde{\mu}\tilde{\theta} = 0, \tag{3}$$

$$\dot{\tilde{\theta}} + \frac{1}{3}\tilde{\theta}^2 + \tilde{\sigma}^{ab}\tilde{\sigma}_{ab} + 4\pi G\tilde{\mu} - \Lambda = 0, \tag{4}$$

where $\tilde{\theta} \equiv \tilde{u}^a{}_{;a}$ is an expansion scalar, and $\tilde{\sigma}_{ab}$ is the shear tensor; for $\tilde{u}_\alpha = 0$ we have no rotation of the four-vector. An overdot with tilde is a covariant derivative along the $\tilde{u}_a$ vector, e.g., $\dot{\tilde{\mu}} \equiv \tilde{\mu}_{,a}\tilde{u}^a$. Combining these equations gives

$$\left(\frac{\dot{\tilde{\mu}}}{\tilde{\mu}}\right)^{\tilde{\cdot}} - \frac{1}{3}\left(\frac{\dot{\tilde{\mu}}}{\tilde{\mu}}\right)^2 - \tilde{\sigma}^{ab}\tilde{\sigma}_{ab} - 4\pi G\tilde{\mu} + \Lambda = 0. \tag{5}$$

Equations (3-5) are valid to all orders, i.e., these equations are fully nonlinear.

3 Correspondence

We consider equations perturbed to the second order in the metric and matter variables. Introduce

$$\tilde{\mu} \equiv \mu + \delta\mu, \quad \tilde{\theta} \equiv 3\frac{\dot{a}}{a} + \delta\theta, \tag{6}$$

where μ and $\delta\mu$ are the background and perturbed energy density, respectively, and $\delta\theta$ is the perturbed part of expansion scalar; we set $\delta \equiv \delta\mu/\mu$. Since our comoving coordinate condition fixes the coordinate system completely, all variables in this coordinate condition are equivalently coordinate invariant to the second order. We *identify* $\mu \equiv \varrho$ to the background, and

$$\delta\mu \equiv \delta\varrho, \quad \delta\theta \equiv \frac{1}{a}\nabla \cdot \mathbf{u}, \tag{7}$$

to the linear and second order perturbations. To the second order, we find that perturbed parts of eq. (3) give

$$\dot{\delta} + \frac{1}{a}\nabla \cdot \mathbf{u} = -\frac{1}{a}\nabla \cdot (\delta\mathbf{u}), \tag{8}$$

$$\frac{1}{a}\nabla \cdot \left(\dot{\mathbf{u}} + \frac{\dot{a}}{a}\mathbf{u}\right) + 4\pi G\mu\delta = -\frac{1}{a^2}\nabla \cdot (\mathbf{u} \cdot \nabla\mathbf{u}) - \dot{C}^{\alpha\beta}\left(\frac{2}{a}u_{\alpha,\beta} + \dot{C}_{\alpha\beta}\right),$$

where $C_{\alpha\beta}$ is the transverse and tracefree tensor-type perturbation (the gravitational wave). Thus, the presence of linear-order gravitational wave can generate the scalar-type perturbation to the second order by generating the shear terms. By combining these equations we have

$$\ddot{\delta} + 2\frac{\dot{a}}{a}\dot{\delta} - 4\pi G\mu\delta = -\frac{1}{a^2}\frac{\partial}{\partial t}\left[a\nabla\cdot(\delta\mathbf{u})\right] + \frac{1}{a^2}\nabla\cdot(\mathbf{u}\cdot\nabla\mathbf{u})$$

$$+\dot{C}^{\alpha\beta}\left(\frac{2}{a}u_{\alpha,\beta} + \dot{C}_{\alpha\beta}\right). \qquad (9)$$

Equations (8-9) are our extension of eq. (2) to the second-order perturbations in Einstein's theory. We will show that, except for the gravitational wave, exactly the same equations also follow from Newton's theory. The linear order gravitational wave contributes to the scalar-type perturbation to the second order. Here, we note the behaviour of the gravitational wave to the linear order. In the super-horizon scale the non-transient mode of $C_{\alpha\beta}$ remains constant, thus $\dot{C}_{\alpha\beta} = 0$, and in the sub-horizon scale, the oscillatory $C_{\alpha\beta}$ redshifts away, thus $C_{\alpha\beta} \propto a^{-1}$. Thus, we anticipate the contribution of gravitational wave to the scalar-type perturbation is not significant to the second order.

In the Newtonian context, the mass conservation, the momentum conservation, and the Poissons' equation give

$$\dot{\delta} + \frac{1}{a}\nabla\cdot\mathbf{u} = -\frac{1}{a}\nabla\cdot(\delta\mathbf{u}), \qquad (10)$$

$$\dot{\mathbf{u}} + H\mathbf{u} + \frac{1}{a}\nabla\delta\Phi = -\frac{1}{a}\mathbf{u}\cdot\nabla\mathbf{u}, \qquad (11)$$

$$\frac{1}{a^2}\nabla^2\delta\Phi = 4\pi G\varrho\delta, \qquad (12)$$

where $\delta\Phi$ is the perturbed gravitational potential. In this Newtonian situation $\mathbf{u}$ is the perturbed velocity and $\delta \equiv \delta\varrho/\varrho$. This completes our proof of the relativistic-Newtonian correspondence to the second order. Although we identified the relativistic density and velocity perturbation variables we *cannot* identify a relativistic variable which corresponds to $\delta\Phi$ to the second order.

4 Summary

We have shown that to the second order, except for the gravitational wave contribution, the zero-pressure relativistic cosmological perturbation equations can be exactly identified with the known equations in Newton's theory. We identified the correct relativistic variables which can be interpreted as the density $\delta\mu$ and the velocity $\delta\theta$ perturbations, and showed the Newtonian hydrodynamic equations remain valid in all cosmological scales including the super-horizon scale. Since the Newtonian system is exact to the second order in nonlinearity, any non-vanishing third and higher order perturbation terms in the relativistic analysis can be regarded as the pure relativistic correction.

Supermassive Black Holes in Galaxies

N. Nowak[1], R.P. Saglia[1], R. Bender[1,2], J. Thomas[2], and R. Davies[1]

[1] Max-Planck-Institut für extraterrestrische Physik, Giessenbachstrasse, 85748 Garching, Germany, nnowak@mpe.mpg.de
[2] Universitäts–Sternwarte, Scheinerstrasse 1, 81679 München, Germany

1 Introduction

Studies of the dynamics of stars and gas in the nuclei of nearby galaxies during the last few years have established that all galaxies with a massive bulge component contain a supermassive black hole closely correlated with the luminosity and the velocity dispersion of the associated bulge [3, 4], which indicates a close link between bulge evolution and black hole growth. We want to investigate whether these relations remain valid or how they change when galaxies with pseudobulges, very low-mass bulges and bulgeless galaxies are considered. This will give us further clues on the connection between central black hole and host galaxy dynamics.

Bulgeless galaxies have not experienced major merger episodes and secular evolution processes also have not yet become significant, therefore they might be the host galaxies of seed black holes in the earliest evolutionary stages. Only very few bulgeless galaxies were studied so far for their central black hole masses (e.g. [5]). The derived upper mass limits do not seem to correlate with disk properties. Although in these galaxies the black hole masses are consistent with zero, at least some should harbour a black hole, because they show nuclear activity. For example the low-luminosity Seyfert galaxy NGC 4395 [6] has no bulge.

Likewise, pseudobulges do not involve mergers but form via secular evolution. The few studies on black holes in pseudobulges [7, 8] indicate that they seem to follow the same $M - \sigma$ relation as classical bulges which would mean that the black hole mass does not depend on the bulge formation process.

The goal of our study is to obtain black hole masses for a representative set of bulgeless galaxies and pseudobulges using stellar kinematics. Therefore we observe the central regions of adequate galaxies with the integral field spectrograph SINFONI [9, 10] at the ESO VLT. This instrument works with adaptive optics delivering diffraction limited spatial resolution and, because it is working in the near infrared, is able to penetrate dust. This makes it particularly suitable to observe late-type spiral galaxies.

A number of galaxies close enough to resolve the black hole sphere of influence were selected from Kormendy & Kennicutt 2004 [11] (pseudobulges), Carollo et al. 1998 [12] and Böker et al. 2002 [13] (disk galaxies). The observation of most of the selected galaxies requires a laser guide star (LGS), therefore additional galaxies with a nearby natural AO guide star (NGS, closer than $15''$) were included in the sample. These are either pure disk galaxies from RC3 (e.g. NGC 3137), galaxies well-known because of their AO star (e.g. NGC 4486a) or galaxies with a nucleus bright enough to serve as AO star (e.g. NGC 1316). So far we observed the three mentioned galaxies in K band using the $3'' \times 3''$ or the $0.8'' \times 0.8''$ field of view. The data reduction was done using the free software package SPRED [14], which was developed specifically for SINFONI data.

2 Analysis of the NGC 4486a Data

NGC 4486a is a low-luminosity elliptical galaxy in the neighbourhood of M87 in the Virgo cluster. It contains an almost edge-on dusty nuclear stellar disk which is ~ 2 Gyr younger than the bulge [15]. A bright AO star is located $\sim 2.5''$ from the centre. We observed NGC 4486a with an exposure time of about 2 hours using the intermediate plate scale of $0.05''$ per spaxel (see Fig. 1 left). The point spread function has a FWHM of about two spaxels, i.e. $0.1''$ and was determined from the observation of the AO star directly after the galaxy. In order to determine the line-of-sight velocity distribution (LOSVD) several kinematic standard stars were observed using the same configuration as for NGC 4486a. A maximum penalized likelihood (MPL) technique [16] was used to obtain the LOSVDs. As a crosscheck they were compared with the LOSVDs obtained with FCQ [17], which were found to be similar. The rotation curve is shown in Fig. 1 right. The dynamical modelling is based on the Schwarzschild axisymmetric orbit

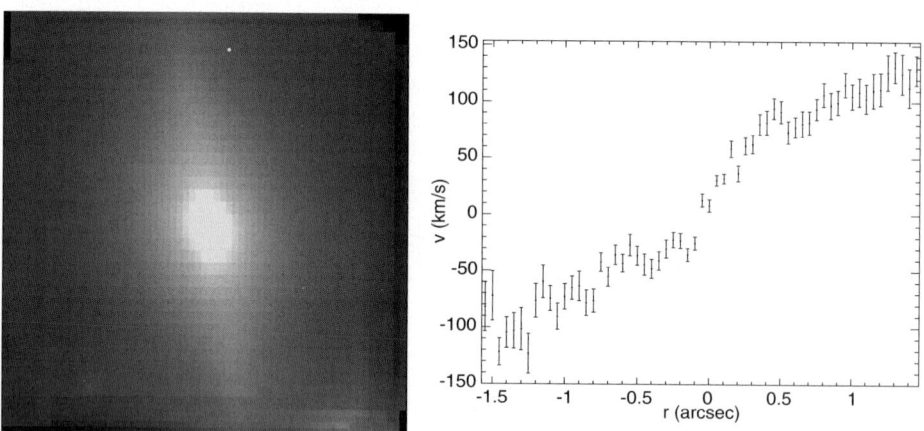

Fig. 1. Left: The combined datacube of NGC 4486a in the K-band. The field of view is $3'' \times 3''$. Right: The rotation curve of NGC 4486a along the major axis.

superposition code of Richstone and Tremaine 1988 [18] in the version of Thomas et al. [19, 20]. The luminosity density, a boundary condition for the dynamical modelling, of both the disk and the bulge component is determined from HST ACS imaging under the assumption that the galaxy is edge-on and axisymmetric. A representative set of orbits is run in every potential calculated from the stellar contribution and the black hole by varying systematically bulge and disk M/L and the black hole mass. The best-fitting kinematic parameters then follow from a χ^2-analysis.

3 Outlook

The modelling of NGC 4486a was successfully tested and is now under way using the entire dataset. The data reduction is finished for all observed galaxies and the modelling of NGC 3137 and NGC 1316 will be started soon. The laser guide star facility PARSEC will be available during the course of this year which allows us now to observe all the pseudobulges and bulgeless galaxies from our sample.

References

1. D. Richstone et al.: Nature **395**, A14 (1998).
2. R. Bender & J. Kormendy: Supermassive Black Holes in Galaxy Centers. In: *Astronomy, Cosmology and Fundamental Physics*, (2003) p. 262.
3. K. Gebhardt et al.: ApJ **539**, L13 (2000).
4. L. Ferrarese & D. Merritt: ApJ **539**, L9 (2000).
5. A.J. Barth, L.C. Ho, R.E. Rutledge & W.L.W. Sargent: ApJ **607**, 90 (2004).
6. A.V. Filippenko & L.C. Ho: ApJ **588**, L13 (2003).
7. J. Kormendy & K. Gebhardt: Supermassive Black Holes in Galactic Nuclei (Plenary Talk). In: *AIP Conf. Proc. 586: 20th Texas Symposium on relativistic astrophysics*, (2001) p. 363.
8. J. Pinkney et al.: ApJ **596**, 903 (2003).
9. H. Bonnet et al.: ESO Messenger **117**, 17 (2004).
10. F. Eisenhauer et al.: ESO Messenger **113**, 17 (2003).
11. J. Kormendy & R.C. Kennicutt, Jr.: ARAA **42**, 603 (2004).
12. C.M. Carollo, M. Stiavelli & J. Mack: AJ **116**, 68 (1998).
13. T. Böker, S. Laine, R.P. van der Marel et al.: AJ **123**, 1389 (2002).
14. J. Schreiber, in: *ASP Conf. Proc.*, Vol. 314, (2004) p. 380.
15. J. Kormendy, K. Gebhardt, D.B. Fisher et al.: AJ **129**, 2636 (2005).
16. K. Gebhardt et al.: AJ **119**, 1157 (2000).
17. R. Bender: A&A **229**, 441 (1990).
18. D. O. Richstone & S. Tremaine: ApJ **327**, 82 (1988).
19. J. Thomas et al.: MNRAS **353**, 391 (2004).
20. J. Thomas et al.: MNRAS **360**, 1355 (2005).

Lagrangian Description for the Cosmic Fluid

T. Tatekawa[1,2,3]

[1] Department of Physics, Ochanomizu University, 2-1-1 Otsuka, Bunkyo, Tokyo, 112-8610, Japan, tatekawa@cosmos.phys.ocha.ac.jp
[2] Department of Physics, Waseda University, 3-4-1 Okubo, Shinjuku, Tokyo, 169-8555, Japan
[3] Advanced Research Institute for Science and Engineering, Waseda University, 3-4-1 Okubo, Shinjuku, Tokyo, 169-8555, Japan

1 Linear Perturbation

The Lagrangian description for the cosmological fluid can be usefully applied to the structure formation scenario. This description provides a relatively accurate model even in a quasi-linear regime. Zel'dovich [1] proposed a linear Lagrangian approximation for a dust fluid. This approximation is called the Zel'dovich approximation (ZA) [1, 2, 3, 4].

In the ZA and its extended models, pressure was ignored. Recently, Lagrangian approximations in which the effect of pressure was taken into consideration have been analyzed. Buchert and Domínguez [5] discussed the effect of velocity dispersion using the collisionless Boltzmann equation They argued that models of a large-scale structure should be constructed for a flow describing the average motion of a multi-stream system. Then they showed that when the velocity dispersion is regarded as small and isotropic it produces effective "pressure" or viscosity terms. Furthermore, they derived the relation between mass density ρ and pressure P, i.e., an "equation of state." Hereafter, we call this model the Euler-Jeans-Newton (EJN) model. Actually, Adler and Buchert [6] have formulated the Lagrangian perturbation theory for a barotropic fluid. Morita and Tatekawa [7] derived the linear perturbative solutions for the polytropic fluid in Einstein-de Sitter Universe model. Then Tatekawa *et al.* [8] showed the solutions in generic Friedmann Universe models.

In the Lagrangian approximation, the displacement of the fluid element from homogeneous distribution is regarded as a perturbative quantity. Using this formalism, the matter density is precisely described. Furthermore we can obtain relatively good description for the density field, because the relation between the density and the displacement is nonlinear.

2 Higher-Order Approximation

ZA solutions are known as perturbative solutions, which describe the structure well in the quasi-nonlinear regime. To improve approximation, higher-order perturbative solutions of Lagrangian displacement were derived. For the EJN model, the higher-order perturbative solutions are derived for only Einstein-de Sitter Universe model.

Table 1 shows the references which derive higher-order perturbative solutions. For other extension which improves Lagrangian description, Tatekawa [4] classified them.

When we continue applying the solutions of ZA or its higher-order approximation after the appearance of caustics, the nonlinear structure diffuses and breaks. For avoidance of the formation of caustics, several modified models have been proposed. For more details, see Tatekawa [4].

Table 1. Higher-order Lagrangian perturbative solutions

order	dust model	the EJN model
2nd	Bouchet *et al.* [9]	Morita and Tatekawa [7]
	Buchert and Ehlers [10]	Tatekawa *et al.* [8]
3rd	Buchert [11]	Tatekawa [15]
	Bouchet *et al.* [12]	Tatekawa [16]
	Catelan [13]	
	Sasaki and Kasai [14]	

3 The Validity of Lagrangian Description

In Sec. 1, we mentioned that the Lagrangian description provides a relatively accurate model even in a quasi-linear regime. Here we check the validity of the Lagrangian description.

We consider the development of the spherical void with Lagrangian perturbation. Here we consider "top-hat" spherical void, i.e., a constant density spherical void in Einstein-de Sitter Universe.

Munshi, Sahni, and Starobinsky [17] derived a perturbative solution up to the third-order. In addition to these, Sahni and Shandarin [18] obtained up to fifth-order, Tatekawa [4] obtained solutions up to eleventh-order. They concluded that ZA remains the best approximation to apply to the late-time evolution of voids. From the viewpoint of the convergence of the series, the conclusion seems reasonable.

4 Future Prospect

According to past studies, it is well-known that the Lagrangian description for cosmic fluid realizes a quasi-nonlinear structure well. Why is the Lagrangian description so accurate? What is the reason? To answer this, several studies have been carried out. Recently, Yoshisato *et al.* [19] discussed the reason. They argued that the Lagrangian description extracts the essence of the gravitational collapse.

Recently, Buchert and Domínguez [20] proposed systematic derivation of the equation of motion for cosmic fluid. They propose systematic generalizations for Newtonian evolution equations. Then they discuss some non-perturbative results for structure formation and try to clarify the phenomenon of stabilization of large-scale structure emerging from gravitational instability. For a reasonable description for high density region or multi-stream region, we should study their generalized approaches.

The Lagrangian perturbation theory seems rather useful until the quasi-nonlinear regime develops. However, the density fluctuation becomes strongly nonlinear in the present Universe. Can we apply the theory to the problem of structure formation? Because a huge simulation has been performed [21], one may think that perturbation theory is no longer useful.

One possibility is analysis of the past structure. Because the density contrast will still be small in the high-z region, we expect that we will be able to discuss the characteristics of the density fluctuation using the Lagrangian description well.

References

1. Ya.B. Zel'dovich, A&A **5**, 84 (1970)
2. S.F. Shandarin and Ya.B. Zel'dovich, Rev. Mod. Phys. **61**, 185 (1989)
3. T. Buchert, A&A **223**, 9 (1989)
4. T. Tatekawa: Recent Res.Devel.Astrophys. **2**, 1-6 (2005)
5. T. Buchert and A. Domínguez, A&A **335**, 395 (1998)
6. S. Adler and T. Buchert, A&A **343**, 317 (1999)
7. M. Morita and T. Tatekawa, MNRAS **328**, 815 (2001)
8. T. Tatekawa *et al.*, Phys. Rev. **D66**, 064014 (2002)
9. F.R. Bouchet *et al.*, ApJ **394**, L5 (1992)
10. T. Buchert and J. Ehlers, Mon. Not. R. Astron. Soc. **264**, 375 (1993)
11. T. Buchert, MNRAS **267**, 811 (1994)
12. F.R. Bouchet *et al.*, A&A **296**, 575 (1995)
13. P. Catelan, MNRAS **276**, 115 (1995)
14. M. Sasaki and M. Kasai, Prog. Theor. Phys. **99**, 585 (1998).
15. T. Tatekawa, Phys. Rev. **D71**, 044024 (2005)
16. T. Tatekawa, Phys. Rev. **D72**, 024005 (2005)
17. D. Munshi, V. Sahni, and A.A. Starobinsky, ApJ **436**, 517 (1994)
18. V. Sahni and S. F. Shandarin, MNRAS **282**, 641 (1996)
19. A. Yoshisato, et al.: ApJ **637**, 555-560 (2006)
20. T. Buchert and A. Domínguez, A&A **438**, 443 (2005)
21. A. E. Evrard *et al.*, ApJ **573**, 7 (2002)

Mach's Principle and a Variable Speed of Light

A. Unzicker

Pestalozzi-Gymnasium München   alexander.unzicker@lrz.uni-muenchen.de

Summary. Ernst Mach (1838-1916) suggested that the origin of gravitational interaction could depend on the presence of all masses in the universe. A corresponding hypothesis of Sciama (1953) on the gravitational constant, $c^2/G = \sum m_i/r_i$, is linked to Dicke's (1957) proposal of an electromagnetic origin of gravitation, a precursor of scalar-tensor-theories. A formula for c depending on the mass distribution is given that reproduces Newton's law of gravitation. This mass distribution allows to calculate a slightly variable term that corresponds to the 'constant' G. The present proposal may supply an alternative explanation to the flatness problem and the horizon problem in cosmology. This is a short form of the author's paper gr-qc/0511038.

1 Riddles in Gravitational Physics

Many puzzling observations have shown up in gravitational physics in the past decades. Since Zwicky's observation of 'missing mass' in galaxy clusters an overwhelming evidence for 'dark matter' has been collected, in particular the flat rotation curves of galaxies. Since 1998, the relatively faint high-redshift supernovae are commonly explained as a manifestation of a new form of matter called 'dark energy'. As [1] comments, these new discoveries have been achieved "at the expense of simplicity"; it is quite disappointing that no counterpart to these forms of matter has been found in the laboratories yet. Thus in the view of recent data the following comment of [3], p. 635, is more than justified:

> 'It is worth remembering...that there is little or no direct evidence that conventional theories of gravity are correct on scales much larger than a parsec or so.'

In the meantime, new incompatibilities with conventional physics have been observed. The detection of an anomalous acceleration from the Pioneer missions [2] indicate that Newtonian gravity may not even be correct at scales within the solar system. Interestingly, this anomaly occurred at the same *dynamic* scale - about $10^{-9}m/s^2$ - as many phenomena indicating dark matter.

Furthermore, the scale $10^{-9}m/s^2$ is the scale that most of the precision measurements of G approach - it is still discussed if the discrepancies between the G measurements in the last decade [6, 12] have a systematic reason.

Besides the distance law, there is another lack of experimental data regarding the mass dependence. The Cavendish torsion balance uses masses of about $1kg$ to determine the mass of the earth, since there is no precise independent geological measure. The two-body treatment of Newtonian gravity tells us that the relative acceleration a_{12} is proportional to $m_1 + m_2$. It may well be that simple form is just an approximation valid for $m_1 >> m_2$. A slightly different mass dependence[1] would remain undetected even by the most precise emphemeris and double pulsar data, since in most cases just the product GM is measured.

In summary, there is quite a big gap between the common belief in the universality of Newton's law of gravitation and the experimental evidence supporting it. So far there is no evidence at all that it still works for accelerations below $10^{-9} m/s^2$.

2 Mach's Principle

Mach's principle is commonly known as follows: The reason for inertia is that a mass is accelerated with respect to all other masses in the universe, and therefore gravitational interaction would be impossible without the distant masses in the universe.

However, from Mach's principle it has been further deduced that the numerical value of the Gravitational G constant must be determined by the mass distribution in the universe, while in Newton's theory it is just an arbitrary constant. Since the square of the speed of light times the radius of the universe divided by the mass of the universe is approximately equal to G, such speculations were first raised by [5], [10] and [4].

3 Sciama's Version of Mach's Principle.

Since Mach never gave a quantitative statement of his idea, it has been implemented in various manners. An idea of a quantitative statement proposes the following functional dependence of Newton's constant G [10]:

$$\frac{c^2}{G} = \sum_i \frac{m_i}{r_i} = \Phi \tag{1}$$

This may be included in the definition of the gravitational potential, thus yielding

$$\phi = -G \sum_i \frac{m_i}{r_i} = -G\Phi = -c^2. \tag{2}$$

Since this suggests a variable c, a similar proposal will be given below.

4 Einstein and a Variable c.

It is worth remembering that a variable speed of light was already considered by Einstein in the years 1907 - 1911 [9].

---

[1] See the interesting measurements by [7]

The only reasonable way to define time and length scales is by means of the frequency f and wave length λ of a certain atomic transition, as it is done by the CODATA units. Thus by definition, $c = f\lambda = 299792458\frac{m}{s}$ has a fixed value *with respect to the time and length scales* that locally however may vary. Einstein had considered just variations of f, with a constant λ. It turns out however, that ([4]; [8]; [11]) all general relativistic effects regarding time and length scales can be described by two hypotheses:

- In the gravitational field frequencies and length scales are lowered by the amount $\frac{\delta f}{f} = \frac{\delta\lambda}{\lambda} = \frac{GM}{rc^2}$, and therefore the velocity of light, seen from outside, is lowered by the double amont.
- During propagation, the frequency f does not change.

5 The Equivalence Principle

Further calculations [11] show that the total energy $E = mc^2$ of a test particle appears as a conserved quantity during the motion in a gravitational field. In first order, we may write

$$\delta(mc^2) = c^2\delta m + m\delta(c^2) = 0. \tag{3}$$

Roughly speaking, the left term describes the increase in kinetic energy due to the relativistic δm and the right term the decrease of potential energy in a region with smaller c^2.

Obviously, *both* terms in eg. 3 are proportional to the mass m of the test particle. There is no reason any more to wonder why inertial and gravitating mass are of the same nature. The deeper reason for this is that gravitation is strictly speaking not an interaction between particles but just a reaction on a changing c.

6 c as a Function of the Mass Distribution in the Universe

For reasons outlined in [11], one of the most simple possibilities is the formula

$$c^2(m_i, r_i) = \frac{c_0^2}{\ln \sum \frac{m_i r_p}{r_i m_p}} =: \frac{c_0^2}{\ln \Sigma}, \tag{4}$$

whereby m_p and r_p stand for mass and radius of the proton. For the acceleration of a test mass then

$$|a| = |\frac{1}{4}\nabla c^2| = \frac{c_0^2}{4(\ln \Sigma)^2} \frac{\sum \frac{m_i}{r_i^2}}{\sum \frac{m_i}{r_i}} \tag{5}$$

holds, yielding the inverse-square law. The gravitational 'constant' can be expressed as

$$G = \frac{c_0^2}{4(\ln \Sigma)^2} \frac{1}{\sum \frac{m_i}{r_i}} = \frac{c^2}{4\ln \Sigma} \frac{1}{\sum \frac{m_i}{r_i}}. \tag{6}$$

which differs by a numerical factor $4\ln \Sigma$ from Sciama's proposal. It should be noted that the term $\sum \frac{m_i}{r_i}$ contained in G is approximately constant. The contributions from earth, sun, and milky way, $9.4 \cdot 10^{17}$, $1.33 \cdot 10^{19}$, and $7.5 \cdot 10^{20}$ (in kg/m), are minute compared to the distant masses in the universe that approximately amount to $1.3 \cdot 10^{27}$. This was first observed by Sciama [10], p. 39. Thus slight variations due to motions in the solar system are far below the accuracy of current absolute G measurements ($\Delta G/G = 1.5 \cdot 10^{-4}$). Sciama's comment on G given on page 39 in the 1953 paper applies here as well:

> '... then, local phenomena are strongly coupled to the universe as a whole, but owing to the small effect of local irregularities this coupling is practically constant over the distances and times available to observation. Because of this constancy, local phenomena appear to be isolated from the rest of the universe.'

7 Visible Matter and Flatness

From (6) and the assumption of an homogeneous universe, elementary integration over a spherical volume yields $\sum \frac{m_i}{r_i} \approx \frac{3m_u}{2r_u}$, and therefore

$$ m_u \approx \frac{c^2 r_u}{6G \ln \Sigma} \tag{7} $$

holds. If one assumes $r_p = 1.2 \cdot 10^{-15}m$, $m_p = 1.67 \cdot 10^{-27}kg$ (the proton values), and taking the recent measurement of $H_0^{-1} = 13.7\ Gyr$ that leads to $r_u = 1.3 \cdot 10^{26}m$, m_u is in approximate agreement with the observations, and corresponds to the fraction of visible matter $\Omega_v = 0.004$. Moreover, there is no more reason to wonder about the approximate flatness, since it is a consequence of the variable speed of light (4), if one uses $\rho = \frac{3m_u}{4\pi r_u^3}$ and $r_u H_0 \approx c$.

Newtonian gravity deduced from a variable speed of light can thus give an alternative explanation for the flatness problem and may not need dark forms of matter (DM, DE) to interprete cosmological data.

8 Outlook

Though the range of the universe we know about since the discovery of GR has increased dramatically, we usually extrapolate conventional theories of gravity to these scales. Physicists have learned a lot from quantum mechanics, but the historical aspect of the lesson was that the 19th-century extrapolation of classical mechanics over 10 orders of magnitude to the atomic level was a quite childish attempt. Under this aspect, the amount of research in cosmology which is done nowadays on the base of an untested extrapolation over 14 orders of magnitude is a quite remarkable phenomenon.

Since it deals with the very basics, it may be to early to try to test the present proposal with sophisticated data. It is merely an attempt to open new roads than a complete solution to the huge number of observational riddles.[2]

---

[2] An excellent overview on the observations modified gravity has to match gives [1].

Thus I consider the quantitative arguments above as preliminary results; besides the reproduction of GR and Newton, the more satisfactory aspect of this proposal seems the motivation for the equivalence principle, the implementation of Mach's principle and the 'removal' of the constant G.

Acknowledgments. I wish to thank Karl Fabian and Hannes Hoff for helpful comments.

References

1. Aguirre et. al.: Class.Quant.Grav. **18**, R223-R232 (2001)
2. Anderson et. al.: Phys.Rev.D. **65**, (2002)
3. Binney, J. and S. Tremaine: *Galactic Dynamics*. Princeton (1994)
4. Dicke, R. H.: Rev.Mod.Phys.**129**, 3, 363 (1957)
5. Dirac, P. A. M.: Proc.Roy.Soc.Lon.A **165**, 199-208 (1938)
6. Gundlach, J. and M. Merkowitz: Phys.Rev.Lett. **85**, 2869-2872 (2000)
7. Holding, S., F. Stacey, and G. Tuck: Phys.Rev.D **33**, 3487 (1986)
8. Puthoff, H.: Found.Phys. **32**, 927-943 (2002)
9. Ranada, A.: *arXiv: gr-qc/0403013* (2004)
10. Sciama, D. W.: (1953). MNRAS **113**, 34-42 (1953)
11. Unzicker, A.: *arXiv: gr-qc/0511038* (2005)
12. Uzan, J.-P.: Rev.Mod.Phys. **75**, 403-455 (2003)

A Century of Cosmology

E.L. Wright

UCLA Astronomy, PO Box 951547, Los Angeles, CA 90095-1547, USA,
wright@astro.ucla.edu

Summary. In the century since Einstein's anno mirabilis of 1905, our concept of the Universe has expanded from Kapteyn's flattened disk of stars only 10 kpc across to an observed horizon about 30 Gpc across that is only a tiny fraction of an immensely large inflated bubble. The expansion of our knowledge about the Universe, both in the types of data and the sheer quantity of data, has been just as dramatic. This talk will summarize this century of progress and our current understanding of the cosmos.

1 Introduction

When the COBE DMR results were announced in 1992, Hawking was quoted in The Times stating that "It was the discovery of the century, if not of all time." But the progress in cosmology in the last century has been tremendous, going far beyond the anisotropy of the cosmic microwave background. A century ago the "Structure of the Universe" meant the patterns of stellar number counts and proper motions that delineated the discus-shaped distribution of observed stars [1]. The true scale of the Milky Way and the nature of the extragalactic nebulae were yet to be determined. As late as 1963 people could say that there were only 2.5 facts in cosmology [2]: 1) the sky is dark at night, 2) the redshifts of galaxies show a pattern consistent with a general expansion of the Universe, and 2.5) the Universe has evolved over time. In 1963 the controversy between the Steady State [3,4] and the Big Bang [5,6] models of cosmology was still quite active, so the last item in the list was only a half-fact.

2 Einstein and λ

Once Einstein developed general relativity, giving a theory for classical gravity, he worked out a cosmological model [7] using what was then known about the Universe. Einstein assumed that the Universe had to be homogeneous, since even if the matter were confined to a finite region initially, the action of gravitational scattering would lead to stars being ejected from the initial distribution. Since the solution of Poisson's equation for a uniform density extending to infinity is not well defined, Einstein considered modifying Newtonian gravity by adding a λ term, giving $\nabla^2\phi - \lambda\phi = 4\pi G\rho$, which has the constant solution $\phi = -4\pi G\rho/\lambda$ for constant density. This modified Newtonian gravity has a short range compared to the infinite range inverse square law behavior of normal gravity. But in General Relativity the λ term had to be multiplied not by ϕ, which is not covariant, but rather by the metric $g_{\mu\nu}$ which contains ϕ as $g_{00} \approx 1 + 2\phi/c^2$ in the Newtonian approximation. Einstein found that for a uniform distribution of matter the geometry of space was that of the 3-sphere S^3 (the surface of a 4 dimensional ball), and that the addition of the λ term could compensate for the tendency for the Universe to collapse.

This static, spherical, homogeneous and isotropic Einstein Universe was not compatible with a solution to Olbers' Paradox. The stars in the Universe were emitting light, and this light would circulate around the spherical Universe and never be lost. As the stars continued to emit light, the Universe would become brighter and brighter. In addition to not solving Olbers' Paradox, Einstein's static Universe was only a unstable equilibrium point between a collapsing model and an infinitely expanding model. After the redshift of distant galaxies was discovered [8], Einstein referred to the introduction of λ as his greatest blunder.

Other cosmological models were developed as well. The de Sitter Universe used only λ and had no matter [9]. It has a redshift growing with distance, consistent with the Hubble Law, and this metric is now recognized as a homogeneous and isotropic Euclidean space ("flat" space means a 3 dimensional Euclidean geometry) that is expanding exponentially with time. The metric of the Steady State model is exactly the same as the de Sitter metric, but since the Steady State model has both matter and the continuous creation of new matter, the λ term was replaced with a C-field that made the matter plus C-field in the Steady State be equivalent to the pure vacuum energy of the de Sitter space.

Friedmann introduced models with matter that expanded from an initial singularity [10]. These models show a redshift proportional to distance which is consistent with the Hubble Law.

3 Big Bang vs. Steady State

After World War II Gamow tried to use the new knowledge about nuclear physics in a cosmological context. He and his students considered first a Universe full of neutrons that expanded and decayed. But they changed to a Universe initially filled with a hot dense medium about equally split between neutrons and protons. As the Universe expanded and cooled, heavier elements would be formed by the successive addition of neutrons. In this model, nuclei with high neutron capture cross-sections would be rapidly converted into heavier nuclei, and would thus be rare in the current Universe. Indeed, a roughly inverse relation between abundance and neutron capture cross section is observed. The time vs temperature during the cooling is related to the matter to radiation ratio in the Universe, and then by estimating the current density of matter, it was possible to estimate the current temperature of Universe [6] as 1 K or 5 K.

In this model, the current Universe is more or less curvature dominated so the ratio $\Omega = \rho/\rho_{crit}$ is $<< 1$ and therefore the age of the Universe is $t_\circ \approx 1/H_\circ = 978 \text{ Gyr}/H_\circ$ in km/sec/Mpc. Since the value of the Hubble constant given by Hubble was ≈ 500 km/sec/Mpc the age of the Universe was about 2 Gyr, which was too short according to the radioactive dating of the Earth. In the Steady State model the scale factor of the Universe is an exponential function of time, $a(t) = \exp(H(t - t_\circ))$, and thus the Hubble constant is actually a constant and the age of the Universe is infinite. But the average age of the matter in the Universe is in fact quite short: $\langle \tau \rangle = 1/3H \approx 700$ Myr for Hubble's value of the Hubble constant. Taking 6 Gyr as a minimum age for the Milky Way based on the radioactive dating of the Earth and adding time needed to form the galaxy and stars, the probability that a random piece of the Universe would be that old or older was only $e^{-9} \approx 10^{-4}$ in the Steady State model but this was still better than the zero probability in the Big Bang model.

4 Discovery and Non-discovery of the CMB

The first evidence [11] for the CMB was a rather inconspicuous interstellar absorption line in the spectrum of the hot, rapidly rotating star ζ Oph. This line was identified with the R(1) line of the cyanogen radical, CN. It was rather unusual, since it arises from a rotationally excited state of CN. Given the low density of the interstellar medium, ions and molecules in the ISM spend almost all of the time in the ground state. The excitation temperature [12] of the rotational transition based on this first CN data was 2.3 K, but its cosmological significance was widely ignored. Nobel Prize winner Herzberg stated: "From the intensity ratio of the lines with K=0 and K=1 a rotational temperature of $2.3°$ K follows, which has of course only a very restricted meaning." But it is not true that the cosmological significance was completely ignored. In fact Hoyle, in a review [13] of a book by Gamow & Critchfield, wrote that "the authors use a cosmological model in direct conflict with more widely accepted results. The age of the Universe is this model is appreciably less than the agreed age of the Galaxy. Moreover it would lead to a temperature of the radiation at present maintained throughout the whole of space much greater than McKellar's determination for some regions within the Galaxy." In making this statement Hoyle ignored the careful and explicit calculations of $T_\circ$ contained in a refereed article [6], which were perfectly compatible with the CN temperature measured by McKellar. I find it remarkable that none of the parties involved thought to follow up this possibility of a decisive test of the Big Bang vs. Steady State.

As a result, the actual discovery of the CMB was left to Penzias & Wilson, who were quite dedicated to finding the source of the excess noise they saw in their low-noise microwave receiver. Within 7 months of the announcement of Penzias & Wilson's result, the brightness temperature of the CMB at the 2.6 mm wavelength of the CN rotational transition had been shown to be the same as that measured at 7.4 cm by Penzias & Wilson. And thus Gamow, Alpher, Herman and Hoyle all missed the Nobel Proze.

Bob Dicke also narrowly missed the Nobel Prize. He invented the Dicke radiometer used in all direct measurements of the CMB spectrum, and during World War II came within a factor of ten of discovering the CMB, even while working from a sea level location in Florida [14]. While building a radiometer with a cold load to specifically search for the CMB, Dicke, Roll, Peebles & Wilkinson heard from Penzias & Wilson about their work.

5 Nucleosynthesis

Since Gamow's motivation for the Big Bang model was the origin of the chemical elements, it is instructive to see how the Big Bang and Steady State models fare on isotopic abundances. One cannot make isotopes heavier than ^4He by the sequential addition of neutrons in the Big Bang because there is no nucleus with atomic weight $A = 5$ that is even slightly stable. Thus the Big Bang, which set out to explain the abundances of the elements from hydrogen to uranium ended up only able to produce the elements from hydrogen to helium, with a sprinkling of lithium.

The Steady State model, on the other hand, proposed that matter was continuously created in the form of hydrogen, and that all heavier elements were created in stars. The triple-α reaction, $^4He + {}^4He + {}^4He \rightarrow {}^{12}C$, can run in stars because conditions of high density and high temperature persist for a long time. Thus all the elements can be produced in stars, starting by fusing hydrogen into helium. Stars produce about 1 gram of elements heavier than helium ("metals" to an astronomer) for each 3 grams of helium. But the average helium to metals ratio is about 12 to 1, and in low metallicity stars the ratio is even higher. Thus the Steady State model fails to produce enough helium, leading to the "helium problem". A proposed solution [15] to this problem was to have the ongoing creation of matter in the Steady State model occur sporadically in a number of "little bangs" that produce a mixture of hydrogen and helium.

The current model uses a combination of Big Bang nucleosynthesis, which produces most of the helium, and stellar nucleosynthesis, which produces the metals and some helium. Nuclear reactions during the Big Bang, starting from a mixture of protons and neutrons in thermal equilibrium at $t < 1$ sec after the Big Bang and $T > 10^{10}$ K, produce the deuterium (D), ^3He, ^4He and ^7Li seen in material that has not been processed through stars. Stars can destroy D and ^7Li, and generate more ^4He. The predicted abundances depend on the number of neutrino species and the baryon to photon ratio. The number of neutrino species primarily controls the ^4He abundance, and appears to be 3, consistent with determinations based on the decay width of the Z boson. The baryon to photon ratio controls the D:H ratio and the ^7Li abundance. The baryon to photon ratio consistent with the D:H ratio seen in high redshift quasar absorption line systems appears to predict a higher ^7Li abundance than that observed in a certain class of stars that has been thought not to have destroyed lithium in their surface layers. But since stars certainly do destroy lithium in their interiors this discrepancy is not too serious.

6 CMB Anisotropy

The CMB was found to be remarkably isotropic. This provided strong evidence that the simple Friedmann-Robertson-Walker metrics, adopted as a useful approximation, were actually quite good representations of the real Universe. While galaxy counts in different directions as a function of brightness had already demonstrated that the Universe was homogeneous and isotropic on large scales, it was still possible in 1967 to propose 10% inhomogeneities leading to 1% anisotropies in the CMB [16]. The first detection of the CMB anisotropy was at the 0.1% level [17,18], and it was soon in textbooks [19] as due to the motion of the Solar System relative to the Universe. The alleged "discovery" of the dipole anisotropy by the U2 experiment [20] was published 6 years after the textbook. Anisotropy other than this dipole term was not detected until 1992, by the COBE DMR [21,22] at the 0.001% level.

The low level of the anisotropy seen by COBE was strong evidence for the existence of dark matter. Dark matter can start to collapse as soon as the matter density exceeds the radiation density, while baryonic matter is frozen to the photons until recombination. Thus there is more growth for structures in dark matter dominated models, and thus the currently observed large scale structure can be generated starting from smaller seeds and hence smaller CMB anisotropies [23]. But the ratio of fluctuations at supercluster scales to the fluctuations at cluster scales required a modest reduction in the small scale power that could be supplied by either an open Universe model (OCDM), a model with a mixture of hot and cold running dark matter (mixed dark matter, or MDM), or by a model dominated by a cosmological constant (ΛCDM) [22].

Detailed calculations of cold dark models (CDM) showed acoustic oscillations in the amplitude of the anisotropy as function of angular scale [24]. These oscillations were caused by an interference between the fluctuations in the dark matter, which has zero pressure and thus zero sound speed, and the baryon-photon fluid which has a sound speed near 170,000 km/sec. The angular scale of the main peak of the angular power spectrum depends on two parameters: whether the Universe is open, flat or closed (Ω_{tot}), and the amount of vacuum energy density (Ω_Λ).

The first observational evidence for the main acoustic peak came from a collection of data from different experiments [25]. By 2003, the *WMAP* experiment [26] had measured the position of the first peak to an accuracy of better than 0.5% [27]. The result requires the Universe to lie along a line segment in the Ω_{tot} vs, Ω_Λ plane, with allowable models lying between a flat ΛCDM model having $\Omega_{tot} = 1$ and $\Omega_\Lambda = 0.73$ and a closed "super-Sandage" model with $\Omega_{tot} = 1.3$ and $\Omega_\Lambda = 0$. This model is referred to as super-Sandage because it has a Hubble constant of $H_\circ = 32$ km/sec/Mpc. The ratio of the first peak amplitude to second peak amplitude and to the valley between the peaks determined the ratio of the dark matter to baryon density and the baryon to photon density. The photon density is well measured by FIRAS on COBE, so the physical matter density $\Omega_m h^2 = 0.135 \pm 10\%$ is determined. Thus a value of $\Omega_m = \Omega_{tot} - \Omega_\Lambda$ also determines a Hubble constant, and the super-Sandage model has $H_\circ = 32$ km/sec/Mpc.

Thus the CMB anisotropy data alone cannot tell whether the Universe is flat or not, and cannot say that the cosmological constant is non-zero. This comes from the fact that the CMB anisotropy power spectrum is generated at recombination, when $z = 1089$, and at high redshifts the cosmological constant is a negligible contribution to the overall density. Other data are needed to verify the existence of the cosmological constant.

7 Supernovae

This other data was provided by observations of Type Ia supernovae. A definite correlation between the decay rate and peak luminosity of Ia SNe was seen [28], and using this calibration it was possible to pin down the acceleration of the Universe. This acceleration is usually denoted by the deceleration parameter, $q_\circ = \Omega_m/2 - \Omega_\Lambda$. If the expansion of the Universe is accelerating, it was slower in the past, and thus a larger time is needed to reach a given expansion ratio or redshift. With the larger travel time comes a larger distance, so distant supernovae appear fainter in accelerating models. Based on the SNe data, the Universe is definitely accelerating, so $q_\circ$ is negative. But the supernova data could be affected by systematic errors. In particular, evolution of the zero-point of the supernova decay rate *vs.* peak luminosity calibration can in principle match any cosmological model. In fact, a very simple exponential in cosmic time evolution in an Einstein - de Sitter Universe matches the supernova data very well, and is actually a slightly better fit to the data than the flat ΛCDM model with no evolution [31]. Since there is no way to rule out evolution with supernova data alone, the existence of the cosmological constant needs to be confirmed using other data or a combination of other data.

There are many other datasets that do confirm the acceleration first seen in the supernova data. The CMB anisotropy in combination with the Hubble constant data require an accelerating, close-to-flat Universe, as does the combination of CMB data and the peak of the large scale structure power spectrum $P(k)$, or the correlation of the CMB temperature fluctuations with superclusters via the late-integrated Sachs-Wolfe effect.

8 Search for Two Numbers

In the February 1970 *Physics Today*, Sandage published an article [32] titled "Cosmology: A search for Two Numbers". At that time, since the cosmological constant had fallen out of favor, the two numbers being sought were the Hubble constant $H_\circ \dot{a}/a$ and the deceleration parameter $q_\circ = -\ddot{a}a/\dot{a}^2$, where $a(t)$ is the scale factor. It is historically interesting that Sandage gave $H_\circ = 80$ km/sec/Mpc$\pm$ a factor of 1.6 for the Hubble constant, in view of the later distance scale controversy. But his value for the deceleration parameter, $q_\circ = 1.2 \pm 0.4$, is far from the currently accepted $q_\circ = -0.6$.

The current cosmological literature is again seeking two numbers, but a different set of two numbers. These are the equation of state parameter $w = P/\rho c^2$ and its time derivative w'. w is exactly -1 for a cosmological constant, but will be different for models involving evolving scalar fields. If w is not exactly -1 then it will interact with matter fluctuations via gravity, and thus the dark energy will be a function of both space and time or redshift. But since the Universe is almost homogeneous and isotropic, the average of w and w' over space should be a good description of the evolution of dark energy.

However, there is a very strong tendency among theorists to assume the Universe is flat when seeking w and w'. This is a logical error, since the evidence for a flat Universe comes from the agreement of the concordance ΛCDM model with all the data. But the concordance ΛCDM model has $w = -1$ and $w' = 0$ exactly. If w and w' are allowed to

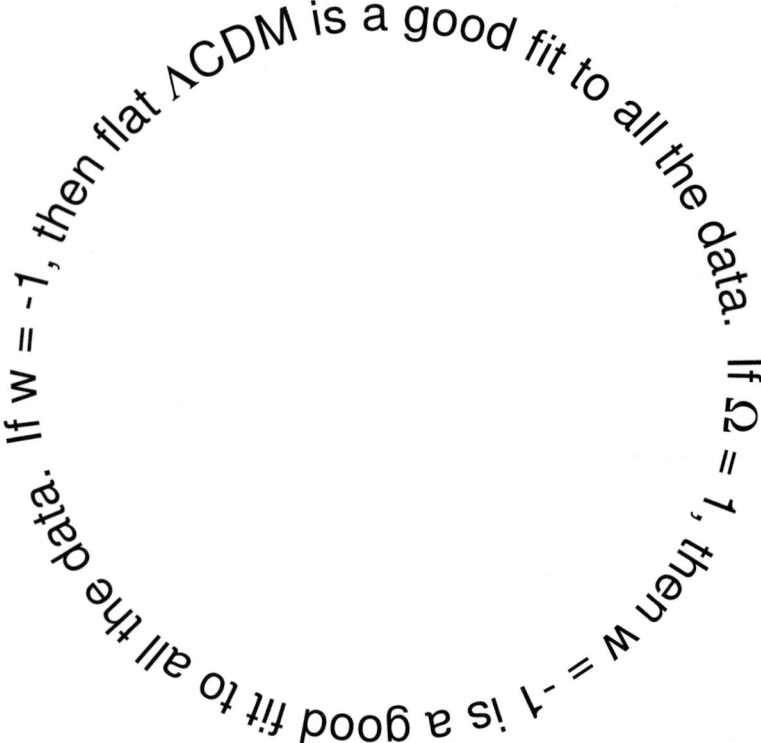

Fig. 1. The circular argument popular among current searches for w and w'. Models fits should always allow Ω_{tot} to be a free parameter.

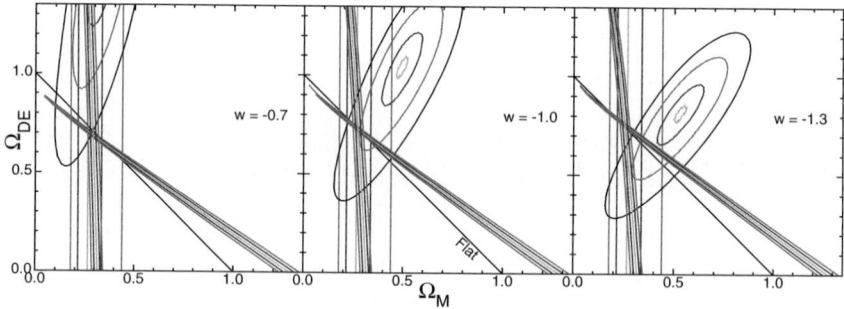

Fig. 2. Constraints on the dark energy density and the matter density from four relatively precise cosmological measurements for three different values of the dark energy equation of state. The ellipses show $\Delta\chi^2 = 0.1,\ 1,\ 4\ \&\ 9$ for my fits to the supernova data [29], while the lines show $-2\sigma,\ -1\sigma,\ 0,\ 1\sigma,\ \&\ 2\sigma$ values for $H_\circ$ [33] (vertical), baryonic oscillations in the SDSS [34] (not quite vertical), and the CMB acoustic peak angular scale [27] (inclined). The concordance at $w = -1$ is gradually lost for other values of w.

vary, then the evidence for a flat Universe must be re-evaluated. Limits on w and w' are only valid when a simultaneous fit for all relevant parameters is done. And when fitting to the CMB data, the spatial variations in the dark energy density should be included even though they are $O(10^{-5})$, since the CMB $\Delta T/T$ is of the same order.

9 Discussion

The progress in cosmology over the past century has been astronomical. We have gone from one fact in 1905 to hundreds of observed facts in 2005.

In terms of the mass of the known Universe the progress is even greater. In 1905 Kapteyn might have given the mass of the Universe as 10^9 $M_\odot$. Today the mass of the Universe is much larger than the mass of the Hubble volume $M = (4\pi/3)\rho_{crit}(c/H_=circ)^3 = 0.5c^3/(GH_\circ) = 4.4 \times 10^{22}$ $M_\odot$ so we can claim to have discovered more than 44 trillion times more of the Universe than was known in 1905. But we have also found that 95% of the density of the Universe is mysterious dark matter or dark energy.

Cosmologists today are working on problems that could hardly have been defined in 1905, but they are fortunate in having a large and growing body of precise observations with which to test their speculative constructs. Further observations of the CMB, large scale supernovae surveys, weak lensing and baryon oscillations will all provide major new datasets in the next century, and future progress in cosmology is assured.

References

1. J. Kapteyn: JRASC **8**, 145 (1914)
2. M. Longair: QJRAS **34**, 157 (1993)
3. F. Hoyle: MNRAS **108**, 372 (1948)
4. H. Bondi and T. Gold: MNRAS **108**, 252 (1948)
5. G. Gamow: Physical Review **74**, 505 (1948)
6. R. Alpher and R. Hermann: Physical Review **75**, 1089 (1949)
7. A. Einstein: Sitzung Derichte per Preussischen Akad. d. Wiss. 1917, 142 (1917)
8. E. Hubble: PNAS **15**, 168 (1929)
9. W. de Sitter: MNRAS **78**, 3 (1917)
10. A. Friedmann: Zeitschrift für Physik **21**, 326 (1922)
11. W. Adams: ApJ **93**, 11 (1941)
12. G. Herzberg: "Spectra of Diatomic Molecules", (New York: Van Nostrand Reinhold) (1950)
13. F. Hoyle: The Observatory **70**, 194-197 (1950)
14. R. Dicke, R. Behringer, R. Kyhl & A. Vane: Physical Review **70**, 340-348 (1946)
15. F. Hoyle & R. Tayler: Nature **203** 1108 (1964)
16. R. Sachs and A. Wolfe: ApJ **147**, 73 (1967)
17. E. Conklin: Nature **222**, 971-972 (1969)
18. P. Henry: Nature **231**, 516 (1971)
19. P. Peebles: "Physical Cosmology", (Princeton: Princeton University Press) (1971)
20. G. Smoot, Gorenstein, M. & R. Muller: PRL **39**, 898 (1977)
21. G. Smoot et al.: ApJL **396**, L1 (1992)

22. E. Wright et al.: ApJL **396**, L13 (1992)
23. P. Peebles: ApJ **263**, L1-L5 (1982)
24. J. Bond & G. Efstathiou: MNRAS **226**, 655-687 (1987)
25. D. Scott, J. Silk, & M. White: Science **268**, 829-835 (1995)
26. C. Bennett et al.: ApJS **148**, 1 (1993)
27. L. Page et al.: ApJS **148**, 233 (2003)
28. M. Phillips: ApJL **413**, L105-L108 (1993)
29. A. Riess et al.: ApJ **607**, 665-687 (2004)
30. S. Perlmutter et al.: ApJ **517**, 565-586 (1999)
31. E. Wright: astro-ph/0201196 (2002)
32. A. Sandage: Physics Today **23**, 34 (1970)
33. W. Freedman et al.: ApJ **553**, 47-72 (2001)
34. D. Eisenstein et al.: ApJ **633**, 560.

Part II

Gravity

The Stochastic Gravitational-Wave Background from Cold Dark Matter Halos

C. Carbone[1,2], C. Baccigalupi[1,2,3], and S. Matarrese[4,5]

[1] SISSA/ISAS, Astrophysics Sector, Via Beirut 4, I-34014, Trieste, Italy,
carbone@sissa.it
[2] INFN, Sezione di Trieste, Via Valerio, 2, 34127, Trieste, Italy
[3] Institut für Theoretische Astrophysik, Universität Heidelberg, Tiergartenstrasse 15, D-69121, Heidelberg, Germany, bacci@ita.uni-heidelberg.de
[4] Dipartimento di Fisica 'Galileo Galilei', Università di Padova, Italy
matarrese@pd.infn.it
[5] INFN, Sezione di Padova, Via Marzolo 8, I-35131 Padova, Italy

1 The Hybrid Approximation and the Most-Probable Halo

The evolution of cosmological perturbations away from the linear regime is characterized by the mode-mixing of different types of fluctuation, which implies that density perturbations can act as a source for tensor metric modes. Moreover, N-body simulations seem to show a departure of the halo density profile from the spherical symmetry and suggest instead a triaxial shape. Consequently, CDM halos around galaxies and galaxy clusters, in the highly non-linear regime, can be potential sources of gravitational waves (GW).

In order to evaluate the halo-induced GW output, we use a hybrid approximation of the Einstein field equations [1], which mixes post-Newtonian (PN) and second-order perturbative techniques. This approach provides, on small scales, a PN approximation to the source of gravitational radiation, and, on large scales, it converges to the first- and second-order perturbative equations, but still describing, on all the cosmologically relevant scales, the dynamics of the involved CDM structures by means of the standard Newtonian equations.

Consequently, we exploit the well known Newtonian continuity, Euler and Poisson equations together with the formula expressing the solution of the inhomogeneous GW equation in the hybrid approximation [1]

$$h^\alpha{}_\beta(\eta, \mathbf{x}) = \frac{4G}{c^4} \frac{1}{ar} \mathcal{P}^\alpha{}_\nu{}^\mu{}_\beta \left[a^3 \int d^3\tilde{x}\, \mathcal{R}^\nu_{\text{eff}\,\mu} \right]_{ret}, \tag{1}$$

where η is the conformal time, a represents the scale factor of the Universe, r is the comoving distance between source and observer and the projection operator is given by $\mathcal{P}^\alpha{}_\beta \equiv \delta^\alpha{}_\beta - x^\alpha x_\beta / r^2$. The subscript "$ret$" means that the quantity has to be evaluated at the retarded space-time point $(\eta - r/c, \tilde{\mathbf{x}})$. Eq. (1) expresses the GW output $h^\alpha{}_\beta$

in terms of integrals over the source "stress distribution" $\mathcal{R}^{\alpha}_{\text{eff }\beta}$, given in terms of the density field, the peculiar velocity field and the peculiar gravitational potential, as

$$\mathcal{R}^{\alpha}_{\text{eff }\beta} = \rho \left(v^{\alpha} v_{\beta} - \frac{1}{3} v^2 \, \delta^{\alpha}{}_{\beta} \right) + \frac{1}{4\pi G a^2} \left(\partial^{\alpha} \varphi \, \partial_{\beta} \varphi - \frac{1}{3} \partial^{\nu} \varphi \, \partial_{\nu} \varphi \, \delta^{\alpha}{}_{\beta} \right) . \quad (2)$$

In order to simulate a CDM halo evolving towards virialization, we use the gravitational collapse of homogeneous ellipsoids as described in Ref. [2] which approximate the evolution of overdensity peaks, in the initial Gaussian density field, via a homogeneous ellipsoid dynamics. Each ellipsoid perturbation is characterized by three parameters: the ellipticity e, the prolateness p and the linear density contrast δ.

In order to evaluate the total energy density $\Omega_{\text{GW}}(f)$ associated with the stochastic halo-induced GW background, we consider CDM structures over a mass range $M = 5 \times 10^9 M_{\odot} - 5 \times 10^{15} M_{\odot}$ and a virialization redshift range $z = 0 - 4$. Moreover, we exploit the unconditional halo mass function (HMF) $n(M, z)$ as determined in Ref. [3], which provides a good fit to N-body simulations of structure clustering in a variety of cosmological models. Assuming that CDM halos emit in all directions, are casually oriented and uniformly distributed around the observer, we approximate each of these structures as the *most-probable* ellipsoid of mass M ($p = 0$, $e = (\sigma/\delta)/\sqrt{5}$, where σ is the linear rms value of the δ distribution), placed at a comoving distance $r(z)$ from the observer, where z is the collapse redshift. Then, using Eqs. (1)–(2) and averaging on directions and orientations, we numerically evaluate the mean power spectral density PSD(f, z, M) for each z and M and integrate the obtained $\Omega_{\text{GW}}(f, z, M) \approx (2\pi^2)/(3H_0^2) f^3 \text{PSD}(f, z, M)$ (weighted by the HMF) over all redshifts and masses to get the total energy density $\Omega_{\text{GW}}(f)$.

2 Results and Conclusions

In this work, we have estimated the GW background from cosmological tensor modes produced by the highly non-linear collapse of CDM density perturbations, i.e. generated during the strongly non-linear stage of CDM halo evolution.

Fig. (1) shows our main result [4], i.e the total energy density $\Omega_{\text{GW}}(f) \approx 10^{-20}$ associated with this stochastic GW background, as a function of the frequency f. Our findings are consistent with previous works in this field [5]. Let us remark that the signal is significant at very low frequencies, $f \approx 10^{-18}$ Hz, due to the cosmological time scales involved in the collapse of CDM halos. The total spectrum of the signal is composed by many single peaks which represent the contribution to the total background from each most-probable halo weighted via the mass function at different redshifts. These peaks are caused by the subset of structures leading to a non-negligible GW signal. In fact, it can be easily shown that the more massive objects give rise to higher values of the GW strain. Consequently, the dominant contribution to this stochastic GW background is likely to be produced by CDM halos corresponding to nearby galaxies and galaxy clusters ($10^{14} - 10^{15} M_{\odot}$) at low redshifts, $z \leq 1$.

Finally and most importantly, the quantity $h^2 \Omega_{\text{GW}}(f) \approx 10^{-20}$ is comparable to the energy density associated with the stochastic background induced by primordial

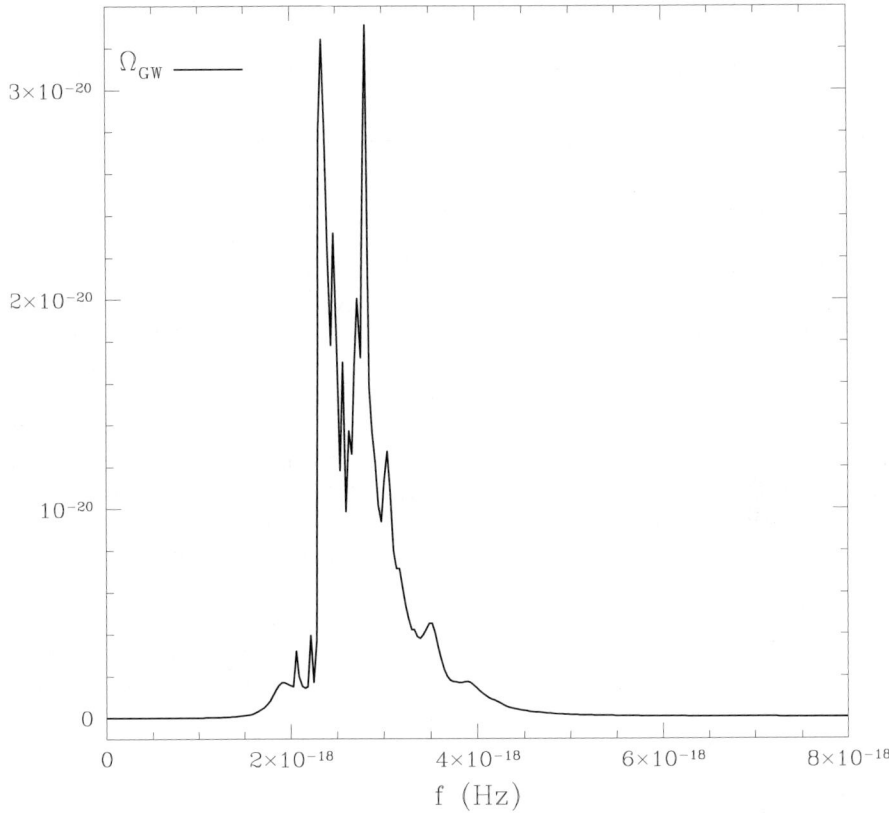

Fig. 1. The total energy density $\Omega_{\mathrm{GW}}(f)$ associated with the stochastic halo-induced GW background as a function of the proper frequency f at observation.

GWs. In fact, if the energy scale of inflation is $V^{1/4} \approx 1 - 2 \times 10^{15} \mathrm{GeV}$, the energy density associated with the primordial stochastic GW background, with a tensor spectral index $n_T \approx 0$, is $\approx 10^{-21} - 10^{-17}$ for frequencies of the order of $10^{-18} - 10^{-17}$ Hz (*e.g* Ref. [6]).

Moreover, due to the cosmological scales involved, the stochastic halo-induced GW background might boost the CMB temperature anisotropies on large angular scales and, on the other hand, the produced temperature quadrupole, being scattered off by the free electrons of the intra-cluster and intra-galactic media, could give rise to secondary E and B polarization modes.

References

1. C. Carbone and S. Matarrese, Phys. Rev. D **71**, 043508-1 (2005)

2. J. R. Bond and S. T. Myers, Astrophys. J. Suppl. Series **103**, 39 (1996)
3. R. K. Sheth, H. J. Mo, G. Tormen, Mon. Not. Roy. Astron. Soc. **323**, 1 (2001)
4. C. Carbone, C. Baccigalupi, S. Matarrese, [arXiv:astro-ph/0403392]
5. V. Quilis, J. M. Ibanez and D. Saez, Astrophys. J. **501**, 21 (1998)
6. M. Maggiore, Phys. Rep. **331**, 283 (2000)

Gravitational Wave from Realistic Stellar Collapse : Odd Parity Perturbation

K. Kiuchi[1], K.-I. Nakazato[2], K. Kotake[2], K.e Sumiyoshi[3], and S. Yamada[4]

[1] Department of Physics, Waseda University, Okubo, 3-4-1, Shinjuku-ku, Tokyo 169-8555, Japan, kiuchi@gravity.phys.waseda.ac.jp
[2] Science & Engineering, Waseda University, 3-4-1 Okubo, Shinjuku, Tokyo, 169-8555, Japan, (nakazato,kkotake)@heap.phys.waseda.ac.jp
[3] Numazu College of Technology, Ooka 3600, Numazu, Shizuoka 410-8501, Japan, sumi@numazu-ct.ac.jp
[4] Science & Engineering, Waseda University, 3-4-1 Okubo, Shinjuku, Tokyo, 169-8555, Japan, shoichi@heap.phys.waseda.ac.jp

1 Introduction

For a future gravitational wave (G.W.) astronomy, it is very important to make theoretical templates of G.Ws. from any kinds of relativistic dynamical system. In this work, we focus on G.W. from a massive stellar core collapse. In many previous works in which the G.Ws from such a kind of source have been studied, it is standard scenario that rotation of stellar core breaks down the spherical symmetry and produce the G.W. as a result. Some works adapted the rapidly rotating core collapse model. But the recent work reported by Heger et al. show that the rotation of the star is not so rapid [1]. Taking account of this result, we consider that there is a possibility that the spherical symmetry preserves under gravitational core collapses. This implies that the linear perturbation approach is efficient to estimate G.Ws. from stellar core collapses. On the other hand, there are many intensive works using this perturbation approach under the dynamical background [2]-[4]. But one cannot say that the templates for detection have been made completely, because the stellar models have been simplified in these works, e.g., dust ball or polytropic EOS. Therefore, our purpose is to analyze G.Ws. from realistic stellar collapses by using the perturbation approach. For simplicity, we only consider the odd parity mode for perturbation. Throughout this paper, we use the geometrical unit $c = G = 1$.

2 Set Up and Numerical Result

For the odd parity perturbation, a basic equation is reduced to a single wave equation in both interior region of a star and exterior one. The interior field equation has a source term in which a perturbation of a four velocity is contained [5, 6]. The exterior equation is the Regge-Wheeler equation [9]. To check a performance of our code, we try the two kinds of test simulations. The first check is the linear gravitational wave propagating in the Minkowski space time [10]. Next we computed the propagation of gravitational waves in the collapse of a homogeneous dust ball [2]. In both test calculation, we confirm that our code works well.

We specify the stellar matter as a perfect fluid and adopt the equation of state developed by Shen et al. [7]. The detail of the hydrodynamic calculation is shown in [8]. We also choose two kinds of stellar collapse as the background. One is the core which has 1.5 solar mass and this model occurs the core bounce. This is the model of supernova explosion. The other core has 13.5 solar mass, which collapses to the black hole, and models population III star. For both models, we set the momentarily static condition on the initial surface. In this work, we only give one kind of matter perturbation for simplicity.

2.1 Model A : type II supernova explosion

We show the numerical results in Fig. 1. Fig. 1(a) shows the gravitational waveform. This waveform is similar to the typical waveform, which is emitted at the core bounce of massive core collapse. We also estimate a frequency of this burst-type gravitational wave. The frequency is about 1 Hz. This value is within the frequency bands of the ground-base detectors, e.g., TAMA, LIGO. Next, we analyze what generates the gravitational wave. Fig. 1(b) shows the trajectories of the mass shell and represents that the core bounce occurred around 0.32s. Fig. 1(c) is the time variation of the lapse function. From this figure, we confirm that the lapse function becomes small at the core bounce, that is the magnitude of the gravity becomes larger. As a result, the gravitational wave generates.

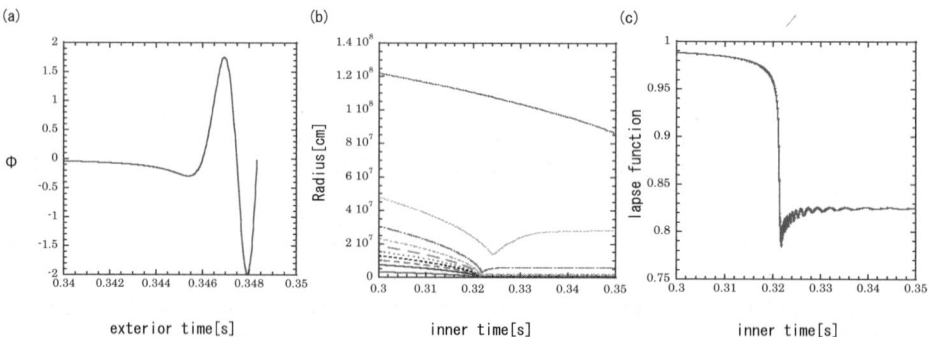

Fig. 1. (a) Gravitational wave for $l = 2$ from the realistic stellar core collapse. We set the momentarily static initial date, (b) the trajectories of the mass shell, and (c) the lapse function.

2.2 Model B : Pop III Star

In the Model B, the background calculation stopped before the gravitational wave reach to the observer because of black hole formation. One of method to avoid this problem is the single null formulation [4]. Using this formulation, one can perform the hydro-dynamic simulation just before black hole forms. But our hydrodynamic code is made to solve the Boltzman equation for neutrino transfer under our metric (Misner-Sharp metric). Therefore, it is a not simple problem to restructure the code for single null formulation. In stead of this method, we plane to perform black hole excision.

3 Conclusion

In this work, we estimate gravitational waves emitted from general relativistic and real-istic stellar collapse. To calculate the wave form, we use the gauge-invariant approach. The results are that in the Model A, the gravitational wave is generated at the core bounce and its frequency is about 1 Hz and in the Model B, black hole formation prevents performing the long-time numerical calculation. Therefore we cannot see the G.W. reaching to the observer.

The future works are following :

1. Black hole excision.

2. The dependence of G.W. for the matter perturbation (initial condition).

3. The dependence of G.W. for initial stellar model.

4. Including the neutrino effect.

References

1. A. Heger, S. E. Woosley and H. C. Spruit, Astrophys. J. **626**, 350 (2005)
2. C. T. Cunningham, R. H. Price and V. Moncrief, Astrophys. J. **224**, 643 (1978); **230**, 870 (1979); **236**, 674 (1979).
3. E. Seidel and T. Moore, Phys. Rev. D **35**, 2287 (1987); **38**, 2349 (1988); **42**, 1884 (1990).
4. T. Harada, H. Iguchi and M. Shibata, Phys. Rev. D **68**, 024002 (2003)
5. U. H. Gerlach and U. K. Sengupta, D **19**, 2268 (1979).
6. J. M. Martin-Garcia and C. Gundlach, Phys. Rev. D **61**, 084024 (2000); **64**, 024012 (2001)
7. H. Shen, H. Toki, K. Oyamatsu and K. Sumiyoshi, Nucl. Phys. A **637** (1998) 435
8. K. Nakazato, K. Sumiyoshi and S. Yamada, arXiv:astro-ph/0509868.
9. T. Regge, Phys. Rev. **108**, 1063 (1957).
10. S. A. Teukolsky, Phys. Rev. D **26**, 745 (1982).

Finding the Electromagnetic Counterparts of Standard Sirens

B. Kocsis[1], Z. Frei[1], Z. Haiman[2], and K. Menou[2]

[1] Institute of Physics, Eötvös University, Pázmány P. s. 1/A, 1117 Budapest, Hungary,
bkocsis@complex.elte.hu
[2] Department of Astronomy, Columbia University, 550 West 120th Street, New York, NY
10027

Summary. The gravitational waves (GWs) emitted during the coalescence of supermassive black holes (SMBHs) will be detectable with the future *Laser Interferometric Space Antenna* (*LISA*). The direction and distance can be determined from the accumulated GW signal with a precision that increases rapidly in the final stages of the inspiral. We find that for $M = (10^5 - 10^7) M_\odot$ near $z = 1$ the angular uncertainty decreases under $1°$ at least several hours before the plunge, allowing a targeted electromagnetic (EM) observation of the final stages of the merger with a wide field instrument. We then calculate the size of the final, three dimensional error volume. Under the plausible assumption that SMBH-SMBH mergers are accompanied by gas accretion leading to Eddington-limited quasar activity, we find that many cases this error volume will contain at most a single quasar for $M = (10^5 - 10^7) M_\odot$ near $z = 1$. This will allow a straightforward test of the hypothesis that GW events are accompanied by bright quasar activity. The identification and observation of counterparts would allow unprecedented tests of the physics of MBH accretion, such as precision–measurements of the Eddington ratio. They would clarify the role of gas as a catalyst in SMBH coalescences, and would also offer an alternative method to constrain cosmological parameters.

1 Introduction

The detection of low frequency gravitational radiation with the future GW space detector *LISA* will open a new window in observational astronomy [6]. Of particular interest is the detection of the coalescence of SMBHs, which is detectable at cosmological distances. A fundamental question is whether the spatial location of the GW event can be localized to within a sufficiently small three-dimensional volume that the EM counterpart can be uniquely identified.

Previous studies associated counterparts with host galaxies and galaxy clusters, and concluded that the identification of such EM counterparts will be difficult because there will be too many counterpart candidates to choose from [20, 9]. In contrast, here we focus on quasars as plausible counterparts, and account for the 3D spatial information by using the redshift of the EM observation obtained from the luminosity distance determined by *LISA*, and restrict the EM Eddington ratio characteristic for quasars [15, 10].

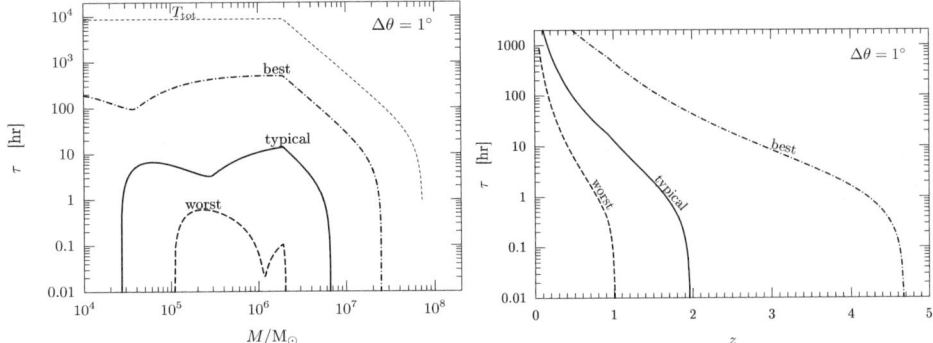

Fig. 1. The amount of time needed before ISCO for the *LISA* angular estimation uncertainty to decrease below $1°$ in three representative cases depending on absolute errors at ISCO. *Left*: total mass, M, is changed while $z = 1$ is kept constant. *Right*: redshift z is changed while $M = 2 \times 10^6\,M_\odot$ is kept constant. The total observation time T is shown with a thin dotted line. Our simple scalings of errors are conservative estimates (see [14] for a detailed analysis).

With these specifications, we demonstrate that in some cases, a specific counterpart can be uniquely determined.

If EM counterparts to *LISA* events exist, they will likely be related to the accretion of gas onto the SMBHs involved in the coalescence. Provided this accretion is not supply-limited, bright quasar counterparts approaching the Eddington luminosity would then be expected. A few additional arguments favor this scenario: galaxy mergers in hierarchical scenarios of structure formation are expected to deliver a significant amount of gas to the central regions of the merging galaxies [4], and this gas may play a catalyst role in driving SMBH coalescence [2, 8, 7]. Ultimately, however, many of the complex processes involved remain poorly understood. For example, [1] have argued that, in the limit of a small mass ratio of the two SMBHs, a prompt and luminous EM signal may be expected during coalescence, while [16] have argued that in the limit of equal mass SMBHs, only a much delayed EM afterglow would be expected. Our working assumption in the present study is that bright quasar activity is a plausible EM counterpart to *LISA* events. Our results show that a joint *LISA* and EM observation allows to test this assumption.

2 The GW Error Volume

A comparison of the gravitational waveform with the anticipated detector noise can be used to estimate the accuracy with which *LISA* will be able to extract the physical parameters of the coalescence events [11, 3, 20, 9]. Due to computational limitations, results are currently only available in a restricted domain of the parameter space. We adopt the results of [20], valid for the single choice of an equal-mass SMBH binary with total mass $M_0 = 2 \times 10^6\,M_\odot$ and redshift $z_0 = 1$ for the uncertainties of luminosity distance δd_L and angle $\delta\theta$. According to the distributions, $\delta d_L/d_L$ is 2, 4, 20 $\times 10^{-3}$ and $\delta\theta$ is $0.10°, 0.17°, 1.7°$ in the best, typical, and worst cases, respectively. For other masses and redshifts we scale the errors with $T^{-1/2} \times (S/N)^{-1}$, where T is the observation time and S/N is the signal to noise ratio of the GW measurement. This scaling roughly accounts for mild correlations between the distance and angular positioning.

Figure 1 shows the observation time τ before the innermost stable circular orbit (ISCO) when the angular uncertainty decreases under $1°$. The results show that a targeted wide field observation of this solid angle could be used to search for EM variability several hours before the merger. In fact, improved estimates on correlated errors suggest that these results corresponds to worst–case estimates, and in some cases much longer warning times (weeks–months) may be possible (Finn & Larson 2005; private communication).

We next consider how to use the three–dimensional spatial localization of the SMBH merger event by *LISA*. The solid angle error box directly yields the two–dimensional angular position error on the sky, in which any EM counterpart will be located. However, an additional step is necessary to convert the luminosity distance, $d_{L,LISA}$, measured by *LISA* into a redshift, z, which is the relevant observable for the EM counterpart. To account for the magnification of weak gravitational lensing, μ, and the peculiar velocity of the source relative to the Hubble flow, v, the standard cosmological distance – redshift relation $d_L(z, p_i)$ is scaled with $\mu^{-1/2}$, and Doppler shift adds a redshift $\Delta z = (1 + z)v/c$ to the fiducial z value. Here $p_i = (\Omega_{DE}, \Omega_M, w, h)$ are the cosmological parameters. Therefore we get the conversion formula $d_L(z; p_i, v, \mu)$.

All of these parameters are uncertain, the average value and the typical variance of these parameters are constrained using other cosmological observations. Therefore in addition to the *LISA* errors δd_L or δz_L, the conversion to redshift introduces additional uncertainty. We assume fiducial parameters are $\langle (\mu, v, \Omega_{DE}, \Omega_M, h, w) \rangle = (1, 0, 0.7, 0.3, 0.7, -1)$ consistent with *WMAP* and SDSS ([19]). Assuming nonzero variance for $\langle \delta \mu^2 \rangle$, $\langle \delta v^2 \rangle$, and $\langle \delta p_i \delta p_j \rangle$, we get

$$\langle \delta z^2 \rangle = \langle \delta z_{LISA}^2 \rangle + \langle \delta z_{cosm}^2 \rangle + \langle \delta z_{pec}^2 \rangle + \langle \delta z_{wl}^2 \rangle, \tag{1}$$

Here the *LISA* term is $\langle \delta z_{LISA}^2 \rangle = (\partial d_L/\partial z)^{-2} \langle \delta d_{L,LISA}^2 \rangle$. The cosmological term is derived from the error matrix $\langle \partial p_i \partial p_j \rangle$ of the cosmological parameters, using the forecasts of two future cosmological probes *Planck* and LSST. We derive the cosmological Fisher matrix following the prescriptions of [12] and [21]. For the peculiar velocity term we use rms errors of $500\,\mathrm{km\,s^{-1}}$. Finally the weak gravitational lensing effects are calculated using the HALOFIT routine of [18] for the matter power spectrum.

The results are shown in Fig. 2. At $z > 0.3$, weak lensing uncertainties greatly exceed all other sources of uncertainties.

3 Search for Quasar Counterparts

To estimate the typical number of quasar counterparts to a specific SMBH merger event, we combine the size of the comoving *LISA* error box with the space density of quasars, by integrating over the quasar luminosity function, $\phi(L, z)$:

$$N = b \left(\frac{\partial^2 V}{\partial z \partial \Omega} \right) \langle \delta z^2 \rangle^{1/2} \langle \partial \Omega^2 \rangle^{1/2} \int_{L_{min}}^{L_{max}} \mathrm{d}L \, \phi(L, z), \tag{2}$$

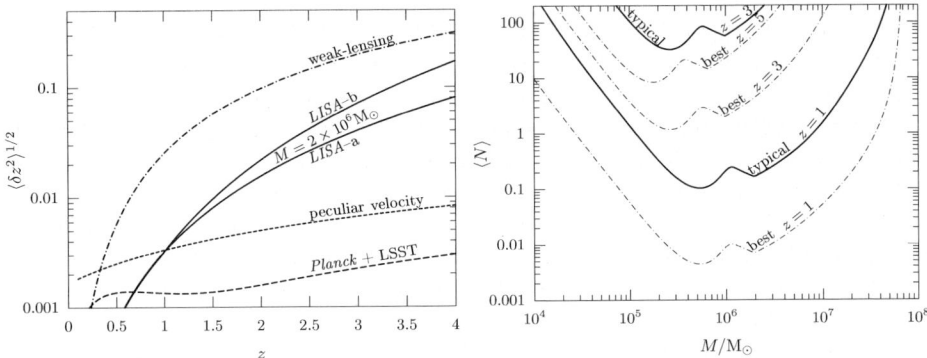

Fig. 2. *Left*: errors on the inferred redshift of an EM counterpart to a *LISA* coalescence event, for $m_1 = m_2 = 10^6 \, M_\odot$. The intrinsic *LISA* error on the luminosity distance, d_L, is shown as a solid line for the two scalings $\delta d_L/d_L \propto SNR^{-1}$ ("*LISA*–a") and $\delta d_L/d_L \propto \Delta T^{-1/2} \times SNR^{-1}$ ("*LISA*–b"). *Right*: the average number of quasars in the three–dimensional *LISA* error volume.

Here b represents the inhomogeneous clustering of quasars, and the second term is the cosmological volume element, $L_{\min}$ and $L_{\max}$ are the minimum and maximum luminosity that can be associated to the quasar counterpart. For b we adopt a double power law profile with exponents 1.20 (1.63) for $s < 25$ ($s > 25$Mpc). For the $\phi(L, z)$ we use the parameteric form of [17] with updated parameter values consistent with recent observations [5, 22]: $L^* = 5.06 \times 10^{10} L_\odot$, $z_s = 1.66$, $\zeta = 2.6$, $\xi = 2.8$, $\beta_l = 1.45$, $\beta_h = 3.31$, and $\Phi_L^* = 1.99 \times 10^{-6} \mathrm{Mpc}^{-3}$. We fix the integration bounds at $L_{\min} = 0.7$ and $L_{\max} = 1.3 \, L_{\mathrm{Edd}}(M)$. Figure 2 shows that the resulting number of possible quasar counterparts are typically less than one for a range of masses and redshifts.

4 Conclusions

We have considered the possibility that SMBH-SMBH mergers, detected as GW sources by *LISA*, are accompanied by gas accretion and quasar activity with a luminosity approaching the Eddington limit. Under this assumption, we have computed the number of quasar counterparts that would be found in the 3D error volume provided by *LISA* for a given GW event. We found that weak lensing errors exceed other sources of uncertainties on the inferred redshift of the EM counterpart and increase the effective error volume by nearly an order of magnitude. Nevertheless, we found that for mergers between $\sim (4 \times 10^5 - 10^7) \, M_\odot$ SMBHs at $z \sim 1$, the error box may contain a single quasar. Also we have shown that the angular uncertainties decrease below $1°$ at least several hours before the merger, enabling a targeted EM observation. This would make the identification of unique EM counterparts feasible, allowing precise determinations of the Eddington ratio of distant accreting SMBHs, and providing an alternative method to constrain cosmological parameters (for detailed discussions see [13]).

Acknowledgments. We acknowledge support from OTKA-T047244 and MRTN-CT-2004-503929.

References

1. Armitage, P. J., & Natarajan, P.: ApJ **567**, 9 (2002)
2. Begelman, M. C., Blandford, R. D., & Rees, M. J.: Nature **287**, 307 (1980)
3. Barack, L. & Cutler, C.: PRD **69**, 082005 (2004)
4. Barnes, J. E., & Hernquist, L.: ARA&A **30**, 705 (1992)
5. Croom, S. M. et al.: MRNAS **356**, 415 (2005)
6. 5th LISA Symposium, Noordwijk, CQG, 22, 10 (2005)
7. Escala, A., et al.: ApJ **607**, 765 (2004)
8. Gould, A., & Rix, H.-W.: ApJ **532**, 29 (2000)
9. Holz, D. E. & Hughes, S. A.: ApJ **629**, 15 (2005)
10. Hopkins, P. F., et al.: ApJS **163**, 1-49 (2006)
11. Hughes, S. A.: MNRAS **331**, 805 (2002)
12. Hu, W., & Haiman, Z.: PRD **68**, 063004 (2003)
13. Kocsis, B., et al.: ApJ **637**, 27 (2006a)
14. Kocsis, B. et al.: in preparation (2006b)
15. Kollmeier, J. A.: ApJ **648**, 128-139 (2006)
16. Milosavljevic, M., & Phinney, E. S.: ApJ **622**, L93 (2005)
17. Madau, P., Haardt, F., & Rees M. J.: ApJ **517**, 648 (1999)
18. Smith, R. E. et al.: MNRAS **341**, 1311 (2003)
19. Tegmark, M.: PRD **69**, 103501 (2004)
20. Vecchio, A.: PRD **70**, 042001 (2004)
21. Wang, S., et al.: PRD **70**, 123008 (2004)
22. Wyithe, J. S., & Loeb A.: ApJ **577**, 68 (2002)

Strong-Field Tests of Gravity with the Double Pulsar

M. Kramer

University of Manchester, Jodrell Bank Observatory, Jodrell Bank, Cheshire SK11 9DL, UK
`michael.kramer@manchester.ac.uk`

Summary. This first ever double pulsar system consists of two pulsars orbiting the common center of mass in a slightly eccentric orbit of only 2.4-hr duration. The pair of pulsars with pulse periods of 22 ms and 2.8 sec, respectively, confirms the long-proposed recycling theory for millisecond pulsars and provides an exciting opportunity to study the works of pulsar magnetospheres by a very fortunate geometrical alignment of the orbit relative to our line-of-sight. In particular, this binary system represents a truly unique laboratory for relativistic gravitational physics.

1 Introduction

This conference, celebrating Albert Einstein's great achievements, showed very clearly that even today hundreds of scientists around the world are deeply involved in searching for the limits up to which his theory of general relativity (GR) can be applied. To date GR has passed all observational tests with flying colours. Still, it is the continued aim of many physicists to achieve more stringent tests by either increasing the precision of the tests or by testing different aspects. Some of the most stringent tests are obtained by satellite experiments in the solar system. However, solar-system experiments are made in the gravitational weak-field regime, while only in strong fields deviations from GR may appear more clearly or even for the first time (see e.g. [1]). This strong-field regime can be explored using radio pulsars.

Pulsars, highly magnetized rotating neutron stars, are unique and versatile objects which can be used to study an extremely wide range of physical and astrophysical problems. Besides testing theories of gravity one can study the Galaxy and the interstellar medium, stars, binary systems and their evolution, plasma physics and solid state physics under extreme conditions. This wide range of applications is exemplified by the first ever discovered double pulsar [2, 3]. In the spirit of the conference, I concentrate here on the gravitational aspects and demonstrate that this rare binary system represents a truly unique laboratory for relativistic gravity.

The presented work is the result of a very active and fruitful collaboration between Jodrell Bank Observatory (Kramer, Lorimer, Lyne, McLaughlin), University of

British Columbia (Stairs, Ferdman), ATNF (Manchester, Sarkissian), Osservatorio Astronomico di Cagliari (Burgay, D'Amico, Possenti), NCRA (Joshi), NAIC (Freire) and
Columbia University (Camilo). Since we can give only a brief summary of the results
here, a complete and detailed report of the timing results is presented elsewhere (Kramer
et al. in prep.).

2 Strong-Field Tests with the Double Pulsar

The double pulsar is a system of two visible radio pulsars with periods of 22.8 ms
(PSR J0737−3039A, simply called "A" hereafter) and 2.8 s (PSR J0737−3039B, simply called "B" hereafter), respectively [3]. Its short and compact (orbital period of
$P_b = 144$ min), slightly eccentric ($e = 0.9$) orbit makes the double pulsar the most
extreme relativistic binary system ever discovered, demonstrated by the system's remarkably high value of periastron advance ($\dot{\omega} = 16.9\,\mathrm{deg\ yr^{-1}}$, i.e. four times larger
than for the Hulse-Taylor pulsar!). In fact, due to the emission of gravitational waves,
the system will shrink and coalesce in only ~ 85 Myr. This boosts the hopes for detecting a merger of two neutron stars with first-generation ground-based gravitational
wave detectors by a factor of several compared to previous estimates [2, 4]. Moreover,
the detection of a young companion B around an old millisecond pulsar A confirms the
evolution scenario proposed for recycled pulsars (e.g. [5, 6]).

Since neutron stars are very compact massive objects, the double pulsar (and other
double neutron star systems) can be considered as almost ideal point sources for testing
theories of gravity in the strong-gravitational-field limit. Tests can be performed when a
number of relativistic corrections to the Keplerian description of an orbit, the so-called
"post-Keplerian" (PK) parameters, can be measured. As A has the faster pulse period,
we can time A much more accurately than B, allowing us to measure a total of five very
precise PK parameters for A's orbit.

The PK parameter, $\dot{\omega}$, is the easiest to measure and describes the relativistic advance
of periastron. It provides an immediate measurement of the total mass of the system.
The parameter γ denotes the amplitude of delays in arrival times caused by the varying
effects of the gravitational redshift and time dilation (second order Doppler) as the
pulsars move in an elliptical orbit at varying distances with varying speeds. The decay
of the orbit due to gravitational wave damping is expressed by the change in orbital
period, $\dot{P}_b$. Two further parameters, r and s, are related to the Shapiro delay caused by
the gravitational field of the companion. Their measurement is possible, since – quite
amazingly – we observe the system almost completely edge-on.

For point masses with negligible spin contributions, the PK parameters in each theory should only be functions of the a priori unknown neutron star masses and the well
measurable Keplerian parameters. With the two masses as the only free parameters, the
measurement of three or more PK parameters over-constrains the system, and thereby
provides a test ground for theories of gravity. These tests can be illustrated in a very el-
egant way [7]: The unique relationship between the two masses of the system predicted
by any theory for each PK parameter can be plotted in a diagram showing the mass of
A on one axis and that of B on the other. We expect all curves to intersect in a single
point if the chosen theory is a valid description of the nature of this system.

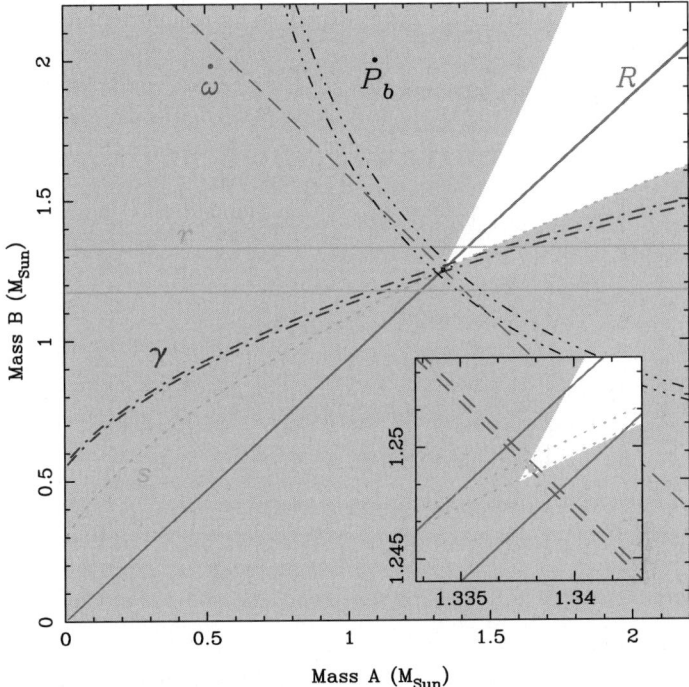

Fig. 1. 'Mass–mass' diagram showing the observational constraints on the masses of the neutron stars in the double pulsar system J0737–3039. The shaded regions are those that are excluded by the Keplerian mass functions of the two pulsars. Further constraints are shown as pairs of lines enclosing permitted regions as given by the observed mass ratio and PK parameters shown here as predicted by general relativity. Inset is an enlarged view of the small square encompassing the intersection of these constraints (see text).

Most importantly, the possibility to measure the orbit of both A and B provides a new, qualitatively different constraint. Indeed, with a measurement of the projected semi-major axes of the orbits of both A and B, we obtain a precise measurement of the mass ratio simply from Kepler's third law, $R \equiv M_A/M_B = x_B/x_A$ where M_A and M_B are the masses and x_A and x_B are the (projected) semi-major axes of the orbits of both pulsars, respectively. We can expect the mass ratio, R, to follow this simple relationship to at least 1PN order [7]. In particular, the R value is not only theory-independent, but also independent of strong-field (self-field) effects which is not the case for PK-parameters. Therefore, any combination of masses derived from the PK-parameters *must* be consistent with the mass ratio derived from Kepler's 3rd law. With five PK parameters already available, this additional constraint makes the double pulsar the most overdetermined system to date where the most relativistic effects can be studied in the strong-field limit. The theory of GR passes this test at the 0.1% level (Kramer et al. in prep.).

3 Further and Future Measurements

In GR, the proper reference frame of a freely falling object suffers a precession with respect to a distant observer, called geodetic precession. As a consequence, the pulsar spins precess about the total angular momentum, changing the relative orientation of the pulsars to one another and toward Earth and hence the observed pulse profile [8]. While GR predicts precession periods of only 75 yr for A and 71 yr for B, The study of the profile evolution of A [9] did not lead to the detection of any profile change yet. This present non-detection greatly simplifies the timing of A but does not exclude the possibility that changes may not happen in the future. In contrast, similar studies of the profile evolution of B [10] reveal a clear evolution of B's emission on orbital and secular time-scales. It is likely that these phenomena are caused by a changing magnetospheric interaction due to geometry variations resulting from geodetic precession.

The precision of the measured timing system parameters increases continuously with time as further and better observations are made. In a few years, we should be able to measure additional PK parameters, including those which arise from a relativistic deformation of the pulsar orbit and those which find their origin in aberration effects and their interplay with geodetic precession (see [7]). On secular time scales, we will even achieve a precision that will require us to consider post-Newtonian terms that go beyond the currently used description of the PK parameters. Indeed, we already achieve a level of precision in the $\dot{\omega}$ measurement where we expect corrections and contributions at the 2PN level. One such effect involves the prediction by GR that, in contrast to Newtonian physics, the neutron stars' spins affect their orbital motion via spin-orbit coupling. This effect modifies the observed $\dot{\omega}$ by an amount that depends on the pulsars' moment of inertia, so that a potential measurement of this effect allows the moment of inertia of a neutron star to be determined for the very first time [11, 3, 12].

4 Summary & conclusions

With the measurement of five PK parameters and the unique information about the mass ratio, the PSR J0737−3039 system provides a truly unique test-bed for relativistic theories of gravity. So far, GR also passes this test with flying colors. The precision of this test and the nature of the resulting constraints go beyond what has been possible with other systems in the past. The test achieved so far is, however, only the beginning of a study of relativistic phenomena that can be investigated in great detail in this wonderful cosmic laboratory.

References

1. T. Damour, G. Esposito-Farèse **58**, 1 (1998).
2. M. Burgay, et al., Nature **426**, 531 (2003).
3. A. G. Lyne, et al. **303**, 1153 (2004).
4. V. Kalogera, et al. **601**, L179 (2004).
5. G. S. Bisnovatyi-Kogan, B. V. Komberg **18**, 217 (1974).

 6. L. L. Smarr, R. Blandford, *Astrophys. J.* **207**, 574 (1976).
 7. T. Damour, J. H. Taylor **45**, 1840 (1992).
 8. T. Damour, R. Ruffini **279**, 971 (1974).
 9. R. N. Manchester, *et al.* **621**, L49 (2005).
10. M. Burgay, *et al.* **624**, L113 (2005).
11. T. Damour, G. Schäfer, *Nuovo Cim.* **101**, 127 (1988).
12. M. Kramer, *et al.*, *Binary Radio Pulsars*, F. Rasio, I. H. Stairs, eds. (Astronomical Society of the Pacific, San Francisco, 2005), pp. 59–65. Astro-ph/0405179.

The Relativistic Time Delay of the Pulsar Radiation in the Non-Stationary Gravitational Field of the Globular Clusters

T.I. Larchenkova[1] and S.M. Kopeikin[2]

[1] Astro Space Center, Lebedev Physical Institute, Moscow, Russia
tanya@lukash.asc.rssi.ru
[2] Department of Physics, University of Missouri-Columbia, USA
kopeikins@missouri.edu

Summary. The propagation of electromagnetic signals of pulsars through the non-stationary gravitational field of the stellar globular clusters formed by an ensemble of arbitrarily distributed stars are discussed. The expression for the relativistic time delay of pulsars radiation in such fields are derived taking into account the negligible aberration corrections. The obtained results are considered in the application to the globular cluster NGC 104 (Tucanae 47) for the cases of the small and large impact parameters.

The problem of propagation of electromagnetic waves in non-stationary gravitational fields is important in relativistic astrophysics. Kopeikin and Schafer (1999) [1] were the first to derive an exact formula for the Shapiro effect produced by a moving massive body and showed that the positions of the light-deflecting gravitating bodies (stars) should be taken not at the pulsar pulse arrival time to the observer, but at the corresponding delayed time due to the finite speed of propagation of the gravitational perturbation [2]. We took into account this effect when deriving an expression for the total relativistic delay time of the pulsar signal in the gravitational field of arbitrarily moving globular cluster stars.

The population of millisecond pulsars in globular star clusters is currently most representative. It is well known that milliseconds can be used as the most stable natural frequency standards. However, a statistically significant estimate of their stability depends considerably on the accurate determination of the parameters of the random low-frequency (LF) noise that is detected in the residuals of time-of-arrivals(TOAs) of pulsar signals observed over long time scales. A prolonged timing of millisecond pulsars has revealed LF uncorrelated (infrared) noise, presumably of astrophysical origin, in the residuals of TOAs of impulse for some of them. In most cases, pulsars in globular clusters show a low-frequency modulation of their rotational phase and spin rate. The relativistic time delay of the pulsar signal in the curved space time of randomly distributed and moving globular cluster stars (the Shapiro effect) is suggested as a possible cause of this modulation.

We concentrate on detailed calculations and analysis of the spectrum of the pulsar noise produced by the stochastic Shapiro effect. Knowledge of the theoretical power spectrum of the stochastic Shapiro effect for a pulsar in a globular cluster will allow us, in the case of its direct measurement, to obtain very important astrophysical information about the structure of the globular cluster core, which is inaccessible to other observational methods, and to analytically extend the Salpeter mass function for globular clusters toward the low-mass stars comparable in mass to Jupiter. Observation of the stochastic Shapiro effect will also allow the mass of the dark matter that is possibly concentrated near the globular cluster cores to be estimated.

Given the smallness of the aberration corrections that arise from the nonstationarity of the gravitational field of the randomly distributed ensemble of stars under consideration, a formula for the Shapiro effect for a pulsar in a globular cluster was found by Larchenkova and Kopeikin [3].

Let us assume that the origin of the coordinate system is at the barycenter of the globular cluster, t_0 – moment of emission of photon, $\boldsymbol{x}$ – barycentric coordinates of the observer, $\boldsymbol{x}_a$ – barycentric coordinates of a-th star, $\boldsymbol{x}_0$ – barycentric coordinates of the pulsar. Then the final expression for the random process produced by the relativistic delay is

$$\varepsilon(t) = \sum_{a=1}^{N} \frac{2GM_a}{c^3} [\ln(1 + 2(t-T)\frac{\boldsymbol{d}_a(T_0)\boldsymbol{v}_a(T_0)}{\boldsymbol{d}_a(T_0)\boldsymbol{x}_a(t_0)})$$
$$- \ln(1 + \frac{(-\boldsymbol{q} - \boldsymbol{K}_0)\boldsymbol{V}_{0a}(T_0)}{Q_{0a} + \boldsymbol{K}_0 \boldsymbol{Q}_{0a}}(t-T))], \tag{1}$$

where G is the gravitational constant, c is the light velocity, T is the beginning of observations, T_0 defines the moment of pulse emission corresponding to the time T, M_a and $\boldsymbol{v}_a$ are the mass and the velocity of a-th star, $\boldsymbol{K}_0$ is the unit vector, $\boldsymbol{V}_{0a} = \boldsymbol{v}_0(T_0) - \boldsymbol{v}_a(T_0)$, $\boldsymbol{Q}_{0a} = \boldsymbol{x}_a(T_0) - \boldsymbol{x}_0(T_0)$, $\boldsymbol{q} = \frac{\boldsymbol{Q}_{0a}}{|\boldsymbol{Q}_{0a}|}$, $\boldsymbol{v}_0$ – the pulsar velocity, $\boldsymbol{d}_a$ – the impact parameter.

The derived formula is used to calculate the autocorrelation function of the low-frequency pulsar noise, the slope of its power spectrum. The autocorrelation function can then be written as

$$\Re(t, \tau) = \int dm_a d\boldsymbol{x}_a d\boldsymbol{v}_a f(m_a, \boldsymbol{x}_a, \boldsymbol{v}_a) \varepsilon(t, m_a, \boldsymbol{x}_a, \boldsymbol{v}_a) \varepsilon(t+\tau, m_a, \boldsymbol{x}_a, \boldsymbol{v}_a), \tag{2}$$

where we assume that the statistical ensemble of stars is defined by uncorrelated parameters. No integration limits are specified in Eq. (2), but we assume that they are known (see below) and specify the range of statistical ensemble parameters. Eq.(2) is integrated in the following limits: $m_a = [0.1 M_\odot - 10 M_\odot]$, $|\boldsymbol{v}_a| = [0 - 4\sigma]$ and $|\boldsymbol{x}_a| = [0 - r_t]$, where σ is the stellar velocity dispersion of the cluster, r_t is the tidal radius and $M_\odot$ is the solar mass. We use the Salpeter function $f(m_a) \sim m_a^{-2.35}$ as the mass function of globular cluster stars and consider two model density distributions of a globular cluster: the model of an isothermal sphere with a core and the King model.

To determine the slope of the power spectrum for the noise process under study, we must calculate the integral in Eq. (2). Analysis indicates that this integral cannot

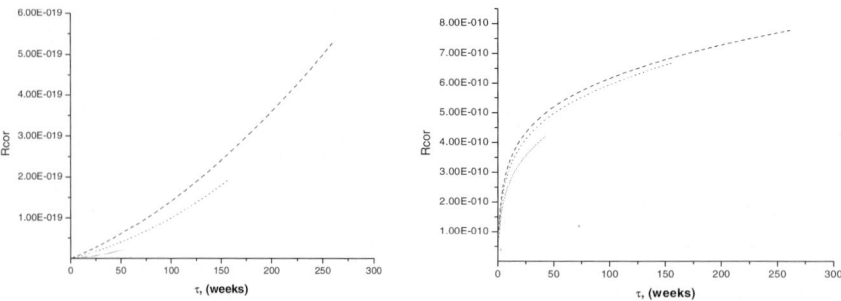

Fig. 1. (a) The autocorrelation function of stochastic Shapiro effect as a function of the time for the isothermal-sphere model. The pulsar is located in the cluster center. (b) The autocorrelation function of stochastic Shapiro effect as a function of time for the isothermal-sphere model in the case of small impact parameters ($|d_a| < 1AU$). The pulsar is located in the cluster center.

be calculated analytically in general form and the effect under study must be simulated numerically, which is done here. The random process attributable to the Shapiro effect can be estimated analytically if we restrict our analysis to star passages far from the line of sight to the pulsar. The random process under study can be separated into long- and short-period parts. The long-period part of the process corresponds to the case where the impact parameter of the pulsar pulse trajectory is large for the cluster stars (analytical case). The short-period part of the process corresponds to the case of small impact parameters.

The numerical simulation shows that the autocorrelation function of the stochastic Shapiro effect can be well approximated by a second-power polynomial in the case of large impact parameters (see figure 1a). The power spectral index is -1.8 as for King model as for the isothermal-sphere model. The autocorrelation function of stochastic Shapiro effect for small impact parameters has a logarithmic behavior and can be interpreted the flickering noise (see figure 1b). The power spectral index is -1.5. So, the rotational phase modulation of a pulsar in a globular star cluster may be attributable to the fluctuations in the relativistic time delay of the pulsar signal caused by the gravitational field variations due to the motion of the cluster stars.

References

1. S.M. Kopeikin, G. Schafer : Physical Review D **60**, 124002 (1999)
2. S. M. Kopeikin : Astrophys. J. **556**, L1 (2001)
3. T.I. Larchenkova & S.M. Kopeikin : Astronomy Letters **32**, 1, 18 (2006)

Relativistic Bose-Einstein Condensation Model for Dark Matter and Dark Energy

T. Fukuyama[1] and M. Morikawa[2]

[1] Department of Physics, Ritsumeikan University, Kusatsu Shiga, 525-8577, Japan
[2] Department of Physics, Ochanomizu University, 2-1-1 Otsuka, Bunkyo, Tokyo, 112-8610, Japan

Summary. We propose a cosmological model in which Bose-Einstein condensation works as Dark Energy. We obtain a novel mechanism of inflation, very early formation of highly non-linear objects, and log-z periodicity in the BEC collapsing time.

1 Introduction

Dynamics of dark matter (DM) and dark energy (DE) is a central problem in the present cosmology. Since this problem is too tough to be solved within the ordinary framework of physics, a variety of possibilities has been proposed so far. Although we do not know most of the matter in the universe, we do not attempt to identify it at present. We would like to emphasize instead the possibility that a well-known mechanism may become essential to understand the DM/DE problem. In this report, we would like to propose the quantum mechanical Bose-Einstein condensation (BEC) in order to explain the structure of DE/DM and the formation of large scale structure [1].

This model does not conflict with the standard LCDM model, which is now phenomenologically established in the linear stage of the density fluctuations, nor with the extreme isotropy of the CMB, which is also observationally and theoretically established. This is because the present model mainly describes local non-linear structure after the photon decoupling era.

There have been many attempts to modify the standard LCDM model in the literature. Most of them yield small deviation and generalization from the LCDM model. However, the newly introduced parameters often fall into the values which coincide with those in the LCDM model [2]. We do not attempt to introduce such small deviation from the LCDM model here, but simply follow a natural scenario of the standard BEC mechanism.

DE should cause the accelerated cosmic expansion and possesses negative pressure. A natural candidate of such matter would be the scalar field. We would like to propose a possibility that this scalar field degrees of freedom comes from the Bose-Einstein condensation (BEC) mechanism of some boson field. The boson may be the axion field with mass scale 10^{-5}eV, or a fermion pair which forms a boson degrees of freedom with the mass of same order. A natural candidate of DM would be then the boson or fermion gas which is free from the condensation.

2 BEC in the Universe

Bose-Einstein condensation is possible if the thermal de Broglie length exceeds the mean separation of particles of the boson gas, $\lambda_{dB} \equiv \left(2\pi\hbar^2/(mkT)\right)^{1/2} > r \equiv n^{-1/3}$. This condition determines the critical temperature $T_{cr} = 2\pi\hbar^2 n^{2/3}/(km)$, which turns out to have the same density dependence as the cosmic temperature for CDM: $T = T_0\,2\pi\hbar^2/m\,(n/n_0)^{2/3}$. Thus, once the temperature becomes lower than the critical temperature and the mass, i.e.$T < T_{cr}$ and $kT < mc^2$, at some moment of cosmic time, then BEC has a chance to initiate all the time after that moment. A rough estimate yields the critical mass about 2eV, below which BEC takes place.

In order to describe the dynamics of BEC, we need a relativistic version of the Gross-Pitaevskii equation, which turns out to be the same form for the complex scalar field

$$\frac{\partial^2\phi}{\partial t^2} - \Delta\phi + m^2\phi + \lambda(\phi^*\phi)\phi = 0, \tag{1}$$

where ϕ here is a classical mean field and the potential is $V \equiv m^2\phi^*\phi + \frac{\lambda}{2}(\phi^*\phi)^2$.

We suppose the attractive interaction in BEC, $\lambda < 0$, which guarantees the local negative pressure $p < 0$ and the instability. Therefore if the BEC energy density exceeds some critical value, BEC immediately collapses into the localized objects such as boson stars or black holes. They act as DM, and form potential inhomogeneity at the smallest scale.

The set of evolution equations in the BEC cosmology becomes

$$\begin{aligned}
H^2 &= \left(\tfrac{\dot{a}}{a}\right)^2 = \tfrac{8\pi G}{3c^2}\left(\rho_g + \rho_\phi + \rho_l\right), \\
\dot{\rho}_g &= -3H\rho_g - \Gamma\rho_g, \\
\dot{\rho}_\phi &= -6H\left(\rho_\phi - V\right) + \Gamma\rho_g - \Gamma'\rho_\phi, \\
\dot{\rho}_l &= -3H\rho_l + \Gamma'\rho_\phi,
\end{aligned} \tag{2}$$

where ρ_g, ρ_ϕ, ρ_l are the energy densities of boson gas, BEC, and localized objects, respectively.

The transition parameter Γ represents the slow steady condensation of the boson gas into BEC. This parameter is controlled by the adiabaticity of the cosmic expansion. Another transition parameter Γ' represents the instantaneous BEC collapse; $\Gamma' = \theta\left(-(\rho + p)\right)c$, where c is a huge positive constant. In other words, $\Gamma' = \infty$ when the BEC violates the weak energy condition, and $\Gamma' = 0$ otherwise.

In this occasion, we point out that the last stage of the evolution is always a novel type of inflation, which is inevitable in our BEC model. This inflation is supported by a balance $\dot{V} = \Gamma\rho_g$, of the condensation speed ($\Gamma\rho_g$) which promotes the condensation ϕ, and the potential force ($\dot{\phi}V'$) which demotes the condensation ϕ. Note that this novel inflation is free from the usual slow-roll conditions.

In the context of BEC collapse, we would like to point out a log(t)-periodicity in the collapsing times. Each BEC collapse takes place when ρ_ϕ ever reaches the critical value $\rho_\phi^{cr} = V_{max} = m^4/(-2\lambda)$, which is almost the moment when the weak energy condition is violated. Just after each collapse, a new BEC initiates from $\phi = 0$. This implies that the occurrence of the BEC collapse is periodic in the logarithm of time $\log(t)$. Especially for not very late stage $z < 1$, and if the cosmic expansion is almost power law in time $a(t) \propto t^{const.}$, log(t)- periodicity means log(a) and log(z) -periodicities. For example in a typical numerical calculation, BEC collapse takes place at $z = 17.0, 10.9, 7.1, 4.1$, which is almost log-periodic.

3 Observational Constraints

Large scale structure:
The collapse of the condensate proceeds in the smallest scale. This collapsing time scale is too short compared to the characteristic time scale of the larger scale fluctuations to grow. In the linear stage of the density fluctuations, therefore, the influence of the BEC collapse does not show up in the larger scales.
Microwave Background Radiation:
The condensation is supposed to collapse well after the decoupling time $t_{dec} \approx 3.8 \times 10^5$[yr], which requires $t_{dec} < \Gamma^{-1}$.

4 Conclusions

We proposed a BEC model, in which multiple BEC collapses take place to form highly non linear objects and a novel type of inflation free from slow-rolling conditions takes place at the end. The BEC collapsing time is almost log-z periodic.

References

1. M. Nishiyama, M. Morita, M. Morikawa, in Proceedings of the XXXIXth Rencontres de Moriond 143 (2004) [astro-ph/0403571]; T. Fukuyama, M. Morikawa, astro-ph/0509789.
2. H. Sandvik, M. Tegmark, M. Zaldarriaga, I. Waga: Phys.Rev. **D69** 123524 (2004)

fi

h   h

T. Nakajima[1] and M. Morikawa[2]

[1] National Astronomical Observatory of Japan, Osawa 2-21-1, Mitaka, 181-8588, Japan
[2] Department of Physics, Ochanomizu University, 2-1-1 Otsuka, Bunkyo, Tokyo, 112-8610, Japan

We propose degenerate fermionic dark matter to explain the flat-top density profile of the cluster A1689 recently observed.

In recent observations of galaxy cluster, the energy density in the central region is very high and it often shows a flat-top profile. This is most prominent in the new observation of the mass concentration in the cluster A1689 [1], using the gravitational lensing technique. This method faithfully extracts the information of the purely gravitating mass. It seems that the popular NFW form [2], which inherently predicts a cusp at the center, cannot describe this flat-top density profile. It would be possible to explain the flat-top profile by considering complicated dynamics of heavy galaxies or by simply introducing repulsive interaction between dark matter particles. However, if this flat-top density profile is general and robust among many clusters, it may indicate that an universal mechanism supports this characteristic density profile.

In this report, we would like to explore such possibility that the dark matter is degenerate at the cluster center in the quantum mechanical sense [3]. The physics is simple and fundamental. It is free from any special interactions and from complicated gravitating many-body systems.

Degenerate structures generally form dense and uniform density profiles, supported by the degenerate quantum pressure. However if the dark matter was composed of bosons, it couldn't support large structures such as clusters; even a boson of mass 10^{-5}eV can only form objects of planetary size. Therefore the dark matter has to be fermionic in our model.

Fermion gas can degenerate if the thermal de Broglie wave length $\left(2\pi\hbar^2 / (mkT)\right)^{1/2}$ exceeds the mean separation of particles: $kT < 2\pi\hbar^2 n^{2/3}/m$. This structure has the characteristic density $\rho = m^4c^3/\hbar^3$. The critical mass beyond which the fermion gas forms black hole is given by $M_{critical} \approx m_{pl}^3/m^2$, where $m_{pl} = (\hbar c/G)^{1/2} = 2.18 \times 10^{-5}$g is the Plank mass. This degenerate structure is general and only determined by the mass of the fermion. If the mass is about one GeV (nucleon), it yields the size of white dwarfs or neutron stars. We would like to consider the size of clusters and then the corresponding mass is about one eV. The detailed structure should be analyzed by the relativistic equation of state and the theory of general relativity (TOV equation[4]).

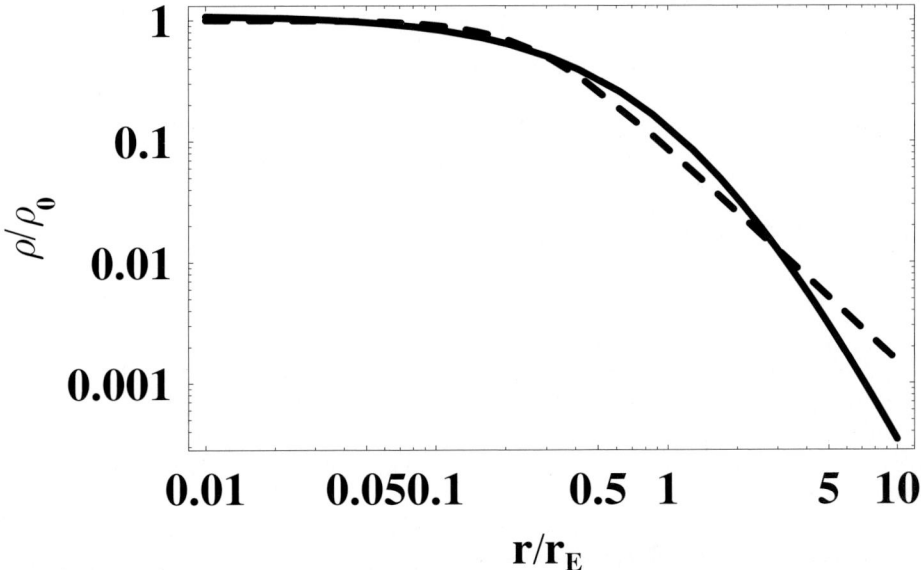

Fig. 1. A numerical solution for the hydrostatic equilibrium using the hybrid equation of state Eq. (1). The solid line is the best fit for A1689 observations, and the broken line is our model. We obtain a fermion mass of $m = 3.8eV$ for $g = 3$ (Majorana), and a central density of $\rho = 2.69 \times 10^{-23}$[g / cm^3].

3 Flat-top Structures at the Centre of A1689 – Hybrid State Equation –

We now consider the mass concentration of the cluster A1689, which is well described by the flat-top profile: $\rho(r) = 1.60 \times 10^{-23}((r/r_E) + 1.28)^{-3.71}hg / cm^3$, where $r_E = 97$kpch^{-1}, $h = 0.71$. We naturally expect that the degenerate fermion exists only at the center where the density is very high, and the normal gas at the outskirts of the cluster. This situation can be phenomenologically expressed by the hybrid state equation:

$$p = p_0 + \frac{2}{3}\frac{GM_r}{r}\rho \tag{1}$$

where $p_0 \propto \hbar^2 \rho^{5/3} \left(m^4 g\right)^{-2/3}$ is the degenerate pressure and the second term is the pressure of the ideal gas, with the temperature evaluated by the gravitational potential assuming a virial relation.

This equation of state and the equation of hydrostatic equilibrium $dp/dr = -\rho G M_r/r^2$ yield an energy density profile. We demonstrated this calculation in Figure 1. The parameters are the central density ρ_{center} and the combination of mass and the degeneracy parameter $mg^{1/4}$. We obtain a fermion mass of $m = 3.8\,\mathrm{eV}$ for $g = 3$ (Majorana), and a central density $\rho_{center} = 2.69 \times 10^{-23}\mathrm{g}/\mathrm{cm}^3$ in the best fit to the observations. These may suggest the possibility of neutrino [5] or sterile neutrinos. Details can be found in our recent paper [6].

Further, since this degenerate structure is quite general and free from any interactions, it can be realized on other scales as well. For example, various masses of the fermion can yield the following characteristic scales.

$$\begin{array}{llll} & m_*[\mathrm{eV}] & M_C[M_\odot] & R_C \\ f_a & 3 & 9.8 \times 10^{16} & 42\mathrm{kpc} \\ f_b & 0.19 \times 10^6 & 2.5 \times 10^7 & 460 R_\odot \\ f_c & 18.2 \times 10^6 & 2.7 \times 10^3 & 0.050 R_\odot \end{array} \tag{2}$$

which may correspond to the characteristic mass scale of black holes and cluster.

References

1. Broadhurst et al.: ApJ, **621**, 53, Broadhurst et al.: ApJ, **619**, L143 (2005).
2. Navarro, J. F., Frenk, C. S. and White, S. D. M.: ApJ, **516**, 591 (1996).
3. Landau and Lifshitz, Statistical Physics, 3rd Ed. (1980)
4. Oppenheimer, J. R. and Volkoff, G. M.: Phys. Rev. **55**, 374, (1939)
   Tolman, R. C.: Relativity, Thermodynamics and Cosmology, Oxford (1934).
5. Shirai, J.: Nucl. Phys. B (Proc. Suppl.), **144**, 286 (2005).
6. Nakajima, T. and Morikawa, M., astro-ph/0506623 (2005).

Hardening in a Stellar Time-Evolving Background: Prospects for *LISA*

A. Sesana[1], F. Haardt[1], and P. Madau[2]

[1] Dipartimento di Fisica e Matematica – Università degli Studi dell'Insubria, via Valleggio 11, 22100 Como, Italy

[2] Department of Astronomy – University of California, 1156 High Street, Santa Cruz, CA 95064, USA

Summary. We use detailed scattering experiments to study the role of 3-body interactions in driving orbital decay of massive black hole binaries (MBHBs) in galactic centers, quantifying also the effect of secondary slingshot on binary shrinking. We find that without invoking other physical mechanisms, such as gas dynamical processes, binaries cannot shrink to the gravitational wave (GW) emission regime in less than a Hubble time, unless they have very small mass ratios. Very unequal mass binaries are therefore a natural target for the planned *Laser Interferometer Space Antenna (LISA)*. The star-binary interactions create a population of hypervelocity stars on nearly radial corotating orbits that is highly flattened in the inspiral plane. Most of the stars are ejected in an initial burst lasting much less then the bulge crossing time.

1 Introduction

It is now widely accepted that the formation and evolution of galaxies and massive black holes (MBHs) are strongly linked: MBHs are ubiquitous in the nuclei of nearby galaxies, and a tight correlation is observed between hole mass and the stellar mass of the surrounding spheroid or bulge [1, 2, 3, 4]. If MBHs were also common in the past (as implied by the notion that many distant galaxies harbor active nuclei for a short period of their life), and if their host galaxies experience multiple mergers during their lifetime, as dictated by popular cold dark matter (CDM) hierarchical cosmologies, then close MBHBs will inevitably form in large numbers during cosmic history [5, 6]. MBH pairs that are able to coalesce in less than a Hubble time will give origin to the loudest GW events in the universe. In particular, a low-frequency space interferometer like the planned *Laser Interferometer Space Antenna (LISA)* is expected to have the sensitivity to detect nearly all MBHBs in the mass range $10^4 - 10^7$, $M_\odot$ that happen to merge at any redshift during the mission operation phase [7, 8]. The coalescence rate of such "*LISA* MBHBs" depends, however, on the efficiency with which stellar and gas dynamical processes can drive wide pairs to the GW emission stage. In the following I briefly review recent results about the role of gravitational slingshot (capture and ejection of stars by the MBHB) in driving the last phase of binary shrinking, after dynamical friction has become inefficient. I also describe the kinematic properties of the stellar population after the interaction with the MBHB. A detailed derivation and discussion of the following results will be found in [9].

2 Star-Binary Interactions

Assume a MBHB of mass $M = M_1 + M_2 = M_1(1 + q)$ $(M_1 > M_2)$ and initial eccentricity e, embedded in a stellar bulge of mass M_B described by a Maxwellian distribution function with velocity dispersion σ. Quinlan [10] showed that energy and angular momentum exchange between the MBHB and the interacting stars is relevant only for binary separation smaller than

$$a_h = \frac{GM_2}{4\sigma^2}. \tag{1}$$

Starting from a_h, the energy transferred to the stars cause the binary shrinking and the stellar population heating.

2.1 Stars: Heating and Ejection

Only a small fraction ($<1\%$) of the bulge stars is in the MBHB loss cone at $t = 0$, nevertheless a significative population of high velocity stars is produced. We define these stars by the condition $v > v_{esc}$, where v_{esc} is the escape velocity from the MBHB influence radius $r_i = GM/2\sigma^2$:

$$v_{esc} = 2\sigma\sqrt{\ln(M_B/M)} \simeq 5\sigma. \tag{2}$$

The second equality comes from the adopted relation $M = 0.0014M_B$ [4].

An example of the effect of slingshot mechanism on the stellar distribution is shown in the left panel of figure 1, where the initial ($t = 0$) velocity distribution of interacting stars is compared to that after loss cone depletion, for an equal mass, circular binary. After the first interaction , a large subset of kicked stars still lies in the (reduced) loss cone of the shrunk MBHB, and the process is iterative. Our calculations show that the high velocity tail depends on the MBHB mass ratio and eccentricity: in general, both a small value of q and a high value of e contribute to extend the tail. Integrating the curves over velocity gives the mass of interacting stars, which turns out to be $\simeq 2M$ for equal mass binary, and scales fairly well with the mass ratio q. The mass ejected from the bulge (i.e. stars with $v > v_{esc}$) is $M_{ej} \sim 0.7\mu$ (μ is the binary reduced mass), nearly independent on e.

In the right panel another interesting effect is shown: the flattening of scattered stars into the MBHB plane. The flattening is a function of ejection velocity: the stronger is the star-binary interaction, the higher is the kinetic energy acquired by the star, and the alignment with the MBHB plane is more pronounced. As a general trend, ejected stars are preferentially moved into the binary orbital plane, on nearly radial, corotating orbits.

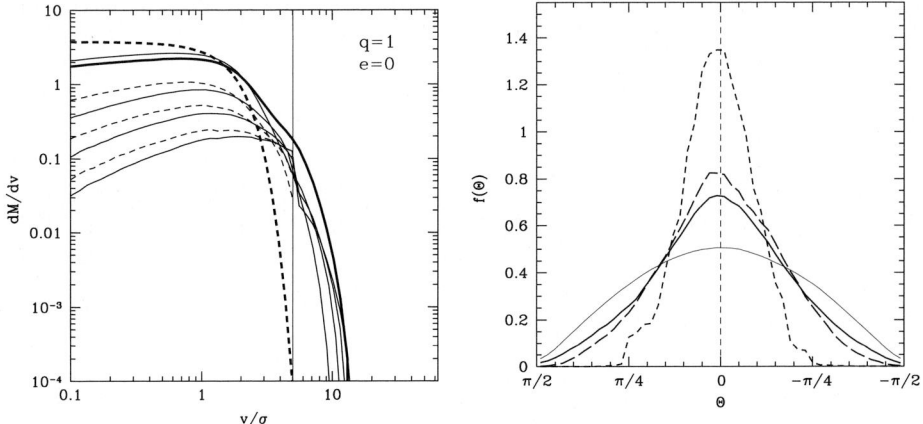

Fig. 1. Left panel: loss cone stellar velocity distribution (normalized to the total mass of interacting stars in units of the binary mass M) for an equal mass, circular binary. the vertical line marks $v_{esc} \simeq 5\sigma$. *Dashed lines*: from top to bottom, shrinking loss cone distributions *before* the 1st, 2nd, 3rd and 4th iteration; the initial distribution is marked with a ticker line. *thin solid lines*; from top to bottom, distribution of stars that have received 1, 2, 3 and 4 kicks. *Thick solid line*: final distribution after loss con depletion is complete. Right panel: distribution $f(\Theta)$ of ejected stars for the same experiment. The MBHB plane is at $\Theta = 0$. *Thin line* isotropic distribution $f(\Theta) \propto \cos\Theta$; *thick solid line*, stars with $v > v_{esc}$; *thick long-dashed line*, stars with $v > 1.5\,v_{esc}$; *thick short-dashed line*, stars with $v > 2\,v_{esc}$.

2.2 Binary: Orbital Decay and Coalescence

From the binary point of view, repeated interactions with the loss cone stars result in a net orbital decay. We compute the MBHB evolution coupling the results of 3-body scattering experiments with an analytical model for the time evolving supply of stars in the loss cone. The results in terms of a_h/a is shown in the left panel of figure 2, where it is also compared to the result of an N-body simulation presented by Milosavljevic & Merritt [11]. The agreement with the simulation is fairly good. The figure shows how the binary hardening declines after few crossing times: this is due to the declining supply of low angular momentum stars from the outer regions of the bulge, once the stars in the central cusp have interacted with the MBHB. We also note that equal mass binaries shrink more than unequal ones, because of the scaling of the mass in stars available for interactions, and of the typical amount of energy exchanged during single 3-body encounters. We find that $d(1/a)/d\ln t \propto q$. Eccentricity plays a marginal role in the orbital evolution of the MBHB. For a given q, the orbital shrinking is larger at most by 10 % for highly eccentric binaries.

We may ask if the derived binary shrinking factors are sufficient to lead the MBHB to GW emission regime and to the final coalescence. The right panel of figure 2 shows the binaries (at two different reference redshifts) in the $M_1 - q$ plane which would be resolved by *LISA* with $SNR > 5$, and binaries (in the same plane) which are going to coalesce within 5 Gyrs after loss cone depletion is completed. The overlap of the

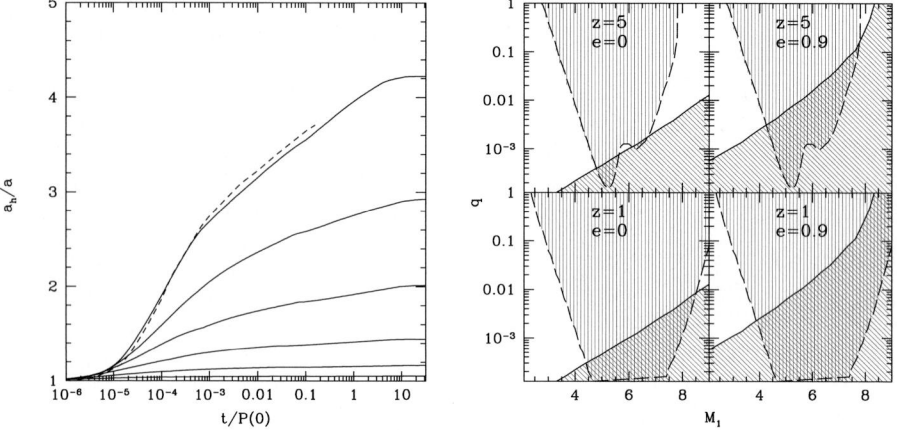

Fig. 2. Left panel: binary orbital decay as a function of time (in terms of the bulge crossing time $P_0 = \sqrt{\pi G M_B / 2\sigma^3}$). *Solid lines* from top to bottom are for circular binaries with q=1, 1/3, 1/9, 1/27, 1/81, 1/243. *Dashed line* is the result of a N-body simulation with a circular equal mass binary performed by Milosavljevic & Merritt [11].Right panel: in the plane $M_1 - q$, the vertical shaded area limits the *LISA* potential targets with $SNR > 5$. The diagonal shaded area on the lower right corner marks binaries that are going to coalesce in 5 Gyrs after loss cone depletion. In each panel, the reference redshift and eccentricity of the MBHBs are labeled.

two areas gives the MBHBs which are potential targets for *LISA*. As expected, the eccentricity is a crucial parameter (the GW shrinking timescale is a steep function of e [12]). Even considering only slingshot of unbound stars as orbital decay driver, very unequal mass binaries are ideal *LISA* targets. It is worth noticing here that we considered only the scattering of unbound stars: the presence of a cusp of stars bound to M_1 could further support the binary decay.

3 Conclusions

We have made a first attempt of modeling the time dependent dynamical evolution of MBH pairs in stellar backgrounds using a hybrid approach, i.e., using results of a 3–body scattering experiment to follow the time evolution of the stellar distribution. Despite several approximations and limitations, we have obtained a number of results that we briefly summarize in the following.

MBHB-stellar interaction causes a significant mass ejection, of the order of $M_{ej} \simeq 0.7\mu$, nearly independent on the binary eccentricity. The properties of ejected stars are well defined. They typically corotate with the binary and they are preferentially ejected in the binary plane in a short burst just after the binary has become hard. The ejected sub-population, if detected in nearby galaxies, would be an unambiguous sign of a relatively recent major merger.

The relevant quantity to assess coalescence is the ratio a_f/a_{GW}, where a_f is the binary separation after loss cone depletion and a_{GW} is the separation at which the binary coalesces in less than a Hubble time. In general a_f is reached in a short time ($t \sim 10^7$ yrs), but is not sufficient to drive the binary to the GW emission stage, unless the binary is very massive ($M > 10^8$ M$_\odot$), and/or very eccentric ($e > 0.7$) and/or formed by very unequal MBHs ($q < 0.01$, in fact the gap $a_h - a_{GW}$ is lower for unequal mass binaries, though the shrinking factor is small). This conclusion is supported by recent high resolution N-body simulations [13, 14]. In terms of *LISA* binaries ($10^4 - 10^7$ M$_\odot$), it is then important to study in details the possible role of other mechanisms, such as the non sphericity of the stellar bulge and the presence of a circumbinary gaseous disk.

References

1. Magorrian J. et al.: AJ **115**, 2285 (1998)
2. Gebhardt K. et al.: ApJ **543**, L5 (2000)
3. Ferrarese L. & Merritt D.: ApJ **539**, L9 (2000)
4. Haring N. & Rix H. W.: ApJ **604**, L89 (2004)
5. Kauffmann G. & Haehnelt M. G.: MNRAS **311**, 576 (2000)
6. Volonteri M., Haardt F. & Madau, P.: ApJ **582**, 599 (2003)
7. Sesana A., et al.: ApJ **611**, 623 (2004)
8. Sesana A., et al.: ApJ **623**, 23 (2005)
9. Sesana A., Haardt F. & Madau P., in prep. (2007)
10. Quinlan G. D.: New A **1**, 35 (1996)
11. Milosavljevic M. & Merritt D.: ApJ **596**, 860 (2003)
12. Peters P. C.: Phys. Rev. B. **136**, 1224 (1964)
13. Makino J. & Funato Y.: ApJ **602**, 93 (2004)
14. Berczik P., Merritt D. & Spurzem R.: ApJ **633**, 680 (2005)

Gravitational Waves for Odd Parity from a Collapsing Dust Ball

H. Sotani[1] and K.-I. Maeda[2]

[1] Advanced Research Institute for Science and Engineering, Waseda University, 3-4-1 Okubo, Shinjuku, Tokyo 169-8555, Japan, sotani@gravity.phys.waseda.ac.jp

[2] Department of Physics, Waseda University, 3-4-1 Okubo, Shinjuku, Tokyo 169-8555, Japan, maeda@gravity.phys.waseda.ac.jp

1 Introduction

In order to detect the gravitational waves directly, gravitational wave projects have been launched in many countries and interferometric gravitational wave detectors have been operating [1]. The wave source is believed to be a system with violent motion of a compact object, such as merger of the binary system, supernova, stellar collapse, and so on. It is considered that the magnetic field plays an important role in the stellar collapse, because of the existence of objects such as magnetars. However although there is some research with respect to the gravitational waves from the stellar collapse using perturbation approach [2, 3, 4, 5], there are a few calculations that take the effect of magnetic field into account. So we focus on the axial gravitational waves from a collapsing dust ball with a magnetic field.

2 Basic Equations

We use the LTB (Lemaître-Tolman-Bondi) spacetime as a background, which describes an inhomogeneuos spherically symmetric dust collapse, and its line element is given by

$$ds^2 = -d\tau^2 + A^2(\tau, \chi)d\chi^2 + R^2(\tau, \chi)d^2\Omega. \tag{1}$$

With respect to the magnetic field, for simplicity we deal with the magnetic field as the perturbation. However it is assumed that the first order perturbation for the metric and the matter is of same order as the second order perturbation for the magnetic field. This assumption is reasonable if the magnetic field becomes stronger by collapsing the dust ball, even if the initial magnetic field is too weak. Under the ideal MHD assumption, the non-zero components of the perturbed field strength tensor $\delta F_{\mu\nu}$ are only δF_{12}, δF_{13}, and δF_{23}, and we can set

$$\delta F_{12} = -b_1 \frac{1}{\sin\theta}\partial_\phi Y_{lm} + b_3\partial_\theta Y_{lm}, \tag{2}$$

$$\delta F_{13} = b_1 \sin\theta\partial_\theta Y_{lm} + b_3\partial_\phi Y_{lm}, \tag{3}$$

$$\delta F_{23} = b_2 \sin\theta Y_{lm}, \tag{4}$$

where b_1, b_2 belong to the axial parity and b_3 belongs to the polar parity. Then from the Maxwell equations we can get the equations for b_1, b_2, and b_3;

$$\partial_\tau b_1 = \partial_\tau b_2 = \partial_\tau b_3 = 0, \tag{5}$$

$$l(l+1)b_1 + \partial_\chi b_2 = 0. \tag{6}$$

On the other hand, under the spherically symmetric background spacetime, 5), we adopt the gauge-invariant perturbation theory [6, 7]. For simplicity we consider only the dipole magnetic field, then we can get the perturbation equations for the axial gravitational waves with $l = 2$, such as

$$\partial_\tau \left[\frac{A}{R^2} \partial_\tau (R^4 \Psi_s) \right] - \partial_\chi \left[\frac{1}{AR^2} \partial_\chi (R^4 \Psi_s) \right] + (l-1)(l+2)A\Psi_s$$

$$= 16\pi(\partial_\tau L_1 - \partial_\chi L_0). \tag{7}$$

$$\partial_\tau (AR^2 L_0) - \partial_\chi \left(\frac{R^2}{A} L_1 \right) + (l-1)(l+2)LA = 0, \tag{8}$$

where Ψ_s, L, L_1, and L_2 are gauge-invariant variables. L, L_1, and L_2 are coupled with the matter perturbation and magnetic field, which are given by

$$L_0 = -\beta\rho, \quad L_1 = -\frac{b_2 b_3}{12\pi R^2}, \quad L = \frac{b_1 b_3}{12\pi A^2}, \tag{9}$$

where β is the perturbation of 4-velocity for axial parity. Thus the magnetic field is determined by equations 5 and 6, and the evolutions of Ψ_s and β can be calculated with equations 7 and 8.

In the case of the (τ, χ) coordinate, the region of numerical calculation is restricted. Thus instead of the (τ, χ) coordinate system, we adopt a single-null coordinate system, i.e., (u, χ'), where we choose $\chi' = \chi$ and u is the outgoing null coordinate. For the case of marginary bound collapse, the evolution equations 7 and 8 become with new coordinate system (u, χ') as follows

$$\frac{d\phi_s}{du} = -\frac{\alpha}{R} \left[3\partial_\chi R + \frac{R}{2}(\partial_\tau R)(\partial_\tau \partial_\chi R) - \frac{5}{4}(\partial_\chi R)(\partial_\tau R)^2 \right] \Psi_s$$

$$+ \frac{8\pi\alpha}{R} \left[\frac{\partial_\tau R}{6\pi R^3} b_2 b_3 + \left(-\frac{\partial_\chi R}{\alpha} \frac{\partial\beta}{\partial u} + \frac{\partial\beta}{\partial\chi'} \right) \rho + \left(-\frac{\partial_\chi R}{\alpha} \frac{\partial\rho}{\partial u} + \frac{\partial\rho}{\partial\chi'} \right) \beta \right]$$

$$- \frac{\alpha}{2} \left[\frac{\partial_\chi^2 R}{(\partial_\chi R)^2} - \frac{2}{R}(1 - \partial_\tau R) \right] \phi_s, \tag{10}$$

$$\frac{\partial\beta}{\partial u} = \frac{-2\alpha}{3(\partial_\chi R)(dF/d\chi)} \left[\frac{\partial_\chi^2 R}{\partial_\chi R} b_2 b_3 - \frac{db_2}{d\chi} b_3 - \frac{db_3}{d\chi} b_2 - 4b_1 b_3 \right], \tag{11}$$

where ϕ_s is defined by

$$\phi_s \equiv R\frac{\partial\Psi_s}{\partial\chi'} + 3(\partial_\chi R)(1 + \partial_\tau R)\Psi_s. \tag{12}$$

3 Conclusion and Discussion

Under the assumption that the stellar magnetic field is so small, we derive the perturbation equations for the axial gravitational waves with the source term including the magnetic filed. As a result, we can see that the perturbation equations for axial gravitational waves have a source term with a coupled magnetic field with the both parities. In other words, if the magnetic field is only axial or polar parities, the magnetic field cannot affect the axial gravitational waves. With regard to the treatment in the exterior region, we have two ideas; one is that we set the thin density distribution in the exterior region and the other is that the exterior region is assumed vacuum. In the case of the former, although we can use the ideal MHD approximation, it may be difficult to set the observer far from the dust ball. On the other hand in the case of the latter, we have to deal with the electro-magnetic field. At any hand, in order to determine how much the magnetic field affects axial gravitational waves, we have to calculate the concrete gravitational waves by numerical simulation.

References

1. B. C. Barish, in *Proceedings of the 17th International Conference on General Relativity and Gravitation*, edited by B. Nolan (World Scientific, 2005).
2. C. T. Cunningham, R. H. Price, and V. Moncrief, Astrophys. J. **224**, 643 (1978); **230**, 870 (1979); **236**, 674 (1980).
3. E. Seidel and T. Moore, Phys. Rev. D **35**, 2287 (1987); E. Seidel, E. S. Myra, and T. Moore, *ibid* **38**, 2349 (1988); E. Seidel, *ibid* **42**, 1884 (1990).
4. H. Iguchi, K. Nakao, and T. Harada, Phys. Rev. D **57**, 7262 (1998); H. Iguchi, T. Harada, and K. Nakao, Prog. Theor. Phys. **101**, 1235 (1999); **103**, 53 (2000).
5. T. Harada, H. Iguchi, and M. Shibata, Phys. Rev. D **68**, 024002 (2003).
6. U. H. Gerlach and U. K. Sengupta, Phys. Rev. D **19**, 2268 (1979); **22**, 1300 (1980).
7. C. Gundlach and J. M. Martín-García, Phys. Rev. D **61**, 084024 (2000); J. M. Martín-García and C. Gundlach, *ibid.* **64**, 024012 (2001).

Part III

Black Holes

Th

h   +

E. Benítez[1], A. Franco-Balderas[1], V. Chavushyan[1,2], and J. Torrealba[1]

[1] Instituto de Astronomía Universidad Nacional Autónoma de México, Apdo. Postal 70-264, 04510 México, D.F., México (erika@astroscu.unam.mx).

[2] Instituto Nacional de Astrofísica, Óptica y Electrónica, Apdo. Postal 51 y 216, 72000 Puebla, Pue, México (vahram@inaoep.mx).

Seyfert galaxies are nearby AGN with host galaxies showing large-scale morphologies that resemble those of normal or inactive galaxies (e.g. McLure et al. 1999). Several studies have concluded that the black hole masses are directly proportional to some properties of the galaxy bulge component like luminosity, mean stellar velocity dispersion of the bulge, etc. (see Kormendy & Gebhardt 2001). It is also known that only a handful of measurements of black hole masses in Seyfert (Sy) galaxies have been obtained using essentially reverberation mapping techniques (e. g. Wandel 2002). However, this method is limited to Sy1 and difficult to apply in some intermediate Sy galaxies mainly because the lack of the broad component in the $H\beta$ emission line. We have recently studied the optical photometric and spectroscopic properties of the Sy galaxy SBS 0748+499. In particular, we have done a surface photometry study that provided us with the R-band total magnitude M_R and the Sérsic-n index for the bulge component. Also, we have done line fitting analysis to the optical spectra of SBS 0748+499 in terms of Gaussian components. We found that we can only see the broad component in the $H\alpha$ emission line (FWHM $=1700$ km s^{-1}) in this object, so we classified it as a Sy1.9 galaxy, and obtained a FWHM $= 450$ km s^{-1} from the narrow lines. In this work we have estimated the M_{BH} using and comparing three different methods that we assumed are valid for intermediate Seyfert objects.

h BH

We have used three different methods to give an estimation of the black hole mass for SBS 0748+499. The first one is based on the absolute R-band magnitude of the bulge component obtained from the photometric data, i.e. from the host galaxy. The second one is based on an estimation of the stellar velocity dispersion of the bulge which was derived indirectly from our spectroscopic data. A third method uses the Sérsic-index obtained from the surface brightness profile analysis done to this object. Calculations were done assuming: $H_0 = 75$ km s^{-1} Mpc^{-1}, $q_0 = 0.5$ and z $= 0.0244$.

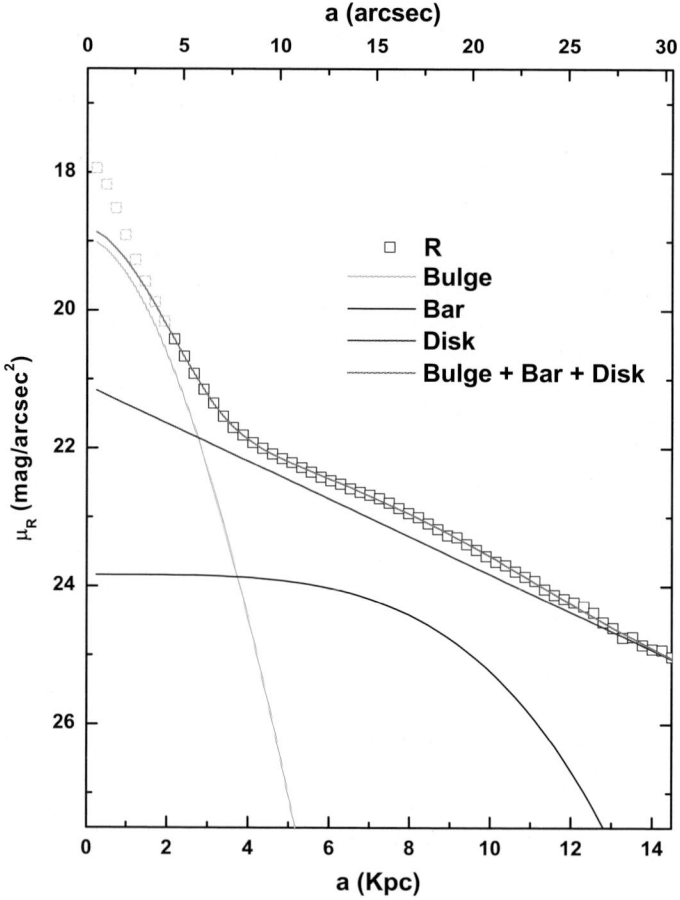

Fig. 1. The light profile of SBS 0748+499 is well described by three components: a bulge and a bar modeled with a Sérsic profile and an exponential disk.

Method 1: Using the total R-band magnitude M_R

The first method is based on the application of the relation between absolute R-band magnitude M_R and M_{BH} given by McLure & Dunlop (2002). We have recalculated this relation for our cosmology as: $\log M_{BH}/M_\odot = -0.50(\pm0.02)M_R - 2.52(\pm0.48)$. Thus, from the luminosity of the bulge component in R-band, i. e. of the host galaxy $L_R = (5.17 \pm 0.21) \times 10^9 L_\odot$, we obtained: $M_R = -19.86(\pm0.05)$ and: $\log M_{BH}/M_\odot = 7.41(\pm0.48)$.

Method 2: Using $\sigma_\star$

The second method needs an indirect inference of the bulge stellar velocity dispersion. The brightness of the nuclear part of SBS 0748+499 does not allow us to detect the absorption lines of CaT in the spectrum (lines commonly used to estimate $\sigma_\star$, e. g. Onken et al. 2004). Therefore, we decided to use the relation found by Nelson (2000): $\sigma_\star = FWHM[OIII]5007 \text{Å}/2.35$ on the assumption that for this galaxy, the forbidden line kinematics is dominated by virial motions in the host galaxy bulge. Also, the author found the following relation for the black-hole mass: $\log M_{BH}/M_\odot = 3.7(\pm0.7)\log\sigma_\star - 0.5(\pm0.1)$. This method has been investigated later by Boroson (2003) using the SDSS spectra of 107 low-z quasars and Sy 1 galaxies. The author confirms the existence of the relation $\sigma_\star$ vs. FWHM [OIII]5007 Å and claimed that this relation can predict the black hole mass in active galaxies to a factor of 5. Thus, measuring from our spectrum the $FWHM[OIII]5007 \text{Å} = 450(\pm14)kms^{-1}$ (see fig. 7 in Torrealba et al. 2006) we obtained that $\sigma_\star = 191(\pm6)kms^{-1}$ which yields a value for the mass of the black hole of: $\log M_{BH}/M_\odot = 7.94(\pm1.6)$.

Method 3: Using the Sérsic-n index

In order to derive the physical parameters for the components of the galaxy, we have modeled the surface brightness profile with an exponential disk plus a Sérsic $r^{1/n}$ profile (Sérsic 1968) with index n. For the R band, we have estimated the index n for the bulge in two cases: (a) n was obtained using the best fitted parameters excluding the innermost 2 arc sec region; (b) n was obtained throughout a seeing-convolved Sérsic $r^{1/n}$ profile. For case (a) $n = 0.58(\pm0.02)$ and for (b) $n = 0.82(\pm0.02)$. We have used the n-index obtained in case (b) to estimate the black-hole mass accordingly to the correlation found by Graham et al. (2003) for R band data: $\log M_{BH}/M_\odot = (2.91 \pm 0.38)\log(n) + 6.37(\pm0.21)$, where n is the Sérsic-index, so for our object we have: $\log M_{BH}/M_\odot = 6.12(\pm0.21)$.

2 Discussion

The values found with methods 1 and 2 differ by a factor of ~3, so we conclude that both results are in agreement. Method 3, which needs an extrapolation of the Sérsic profile towards the nucleus of SBS 0748+499, provides an estimation of the M_{BH} that differs by a factor of ~20 and ~ 66 with respect to methods 1 and 2, respectively. It is important to note that the shape of the profile shown in Fig. 1 would be impossible to fit with a n-index > 1 (needed to have an estimation of $\log M_{BH}/M_\odot$ that would be closer to the one obtained with methods 1 and 2). In a future work we plan to study a sample of intermediate Seyfert galaxies in order to test the validity of these methodologies.

References

1. T. A. Boroson: ApJ **585**, 647 (2003)

114   E. Benítez et al.

2. A. W. Graham, P. Erwin, N. Caon, I. Trujillo: RMxAC   , 196 (2003)
3. J. Kormendy & K. Gebhardt: In: *Conference Proceedings 586, 20th Texas Symposium*, vol 3, edited by J. C. Wheeler and H. Martel (American Institute of Physics, Austin 2001) pp 363-379
4. R. J. Mc Lure & J. S. Dunlop: MNRAS   , 795 (2002)
5. R. J. Mc Lure, M. J. Kukula, J. S. Dunlop et al: MNRAS,   , 377 (1999)
6. C. H. Nelson: ApJ   , L91 (2000)
7. C. A. Onken, L. Ferrarese, D. Merritt, D. et al. : ApJ   , 645 (2004)
8. J. L. Sérsic, Atlas de Galaxias Australes, (Obs. Astronómico, Córdoba 1968)
9. J. Torrealba, E. Benítez, A. Franco-Balderas & V. Chavushyan: RMxAA   , 3 (2006)
10. A. Wandel : ApJ   , 762 (2002)

A   h

J. Cuadra[1], S. Nayakshin[1,2], V. Springel[1], and T. Di Matteo[1,3]

[1] Max-Planck-Institut für Astrophysik, D-85741 Garching, Germany
   jcuadra@mpa-garching.mpg.de
[2] University of Leicester, LE1 7RH, UK
[3] Carnegie Mellon University, Pittsburgh, PA 15213, USA

We report a 3-dimensional numerical study of the accretion of stellar winds onto Sgr A*, the super-massive black hole at the centre of our Galaxy. Compared with previous investigations, we allow the stars to be on realistic orbits, include the recently discovered slow wind sources, and allow for optically thin radiative cooling. We find that the slow winds shock and rapidly cool, forming cold gas clumps and filaments that coexist with the hot X-ray emitting gas. The accretion rate in this case consists of two components: the hot quasi steady-state one, and the cold one that is highly variable on time-scales of tens to hundreds of years. Such variability can in principle lead to a strongly non-linear response through accretion flow physics not resolved here, making Sgr A* an important energy source for the Galactic centre.

Sgr A* is identified with the $M_{\mathrm{BH}} \sim 3.6 \times 10^6 M_\odot$ super-massive black hole (SMBH) at the centre of our Galaxy [6]. By virtue of its proximity, Sgr A* may play a key role in the understanding of Active Galactic Nuclei (AGN). Indeed, this is the only AGN where observations detail the origin of the gas in the vicinity of the SMBH. This information is absolutely necessary for the accretion problem to be modelled self-consistently.

One of Sgr A* puzzles is its low luminosity with respect to estimates of the accretion rate. From *Chandra* observations, one can measure the gas density and temperature around the inner arcsecond[4] and then estimate the Bondi accretion rate [1]. However, hot gas is continuously created in shocked winds expelled by tens of young massive stars near Sgr A*, so the situation is far more complex than in the idealised, symmetric and steady state, Bondi model.

An alternative approach is to model the gas dynamics of stellar winds, assuming that the properties of the wind sources are known [2, 10, 12]. Here we present the first numerical simulations of wind accretion onto Sgr A* that include optically thin radiative cooling and allow the wind-producing stars to be on Keplerian orbits.

---

[4] One arcsecond (1″) correspondes to $\sim 10^{17}$cm or $\sim 10^5 R_{\mathrm{S}}$ for Sgr A*.

2 Method and Initial Conditions

A full account of our numerical method along with validation tests was presented elsewhere [4]. Here we describe only briefly the method. We use the SPH/N-body code
GADGET-2 [14] to simulate the dynamics of stars and gas in the gravitational field of
the SMBH. The SMBH itself is modelled as a "sink" particle [13], with all the gas passing within a given distance from it ($0.07''$ in the present simulation) disappearing from
the computational domain. To model the stellar winds, new gas particles are continously
created around the stars.

The initial conditions are chosen to broadly reproduce the observed stellar population and dynamics [5, 8]. We introduce two stellar types: luminous blue variable
candidates (LBVs) with wind velocity $v_w = 300$ km s^{-1}, and Wolf-Rayet stars (WRs)
with $v_w = 1000$ km s^{-1}. We distribute the stars in two perpendicular discs, with radial
extents 2–5$''$ and 4–8$''$. We use 20 wind sources, 6 LBVs and 2 WRs stars in the inner
disc, plus 3 LBVs and 9 WRs in the outer disc. Each star has the same mass loss rate of
$\dot{M}_* = 4 \times 10^{-5} \mathrm{M}_\odot \mathrm{yr}^{-1}$.

3 Results

Two-phase gas

Figure 1 shows the resulting morphology of the gas 2,450 yr into the simulation. Cool
dense regions in the gas distribution (shown as bright yellow regions) are mainly produced by winds from LBVs. When shocked, these winds attain a temperature of only
around 10^6 K, and, given the high pressure environment of the inner parsec of the GC,
quickly cool radiatively [3]. LBV winds form bound clouds of gas, often flattened into
filaments due to the SMBH potential and the symmetry of the problem. As more filaments are formed in the inner region, they start overlapping and eventually form a disc
that lies almost at the plane $z = 0$. The orientation of the disc plane is very close to
that of the inner stellar disc. This happens because most of the winds coming from the
LBVs in the outer disc are unbound and escape the region, without influencing the inner
arcsecond.

On the other hand, the WRs by themselves do not produce much structure. The fast
winds they emit have temperatures $> 10^7$ K after shocking, and do not cool fast enough
to form filaments. This temperature is comparable to that producing X-ray emission
detected by *Chandra*. Gas cooler than that would be invisible in X-rays due to the finite
energy window of *Chandra* and the huge obscuration in the Galactic plane.

The prominent $\sim 1''$ disc-like structure that we find in the simulation appears to be
inconsistent with current observations. The most likely reason for the discrepancies may
be ascribed to the specific initial conditions we used. In particular, placing many LBVs
in the same plane at a short distance from the SMBH may be particularly favourable
for the development of a disc. Newer observations show that LBVs and WRs are more
evenly distributed between both discs [9]. To test the influence of the stellar distribution
on the results we ran simulations where the stellar orbits are oriented randomly [4]. The
distribution of angular momenta vectors is obviously isotropic in this case, so there is
no obvious preference for a particular plane and no disc was formed.

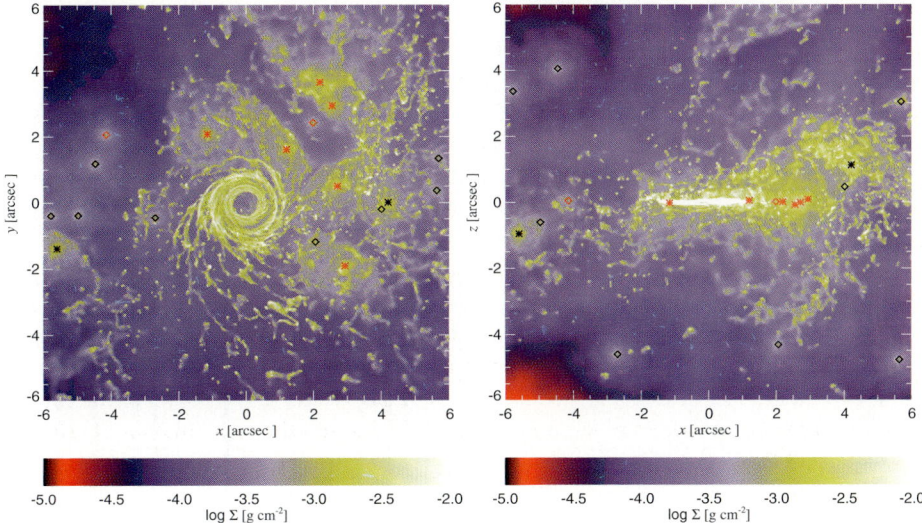

Fig. 1. Left panel: Column density of gas in the inner $6''$ of the computational domain. Asterisks mark slow wind stars (LBVs), whereas diamonds mark stars producing fast winds (WRs). The inner stellar disc is face-on, rotating clock-wise, while the outer one is seen edge-on (plane $y = 0$). Right panel: same as on the left, but the outer stellar disc is now seen face-on and rotating counter clock-wise, while the inner one appears edge-on ($z = 0$).

Variable accretion

Figure 2 shows the accretion rate onto the SMBH as a function of time. The accretion of hot gas proceeds at a fairly constant rate, at a value $\dot{M}_{\mathrm{BH}} \approx 3 \times 10^{-6} \mathrm{M}_\odot \ \mathrm{yr}^{-1}$, consistent with the *Chandra* estimates [1]. In addition to this component, the accretion of cold clumps occurs in short bursts, producing a highly variable transfer of mass into the inner $0.07''$. The actual rate of accretion onto the SMBH has still to be determined by the accretion flow physics that we cannot resolve here, but it is expected to maintain most of the variability in time-scales longer than ~ 100 yr.

There are additional sources of variability that have not been included in our treatment. Some of the stars with high outflow rates probably have orbits with a non-negligible eccentricity. Then the fraction of winds that can be captured from a given star by the black hole changes with time. Preliminary tests where the stellar cluster IRS13E is included with an eccentric orbit [9] show that the accretion rate can go up by a factor ~ 20 during its closest approach to Sgr A*.

In the extreme sub-Eddington regime of Sgr A* accretion, the dependence of the luminosity on the accretion rate is very non-linear [15]. Thus, even small changes in the accretion rate could result in a strongly enhanced X-ray emission. The results from our simulations, the observational evidence for higher luminosity in the recent past [11], and the idea of star formation in an AGN-like accretion disc a few million yr ago [7],

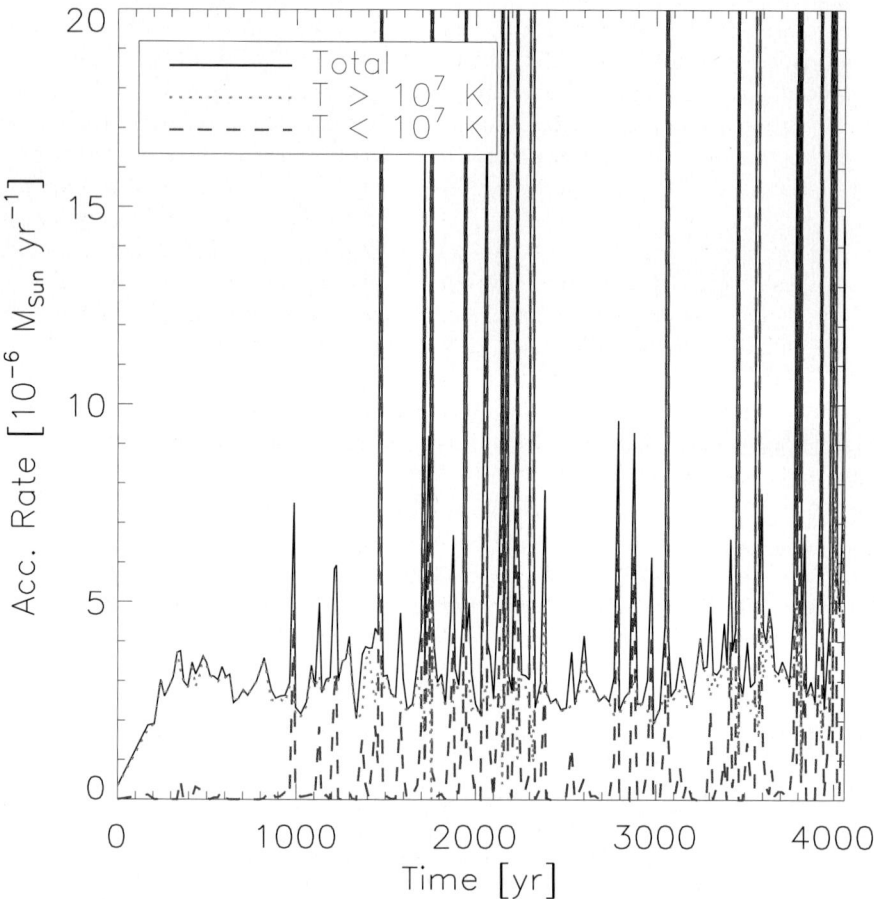

Accretion rate onto the SMBH as a function of time (black line). This rate is then divided into that of hot gas (dotted) and that of low temperature gas (dashed). The time-averaged accretion is dominated by the hot component that is quasi-constant; the accretion of cold gas is episodic and highly variable, but smaller overall.

all suggest that on long time-scales Sgr A* is an important energy source for the inner Galaxy.

Here we presented our new numerical simulations of wind accretion onto Sgr A*. Compared with previous works, our methodology includes a treatment of stellar orbital motions and of optically thin radiative cooling. While the results depend on the assumptions about stellar mass loss rates, orbits, and wind velocities, some relatively robust conclusions can be made.

Unless mass loss rates of narrow line stars are strongly over-estimated, the gas at $r \sim 1''$ distances from Sgr A* has a two-phase structure, with cold filaments immersed into hot X-ray emitting gas. Depending on the geometry and orbital distributions in the mass-loosing star cluster, the cold gas may be settling into a coherent structure such as a disc, or be torn apart and heated to X-ray emitting temperatures in collisions. Both the fast and the slow phase of the winds contribute to the accretion flow on to Sgr A*. The accretion flow is rotating rather than free-falling in sub-arcsecond regions, with a circularisation radius of order $10^4 R_S$. The accretion rates we obtain are of the order of $3 \times 10^{-6} M_\odot$ year^{-1}, consistent with the *Chandra* estimates. The accretion of cooler gas proceeds separately via clump infall and is highly intermittent, although the average accretion rate is dominated by the quasi-constant inflow of hot gas.

A generic result of our simulations is the large variability in the accretion rate on to Sgr A* on time scales of the order of the stellar wind sources orbital times (hundreds to thousands years). This implies that the current very low luminosity state of Sgr A* may be the result of a relatively unusual quiescent state. It also means that the real time-averaged output of Sgr A* in terms of radiation and mechanical jet power may be orders of magnitude higher than what is currently observed. The role of Sgr A* for the energy balance of the inner region of the Galaxy may therefore be far more important than its current meager energy output would suggest. Observations of γ-ray/X-ray echos of past activity of Sgr A* [11] seem to confirm these suggestions from our simulations.

References

1. Baganoff F. K., Maeda Y., Morris M., et al.: ApJ **591**, 891 (2003)
2. Coker R. F., Melia F.: ApJ **488**, L149 (1997)
3. Cuadra J., Nayakshin S., Springel V., Di Matteo T.: MNRAS **360**, L55 (2005)
4. Cuadra J., Nayakshin S., Springel V., Di Matteo T.: MNRAS **366**, 358 (2006)
5. Genzel R., Schödel R., Ott T., et al.: ApJ **594**, 812 (2003)
6. Eisenhauer F., Genzel R., Alexander T., et al.: ApJ **628**, 246 (2005)
7. Nayakshin S., Cuadra J.: A&A **437**, 437 (2005)
8. Paumard T., Maillard J. P., Morris M., Rigaut F.: A&A **366**, 466 (2001)
9. Paumard T., Genzel R., Martins F., et al.: ApJ **643**, 1011 (2006)
10. Quataert E.: ApJ **613**, 322 (2004)
11. Revnivtsev M. G., Churazov E. M., Sazonov S. Y., et al.: A&A **425**, L49 (2004)
12. Rockefeller G., Fryer C. L., Melia F., Warren M. S.: ApJ **604**, 662 (2004)
13. Springel V., Di Matteo T., Hernquist L.: MNRAS **361**, 776 (2005)
14. Springel V.: MNRAS **364**, 1105 (2005)
15. Yuan F., Markoff S., Falcke H.: A&A **383**, 854 (2002)

Winds Driven by Line Opacity near Neutron Stars and Black Holes

A.V. Dorodnitsyn[1,2]

[1] Max-Planck-Institute for Nuclear Physics, Postfach 10 39 80, D-69029 Heidelberg, Germany
`dora@mpi-hd.mpg.de`
[2] Space Research Institute, 117997, Profsoyuznaya st.84/32, Moscow, Russia

1 Introduction

It is widely accepted that radiation pressure can play an important role in formation of fast plasma outflows from compact objects (BHs, NS). In this paper we focus on the scenario, when momentum is extracted from the radiation field via transitions in lines of abundant elements, particularly studying the situation when the influence of the strong gravitational field on the formation of such a flow cannot be neglected.

Accretion discs are believed to be common ingredients in many astrophysical cases. Radiation from these disks are probably responsible for the formation of fast radiation-driven winds in AGNs, cataclismic variables and young stellar objects.

Although the opacity in spectral lines is high in comparison with that of the continuum, in a static atmosphere the acceleration due to absorption in a line transition is not efficient. The reason for that is because of the large optical depth in a line, the radiation flux in the frequency band of the line is reduced: $e^{-\tau_l}$. It was in the pioneering paper by Sobolev [9] where it was recognized that in the gradually accelerating medium, the process of transfer in a line is simplified drastically in comparison with the same process in a static atmosphere. This is because a line with a rest frequency ν_0 will be redshifted if seen from adjacent downstream regions of the flow. Thus after emission in a line transition a probability of a photon to escape to infinity will be greater as the greater will be the velocity gradient.

In the papers by [7] and [1], (hereafter CAK), the importance of the line opacity for the formation of fast radiation-driven winds from hot stars has been understood. In the latter paper the authors succeeded in building of what is now referenced "standard theory of line-driven winds".

In the paper by Dorodnitsyn [2] (hereafter **D1**) a mechanism was proposed when line-driven acceleration occurs in the vicinity of compact object so that the the gravitational redshifting can play an important role.

The detection of a gravitationally redshifted absorption from highly ionized species will give an important piece of evidence of the interaction of radiation and plasma in the vicinity of the compact object. However in case of quasars there are still only a few cases of such an interpretation. In the case of the high-luminosity ($L(2 \div 10\mathrm{keV}) =$

$3 \div 4 \cdot 10^{45}$ergs/s) quasar E 1821+643, an observed absorption feature was attributed to resonant absorption in highly ionized Fe [10]. A signature of a redshifted line in the quasar Q0056-363 was reported in [5] and for quasar PG 1211+143 in [8]. Although the quality of the data accepts alternative explanations, a superposition of gravitationally redshifted absorption and blue/redshifted emission seems to be the most likely explanation in these cases. The characteristic width of the line in case of quasar Q0056-363 is $\Delta E \simeq 50$eV, if we assume that the line broadening is of purely gravitational origin we find the characteristic length scale $\delta l / r_g = 2x^2 \Delta E / E_0$, which gives $\delta l / r_g \sim 1$, if the emission is seen from the wind/blob from $10 r_g$. In our calculations of the optical depth and radiation force we assume the Sobolev approximation for the interration of photons and matter. That means we accept that while an emitted photon has a certain probability to interact either in the vicinity of this point or not at all. As it was derived in **D1** in such a case interration occurs in the resonant layer of the thickness: $l_{res} = v_{th} / \left(dv/dr + \dfrac{1}{c} d\varphi/dr \right)$ in case of pure gravitational redshift we have: $l_{res}/r_g = 8.56 \cdot 10^{-5} \sqrt{T_4/A}$. A photon emitted at a given radius will suffer a continuous redshifting both gravitational and Doppler and may become resonant with a line transition at some point downstream. We restrict ourself to the *radially streaming photons* only and assume that they are emitted from a point source.

Optical depth between points s_0 and s, as measured along the photon's trajectory is calculated as follows:

$$t = \int_{s_0}^{s} \chi_{lab}^{l} \Psi(y_0) \, ds \simeq \int_{y_0(s_0)}^{y_0(s)} \chi_{com}^{l} \frac{\Psi(y_0)}{\left(\dfrac{dy_0}{ds} \right)} \, dy_0, \tag{1}$$

where the usual procedure of transformation from space to frequency variable was adopted.

The opacity is transformed according to the relation: $\chi_{lab}^{l} = \chi_{com}^{l} \tilde{\nu}/\nu_{lab}$, where χ_{com}^{l} is the absorption coefficient measured by the co-moving observer. A probability for a photon of frequency ν, to be emitted at some point s_0 being in the frequency range ν, $\nu + d\nu$, and solid angle range Ω, $\Omega + d\Omega$: $dP_e = (4\pi)^{-1} \Psi(y_0) \, d\nu_e \, d\Omega$, where $y_0 \equiv (\tilde{\nu} - \nu_o)/\Delta \nu$ is the non-dimensional frequency displacement variable in the co-moving frame of the fluid. Note that in the current studies we retain only terms of the order $O(v/c)$. It is also convenient to introduce the frequency displacement variable $y \equiv (\nu - \nu_0)/\Delta \nu$, so that $\tilde{y} = y(\tilde{\nu})$, $y^{\infty} \equiv y(\nu^{\infty})$ - a readshift as measured by the observer at infinity: $y^{\infty} \equiv (\nu^{\infty} - \nu_0)/\Delta \nu$, where ν^{∞} is the frequency of the photon at infinity.

In the co-moving frame of the fluid a photon will have a following frequency:

$$\tilde{\nu}(s) = \nu \left(1 - \mu \beta - \frac{\phi(s) - \phi(s_0)}{c^2} \right). \tag{2}$$

In the case when the Schwarzschild black hole is considered the optical depth was derived in [4]:

$$t = \int_{r_d}^{\infty} \Psi(\tilde{\nu} - \nu_0)\chi_{l\nu} \frac{dr}{\sqrt{h}} = \frac{\chi_l v_{th}}{\left| \sqrt{h}\frac{dv}{dr} + cw \right|}, \tag{3}$$

where $\chi_{l\nu}(\mathrm{Hz \cdot cm^{-1}}) = \Delta\nu_D \chi_l$ and $h \equiv g_{00}.$, and $\mathrm{w} = \frac{d}{dr}\sqrt{h}$. If there is no gravitational redshifting taken into account then the Sobolev optical depth is obtained: $\tau_{sob} = \frac{\chi_l v_{th}}{|dv/dr|} = \chi_l l_{sob}$ The radiation pressure can be calculated when summing over the ensemble of optically thin and optically thick lines. Using the parameterization law of CAK we obtain the result by [4]:

$$g_l = \sum_l G_l = \frac{F\sigma_e}{c} k \left(\frac{\sigma_e \rho_0 v_{th}}{\sqrt{h}\frac{dv}{dr} + cw} \right)^{-\alpha}, \tag{4}$$

where σ_e is the electron scattering opacity per unit mass and F is the radiation flux.

2 Gravitationally Exposed Flow

The equation of motion for stationary, spherically-symmetric, isothermal wind reads:

$$\frac{P+\rho}{h\rho_0} \left(vh\frac{dv}{dr} + \frac{GM}{r^2} \right) + \frac{1}{\rho_0}\frac{dP}{dr} - \frac{\sigma_e}{\sqrt{h}}\frac{L}{4\pi r^2 c}$$

$$- \frac{\sigma_e}{\sqrt{h}}\frac{L}{4\pi r^2 c} k \left(\frac{4\pi}{\sigma v_{th}\dot{M}} \right)^{\alpha} \times$$

$$\left\{ \sqrt{h}\, vr^2 \left[\sqrt{h}\frac{dv}{dr} + cw \right] \right\}^{\alpha} = 0, \tag{5}$$

Here we retain only terms of the order $O(v/c)$. We adopt the equation of state for the ideal gas: $P = \rho_0 \mathcal{R}T$, $E_i = 3/2\,\mathcal{R}T$, where $\mathcal{R} = k/m_p$ is the gas constant. For a given position of the critical point r_c we can calculate the value of the velocity and velocity gradient in the critical point. Adjusting the position of the critical point r_c we integrate the equation of motion inward, looking for the solution that satisfies the inner boundary condition. The qualitative picture that has been obtained in **D1** is confirmed throughout our calculations. We detect a considerable gain in terminal velocity both in comparison with CAK case and between fully relativistic calculations presented here, and semi-classical treatment of **D1**.

3 Discussion

The problem of direct detection of line emmision from regions close to compact objects are obviously of greate importance. However only in a few cases where observational evidence of formation of spectral lines in outflowing/inflowing plasma (continuous or

Table 1. Comparison of GEF solution and SLDW solution. $x_c = r_c/r_g$, $\Delta_c^{\mathrm{GEF}} = x_c^{CAK} - x_c^{\mathrm{GEF}}$. $\Delta_{\mathrm{CAK}}^{\infty} = (v_{\mathrm{GEF}}^{\infty} - v_{\mathrm{CAK}}^{\infty})/v_{\mathrm{CAK}}^{\infty}$

Model	x_c^{GEF}	Δ_c^{CAK}	$\Delta_{\mathrm{CAK}}^{\infty}$
s_1	15	7.47	0.57
s_2	20	9.95	0.51
s_3	50	25	0.37
s_4	100	50	0.2

blob) near BH in AGNs is established [10], [8], [5]. Two issues arise when one considers an outflow, driven by radiation pressure in lines. The first one is the problem of acceleration and the second is the observational signature of such a flow. Here the first issue has been adressed. The problem of formation of spectral lines in a flow in a strong gravitational field is addressed in [3]. We have shown that it is indeed of significant importance to include gravitational redshift when calculating the radiation force.

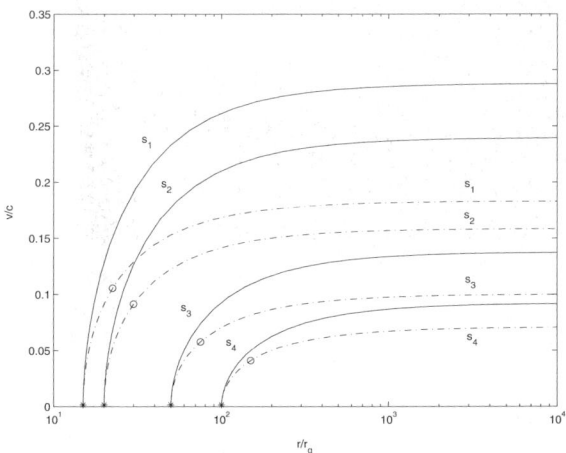

Fig. 1. Solutions of the equation of motion. Solid line - "Gravitationally Exposed Flow" (GEF) solution, dashed line - standard line-driven wind (SLDW) regime. Stars indicate GEF critical points, circles - SDLW critical points. cf. Figure 1. of Dorodnitsyn (2003) s_1 - s_4 correspond to different sets of solutions.

References

1. J. Castor, D. Abbott, R. Klein, R.: ApJ **195**, 157 (1975)
2. A. Dorodnitsyn: MNRAS **339**, 569 (2003)
3. A. Dorodnitsyn: in preparation (2006)
4. A. Dorodnitsyn, I. Novikov: ApJ **621**, 932D, (2005)

5. G. Matt, D. Porquet, S. Bianchi, S. Falocco, R. Maiolino, J. Reeves, L. Zappacosta: A&A **435**, 857 (2005)
6. R. Mushotzky, P. Solomon, P. Strittmatter: ApJ **174**, 7 (1972)
7. L. Lucy, P. Solomon: ApJ **159**, 879 (1970)
8. J. Reeves, K. Pounds, P. Uttley, S. Kraemer, R. Mushotzky, T. Yaqoob, M. George, T. Turner: ApJ **633**, L81 (2005)
9. V.V. Sobolev: Moving envelopes of stars. (Cambridge: Harvard University Press 1960)
10. T. Yaqoob, P. Serlemitsos: ApJ **623**, 112 (2005)

Inspiral of Double Black Holes in Gaseous Nuclear Disks

M. Dotti[1], M. Colpi[2], and F. Haardt[1]

[1] Università dell'Insubria, Como, Italy, `(dotti,haardt)@mib.infn.it`
[2] Università di Milano-Bicocca, Milano, Italy, `colpi@mib.infn.it`

Summary. We study the inspiral of double black holes orbiting inside a massive rotationally supported gaseous disk with masses in the Laser Interferometer Space Antenna (*LISA*) window of detectability. Using high–resolution SPH simulations, we follow the black hole dynamics in the early phase when gas–dynamical friction acts on the black holes individually, and continue our simulation until they form a close binary. We find that in the early sinking the black holes lose memory of their initial orbital eccentricity, if they co–rotate with the gaseous disk. As a consequence the massive black holes *form a binary with very low eccentricity*. During the inspiral, gravitational capture of gas by the black holes occurs mainly when they move on circular orbits and may ignite AGN activity: eccentric orbits imply instead high relative velocities and weak gravitational focusing.

1 Introduction

Close massive black hole (MBH) binaries are natural, powerful sources of gravitational radiation, whose emission is one of the major scientific targets of *LISA* (see, e.g., Haehnelt 1994, Jaffe & Backer 2003, Sesana et al. 2005). How can MBHs reach sub–parsec distance scales and coalesce when resulting from the collision and merger of galaxies? Recently, Kazantzidis et al. (2005) explored the effect of gaseous dissipation in mergers between gas–rich disk galaxies with central MBHs, using high resolution N–Body/SPH simulations. They found that the interplay between strong gas inflows, cooling processes, and star formation, leads naturally to the formation of a close MBH pair and of massive nuclear gaseous disks, on a scale of ~ 100 pc, close to the numerical resolution limit (updated simulations are presented by Mayer et al. in these proceedings). On smaller scales, Escala et al. (2005) have studied the role of gas on the orbital evolution of MBH binaries as a function of disk clumpiness, MBH to gas mass ratio, and orbital inclination angle. In the same context, we studied the evolution of the eccentricity, a key parameter for assessing the role played by gravitational wave emission in the orbital decay of MBH binaries in the *LISA* mass range.

2 The Simulations

We perform our simulations with MBHs embedded in a spheroidal component (bulge) modeled initially as a Plummer sphere, and in a Mestel gaseous disk. We evolve the system using the N–Body/SPH code GADGET (Springel, Yoshida & White 2001).

We show the case of two MBHs of mass $10^6 \, M_\odot$ and $5 \times 10^6 \, M_\odot$ orbiting in the disk plane. We allow the heavier MBH (MBH1) to reach the center before the sinking process of the light one (MBH2) takes place. MBH2 is initially moving on a prograde eccentric orbit ($e = 0.95$). Figure 1, upper panel, shows the MBH distances from the center of mass, as a function of time. We find that the sinking time of MBH2 is $\simeq 10^7$ yr, and that there is a pronunced circularization of its orbit. The lower panel shows the gas mass collected by the two MBHs during their orbital evolution. Only when the orbit of MBH2 becomes circular gathering of gas can occur. Figure 2 shows, at two selected times, the face-on projection of the gas density together with the orbit of MBH2: close to pericentre the MBH has a speed larger, in modulus, than the local gas speed so that

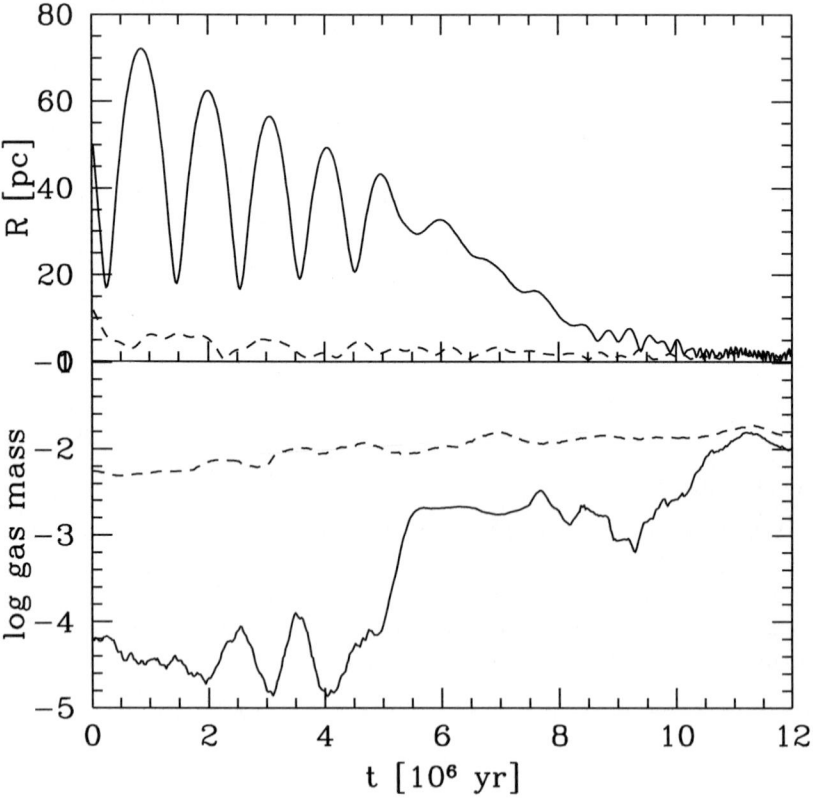

Fig. 1. Upper panel: Solid (dashed) line shows the distance R (pc) of MBH2 (MBH1) from the center of mass as a function of time. Lower panel: Solid (dashed) line shows the mass of the over-density corresponding to MBH2 (MBH1) as a function of time.

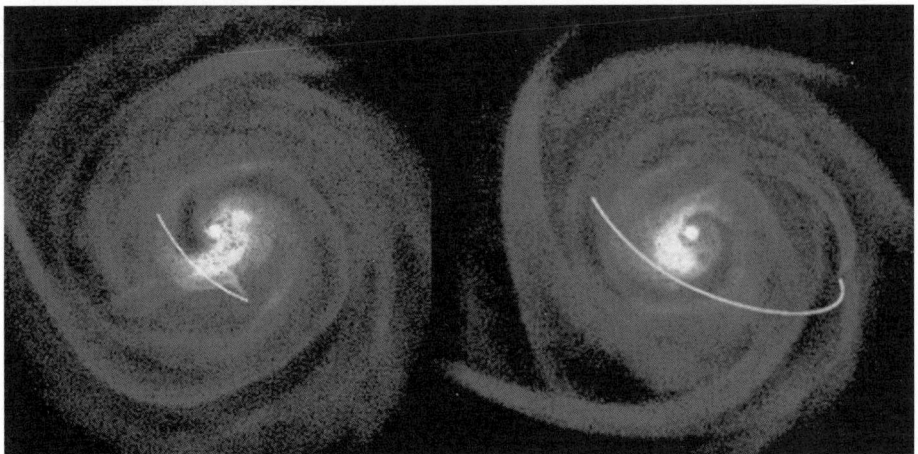

Fig. 2. Time sequence of the sinking of the MBH2. The intensity indicates the z–averaged gas density (in logaritmic scale), and the white line traces the MBH2 counterclockwise prograde orbit. In the left panel the over–density created by MBH2 is behind its current trail, while in the right panel, the BH finds its own wake in front of its path. The wake is dragged by the faster rotation of the disk.

the wake is excited behind its trail, eroding the radial component of the velocity (left panel). On the other hand, near to apocentre, the MBH tangential velocity is slower than the local rotational gas velocity, so that the wake is dragged in front of the MBH, increasing its angular momentum (right panel). The net effect is the circularization of the initially eccentric MBH orbit. New simulations with pure stellar disks present the same effect.

3 Conclusion

Our main findings are:

- dynamical friction due to the MBH–gas interaction is effective down to pc distence scales
- the MBH–disk interaction circularizes prograde orbits
- substantial gas mass can be gathered along circular MBH orbits: binary coalescence could be accompanied by (double) AGN activity.

For full details and references please see: Dotti, Colpi, Haardt: MNRAS.

References

1. Dotti M., Colpi M., Haardt F.: MNRAS **367**, 103 (2006)
2. Escala A., Larson R.B., Coppi P.S., Maradones D.: ApJ **630**, 152 (2005)

3. Haenelt M.G.: MNRAS **269**, 199 (1994)
4. Jaffe A.H., Backer D.C.: MNRAS **583**, 616 (2003)
5. Kazantzidis S., Mayer L., Colpi M., Madau P., Debattista V.P., Wadsley J., Stadel J., Quinn T., Moore B.: ApJ **623**, L67 (2005)
6. Sesana A., Haardt F., Madau P., Volonteri M.: ApJ **623**, 23 (2005)
7. Springel V., Yoshida N., White S.D.M.: NewA **6**, 79 (2001)

The Cosmogony of Super-Massive Black Holes

W.J. Duschl[1,2] and P.A. Strittmatter[2]

[1] Institut für Theoretische Astrophysik, Zentrum für Astronomie der Universität Heidelberg, Albert-Ueberle-Str. 2, 69120 Heidelberg, Germany, wjd@ita.uni-heidelberg.de
[2] Steward Observatory, The University of Arizona, 933 N. Cherry Ave., Tucson, AZ 85721, USA, pstrittmatter@as.arizona.edu

1 Introduction and Motivation

Two recently found properties of (super-massive) black holes (SMBHs) in the centers of active galaxies shed new light on their formation and growth:

- The luminosities of the quasars with the largest redshifts indicate that central BH masses of $> 10^9 \, M_\odot$ were already present when the Universe was less than 10^9 years old (Barth et al. 2003). These masses are lower limits as they are based on the assumption that the BHs accrete at their Eddington limit. There is no indication that the (majority of the) sample of highest-redshift-quasar luminosities is afflicted by amplification by gravitational lensing (White et al. 2005).
- Surveys in the X-ray regime (Hasinger et al. 2005) and in the optical/UV regime (Wolf et al. 2003) show a strong luminosity dependence of the redshift at which the active galactic nuclei (AGN) space density peaks: The lower the AGN luminosity, i.e., the smaller the BH mass, the later in the evolution of the Universe the co-moving space density of these AGN peaks. In other words, it takes BHs of a lower *final* mass longer to reach that mass than BHs of a larger final mass. This is also supported by a comparison of the local and the derived accretion mass function of SMBHs (Shankar et al. 2004).

In this contribution we report results of a project investigating the growth of SMBHs by disk accretion. We find that both above-mentioned phenomena can be explained in the framework of such a model.

2 Black Hole Formation and Growth in Galactic Centers

Our model involves two basic processes, namely: (a) galaxy mergers resulting in aggregation of material into a self-gravitating disk around the galactic center (GC); and (b) subsequent disk accretion onto a central BH. The issue in the past has been whether enough disk material could be accreted in the available time, a problem that has become steadily more acute as increasingly luminous quasars have been found at great redshifts.

In regard to (a), we envisage that the interaction between two (gas rich) galaxies leads to the rapid formation of a circum-nuclear gaseous disk. This process is expected to occur on a dynamical time scale. The mass and spatial extent of this disk will depend on the mass and gas content of the interacting galaxies and on the impact parameters of the collision. The process will thus result in a range of disk masses and extents.

For our subsequent reasoning it is important to note that fairly early in the evolution of the Universe ($z > 1.5 \cdots 3$) massive galaxies were already in place (Chen & Marzke 2004, Glazebroke et al. 2004, van Dokkum & Ellis 2003). Moreover, there is mounting evidence (e.g., Dunlop et al. 2003, Sánchez et al. 2004) that interactions between and mergers of galaxies trigger nuclear activity (e.g., McLeod & Rieke 1994 a+b, Bahcall et al. 1997, Canalizo & Stockton 2001, Sánchez et al. 2005, Sanders 2004).

Thus for a *major merger* the resulting gaseous disk mass may well contain $> 10^{10} \, M_\odot$ within a few hundred parsecs of the GC. This basic process has been demonstrated by numerical simulation (e.g., Barnes 2002, Barnes & Hernquist 1996+ 1998, Iono, Yun & Mihos 2004) and provides the essential initial conditions for our analysis. The resulting nuclear disk will, of course, be subject to viscous dissipation causing an inward flow of material towards the GC where it is potentially available for accretion onto a BH. With the initial mass and radius parameters discussed above, such a disk must inevitably be self-gravitating, at least initially. In these circumstances, the disk accretion time scale and, hence, the growth times and limiting mass of the putative BH, depend on both the mass and extent of the initial disks.

We also note that, initially, the disk may provide mass to the BH at a rate higher than is allowed by the Eddington limit. Thus in such a case initially the BH growth rate is defined by the Eddington limit, with the rest of the material presumably driven from the system (or at least the proximity of the BH) by radiation pressure. The peak luminosity will occur roughly when the rate of supply of material from the disk equals the Eddington limit for the current BH mass (higher for higher accretion rate).

3 Evolution of Self-Gravitating Accretion Disks and the Growth of Black Hole Masses

We carried out numerical simulations modelling the evolution of initially self-gravitating accretion disks and the ensuing growth of the central BH. Our model disks are geometrically thin and rotationally symmetric, with the following modifications with respect to standard accretion disk models:

- We allow for a disk mass which is not necessarily small compared to the mass of the central BH, i.e., we do not assume a Keplerian rotation curve in the disk but solve Poisson's equation.
- We use the generalized viscosity prescription by Duschl et al. (2000; β-viscosity).
- We take the Eddington limit into account. Mass flow above the Eddington limit is assumed to be lost from the system.

Our numerical code is based on an explicit finite-difference scheme. For further details of the modelling, we refer the reader to an upcoming paper (Duschl & Strittmatter, *in prep.*)

3.1 A Reference Model

As a reference model, we define an accretion disk with the following set of parameters:

- Inner radius of the disk: $s_i = 10^{16.5}$ cm
- Outer radius of the disk: $s_o = 10^{20.5}$ cm $\approx 10^2$ pc
- Initial disk mass: $M_{d,0} = 10^{10}\,M_\odot$
- Initial surface density distribution: $\Sigma_0(s) \propto s^{-1}$.
- Seed black hole mass: $M_{BH,0} = 10^6\,M_\odot \ll M_{d,0}$
- Viscosity parameter: $\beta = 10^{-3}$.

3.2 The Evolution of the Reference Model

The evolution of the mass flow rate of the reference model is shown in the left panel of Fig. 1. The right panel shows the corresponding evolution of the BH mass. The zero-point of the time is the point at which, as a consequence of a galaxy-galaxy interaction, a massive nuclear accretion disk has been established. One can clearly discern two phases of the accretion process: From the beginning of the evolution to $t_{Edd} \sim 2.7 \cdot 10^8$ years the evolution is dominated by the Eddington limit: The disk delivers mass at a larger rate (broken line; $\dot M_d$) than the BH can accrete due to the Eddington limit (dash-dot-dotted line; $\dot M_{Edd}$). For times $t < t_{Edd}$ the growth rate of the BH, $\dot M_{BH}$, is subject to the Eddington limit, i.e., $\dot M_{BH}(t < t_{Edd}) = \dot M_{Edd}$. For $t > t_{Edd}$, however, both the mass of the BH has become so large and the mass flow rate from the disk has dropped by so much that $\dot M_d$ now has fallen below $\dot M_{Edd}$ and all the mass delivered by the disk can be accreted: $\dot M_{BH}(t > t_{Edd}) = \dot M_d$. For the following few 10^8 years the free accretion rate, however, is still large enough to make the BH grow at a fast rate. This is slowed down considerably only after another $\sim 3.5 \cdot 10^8$ years by when the accretion rate has fallen by approximately one and a half orders of magnitude. From now on the BH grows only slowly; it has almost reached its *final* mass of $2.1 \cdot 10^9\,M_\odot$ (broken line in the right panel).

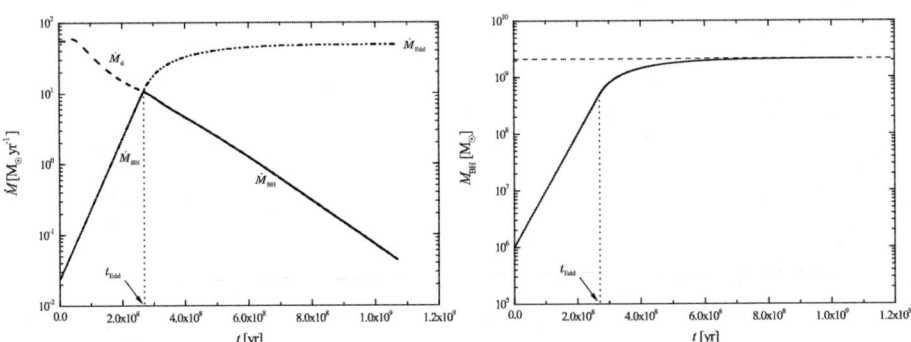

Fig. 1. The evolution of the mass flow rate (left panel) and the BH mass (right panel) for the reference model.

3.3 Variations of the Initial Physical Setup: The Black Hole Growth Time

As an example of the influence of the variation of the initial physical setup, we show in Fig. 2 the growth time scale $t_{0.5}$ of BHs where the initial disk mass and the inner disk radius have been changed, while all the other parameters of the reference model remained unaltered. $t_{0.5}$ is defined as the time at which the BH has reached half its final mass. In all our models, at this time the accretion rate, and thus the accretion luminosity, have already fallen considerably below their maximum value. It is noteworthy that for BH masses in the realm of our GC, the growth times reach values comparable to the Hubble time.

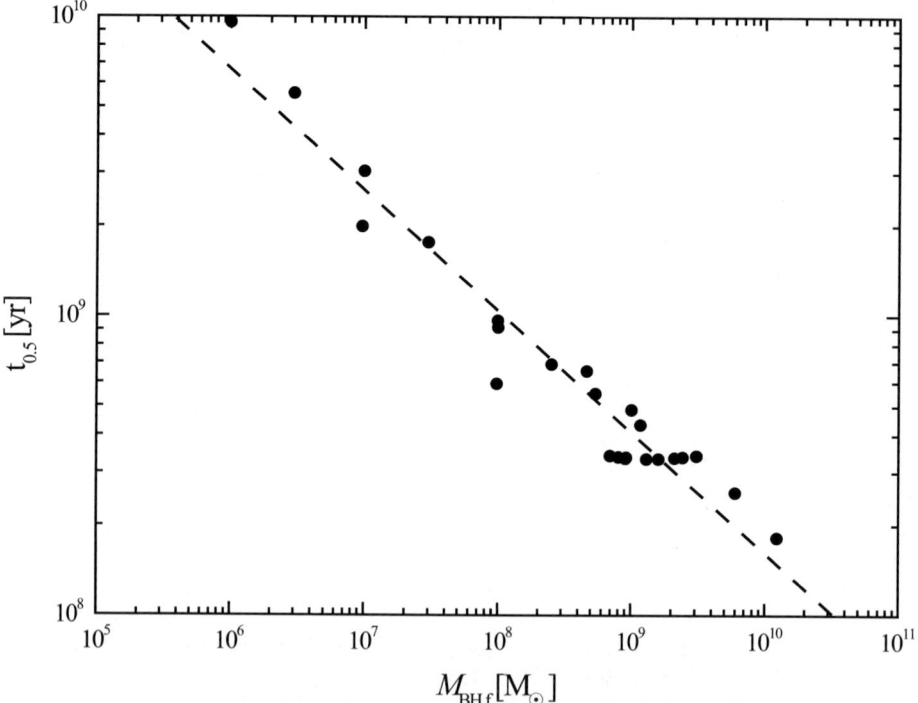

Fig. 2. The BH growth time scale $t_{0.5}$ as a function of the final BH mass M_{BH}.

4 Discussion and Outlook

Massive accretion disks seem to have the required properties to explain the observations described at the beginning of this contribution: The BH mass growth is quick enough to account for the inferred masses in the highest-redshift quasars, and the evolution time is an inverse function of the final BH mass[3]. We expect that the evolution of the Universe as a whole will even emphasize the latter effect: In the early Universe, galaxy mergers and collisions were much more frequent than they are in today's Universe making high initial disk masses more likely at higher redshifts. For a detailed comparison to the observed luminosity functions (e.g., Hasinger et al. 2005) this cosmological evolution of the initial conditoions has to be taken into account.

Acknowledgments. We benefitted very much from discussions of the topic with G. Hasinger and S. Komossa (Garching) and A. Burkert and T. Naab (Munich). This work is partially supported by the German Science Foundation DFG through the Collaborative Research Center SFB439 *Galaxies in the Young Universe*.

References

1. J.N. Bahcall, S. Kirhakos, D.H. Saxe, & D.P. Schneider: ApJ **479**, 642 (1997)
2. J.E. Barnes: MNRAS **333**, 481 (2002)
3. J.E. Barnes & L. Hernquist: ApJ **471**, 115 (1996)
4. J.E. Barnes & L. Hernquist: ApJ **495**, 187 (1998)
5. A.J. Barth, P. Martini, Ch.H. Nelson, & L.C. Ho: ApJ **594**, L95 (2003)
6. G. Canalizo & A. Stockton: ApJ, **555**, 719 (2001)
7. H.-W. Chen & R.O. Marzke: ApJ **615**, 603 (2004)
8. J.S. Dunlop, R.J. McLure, M.J. Kukula, *et al.*: MNRAS **340**, 1095 (2003)
9. W.J. Duschl W.J., P.A. Strittmatter & P.L. Biermann: A&A **357**, 1123 (2000)
10. K. Glazebroke, R.G. Abraham, P.J. McCarthy, *et al.*: Nature **430**, 181 (2004)
11. G. Hasinger, T. Miyaji, & M. Schmidt: A&A **441**, 417 (2005)
12. K.K. McLeod & G.H. Rieke: ApJ **420**, 58 (1994a)
13. K.K. McLeod & G.H. Rieke: ApJ **431**, 137 (1994b)
14. S.F. Sánchez, K. Jahnke, L. Wisotzki, *et al.*: ApJ **614**, 586 (2004)
15. S.F. Sánchez, T. Becker, B. Garcia-Lorenzo B., *et al.*: A&A **429**, L21 (2005)
16. D.B. Sanders: AdvSpR **34**, 535 (2004)
17. F. Shankar, P. Salucci, G.L. Granato, G. de Zotti, L. Danese: MNRAS **354**, 1020 (2004)
18. P.G. van Dokkum & R.S. Ellis: ApJ **592**, L53 (2003)
19. R.L. White, R.H. Becker, X. Fan, M.A. Strauss: AJ **129**, 1202 (2005)
20. C. Wolf, L. Wisotzki, A. Borch, *et al.*: A&A **408**, 499 (2003)

---

[3] For a presentation of the entire set of model calculations and for a more exhausting discussion of their results, we refer the reader to an upcoming paper (Duschl & Strittmatter, *in prep.*).

The Flare Activity of Sagittarius A*

A. Eckart[1], R. Schödel[1], L. Meyer[1], T. Ott[2], S. Trippe[2], and R. Genzel[2,3]

[1] I. Physikalisches Institut, Universität Köln of Cologne, Zülpicher Str.77, 50937 Köln, Germany eckart@ph1.uni-koeln.de
[2] Max Planck Institut für extraterrestrische Physik, Giessenbachstraße, 85748 Garching, Germany
[3] Department of Astronomy and Radio Astronomy Laboratory, University of California at Berkeley, Le Conte Hall, Berkeley, CA 94720, USA

1 Introduction

Over the last decades, evidence has been accumulating that most quiet galaxies harbor a massive black hole (MBH) at their centers. Especially in the case of the center of our Galaxy, progress has been made through the investigation of the stellar dynamics (Eckart & Genzel 1996, Genzel et al. 1997, 2000, Ghez et al. 1998, 2000, 2003a, 2003b, 2005, Eckart et al. 2002, Schödel et al. 2002, 2003, Eisenhauer 2003, 2005) which has revealed the presence of a massive $(3.6 \pm 0.3) \times 10^6 M_\odot$ black hole (MBH) at the Galactic Center. Its position coincides with that of the compact radio source SgrA*. At a distance of only 7.6 ± 0.3 kpc (Eisenhauer et al. 2005) the Galactic Center is the closest nucleus and allows for detailed investigations of the region.

Additional compelling evidence[4] for a massive black hole at the position of Sgr A* is provided by the observation of variable emission from that position both in the X-ray and recently in the near-infrared (Baganoff et al. 2001, 2003, Eckart et al. 2004, 2006a, Porquet et al. 2003, Goldwurm et al. 2003, Genzel et al. 2003, Ghez et al. 2004a, Eisenhauer et al. 2005, Belanger et al. 2005, and Yusef-Zadeh et al. 2006). The close temporal correlation between rapid variability of the near-infrared (NIR) and X-ray emission (Eckart et al. 2004, Eckart et al. 2006a) suggests that the emission with 10^{33-34} erg/s flares arises from a compact source within a few ten Schwarzschild radii (R_S) of the MBH. This points to a common physical origin of the phenomena and may be linked to the variability at radio through sub-millimeter wavelengths (Herrnstein et al. 2004, Mauerhan et al. 2005 and references therein).

---

[4] Based on observations with CHANDRA and observations at the Very Large Telescope (VLT) of the European Southern Observatory (ESO) on Paranal in Chile; Program: 271.B-5019(A) and 271.B-5019(A).

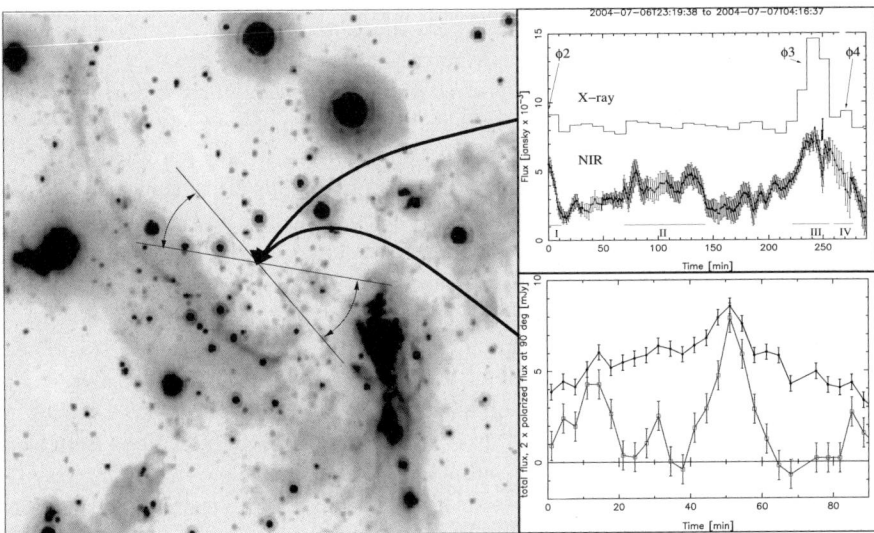

Fig. 1. Left: An image of the central 0.5×0.5 square parsec (1 arcsec ~ 0.039 parsecs) of the Galactic Center at a wavelength of $3.8 \mu m$ using the NACO adaptive optics system at the VLT. In addition to the stars the dust emission from the mini-spiral is seen. The two lines are centered on the position of SgrA* in close the range over which the polarization angle varied on the sky. **Top right:** The X-ray and NIR light curves plotted with a common time axis. **Bottom right:** Polarized emission from the NIR counterpart SgrA*. The total de-reddened flux density light curve (black) and the de-reddened flux density at polarization angle 90° (East over North; grey) corrected for the flux density measured at a PA of 0° at which the sub-flares cannot be seen.

2 NIR/X-Ray Correlation

Simultaneous observations of SgrA* across different wavelength regimes are indispensable, since they provide information on the emission mechanisms responsible for the radiation from the immediate vicinity of the central black hole. The first observation of SgrA* detecting an X-ray flare simultaneously in the near-infrared was presented by Eckart et al. (2004). They detected a weak 6×10^{33} erg/s X-ray flare and covered its decaying flank simultaneously in the NIR.

Variability at radio through submillimeter wavelengths has been studied extensively, showing that variations occur on time scales from hours to years (e.g. Bower et al. 2002, Herrnstein et al. 2004, Zhao et al. 2003). Some of the radio variability may be due to interstellar scintillation. The connection to variability at NIR and X-ray wavelengths has not been clearly elucidated. Zhao et al. (2004) showed a probable link between the brightest X-ray flare ever observed and flux density at mm- and short cm-wavelengths on a timescale of less than 24 hours (see also Mauerhan et al. 2005).

Eckart et al. (2006a) present new simultaneous NIR/ sub-millimeter/ X-ray observations of the SgrA* counterpart associated with the massive $3-4 \times 10^6 M_\odot$ black hole at the Galactic Center. They investigate the physical processes responsible for the variable emission from SgrA*. The observations have been carried out using the NACO adaptive optics (AO) instrument at the European Southern Observatory's Very Large

Telescope and the ACIS-I instrument aboard the *Chandra X-ray Observatory* as well as the Submillimeter Array SMA[5] on Mauna Kea, Hawaii, and the Very Large Array[6] in New Mexico. Eckart et al. (2006a) detected one moderately bright flare event in the X-ray domain and 5 events at infrared wavelengths. The X-ray flare had an excess 2 - 8 keV luminosity of about 33×10^{33} erg/s. The duration of this flare was completely covered in the infrared and it was detected as a simultaneous NIR event with a time lag of ≤ 10 minutes. Simultaneous infrared/X-ray observations are available for 4 flares. All simultaneously covered flares, combined with the flare covered in 2003, indicate that the time-lag between the NIR and X-ray flare emission is very small and in agreement with a synchronous evolution (see Fig.1). There are no simultaneous flare detections between the NIR/X-ray data and the VLA and SMA data. The excess flux densities detected in the radio and sub-millimeter domain may be linked with the flare activity observed at shorter wavelengths. The flaring state can be explained with a synchrotron self-Compton (SSC) model involving up-scattered sub-millimeter photons from a compact source component. This model allows for NIR flux density contributions from both the synchrotron and SSC mechanisms. Indications for an exponential cutoff of the NIR/MIR synchrotron spectrum allow for a straightforward explanation of the variable and red spectral indices of NIR flares (see also Gillessen et al. 2006).

3 NIR Polarization Measurements

Eckart et al. (2006bin preparation) report new polarization measurements of the variable near-infrared emission of the SgrA* counterpart associated with the massive 3–$4 \times 10^6 M_\odot$ black hole at the Galactic Center. The authors investigate the physical processes responsible for the variable emission from SgrA*. The observations have been carried out using the NACO adaptive optics (AO) instrument at the European Southern Observatory's Very Large Telescope Eckart et al. (2006b) find that the variable NIR emission of SgrA* is highly polarized and consists of a contribution of a non- or weakly polarized main flare with highly polarized sub-flares (see Fig.1). The flare activity shows a quasi-periodicity of 20 ± 3 minutes consistent with previous observations (Genzel et al. 2003). The highly variable and polarized emission clearly shows that the NIR emission is non-thermal. The observations cannot discriminate between a jet or temporary disk model. In the disk model the quasi-periodic flux density variations can be explained due to spots on relativistic orbits around the central MBH. The variable polarization could also be due to a helical magnetic field along a short jet and the variable sub-flare emission could be explained by temporal instabilities in the jet. However, near the last stable orbit (LSO) a short jet emerging from a disk may likely look almost indistinguishable from a case involving a pure disk or orbiting spots (see also Fig.4 in

---

[5] The Submillimeter Array is a joint project between the Smithsonian Astrophysical Observatory and the Academia Sinica Institute of Astronomy and Astrophysics, and is funded by the Smithsonian Institution and the Academia Sinica.

[6] The VLA is operated by the National Radio Astronomy Observatory which is a facility of the National Science Foundation operated under cooperative agreement by Associated Universities, Inc.

Bower et al. 2004). Alternative explanations for the high central mass concentration involving boson or fermion balls are increasingly unlikely.

References

1. Baganoff, F.K., Bautz, M.W., Brandt, W.N., et al.: Nature **413**, 45 (2001)
2. Baganoff, F. K., Maeda, Y., Morris, M., et al.: ApJ **591**, 891 (2003)
3. Belanger, G., et al.: ApJ **635**, 1095 (2005)
4. Bower, G.C., Falcke, H., Sault, R.J. and Backer, D.C.: ApJ **571**, 843 (2002)
5. Eckart, A., Genzel, R., Ott, T. and Schödel, R.: MNRAS **331**, 917 (2002)
6. Eckart, A. and Genzel, R.: Nature **383**, 415 (1996)
7. Eckart, A., et al.: A&A **427**, 1 (2004)
8. Eckart, A., et al.: A&A **450**, 535 (2006)
9. Eckart, A., Schödel, R., Meyer, L., Trippe, S., Ott, T., Genzel, G.: A&A **455**, 1 (2006)
10. Eisenhauer, F., et al.: ApJ **597**, L121 (2003)
11. Eisenhauer, F. et al.: ApJ **628**, 246 (2005)
12. Genzel, T., Eckart, A., Ott, T., Eisenhauer, F.: MNRAS **291**, 219 (1997)
13. Genzel, R., Pichon, C., Eckart, A., Gerhard, O.E., Ott, T. 2000, MNRAS 317, 348
14. Genzel, R., Schödel, R., Ott, T., et al.: Nature **425**, 934 (2003)
15. Ghez, A., Klein, B.L., Morris, M., Becklin, E.E.: ApJ **509**, 678 (1998)
16. Ghez, A., Morris, M., Becklin, E.E., Tanner, A., Kremenek, T.: Nature **407**, 349 (2000)
17. Ghez, A. M., Duchéne, G., Matthews, K., et al.: ApJ **586**, L127 2003a
18. Ghez, A.M., et al., 2003b, ApJ, submitted, astro-ph/0306130
19. Ghez, A.M., Wright, S.A., Matthews, K., et al.: ApJ **601**, 159 (2004)
20. Ghez, A.M., et al.: ApJ **620**, 744 (2005)
21. Gillessen, S., et al.: ApJ **640**, L163 (2006)
22. Goldwurm, A., Brion, E., Goldoni, P., et al.: ApJ **584**, 751 (2003)
23. Herrnstein, R.M., Zhao, J.-H., Bower, G.C., Goss, W.M.: AJ **127**, 3399 (2004)
24. Mauerhan, J.C.; Morris, M.; Walter, F.; Baganoff, F.K.: **ApJ** 623, L25 (2005)
25. Porquet, D., Predehl, P., Aschenbach, et al.: A&A **407**, L17 (2003)
26. Schödel, R., Ott, T., Genzel, R. et al.: Nature **419**, 694 (2002)
27. Schödel, R., Genzel, R., Ott, et al.: ApJ **596**, 1015 (2003)
28. Yusef-Zadeh, F., et al.: ApJ **644**, 198 (2006)
29. Zhao, J., et al.: ApJ **586**, L29 (2003)
30. Zhao, J.-H., et al.: ApJL **603**, L85 (2004)

Mass Function of Remnant Black Holes in Nearby Galaxies

M.E. Gáspár[1], Z. Haiman[2], and Z. Frei[1]

[1] Institute of Physics, Eötvös University, Pázmány P. s. 1/A, 1117 Budapest, Hungary,
frei@alcyone.elte.hu
[2] Department of Astronomy, Columbia University, 550 West 120th Street, New York, NY 10027

1 Overview

Observations have shown that most nearby spiral galaxies harbor massive dark objects at their centers [1], that are possibly black holes (BHs). The masses of these objects ($10^6 - 10^9 M_\odot$) are consistent with the density of quasar remnants expected from energy arguments [2]. Therefore, massive BHs are imprints of the evolution of galaxies. A crucial step in understanding this process is to measure the mass function of these objects. Recently the data analysis of millions of galaxies became possible. The calculation of the mass function is based on empirical correlations between the central BH mass and the measured physical parameters of the host galaxy spheroidal (bulge) component. The precision of estimates are ultimately limited by these correlations. Therefore, every possible improvement in measured bulge parameters is a crucial step forward.

Three empirical correlations can be utilized. Velocity dispersion of the bulge [3] correlates best with the central BH mass, however only high resolution spectra allow the calculation of the velocity dispersion. Second, the bulge luminosity correlates (not too strongly [4, 5]) with the BH mass, and here only photometric data is needed. Third, the concentration index correlates quite well with the BH mass [6]. The concentration index is a purely photometric quantity requiring no calibration. However, the point spread function distorts the intensity profile; therefore we conclude that for small or distant galaxies the bulge luminosity is the only quantity that we can measure well.

Half of the total mass of central BHs in faint galaxies comes from BHs in centers of bulges in spirals, therefore precise bulge-disk decomposition is required to estimate the total mass contribution (of BHs in the range of $10^6 - 10^7 M_\odot$).

2 Previous results

Proper calculation of the bulge luminosity requires bulge-disk decomposition for spiral galaxies, which is an additional time consuming task in data processing. Previous works [2, 7] usually omit this step. In the common treatment in calculating the spiral part of the

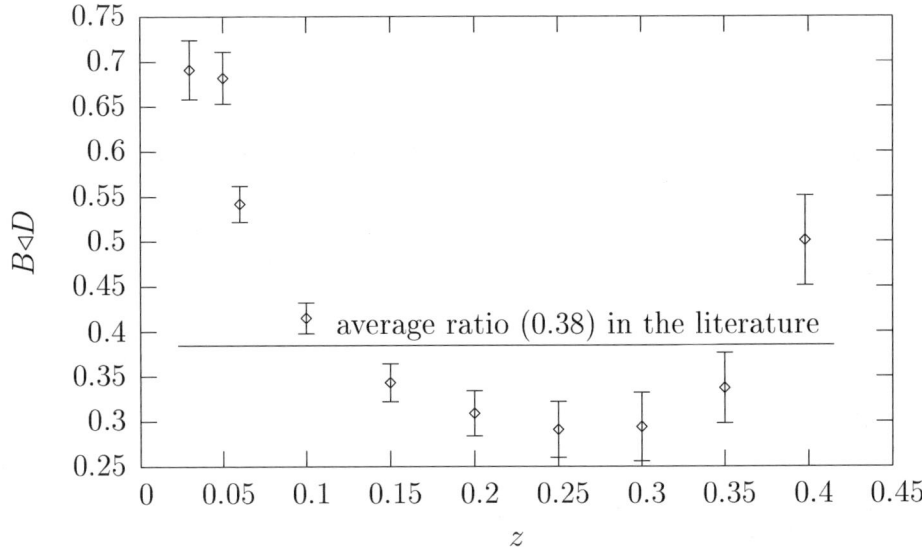

Fig. 1. Bulge-to-disk ratio for various redshifts.

mass function it is customary to divide the spiral sample in morphological subsets and do the conversion on each individual subset separately, considering only the average bulge-to-disk ratio for each subset. Using this method new dispersion is introduced into the data. Spiral galaxies dominate the lower end of the mass function. Our work, in which we properly perform the bulge-disk decomposition is very important, since the morphological division can become redundant this way.

3 Our sample

We process SDSS galaxies which have ellipticities less than 0.7. This constraint eliminates those edge-on spiral galaxies in which dust dims the bulge (we apply statistical corrections to make up for this omission).

4 Preliminary results

We have developed a fast and fully automatic software in IDL that performs bulge-disk decomposition using a one-dimensional brightness profile along the major axis. We plan to do a better, two-dimensional fit in the future.

All galaxies between $0.04 < z < 0.05$ have been processed. Error bars in Fig. 2 show errors of the fitted model parameters. Figure 1 shows quantitatively the well-known correlation: bulges are redder than disks.

In Fig. 1 each point corresponds to hundreds of galaxies within a narrow z-band. On average, half of the galaxies have fitted well to our model. Results are consistent with

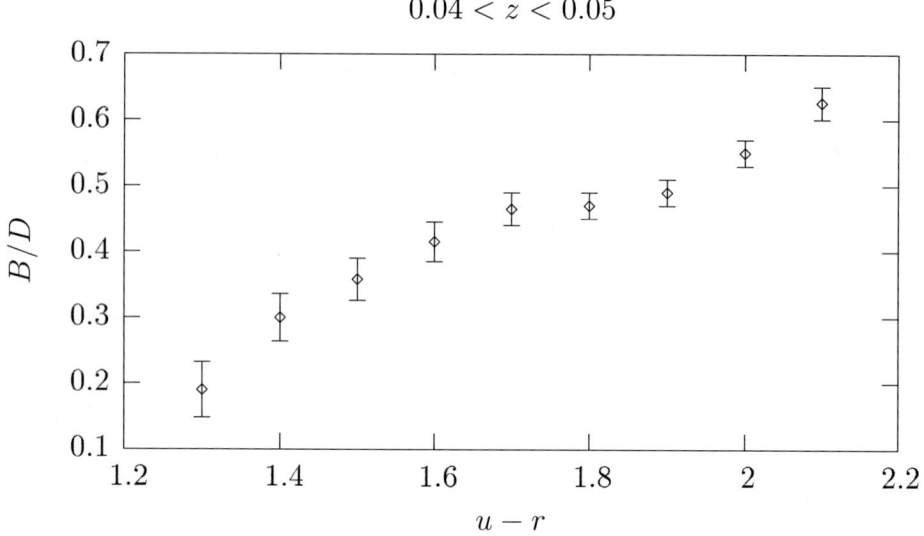

Fig. 2. Bulge-to-disk ratio vs. color.

[8], however the comparison is difficult. It is important to note that these results suffer from the selection bias related to the SDSS luminosity range.

Acknowledgments. We acknowledge support from OTKA through grant nos. T37548 and T47042 and from EU through grant no. MRTN-CT-2004-503929.

References

1. Ferrarese, L., & Ford, H.: Space Sci. Rev. **116**, 523 (2005)
2. Yu, Q., & Tremaine, S.: MNRAS **335**, 965 (2002)
3. Tremaine S. et al.: ApJ **574**, 740 (2002)
4. McLure R. J., & Dunlop. J. S.: MNRAS **331**, 795 (2002)
5. Marconi A., & Hunt L.: ApJL **589**, L21 (2003)
6. Graham A.W., Erwin P., Caon N., & Trujillo I.: ApJL **563**, 11 (2001)
7. Shankar F. et al.: MNRAS **354**, 1020 (2004)
8. Fukugita M., Hogan C. J., & Peebles P. J. E.: ApJ **503**, 518 (1988)

Tidal Capture by a Black Hole and Flares in Galactic Centres

A. Gomboc[1], A. Čadež[1], M. Calvani[2], and U. Kostič[1]

[1] Department of Physics, Faculty of Mathematics and Physics, University in Ljubljana, Slovenia, andreja.gomboc@fmf.uni-lj.si
[2] INAF, Astronomical Observatory of Padova, Padova, Italy calvani@pd.astro.it

1 Introduction

The centre of our Galaxy may harbour the nearest (8 kpc) massive black hole. Its proximity allows us to study the environment of massive black holes in detail, including the effects of black hole's gravity on stellar systems in its vicinity. Recent stellar orbits determinations reveal a central dark mass of $(3.7 \pm 0.2) \times 10^6 \, [R_0/(8\text{kpc})]^3 \, M_\odot$, where R_0 is the distance to Galactic centre [1]. We would like to point out that eccentricities of all these orbits (except one) are close to 1 (see Table 3 in [1]). In recent years it has been reported that the S0-2 star skimmed the Sgr A* at 17 light hours at the periastron [2], which corresponds to $\sim 3000 \, r_g$ (where r_g is the gravitational radius of the black hole: $r_g = GM_{bh}/c^2$) and the S0-16 came even closer to 45 AU, corresponding to 6.2 light hours [1] or $\sim 1200 r_g$.

The first rapid X-ray flaring from the direction of Sgr A* was observed in October 2000 with the duration of about 10 ks [3]. In September 2001, an early phase of a similar X-ray flare was observed in which the luminosity increased by ≈ 20 in about 900 s [4]. The brightest (observed so far) X-ray flare reaching a factor of 160 of the Sgr A* quiescent value was detected in October 2002 and had a duration of 2.7 ks [5]. In addition, in May and June 2003 four infrared flares from Sgr A* were observed with the duration from ≤ 0.9 ks to 5.1 ks and reaching a variability factor of 5 [6].

2 Flares from a tidal disruption of a Solar type star by a $10^6 M_\odot$ black hole

To shed some light on phenomena, which may produce bright flares in galactic centres, we investigate the tidal interaction between a star and massive black hole during a close encounter. A star approaching the massive black hole will probably follow a highly eccentric orbit. Once the star plunges deep through the Roche radius:

$$R_{\mathcal{R}} = 50 \times (\varrho_\odot/\varrho_*)^{1/3}(10^6 M_\odot/M_{bh})^{2/3} r_g, \tag{1}$$

(where ϱ_* and $\varrho_\odot$ are star's and Solar density, respectively), it experiences an enormous work done by tidal forces (reaching as high as $\sim 0.1 m_* c^2$) and is disrupted on a timescale of $\sim 50 r_g/c \sim 250 s \times (M_{bh}/10^6 M_\odot)$. As the outer layers of the star are stripped off and the hot core is exposed, the luminosity rises dramatically. The estimates from our numerical simulations show that the rise in luminosity could be as high as $10^{11} - 10^{13} L_\odot$ (for details see [7] and [8]). As the stellar debris is scattered and they cool, the luminosity decreases. Such event would therefore be observed as a bright flare coming from a galactic centre. The exact duration of bright phases depends on the cooling mechanisms and hydrodynamics.

In the case of stars in vicinity of Sgr A*, both S0-2 and S0-16 are at their periastrons still safely outside the Roche radius (1), which for a Solar type star in the Galactic centre lies at $R_\mathcal{R} \sim 20 r_g \sim 7$ light minutes, and therefore do not get tidally disrupted.

3 The time scale puzzle of flares in Sagittarius A* and tidal disruption and infall of a comet or asteroid

In analyzing the flares observed in Sgr A*, we were puzzled by the fact that the characteristic rise and switch-off times of all flares are very similar, about 900 sec. Could such a unique time scale suggest common origin for these flares observed at quite different wavelengths?

If the timescale is due to the sources' characteristics, they should have almost exactly the same mass. We find this explanation highly unlikely and suggest that the timing is not so much due to sources themselves, but to the space-time of the central Galactic black hole they are moving in.

We explore the idea that observed flares are produced by small objects, e.g. cometary or planetary ones, which are heated by resonant tides on their way down to the black hole.

In the first phase of such a scenario, the stars moving close to the Galactic centre black hole are gradually being stripped off their comets, asteroids, planets. In the process, the remaining stellar system is loosing its orbital angular momentum, making the stellar system orbit more and more elliptical.

In the second phase, a stripped asteroid (with mass M) is likely to move on a highly eccentric orbit, reaching deep into the potential well of the black hole. Each periastron passage produces an increasing tidal wave and reduces the orbital angular momentum and the orbital energy in such a way that the orbit is becoming more and more eccentric (parabolic) with the angular momentum slowly approaching the angular momentum of tidal capture $l_{crit} = 4 M M_{bh} c$. The last tidal kick, that occurs just before capture, releases up to $\Delta E \sim 0.1 M c^2 = 10^{41} erg$ of tidal energy to the asteroid, which is more than enough to evaporate it and heat it to X-ray temperatures. The result is the formation of a comet-like tidal tail with the length of the circumference of the last circular orbit. The luminosity increases with the characteristic rise time determined by the black hole's gravity: $\sim 200 s \times (M_{bh}/10^6 M_\odot)$. We assume that a distant observer is located close to the orbital plane, so that the asteroid's light curve is further modulated by black hole's gravity: as the brightening object is making the last turns before its final demise down the black hole, the Doppler effect, aberration of light, and light bending will produce luminosity peaks with the quasi-period of the last circular orbit - see our fit in Figure 1. The luminosity in our model decreases, as the object with its tidal tail falls behind the horizon.

We estimate the capture rate of asteroids as: stellar capture rate × number of asteroids per star, yielding $\sim (10^{-4}\mathrm{y}^{-1}) \times 10^5 = 10\mathrm{y}^{-1}$.

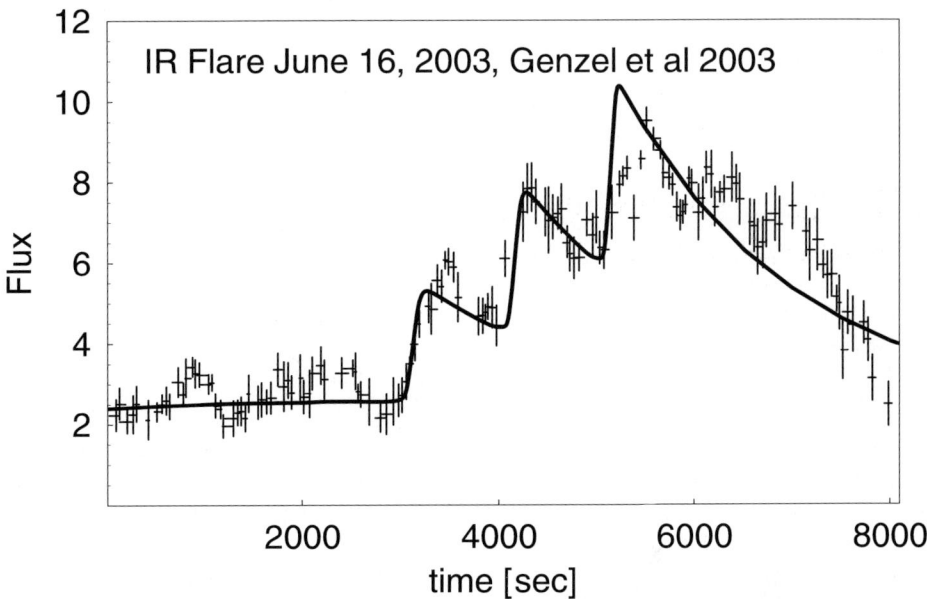

Fig. 1. Flare observed in Sgr A* and our fit (line) to the observed light curve obtained with a very rudimentary model, assuming that asteroid's luminosity is increasing exponentially with time and the luminosity of its tidal tail is decreasing exponentially with the distance from the asteroid's core.

References

1. A. M. Ghez, et al.: Astrophys. J. **620**, 744 (2005)
2. R. Schödel, et al.: Nature. **419**, 694 (2002)
3. F. K. Baganoff, et al.: Nature. **413**, 45 (2001)
4. A. Goldwurm, et al.: Astrophys. J. **584**, 751 (2003)
5. D. Porquet, et al.: A&A. **407**, L17 (2003)
6. R. Genzel, et al.: Nature. **425**, 934 (2003)
7. A. Gomboc, A. Čadež: Astrophys. J. **625**, 278 (2005)
8. A. Gomboc: Rapid Luminosity Changes Due to Interaction with a Black Hole. PhD Thesis, University of Ljubljana, Ljubljana (2001)

Low-Rate Accretion onto Isolated Stellar-Mass Black Holes

S. Karpov[1] and G. Beskin[1]

Special Astrophysical Observatory of Russian Academy of Sciences, Zelenchukskaya, Karachaevo-Cherkesia, Russia, karpov@sao.ru

Summary. Magnetic field behaviour in a spherically-symmetric accretion flow for parameters typical of single black holes in the Galaxy is discussed. It is shown that in the majority of the Galaxy volume, accretion onto single stellar-mass black holes will be spherical and have a low accretion rate ($10^{-6} - 10^{-9}$ of the Eddington rate). An analysis of plasma internal energy growth during the infall is performed. Adiabatic heating of collisionless accretion flow due to magnetic adiabatic invariant conservation is 25% more efficient than in the standard non-magnetized gas case. It is shown that magnetic field line reconnections in discrete current sheets lead to significant nonthermal electron component formation. In a framework of quasi-diffusion acceleration, the "energy-radius" electron distribution is computed and the function describing the shape of synchrotron radiation spectrum is constructed. It is shown that nonthermal electron emission leads to formation of a hard (UV, X-ray, up to gamma), highly variable spectral component in addition to the standard synchrotron optical component first derived by Shvartsman generated by thermal electrons in the magnetic field of accretion flow. For typical interstellar medium parameters, a black hole at 100 pc distance will be a 16-25^m optical source coinciding with the highly variable bright X-ray counterpart, while the variable component of optical emission will be about 18-27^m. The typical time scale of the variability is 10^{-4} sec, with relative flare amplitudes of 0.2-6% in various spectral bands. Possible applications of these results to the problem of search for single black holes are discussed.

Even though more than 60 years have passed since the theoretical prediction of black holes as an astrophysical objects in some sense they have not been discovered yet. The detection of the event horizon is a very complicated task that cannot be easily performed in x-ray binaries and AGNs due to high accretion rate and so the high optical depth of the accreting gas. At the same time, single stellar-mass black holes, which accrete interstellar medium of low density ($10^{-2} - 1 \mathrm{cm}^{-3}$), are the ideal case for the detection and study of the event horizon.

The analysis of existing data on possible black hole masses and velocities in comparison with the interstellar medium structure shows that in the majority of cases in the Galaxy ($> 90\%$) the accretion rate $\dot{m} = \dot{M}c^2/L_{edd}$ can't exceed $10^{-6} - 10^{-7}$.

Black holes usually move supersonically (at Mach numbers 2-3). And only in cold clouds of the interstellar hydrogen ($n \sim 10^2 - 10^5$, $T \sim 10^2$) and at low velocities of motion (< 10 km/s) the accretion rates may be high and black hole luminosities may reach $10^{38} - 10^{40}$ erg/c. For typical interstellar medium inhomogeneity the captured specific angular momentum is smaller than that on the BH last stable orbit, and the accretion is always (with the exception of the case of a black hole in a dense molecular cloud) spherically-symmetric.

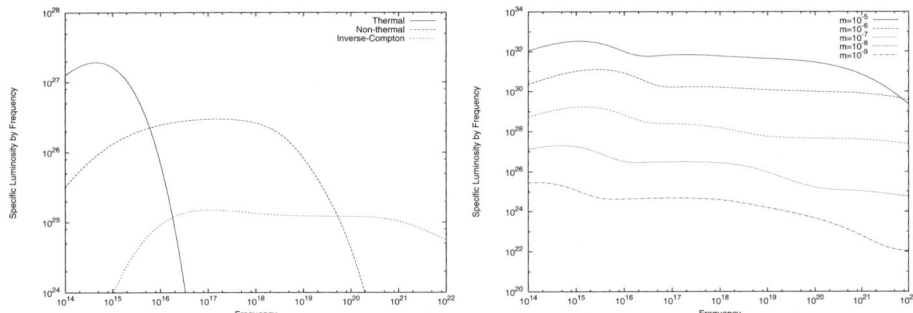

Fig. 1. Left pane is a decomposition of a single black hole (with the mass $10M_\odot$) emission spectrum into thermal and nonthermal parts. The accretion rate is $1.4 \cdot 10^{10}$ g/s, which corresponds to $\dot{m} = 10^{-8}$. Right pane shows the accreting $10M_\odot$ black hole spectra for different accretion rates.

The accreting plasma is initially collisionless, and it remains so until the event horizon. The electron-electron and electron-ion free path $\lambda \sim 2.4 \cdot 10^3 T^2 n^{-1}$ even at the capture radius is as high as $\sim 10^{12}$ cm. Only the magnetic fields trapped in plasma (the proton Larmor radius at r_g is 10 cm) make it possible to consider the problem as a quasi-hydrodynamical one; it is only due to the magnetic field that the particle's momentum is not conserved, allowing particles to fall towards the black hole. In addition, the magnetic field effectively "traps" particles in a "box" of variable size, which allows us to consider its adiabatic heating during the fall; a correct treatment of such a process shows that for magnetized plasma such heating is 25% more effective than for ideal gas due to the conservation of magnetic adiabatic invariant $I = \frac{3cp_t^2}{2eB}$, where p_t is the tangential component of electron momentum. Therefore, the plasma temperature in the accretion flow grows much faster and electrons become relativistic earlier – $R_{rel} \approx 6000$ in contrast to $R_{rel} \approx 1300$ in [6] and $R_{rel} \approx 200$ in [10]. The accretion flow is much hotter, and our estimation of "thermal" luminosity

$$L = 9.6 \cdot 10^{33} M_{10}^3 n_1^2 (V^2 + c_s^2)_{16}^{-3} \text{ erg/s} \qquad (1)$$

is significantly higher than those of [10]

$$L_{\text{IP}} = 1.6 \cdot 10^{32} M_{10}^3 n_1^2 (V^2 + c_s^2)_{16}^{-3} \text{ erg/s} \qquad (2)$$

and [6]

$$L_{\text{BKR}} = 2 \cdot 10^{33} M_{10}^3 n_1^2 (V^2 + c_s^2)_{16}^{-3} \text{ erg/s}, \qquad (3)$$

while the optical spectral shape is nearly the same.

The basis of our analysis is the assumption of the energy equipartition in the accretion flow (Shvartsman' theorem [18]). The straight consequence of this assumption is the necessity of exceeding magnetic energy dissipation with the rate

$$\frac{dE}{dV\,dt} = 4\frac{v}{r}\frac{B^2}{8\pi} - \frac{5}{2}\frac{v}{r}\frac{B^2}{8\pi} = \frac{3}{2}\frac{v}{r}\frac{B^2}{8\pi}.$$

(4)

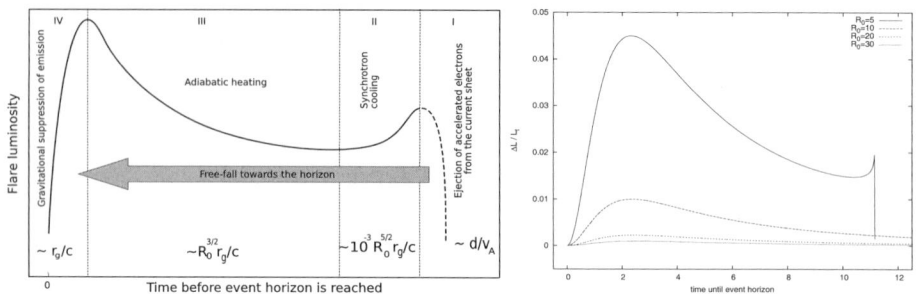

Fig. 2. Left pane – internal structure of a flare as a reflection of the electron cloud evolution. Right pane – light curves of separate non-thermal flares – beams of accelerated electrons ejected at R_0 – for $\dot{m} = 10^{-8}$. Time is measured in units of r_g/c, luminosity in units of total thermal luminosity.

In the previous accretion models the dissipation goes continuously[7] in the turbulent flow and its mechanism is not examined in detail. We considered conversion of the magnetic energy in the discrete turbulent current sheets [15] as a mechanism providing such dissipation. At the same time, various modes of plasma oscillations are generated (ion-acoustic and Lengmur plasmons mostly), magnetic field lines reconnect and are ejected with plasma from the current sheet and electrons are accelerated. The latter effect is very important for the observational appearances of the whole accretion flow. The beams of the accelerated electrons emit its energy due to motion in the magnetic field and generate an additional (in respect to synchrotron emission of thermal particles) nonthermal component. These electrons have power-law energy distribution and its emission spectrum is flat up to gamma band (Fig.1). An important property of the nonthermal emission is its flaring nature – the electron ejection process is discrete, typical light curves of single beams are shown in Fig. 2. The light curve of each such flare has a stage of fast intensity increase and sharp cut-off, its shape reflects the properties of the magnetic field and space-time near the horizon.

The presence of a discrete set of current sheets complicates significantly the structure of the accretion flow. In essence, the accretion flow is a complex dynamical system with nonlinear feedbacks. The latter are ensured by the plasma oscillations generated in each reconnection event, beams of the accelerated electrons and clouds of the magnetized plasma ejected from current sheets. All these agents may act as triggers for already "prepared" inhomogeneities which turn on magnetic energy dissipation processes. This situation seems to be similar to the Solar one determining its flaring activity, and also to the case of UV Cet stars and maybe accretion disks of X-ray binaries and active galactic nuclei. All these non-stationary processes are characterized by power-law scalings of flare energies with similar slopes of 1.5-2 at a very wide range of energies – from 10^{23} erg/s for the Sun till 10^{45} erg/s for quasars. Universality of these processes may be interpreted in the framework of fractal approach as it has been made in [1, 2, 12, 13, 15].

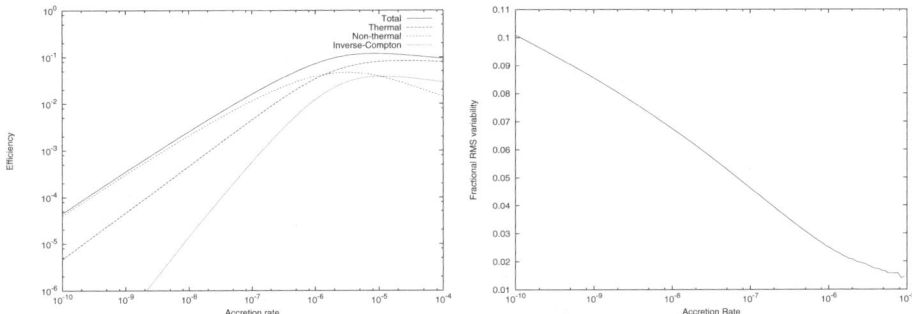

Fig. 3. Left pane - efficiencies of the synchrotron emission of thermal and non-thermal electron components of the accretion flow. Right pane - fractional RMS variability of the flaring nonthermal emission as a function of dimensionless accretion rate $\dot{m}$.

This means the realization (at least in active phases) of some collective state, sometimes called as "self-organized criticality" [2], which is characterized by the same behaviour of the parameters on all scales. These are percolation processes. There are evidences that the accretion flow is in this state, and so its observational appearances (at least those related to non-stationary processes) may be predicted and interpreted in the framework of this approach. Initial steps in this direction have been done [5], but, of course, it needs to be studied much deeper and wider.

It is clear that the search for a black hole strategy may be modified in accordance with such results. Optical high time resolution studies of X-ray sources may be very important. Single black holes may be contained inside the known stationary gamma sources [8] as well as objects causing long microlensing events [14]. Thus it is very important to look for X-ray emission as well as for fast optical variability of these objects. Sample observations of the longest microlensing event MACHO 1999-BLG-22[3], a stellar-mass black hole candidate, have been performed at the Special Astrophysical Observatory of RAS in the framework of the MANIA experiment[4] in 2003-2004.

The best evidence will be provided by the synchronous high time resolution observations in optical and X-ray ranges.

Detection of the event horizon signatures cannot result from statistical studies. A detailed study of each object is needed to detect its specific appearance.

Acknowledgments. This work has been supported by the Russian Foundation for Basic Research (grants No. 01-02-17857 and 04-02-17555) and by the grant in the framework of the CNR (Italy) – RAS (Russia) agreement on scientific collaboration. S.K. thanks the Russian Science Support Foundation for support. G.B. thanks the Cariplo Foundation for the scholarship and Merate Observatory for hospitality.

References

1. Anastasiadis, A., et al.: ApJ **489** 367 (1997)
2. Bak, P., Tang, C., & Weisenfeld, K.: Phys.Rev. Letters **59** 381 (1987)

3. Bennett, D.P., et al.: ApJ **579**, 639 (2002)
4. Beskin, G.M., et. al.: ExA **7** 413 (1997)
5. Beskin, G.M., & Karpov, S.V.: Gravitation and Cosmology Suppl. Ser. **8** 182 (2002a)
6. Bisnovatyi-Kogan, G.S., & Ruzmaikin, A.A.: Ap& SS **28** 45 (1974)
7. Bisnovatyi-Kogan, G.S., & Lovelace, R.V.E.: ApJ **486** L43 (1997)
8. Gehrels, N., et al.: Nature **404** 363 (2000)
9. Ipser, J.R., & Price, R.H.: ApJ **216** 578 (1977)
10. Ipser, J.R., & Price, R.H.: ApJ **255** 654 (1982)
11. Kawaguchi, T., & Mineshige, S.: PASP **52** L1 (1999)
12. Lu, E.T., & Hamilton, R.J.: APJ **380** 89 (1991)
13. Lu, E.T., et al.: APJ **412** 841 (1993)
14. Paczynski, B.: ApJ **371** 63 (1991)
15. Pustilnik, L.A.: ApSS **252** 325 (1997)
16. Shapiro, S.L.: ApJ **185** 69 (1973b)
17. Shapiro, S.L., & Teukolsky, S.A.: "Black holes, white dwarfs, and neutron stars: The physics of compact objects", New York, Wiley-Interscience (1973)
18. Shvartsman, V.F.: AZh **48** 438 (1971)

Clumps of material orbiting a black hole and the QPOs

U. Kostić, A. Cadež, and A. Gomboc

Department of Physics, Faculty of Mathematics and Physics, University of Ljubljana, Jadranska 19, 1000 Ljubljana, andrej.cadez@fmf.uni-lj.si

1 Succesive passages of an asteroid about a black hole

Clumps of material orbiting a black hole may be distributed, somewhat like comets in the Kuiper belt. Some clumps are perturbed to relatively small periastron high eccentricity orbits. During each periastron passage tides do work and so change orbital parameters to make the orbit more eccentric with less orbital angular momentum and lower periastron. The periastron crossing time decreases accordingly. When the periastron touches the Roche radius, tides resonate with the fundamental quadrupole mode of the asteroid [1]. The transfer of orbital angular momentum to internal tidal modes is high and the work done by tides is large enough to heat the asteroid to high temperatures and to break loose all parts that are no more than gravitationaly bound to the asteroid. Each next periastron passage has a shorter characteristic time and excites higher mechanical modes of the asteroid. The orbital energy keeps decreasing and the transferred energy is used to heat it to higher and higher temperatures and to accelerate more breaking-away parts. Soon the asteroid is split into small pieces that are no longer kept together by gravity, but by solid state forces. Sufficiently small bodies become stable against tidal destruction. They are excited, heated and spun up by tidal waves at the periastron. Tides and the tidal energy grow at each next periastron passage. When the periastron reaches close to the last circular orbit, the overwhelming tidal force crushes all solids that are larger than about $0.01 GM/c^2$ (here M is the mass of the black hole; $GM_\odot/c^2 = 1500m$) and produces supersonic relativistic tides, which do enough work on the body to heat it to hard X-ray temperatures.

The work is taken out of orbital energy, so that the body can no longer return to the far apastron, but is caught by the black hole. After making a small number of very flashy and characteristic turns about the black hole, it disappears behind its horizon. The overwhelming tidal force is very rapid with respect to the frequency of fundamental modes of the asteroid, so that the work done by tides can be as high as $0.1mc^2$, where m is the mass of the asteroid. Even a small part of this energy is more than enough to make the asteroid explode. Yet, the shock wave is still slow with respect to asteroid's relativistic speed, so that it has no time to grow large with respect to the black hole. In making the few last turns to the black hole it accelerates toward and away from the observer so that the Doppler effect, aberration of light and possibly gravitational lensing make a characteristic bright chirp in the light curve of this last orbit.

We assume that a black hole is fed by a large number of small debris and calculate light curves from such events and their Fourier spectra.

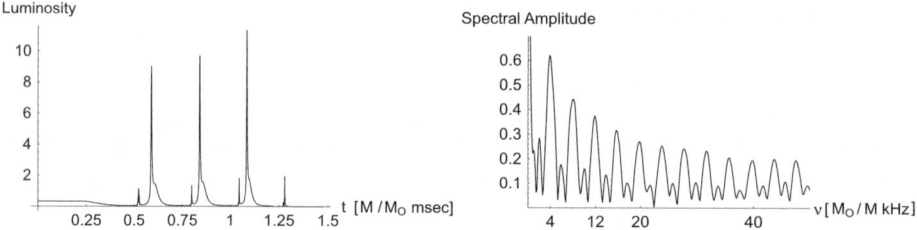

Fig. 1. *Left*: A light curve of a typical chirp produced by a small ($r = 0.01GM/c^2$) solid body on its way to the black hole following a parabolic orbit which makes three turns before crossing the horizon. In this example the body's velocity, when far from the black hole, subtends an angle $24°$ with the line of sight and its orbit is inclined $85°$ to the observer. The sharp peaks in observed luminosity are the effect of gravitational lensing. The wide bumps following the peaks are due to aberration of light and Doppler shift.
Right: The Fourier spectrum of the chirp on the left. Note that the frequency scale is in kHz, divided by the mass of the black hole in solar units; for example, a chirp from a 10 solar mass black hole would have the lowest peak frequency at $\nu \sim 400$Hz. This frequency corresponds to the orbital period on the last circular orbit about the black hole.

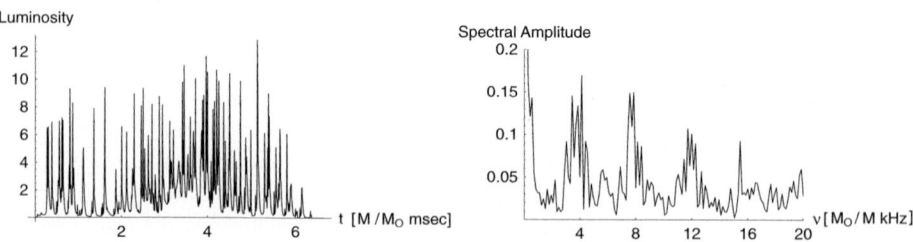

Fig. 2. A light curve (left) and the Fourier spectrum (right) of 24 chirps produced by 24 particles orbiting in a plane inclined $85°$ with respect to the observer. The particles are distributed randomly in time and with respect to the angle of the line of nodes.

2 Conclusions

Accretion of solid asteroid-like material onto a black hole can produce signals with a signature similar to that of QPOs. Small solid particles are the only form of matter that can survive huge tidal forces deep in the gravitational field of a stellar mass black

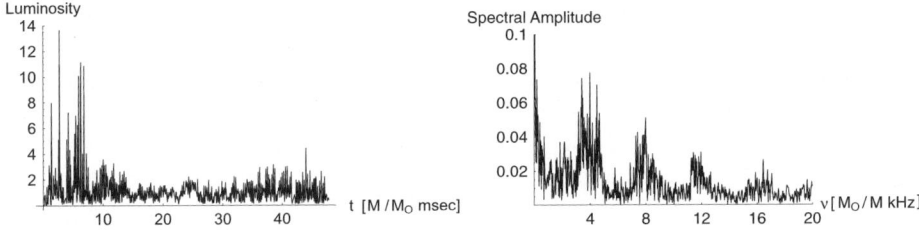

Fig. 3. Left: A signal produced by 178 objects falling onto the black hole from random directions and at random times. Right: Fourier spectrum of this signal. The broad peaks appear at harmonics of the frequency corresponding to the period of the last circular orbit, but their amplitudes are again modulated by random factors determined by the relative timing of contributing chirps. Another random series of chirps would produce a similar, but not the same power spectrum. For example, the small and narrow peak at $\nu \sim 6\text{kHz}[M_\odot/M]$ may become quite comparable with the peak at $\nu \sim 4\text{kHz}[M_\odot/M]$.

hole. They can enter the last circular orbit almost intact and transform the mounting tidal force into internal energy. Since the theoretical upper limit for this energy increase is $\sim 0.1\text{mc}^2$, it seems plausible that the body heats to hard X-ray temperatures when making a few last turns before plunging behind the horizon. Such a signal is modulated by Doppler shift, aberration of light and gravitational lensing and produces a characteristic chirp. All chirps have the same characteristic frequency $1/50[c^3/GM]$, but the higher frequencies depend somewhat on the orientation of the orbit with respect to the observer. Many chirps produce a signal which is very similar to that of a QPO [2, 3].

References

1. A. Gomboc, A. Čadež: ApJ **625**, 278 (2005)
2. G. L. Israel, et al.: ApJL **628**, L53 (2005)
3. J. Homan, et al.: ApJ **623**, 383 (2005)

Multi-Scale Simulations of Merging Galaxies with Supermassive Black Holes

L. Mayer[1], S. Kazantzidis[2], P. Madau[3], M. Colpi[4], T. Quinn[5], and J. Wadsley[6]

[1] Institut für Astronomie, ETH Zürich, Switzerland, lucio@phys.ethz.ch
[2] KICP, University of Chicago, USA, stelios@cfcp.uchicago.edu
[3] University of California, Santa Cruz, USA, pmadau@ucolick.org
[4] Universita' Milano-Bicocca, Milano, Italy, Monica.Colpi@mib.infn.it
[5] University of Washington, Seattle, USA, trq@astro.washington.edu
[6] McMaster University, Canada, wadsley@mcmaster.ca

Summary. We present the results of the first multi-scale N-body+hydrodynamical simulations of merging galaxies containing central supermassive black holes (SMBHs) and having a spatial resolution of only a few parsecs. Strong gas inflows associated with equal-mass mergers produce non-axisymmetric nuclear disks with masses of order $10^9 M_\odot$, resolved by about 10^6 SPH particles. Such disks have sizes of several hundred parsecs but most of their mass is concentrated within less than 50 pc. We find that a close pair of supermassive black holes forms after the merger, and their relative distance then shrinks further owing to dynamical friction against the predominantly gaseous background. The orbits of the two black holes decay down to the minimum resolvable scale in a few million years after the merger for an ambient gas temperature and density typical of a region undergoing a starburst. The conditions necessary for the eventual coalescence of the two holes as a result of gravitational radiation emission appear to arise naturally from the merging of two equal-mass galaxies whose structure and orbits are consistent with the predictions of the LCDM model. Our findings have important implications for planned gravitational wave detection experiments such as *LISA*.

1 Introduction

Recent observations of molecular gas in the nuclear region of candidate merger remnants such as starbursting ultraluminous infrared galaxies (ULIRGs) reveal the presence of rotating gaseous disks which contain in excess of $10^9 M_\odot$ of gas within a few hundred parsecs ([3],[5]). Some of these galaxies, such as Mrk 231, also host a powerful AGN. The high central concentration of gas is likely the result of gaseous inflows driven by the prodigious tidal torques and hydrodynamical shocks occurring during the merger ([1], [7]), and possibly provides the reservoir that fuels the SMBHs. If both the progenitor galaxies host a SMBH the two holes may sink and eventually coalesce as a result of dynamical friction against the gaseous background. The sinking of two holes with an initial separation of 400 pc in a nuclear disk described by a Mestel model has been studied by [6] and [4]. Here we avoid any assumption on the structure of the nuclear gaseous disk and initial separation of the pair: rather, we follow the entire merging process starting from when the cores of the two galaxies are hundreds of kiloparsecs apart up to the point where they merge, produce a nuclear disk, and leave a pair of SMBHs separated by the adopted force resolution of 10 pc. Previous simulations of merging galaxies with SMBHs have not followed the evolution of the central region below a scale of a few hundred parsecs due to their limited mass and force resolution ([11], [7]). We are able to extend the dynamic range of previous works by orders of magnitude using the technique of particle splitting.

2 The Numerical Simulations

Our starting point are the high-resolution simulations presented in [7] which, as the new simulations presented here, were performed with the parallel tree+SPH code GASO-LINE [13]. In particular we refer to those simulations that followed the merger between two equal-mass, Milky-Way sized early type spirals having 10% of the mass of their exponential disk in a gaseous component and the rest in stars. The structural parameters of the two disks and their NFW dark matter halos, as well as their initial orbits, are motivated by the results of cosmological simulations. We apply the static splitting of SPH particles [8] to increase the gas mass resolution in such calculations (the stars and dark matter resolution remain the same), and reduce the gravitational softening accordingly as we increase the mass resolution. We select an output about 50 Myr before the merger is completed, when the cores of the two galaxies are still $\sim$ 6 kpc away, and split each SPH particle into 8 children, reaching a mass resolution of about 3000 $M_\odot$ in the gas component (Figure 1). The gravitational softening of the gas particles is decreased from 200 pc to either 40 pc, 10 pc, or 2 pc (one run is performed for each of the three different softenings). With the new mass resolution even for the smallest among the gravitational softenings considered the number of SPH particles within a sphere of radius equal to the local Jeans length is much larger than twice the number of SPH neighbors (=32), thus avoiding spurious fragmentation ([2]). The two SMBHs are point masses with a softening length set equal to 10 pc and a mass of $3 \times 10^6 M_\odot$.

We have run a suite of simulations with different prescriptions for the gas thermodynamics, and show here the results of two runs in which radiative cooling and heating processes are not included directly, rather an adiabatic equation of state with either $\gamma = 7/5$ or $\gamma = 5/3$ is adopted (irreversible shock heating, which is important during the merging phase, is included via an artificial viscosity term in the energy equation). According to the radiative transfer calculations of [10] the case $\gamma = 7/5$ approximates quite well the balance between radiative heating and cooling in a starburst galaxy (in [7] a central starburst indeed does occur in the final phase of the merger that we are considering here). A stiffer equation of state such as that with $\gamma = 5/3$ might instead be relevant when an additional strong heating source, for example AGN feedback, comes into play ([11]).

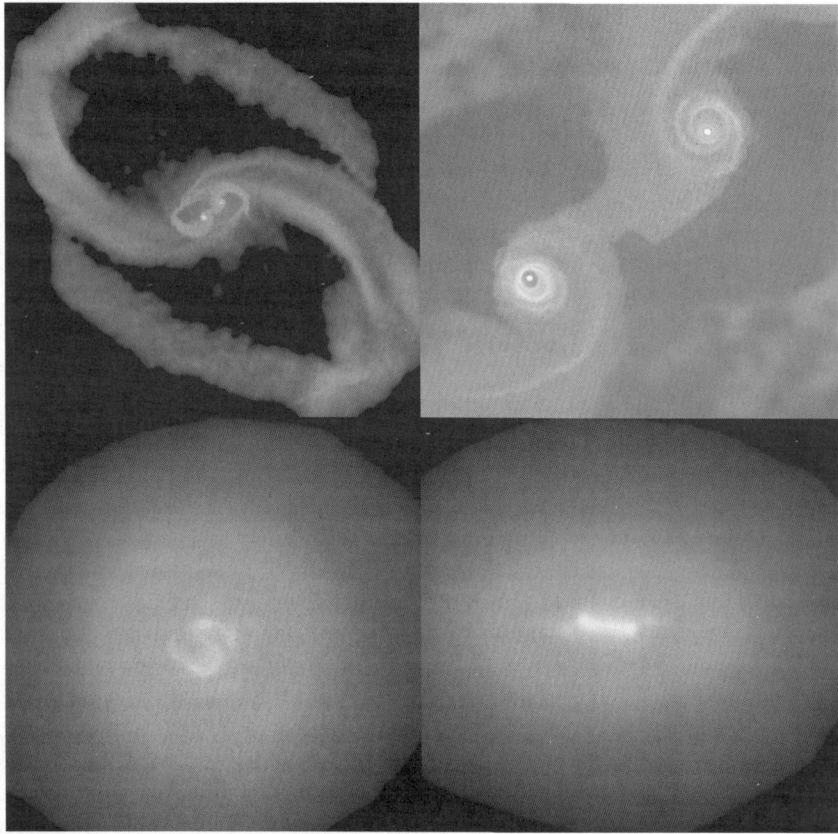

Fig. 1. Intensity coded density map of the nuclear region 50 Myr before the merger (top panel) and just after the merger (bottom panels). The top panel shows a box 30 kpc on a side (left) and a zoom-in within the inner 6 kpc (right). The bottom panels show the inner 300 pc, with the disk seen face-on (left) and edge-on (right). An adiabatic equation of state with $\gamma = 7/5$ was used in this run.

3 Gas Inflows and the Structure of the Nuclear Disks

About 80% of the gas of the two galaxies is funneled to the central kiloparsec during the last stage of the merger and settles into two rotationally supported disks. When the cores of the two galaxies finally merge, the two disks also merge into a single gaseous core which rapidly becomes rotationally supported as radial motions are largely dissipated in shocks. The disk however remains non-axisymmetric, with evident bar-like and spiral patterns (Figure 1). A coherent thick disk forms independently on the relative inclination of the initial galactic disks, albeit the orientation of its angular momentum vector relative to the global angular momentum will change depending on those initial parameters ([7]). In the run with $\gamma = 7/5$, that was designed to reproduce the thermodynamics of a starburst region, the disk has a vertical extent of about 20 pc and $v_{rot}/\sigma > 1$ out to about 600 pc. The thickness is about 5 times higher in the run with

$\gamma = 5/3$. The scale height in the $\gamma = 7/5$ run is comparable to that of the disks in the multi-phase simulations of a 1 kpc-sized nuclear region performed by [12]. The simulations of [12] include radiative cooling and resolve the turbulence generated from supernovae explosions as well as gravitational instability, suggesting that our equation of state yields a characteristic pressure scale that accounts for the combined thermal and turbulent pressures.

Gravitational torques and the balance between gravity and the thermodynamical pressure at small scales depend ultimately on the adopted gravitational softening. We verified that the mass inflow seen in the simulations for a given value of γ converges as the softening approaches 10 pc. Convergence in the disk vertical extent is also observed at such a spatial resolution. At high resolution more than 60% of the mass piles up within as little as 30 pc.

4 Sinking SMBHs

Once the merging between the two galactic cores is complete, a close but unbound pair of SMBHs is delivered. The pair is separated by about 100 pc and is embedded within the newly formed nuclear gaseous disk. Up to this point the orbital decay of the two SMBHs was driven by dynamical friction of the surrounding galactic cores within the *collisionless* background of stars and dark matter. Once inside the massive gaseous disk the orbital decay of the two holes is dominated by dynamical friction in a *gaseous* background ([9]). The intensity of the drag is then higher or lower depending on whether the black holes move supersonically or subsonically with respect to the background, and increases also as the the characteristic density of the background increases. The run with $\gamma = 7/5$ falls in the supersonic regime (the orbital velocity of the black holes is of order 300 km/s, which corresponds to a temperature of about 10^6 K), while the black holes move slightly subsonically in the run with $\gamma = 5/3$. In the latter run the average background density is also a factor ~ 2 lower compared to the run with $\gamma = 7/5$. These differences in the thermal and density structure of the remnants explain why in the run with $\gamma = 7/5$ the two black holes reach a separation comparable to the force resolution limit (10 pc) in about 10 Myr whereas they remain separated by a distance larger than 100 pc in the other run (Figure 3). In neither case the SMBHs form a bound binary by the end of the simulation. However, with an even higher force resolution the formation of a bound pair is likely in the $\gamma = 7/5$ since the orbital energy of the binary is only marginally positive at $t = 5.128$ Gyr. Instead in the simulation with $\gamma = 5/3$ the binding of the two SMBHs will be aborted because their orbital decay time is longer than the Hubble time on the last few orbits.

5 Conclusions

We have shown that dense, rotationally supported nuclear disks are the natural outcome of dissipative mergers starting from cosmologically motivated initial conditions. The orbital evolution of the central SMBH delivered at the center of the merger remnant is then dominated by dynamical friction against the surrounding gaseous medium. The

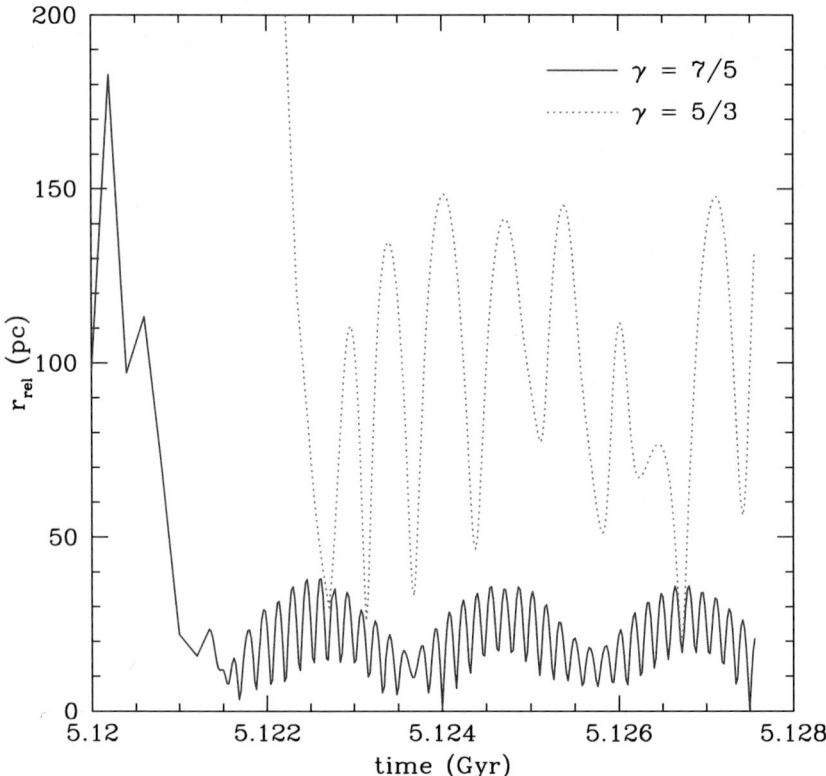

Fig. 2. Orbital separation of the two black holes as a function of time. Two runs with different values of γ and with a gas gravitational softening of 10 pc are shown. The curves start from the time at which the galaxy cores have already merged.

details of this process are extremely sensitive to thermodynamics of the gas. For the black holes to form a loose binary and eventually coalesce the equation of state can hardly be stiffer than that expected during a major starburst. This suggests that either AGN feedback has a minor impact on the gas in the disk or that strong AGN feedback has to be delayed for several million years after the galaxy merger is completed. In the second case the coalescence of the two black holes will occur when the merger remnant is a powerful starburst, such as a ULRIG, rather than a powerful AGN.

References

1. Barnes, J. & Hernquist, L.: ApJ **471**, 115 (1996)
2. Bate, M. & Burkert, A.: MNRAS **288**, 1060 (1997)
3. Davies, R.I., Tacconi, L.J. & Genzel, R.: ApJ **613**, 781 (2004)
4. Dotti, M., Colpi M. & Haardt F.: MNRAS **367**, 103 (2006)

5. Downes D. & Solomon, P.M.: ApJ **507**, 615 (1998)
6. Escala A., Larson, R.B., Coppi, P.S. & Mardones, D.: ApJ **630**, 152 (2005)
7. Kazantzidis, S., Mayer. L., Colpi, M., Madau, P., Debattista, V., Quinn, T., Wadsley, J. & Moore, B.: ApJ **623**, L67 (2005)
8. Kitsionas, S. & Whitworth, S.: MNRAS **330**, 129 (2002)
9. Ostriker, E.: ApJ **513**, 252 (1999)
10. Spaans, M. & Silk: MNRAS **538**, 115 (2000)
11. Springel, V., Di Matteo, T. & Hernquist, L.: MNRAS **361**, 776 (2005)
12. Wada, K. & Norman, C.: ApJ **566**, L21 (2002)
13. Wadsley, J., Stadel, J. & Quinn, T.: New Astronomy **9**, 137 (2004)

The Parallel Lives of Supermassive Black Holes and their Host Galaxies

A. Merloni, G. Rudnick, and T. Di Matteo

[1] MPI für Astrophysik, K.Schwarzschildstr.1, 85741 Garching, Germany
[2] NOAO, North Cherry Ave., Tucson, AZ 85721, USA
[3] Carnegie Mellon University, Department of Physics, 5000 Forbes Ave., Pittsburgh, PA 15213, USA

Summary. We compare all the available observational data on the redshift evolution of the total stellar mass and star formation rate density in the Universe with the mass and accretion rate density evolution of supermassive black holes, estimated from the hard X-ray selected luminosity function of quasars and active galactic nuclei (AGN). We find that on average black hole mass must have been higher at higher redshift for given spheroid stellar mass. Moreover, we find negative redshift evolution of the disk/irregulars to spheroid mass ratio. The total accretion efficiency is constrained to be between 0.06 and 0.12, depending on the exact value of the local SMBH mass density, and on the critical accretion rate below which radiatively inefficient accretion may take place.

1 Introduction

Observational evidence indicates that the mass of supermassive black holes (SMBH) is correlated with the luminosity (Marconi & Hunt, 2003 and references therein) and velocity dispersion (Tremaine et al., 2002 and references therein) of the host spheroids, suggesting that the process that leads to the formation of galaxies must be intimately linked to the growth of the central SMBH. Studying low redshift AGN, Heckman et al. (2004) have shown that not only does star formation directly trace AGN activity, but also that the sites of SMBH growth must have shifted to smaller masses at lower redshift, thus mimicking the "cosmic downsizing scenario" first put forward to describe galaxy evolution by Cowie et al. 1996. Such a scenario has recently received many independent confirmations, both for the evolution of SMBH as traced directly by X-ray and radio luminosity functions (LF) of AGN (Marconi et al. 2004; Merloni 2004; Hasinger et al. 2005), and for that of star forming galaxies, thanks to large surveys such as SDSS, GDDS, COMBO-17, GOODS, etc. (see, e.g. Heavens et al. 2004; Juneau et al. 2005; Pérez-González et al. 2005; Feulner et al. 2005).

Following Merloni, Rudnick and Di Matteo (2004; MRD04), here we discuss a *quantitative* approach to the study of the posited link between star formation and SMBH growth, based on a detailed comparison of the redshift evolution of integral quantities, such as the total stellar mass, black hole mass and star formation rate densities.

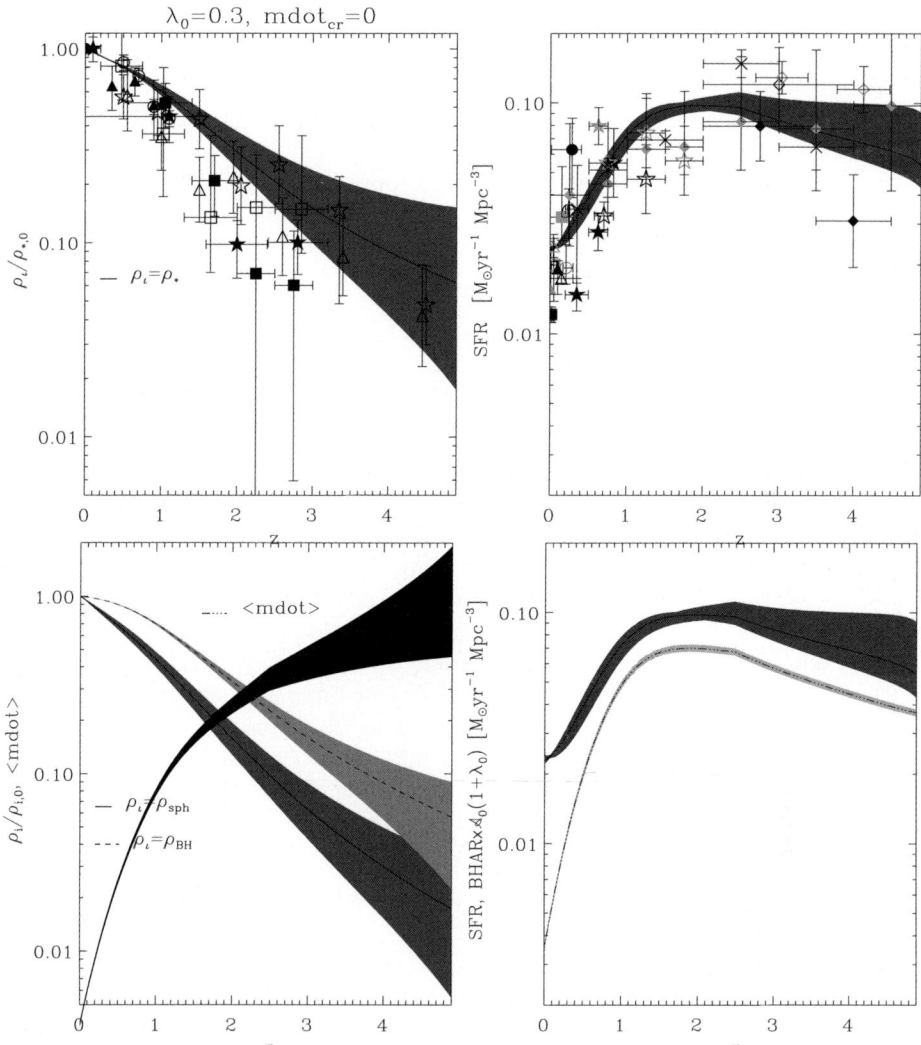

Fig. 1. The upper left panel shows the evolution of the stellar mass density as a function of redshift, where the density is given as a ratio to the local value, $\rho_{*,0} = 5.6 \times 10^8 M_\odot \text{ Mpc}^{-3}$, the upper right panel shows the SFR density. Our best-fit model is also shown in each panel. Values of $\lambda_0 = 0.3$ and $\rho_{\text{BH},0} = 2.5 \times 10^5 M_\odot \text{ Mpc}^{-3}$ and $\dot{m}_{\text{cr}} = 0$ are adopted here. Shaded areas represent 1-sigma confidence intervals of the model fits. The lower left panel shows a direct comparison between best-fit normalized mass density of spheroids (solid line, red / grey shaded area) and black holes (dashed line, light red / light grey shaded area). Also shown is the evolution of the average Eddington scaled accretion rate $\dot{m}(z)$ (dotted line, dark red / light grey shaded area). Finally, the lower right panel shows a direct comparison between the best-fit SFR and BHAR (rescaled by a factor $\mathcal{A}_0(1 + \lambda_0)$) densities.

2 SMBH as Tracers of Galaxy Evolution

Under the standard assumption that black holes grow mainly by accretion, their cosmic evolution can be calculated from the luminosity function of AGN $\phi(L_{\mathrm{bol}}, z) = dN/dL_{\mathrm{bol}}$, where $L_{\mathrm{bol}} = \epsilon \dot{M} c^2$ is the bolometric luminosity produced by a SMBH accreting at a rate of $\dot{M}$ with a *radiative* efficiency ϵ (Soltan 1982). Following the discussion in MRD04, we will assume that the absorption corrected 2-10 keV luminosity function of AGN, $\phi(L_{\mathrm{X}}, z)$, (La Franca et al. 2005) best describes the evolution of the *entire* accreting black holes population, yielding:

$$\frac{\rho_{\mathrm{BH}}(z)}{\rho_{\mathrm{BH},0}} = 1 - \int_0^z \frac{\Psi_{\mathrm{BH}}(z')}{\rho_{\mathrm{BH},0}} \frac{dt}{dz'} dz', \tag{1}$$

where the black hole accretion rate (BHAR) density is given by:

$$\Psi_{\mathrm{BH}}(z) = \int_0^\infty \frac{(1-\epsilon)L_{\mathrm{bol}}(L_{\mathrm{X}})}{\epsilon c^2} \phi(L_{\mathrm{X}}, z) dL_{\mathrm{X}} \tag{2}$$

L_{X} is the X-ray luminosity in the rest-frame 2-10 keV band, and the bolometric correction function $L_{\mathrm{bol}}(L_{\mathrm{X}})$ is given by eq. (21) of Marconi et al. (2004). The exact shape of $\rho_{\mathrm{BH}}(z)$ and $\Psi_{\mathrm{BH}}(z)$ then depends only on the local black holes mass density $\rho_{\mathrm{BH},0}$ and the (average) radiative efficiency ϵ. This, in turn, is given by the product of the total accretion efficiency $\eta(a)$, itself a function of the inner boundary condition and thus of the black hole spin parameter a, and a function f of the Eddington scaled dimensionless accretion rate $\dot{m} \equiv L_{\mathrm{bol}}/L_{\mathrm{Edd}}$. Below a critical rate, $\dot{m}_{\mathrm{cr}}$, accretion does not proceed in the standard optically thick, geometrically thin fashion for which $\epsilon = \eta$. Its radiative efficiency, instead, critically depends on the nature of the flow: if powerful outflows/jets are capable of removing the excess energy which is not radiated, as, for example, in the ADIOS scenario (Blandford & Begelman 1999), then $f = 1$, and black hole are always efficient radiators *with respect to the accreted mass* ("black holes are green!", Blandford 2005). On the other hand, if advection across the event horizon is the dominant process by which energy is disposed of (ADAF, Narayan & Yi 1995), we have:

$$\epsilon \equiv \epsilon(a, \dot{m}, \dot{m}_{\mathrm{cr}}) = \eta(a)f(\dot{m}, \dot{m}_{\mathrm{cr}}) = \eta(a) \begin{cases} 1, & \dot{m} \geq \dot{m}_{\mathrm{cr}} \\ \dot{m}/\dot{m}_{\mathrm{cr}}, & \dot{m} < \dot{m}_{\mathrm{cr}} \end{cases} \tag{3}$$

We use the total BHAR and mass densities, $\Psi_{\mathrm{BH}}(z)$ and $\rho_{\mathrm{BH}}(z)$ respectively, to estimate the redshift evolution of the average global Eddington scaled accretion rate $\dot{m}(z) \propto \Psi_{\mathrm{BH}}(z)/\rho_{\mathrm{BH}}(z)$ (lower left panel of Fig. 1) and the corresponding radiative efficiency according to eq. (3). This allows us to identify the redshift at which the transition occurs in the global accretion mode of growing SMBH. Depending on the assumed value of $\dot{m}_{\mathrm{cr}}$, this transition redshift is 0, if black holes are always efficient accretors, i.e. $\dot{m}_{\mathrm{cr}} = 0$, or $z \approx 0.6$ if $\dot{m}_{\mathrm{cr}} \approx 0.05$. This also implies that radiatively inefficient accretion could contribute to only a small fraction of the total black hole mass density (see also Yu and Tremaine 2002; Merloni 2004; Hopkins et al. 2006 and references therein).

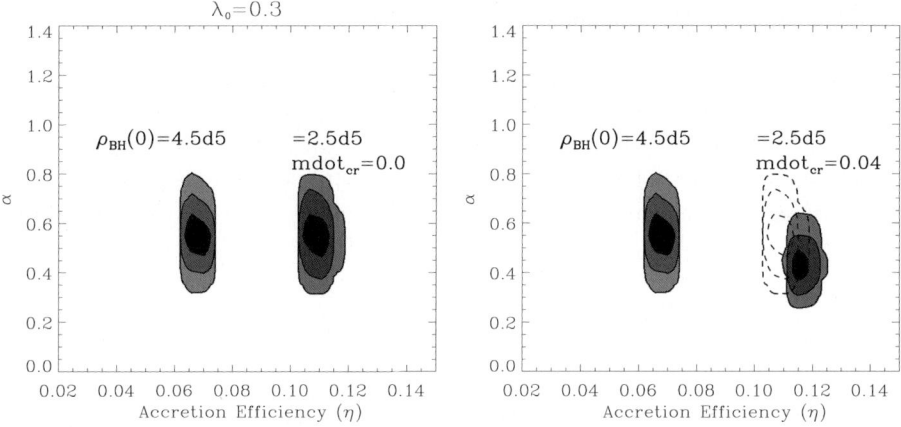

Fig. 2. 1,2 and 3 sigma confidence contours for the average accretion efficiency η and the index α describing the evolution of the stellar spheroid to black hole mass density ratio. In each panel, the leftmost and rightmost set of contours show the results obtained assuming $\rho_{BH,0} = 4.5$ (Marconi et al. 2004), and $2.5 \times 10^5 M_\odot$ Mpc^{-3} (Yu and Tremaine 2002), respectively. Also shown is the dependence of such constraints on the assumed value of the critical rate between radiatively efficient and inefficient accretion (ADAF-like), with the dashed contours in the right panel showing the $\dot{m}_{cr}$=0 reference point.

2.1 The Parallel Evolution

Our goal is to link the growth of SMBH from eq. (1) to the growth of stellar mass in galaxies. Because local SMBH are observed to correlate with spheroids only, we introduce the parameter $\lambda(z)$, the ratio of the mass in disks and irregulars to that in spheroids at any redshift, so that the total stellar mass density can be expressed as: $\rho_*(z) = \rho_{\rm sph}(z) + \rho_{\rm disk+irr}(z) = \rho_{\rm sph}(z)[1 + \lambda(z)]$. We then assume that $\lambda(z)$ evolves according to $\lambda(z) = \lambda_0(1 + z)^{-\beta}$, where λ_0 is the value of the disk to spheroid ratio in the local universe. Also we assume that the mass density of spheroids and supermassive black holes evolve in parallel, modulo a factor $(1 + z)^{-\alpha}$, obtaining a prediction for the observable stellar mass density evolution as traced by SMBH growth:

$$\rho_*(z) = \mathcal{A}_0 \rho_{BH}(\epsilon, z)(1 + z)^{-\alpha}[1 + \lambda_0(1 + z)^{-\beta}] \qquad (4)$$

where $\mathcal{A}_0$ is the constant of proportionality in the Magorrian relation. By taking the derivative of (4), accounting for stellar mass loss, an expression is also found for the corresponding star formation rate (SFR) density evolution (see eq. (7) of MRD04).

With these expressions we obtain statistically acceptable simultaneous fits to all available observational data points (see MRD04 for a complete list of references) of both $\rho_*(z)$ and SFR(z). For each choice of $\rho_{BH,0}$, λ_0, and of the critical accretion rate $\dot{m}_{cr}$, the fitting functions depend only on three parameters: α, β and the accretion efficiency η. One example of such fits is shown in Fig. 1 for the specific case $\rho_{BH,0} = 2.5 \times 10^5 M_\odot$ Mpc^{-3}, $\lambda_0 = 0.3$ and $\dot{m}_{cr} = 0$. Because the drop in the AGN integrated luminosity density at low z is apparently faster than that in SFR density (see lower right panel of Fig. 1), the average black hole to spheroid mass ratio must evolve with lookback time ($\alpha > 0$; see lower left panel of Fig. 1). This result is independent from the local black hole mass density, or from λ_0, and is not strongly affected by the choice of the value for the critical accretion rate. This is shown also in Fig. 2, where the constraints on the fit parameters are shown as confidence contours in the accretion efficiency–α plane for various choices of $\dot{m}_{cr}$. As SMBH grow most of their mass at high $\dot{m}$, the global constraints on η are not strongly affected by any reasonable choice of $\dot{m}_{cr}$.

3 Conclusions

We have made quantitative comparisons between the redshift evolution of the integrated stellar mass in galaxies and the mass density of SMBH. Although clearly correlated with the stellar mass density, SMBH accretion does not exactly track either the spheroid nor the total star assembly: irrespective of the exact mass budget in spheroids and disks + irregulars, the ratio of the total or spheroid stellar to black hole mass density was lower at higher redshift. Our results also suggest that the fraction of stars locked up into the non-spheroidal components of galaxies and in irregular galaxies should increase with increasing redshift ($\beta < 0$, see MRD04 for a discussion of this point). Our version of the Soltan argument yields a well defined constraint on the average radiative efficiency, and a corresponding one for the total accretion efficiency (i.e. on the mean mass-weighted SMBH spin), only weakly dependent on the uncertain physics of low $\dot{m}$ accretion flows.

References

1. R.D. Blandford: A Black Hole Manifesto. In: *Growing Black Holes*, eds A. Merloni, S. Nayakshin and R. Sunyaev (Springer 2005) pp 477
2. R.D. Blandford & M.C. Begelman, MNRAS, **303**, L1 (1999)
3. L.L. Cowie et al., AJ, **112**, 839 (1996)
4. G. Feulner et al., ApJ, **633**, L9 (2005)
5. G. Hasinger, T. Miyaji & M. Schmidt, A&A, **441**, 417 (2005)
6. A. Heavens, B. Panter, R. Jiménez, & J. Dunlop, Nature, **428**, 625, 2004
7. T.M. Heckman et al., ApJ, **613**, 109 (2004)
8. P.E. Hopkins, R. Narayan & L. Hernquist, ApJ, **643**, 641-651 (2006)
9. S. Juneau et al., ApJ, **619**, L135 (2005)
10. F. La Franca, et al., ApJ, **635**, 864 (2005)
11. A. Marconi & L.K. Hunt, ApJL, **589**, L21 (2003)
12. A. Marconi et al., MNRAS, **351**, 169 (2004)
13. A. Merloni, MNRAS, **353**, 1035 (2004)
14. A. Merloni, G. Rudnick & T. Di Matteo, MNRAS, **354**, L37 (2004) MRD04
15. R. Narayan & I. Yi, ApJ, **452**, 710 (1995)
16. P.G. Pérez-González, et al., ApJ, **630**, 82 (2005)
17. A. Soltan, MNRAS, **200**, 115, (1982)
18. S. Tremaine, et al., ApJ, **574**, 554 (2002)
19. Q. Yu & S. Tremeaine, MNRAS, **335**, 965 (2002)

The Polarization Properties of Sgr A* at Submillimeter Wavelengths

J.M. Moran[1], D.P. Marrone[1], J.-H. Zhao[1], and R. Rao[2]

[1] Harvard-Smithsonian Center for Astrophysics, 60 Garden Street, Cambridge, MA 02138,
    USA, {jmoran,dmarrone,jzhao}@cfa.harvard.edu
[2] rrao@sma.hawaii.edu

Summary. We have placed an upper limit on the magnitude of the rotation measure of 7×10^5 rad m^{-2} for the putative Faraday screen (magnetized plasma) in front of Sgr A*, the radio source associated with the black hole in the Galactic Center. There is evidence that the actual rotation measure is about -5×10^5 rad m^{-2}. With a simple model of equipartition of energy and reasonable inner radius for the screen, the accretion rate is estimated to be less than $10^{-6} M_\odot \mathrm{yr}^{-1}$. In addition, we have detected, for the first time, intra-day variability in the polarization of Sgr A*, which may be due to either intrinsic variations in Sgr A* or variations in the composition of the Faraday screen.

1 Introduction

The black hole in the center of our galaxy gives rise to a radio source known as Sgr A*, which radiates via the synchrotron process. Its image is broadened by refraction and diffraction effects in the turbulent ionized interstellar medium along the line of sight. The apparent size of the source is about $0.8\nu^{-2}$ arcseconds, where ν is the frequency in GHz. Observations with very long baseline interferometry(VLBI) suggests that the intrinsic size of the source is about 240 μas, or about 2.7×10^{13} cm at 43 GHz (Bower et al. 2004) and 1.5×10^{13} cm at 86 GHz (Shen et al. 2005), for a distance of 7.6 kpc (Eisenhauer et al. 2003; Ghez et al. 2005). Since the Schwarzschild radius (r_s) for the 3.5×10^6 solar mass black hole is 10^{12} cm, the radius of the source is only about $8r_s$ at the higher frequency. The accretion envelope surrounding and feeding the black hole has never been detected directly, but is expected to contain a hot, magnetized plasma, which can also affect the image of the radio source. If the intrinsic source (SgrA*) is linearly polarized, then the position angle of the radiation will be changed by the plasma due to the effect of Faraday rotation. The observed polarization angle is given by the relation

$$\chi(\nu) = \chi_o + (c/\nu)^2 RM, \qquad (1)$$

where χ_o is the intrinsic polarization position angle (PA) of the source, c is the speed of light and ν is the frequency. RM, called the rotation measure, is given by,

$$RM = (8.1 \times 10^5) \int n_e \, B \, dr \quad (\text{rad m}^{-2}),\tag{2}$$

where n_e is the electron density in cm^{-3}, B is the magnetic field strength along the line of sight in Gauss, and dr is the path length in parsecs. Such a Faraday screen, when it lies in front of the source, and has no lateral variations in n_e and B, changes the position angle, but not the degree, of polarization.

Sgr A* is less than 1 percent linearly polarized at 22 and 43 GHz (Bower et al. 1999). Recently, polarization of about 2 percent has been seen at 86 GHz (Macquart et al. 2006). The polarization percentage at 220 GHz is typically 5 percent (Bower et al. 2005) and may rise with increasing frequency (Aitken et al. 2000). We present polarization measurements made with the Submillimeter Array at 340 GHz, which place new constraints on the parameters of the accretion environment.

2 The Submillimeter Array (SMA)

The Submillimeter Array, a joint project of the Smithsonian Astrophysical Observatory and the Academia Sinica Institute of Astronomy and Astrophysics, is a new imaging instrument that currently operates in the frequency range of 180 to 690 GHz (Ho, Moran, & Lo 2004; Blundell 2005). It is the only instrument currently available to make subarcsecond resolution images at submillimeter wavelengths ($\nu > 300$ GHz). Such resolution is important for observations of Sgr A* in order to be able to separate its emission from that of the surrounding dust which is also polarized. The SMA, shown in Figure 1, is located near the summit of Mauna Kea at an altitude of 4000m, which lessens the effects of absorption by atmospheric water vapor. The SMA has eight 6-m diameter elements, which can be configured on baselines from 8 to 508 meters. This

Fig. 1. The Submillimeter Array near the summit of Mauna Kea in its nominal compact configuration, similar to the one used for the observations described here. The vehicle on the right is the antenna transporter, which is used to move the antennas among various configurations. The cylindrical dome on the left is the James Clerk Maxwell Telescope, which will be included as an element in the SMA, along with the CSO, later in 2006.

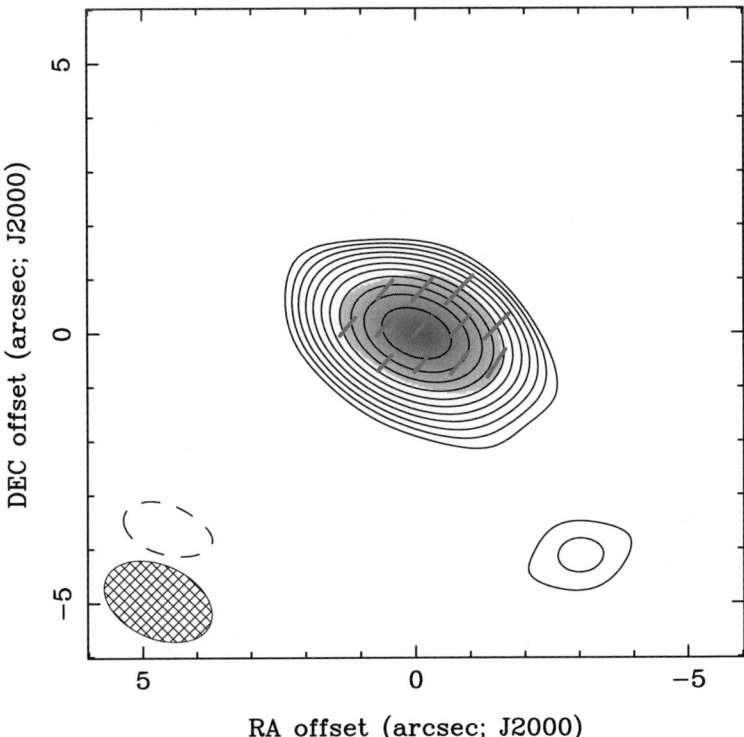

Fig. 2. Image of Sgr A* made with the SMA on 25 May 2004. The source is unresolved in the 2 arcsecond beam of the Array (shown in the lower left). The contour levels are -3.5, 3.5, 5, 7, 10, 14, 20, 28, 40, 57 80, 113 times the rms noise level of 26 mJy. The position angle of the linear polarization is shown by the vectors.

corresponds to a range in resolution of 0.3 to 22 arcseconds at 340 GHz. In the continuum mode with a bandwidth of 2 GHz, and reasonably good weather conditions with opacity of 0.2, the sensitivity in 4 hours is about 3 mJy (see SMA Observer Center site, http://sma1.sma.hawaii.edu/beamcalc.html).

In about a year's time, the SMA will be equipped with dual orthogonally polarized [right circular (R) and left circular(L)] receivers. This will allow the measurements of the full set of Stoke parameters of the image, constructed from the RR, LL, RL, and LR combinations of circular polarization on each baseline. In the meantime, one of us (Marrone) has constructed an interim system that measures all Stokes parameters with only one receiver per antenna as part of his PhD thesis research at Harvard University. The system works in the following way. A quarter waveplate is positioned in the beam of each antenna and switched between two positions differing by 90 degrees to convert the linear polarization to either right or left circular polarization. The polarization state of each antenna is controlled by a series of Walsh functions so that all possible combinations of polarizations among all antennas can be sampled in 7 minutes with a 20 second dwell time. In the absence of leakage, the cross-circular polarization correlation

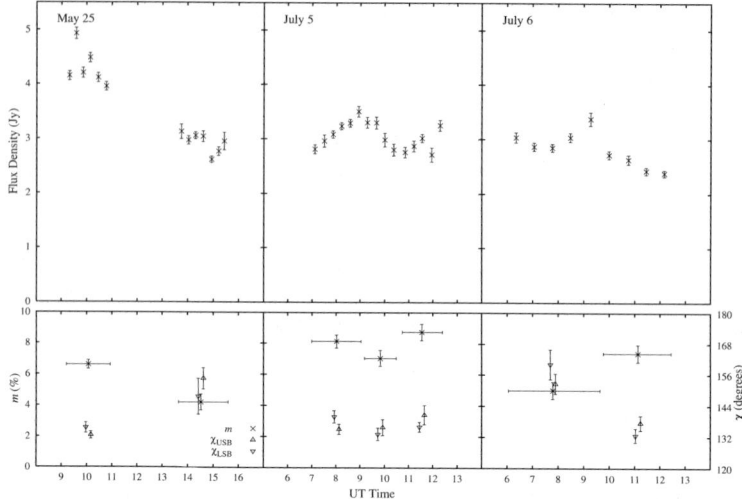

Fig. 3. Variability of Sgr A* at 340 GHz shown for the three most sensitive epochs of data (25 May, 5 and 6 July, 2004). (top) Flux density versus time. (bottom) Polarization data binned into two periods for 25 May, and three periods for 5 & 6 July, as shown by the horizontal bars on the polarization fraction (m) data points. The polarization position angle for the 342 GHz (USB) and 332 GHz (LSB) channels are shown separately and offset in time for clarity.

for any baseline responds only to the linearly polarized signal component (Thompson, Moran & Swenson 2002). The leakage components were typically less than 4 percent, and calibrated to better than 1 percent, for these observations.

3   Observations and Results

We observed Sgr A* on six nights in 2004: 25-26 May, 5-7 July, and 14 July (see Marrone et al. 2006, for more details). Observations were made in the two sidebands of the double sideband receivers centered at 332 and 342 GHz. A typical image of Sgr A* in total intensity, with the linear polarization vectors superimposed, is shown in Figure 2. The snr is greater than 100:1. The results of the three epochs for which the opacity at 340 GHz was lowest (0.11-0.16) are shown in Figure 3. The flux density scale was derived from observations of Neptune. The reliability of the calibration was checked by frequent observations of nearby-in-angle quasars, including NRAO530, 1741-038, 1749+096 and 1921-293.

We obtained the following results (see Marrone et al. 2006, for more details):

- Inter-night variability was detected in the flux density in the range of 2.7 to 3.8 Jy. Similar levels of fluctuations have been reported at 230 by Zhao et al. (2003) and Bower (2005), and at 86 GHz by Mauerhan et al. (2005).
- The degree of polarization and the polarization position angle varied within the ranges of 2.3 to 8.5 percent and 137 to 156°, respectively.

- Intra-day variability was detected with a range (peak-to-peak) of 2.3 Jy, 2.5% and 20°, in total intensity, polarized fraction, and position angle, respectively. This is the first detection of intra-day variability in the polarized component of the radio emission.
- The difference in position angle of the polarization between the two sidebands (332 and 342 GHz) was less than 1.8° on the day with the most sensitive measurements (6 July).
- No circular polarization was detected at a 1-σ level of 0.5 percent based on an average of all of the data. This is the most stringent limit on circular polarization above 90 GHz. Sgr A* has a circular polarization at cm wavelengths of about 21, 31, 34 and 62 percent at 1.4, 4.8, 8.4 and 15 GHz, respectively (Bower et al. 2002).

4 Discussion

The lack of change in the position angle in the linear polarization between 332 and 342 GHz can be used to put a limit on the Faraday rotation, and with certain assumptions, on the accretion rate. The limit on rotation measure, implied by a limit in PA difference over 10 GHz of 1.8° is, from Equation 1, equal to 7×10^5 rad m^{-2}. To relate this limit to accretion rate we consider a simple phenomenological model. We assume that the electron density scales as a power law, i.e., $n_e(r) \sim r^{-\beta}$. We further assume equipartition of magnetic and gravitational energies, so that $B \sim r^{(\beta+1)/2}$. Integration from an inner radius, r_{in}, to infinity gives the relation

$$RM = 3.4 \times 10^{19} \left(\frac{2}{3\beta - 1} \right) \left(\frac{r_{in}}{r_s} \right)^{-7/4} \dot{M}_\odot^{3/2} \quad (\text{rad m}^{-2}), \qquad (3)$$

where $\dot{M}_\odot$ is the mass accretion rate in $M_\odot \text{yr}^{-1}$. The quantity r_{in} corresponds to the radius where the plasma becomes relativistic, within which the Faraday rotation effect is greatly diminished (Quataert & Gruzinov 2000a). β is 3/2 for free fall, but can range from 1/2 for the CDAF model (Quataert & Gruzinov 2000b) to 3/2 for the ADAF model (Narayan & Yi 1994). For ranges of r_{in} from 30 to 300 r_s, and ranges of β from 1/2 to 3/2, the mass accretion rate varies from 10^{-9} to $10^{-6} M_\odot \text{yr}^{-1}$. The expected suppy of material to the accretion flow based on X-Ray observations is $10^{-5} M_\odot \text{yr}^{-1}$ (Baganoff et al. 2003). This suggests that the accretion rate may decrease with decreasing radius.

In studying the temporal behavior of the postion angles of the polarization in our data at 340 GHz and that of Bower et al.(2005) at 230 GHz, we notice that there is a tendency for the them to cluster about the values of 140 and 111°, respectively. We speculate that these may correspond to "quiescent" values of the polarization. If this difference is due to Faraday rotation, then the RM would be -5.1×10^5 rad m^{-2}. An approximate set of values for parameters for the case of free fall, $\beta = 3/2$, where most of the Faraday rotation would be confined to a small annular region, and an inner effective radius of $30r_s$ (3×10^{13} cm) would be $\dot{M} = 4.6 \times 10^{-8} M_\odot \text{yr}^{-1}$, $n_e = 3 \times 10^4$ cm^{-3}, B = 4 G, and a screen thickness, $\delta r = 16r_s$ or 1.6×10^{13} cm. The intrinsic polarization angle would be 160°. We note that the polarized infrared flares, which should suffer negligible Faraday rotation, has a position angle of about 40° (Eckart et al

2006b). Also note that none of the IR flares has been identified with significant events in the submillimeter band (Eckart et al. 2006a)

The origin of the temporal changes in position angle is not clear. If changes in the Faraday screen are responsible, then the relation between the change in RM for a change in χ at 340 GHz would be, from Equation (1),

$$\delta RM = (2.2 \times 10^4)\delta\chi, \tag{4}$$

for $\delta\chi$ in degrees. Hence, our maximum change of $19°$ in PA would correspond to a change in rotation measure of 4×10^5 rad m^{-2}. This value is about equal to the inferred rotation measure. Alternatively, the position angle changes could be intrinsic to the source. The distinction between these interpretations requires simultaneous polarization measurements at substantially different frequencies.

5 Future Observations

We have installed new optical plates (Marrone 2006) that act as quarter wave plates at 230 GHz and 3/4 wave plates at 690 GHz so that we can make simultaneous polarization measurements at these frequencies, and thereby determine the Faraday rotation without concern about comparing position angles measured at different times. In 2005 July, such measurements, interleaved on alternate days with 340 GHz measurements, were made. Their analysis is underway. Additional measurements are scheduled for the spring of 2006. In 2007 the full dual channel polarization system will be available with a forty percent improvement in sensitivity and simultaneous measurements of all polarization parameters.

References

1. D.K. Aitken, J. Greaves, A. Chrysostomou, T. Jenness, W. Holland, J.H. Hough, D. Pierce-Price, & J. Richer: Ap.J.(Lett.) **534**, L173 (2000)
2. F.K. Baganoff, Y. Maeda, M. Morris et al: Ap.J. **591**, 891 (2003)
3. R. Blundell: The Submillimeter Array Antennas and Receivers. In: *Proceedings of 15th International Symposium on Space Terahertz Technology*, ed by G. Narayanan (Amherst: U. Massachusetts Press 2004) astroph/0508492
4. G.C. Bower, H. Falcke, R.J. Sault, D.C. Backer: Ap.J. **571**, 843 (2002)
5. G.C. Bower, M.C.H. Wright, D.C. Backer, H. Falcke: Ap.J. **527**, 851 (1999)
6. G.C. Bower, H. Falcke, R.M. Herrnstein et al: Science **304**, 704 (2004)
7. G.C. Bower, H. Falcke, M.C.H. Wright, D.C. Backer: Ap.J. (Lett.) **618**, L29 (2005)
8. A. Eckart, F.K. Baganoff, R. Schoedel et al: A&A **450**, 2, 535-555 (2006a)
9. A. Eckart, R. Schodel, L. Meyer, et al: JPhCS **54**, 1, 391-398 (2006b)
10. F. Eisenhauer, R. Schödel, R. Genzel, et al: Ap.J.(Lett) **597**, L121 (2003)
11. A.M. Ghez, S. Salim, S.D. Hornstein, et al: Ap.J. **620**, 744 (2005)
12. P.T.P. Ho, J.M. Moran, K.-Y. Lo: Ap.J.(Lett.) **616**, L1 (2004)
13. J.P. Macquart, G.C. Bower, M.C.H. Wright, D.C. Backer, H. Falcke: Ap.J. **646**, L111 (2006)
14. D.P. Marrone: PhD Thesis, Harvard University (2006)
15. D. P. Marrone, J. M. Moran, J.-H. Zhao, R. Rao: Ap.J. **640** 308 (2006)

16. J.C. Mauerhan, M. Morris, F. Walter, F.K. Baganoff: Ap.J.(Lett.) **623**, L25 (2005)
17. R. Narayan, I. Yi: Ap.J.(Lett.) **428**, L13 (1994)
18. E. Quataert, A. Gruzinov: Ap.J. **545**, 842 (2000a)
19. E. Quataert, A. Gruzinov: Ap.J. **539**, 809 (2000b)
20. Z.Q. Shen, K.Y. Lo, M.C. Liang, P.T.P. Ho, J.H. Zhao: Nature, **438**,62 (2005)
21. A.R. Thompson, J.M. Moran & G.W. Swenson, Jr.: *Interferometry and Synthesis in Radio Astronomy*, 2nd edn (John Wiley & Sons New York 2002)
22. J.-H. Zhao, K.H. Young, R.M. Herrnstein et al: Ap.J.(Lett.) **586**, L29 (2003)

Highlights of XMM-Newton Observations of Black Holes

N. Schartel

XMM-Newton SOC, ESAC, ESA, Apartado 50727, E-28080 Madrid, Spain,
Norbert.Schartel@sciops.esa.int

1 Introduction

From a scientific historic viewpoint the theory of general relativity is unique in many aspects. In 1916 Einstein formally completed its formulation with his famous publication, "Die Grundlage der allgemeinen Relativitätstheorie"[1], in the Annalen der Physik [1]. In the same year, Einstein could communicate the radial symmetric solution of the field equations found by Schwarzschild [2]. This Schwarzschild solution describes a black hole without charge and angular momentum. Almost 60 years had to pass before in 1963 Kerr found the description of a black hole with angular momentum.

We might consider 1975, the year of the famous bet between Hawking and Thorn on the nature of Cygnus X-1 [3], as the time when black holes were entering the horizon of observation. Consequently we can take 1990, the year when Hawking conceded, as the date when most astronomers were convinced of black holes' existence. But black hole observations were based on indirect evidence rather than direct proof.

In a typical argumentation chain, first the total mass within a certain volume was determined, e.g. through measurement of velocity dispersion or through observation of a visible companion, and second it was shown that the objects found within the studied volume could not account for the determined mass, which then implied the postulation of a black hole. Similar considerations were based on variability, i.e. the detected objects were unable to explain an observed luminosity change, where an accreting black hole was required. In general the claim of a black hole followed the lack of objects, which could explain a determined total mass or a measured luminosity change.

The basic problem was that all the observations were taken far away from the event horizon of the black hole such that only the weak gravitational field is acting. The comparison between a neutron star with a radius of about 10 Schwarzschild radii and a black hole of the same mass illustrates how narrow the space is for observing black holes and for studying the strong gravitational field.

In 1995 the situation changed with the observation of a gravitationally redshifted iron emission line by Tanaka et al. [4], which implied an accretion disk around the

---

[1] The basis of the general theory of relativity

massive black hole in MGC-6-30-15. With this measurement, for the first time, the strong gravitational field could be directly observed.

2 Birth of Black Holes

By observing gamma ray bursts (GRB) we are most probably witnessing the birth of black holes. XMM-Newton observes, as targets of opportunity, X-ray afterglows of GRBs that are detected by other missions, like ESA's Integral or NASA's Swift. A burst with a gamma ray emission duration of more than one second indicates the creation of a stellar-mass black hole in connection with a supernova explosion. Measurements with XMM-Newton were the first that provided observational support for such a scenario: Reeves et al. [5] reported about the X-ray spectrum of the afterglow of GRB011211, which shows emission lines of magnesium, silicon, sulphur, argon and calcium, arising in metal-enriched material with an outflow velocity of the order of one-tenth of the speed of light. These findings strongly support models where a supernova explosion from a massive stellar progenitor precedes the burst event and is responsible for the outflowing matter.

3 The Growing of Black Holes

Outbursts and transient behaviour are the most acknowledged tracers of a black holes' growth. During outburst, Galactic black holes are among the brightest X-ray sources in the sky. XMM-Newton observed outbursts of Galactic black holes as targets of opportunity, like XTE J1650-500 in 2002 [6] or GX 339-4 in 2004 [7]. The spectra of both show broad Fe K_α line emission which characterizes the outburst as a violent form of accretion via a disk.

Komossa et al. [8] measured a factor of ~ 200 drop in luminosity of the nonactive galaxy RX J1242.6-1119A using data of ROSAT, Chandra and XMM-Newton observations. In this case the favoured outburst scenario is the tidal disruption of a star by a supermassive black hole. The post-outburst "low-state" emission has properties, which are related to the flare and show a "hardening" compared to the outburst spectrum. The inferred black hole mass, the amount of liberated energy, and the duration of the event favour an accretion event of the form expected from the (partial or complete) tidal disruption of a star.

An unexpected example for the normal continuous growth of supermassive black holes was reported by Streblyanska et al. [9]. The stacked spectra of the faint Lockman hole Active Galactic Nuclei (AGNs) (53 type 1 and 41 type 2) observed for 770 ks show a broad iron line and therefore the clear signature for an accretion disk.

4 Close to the Event-Horizon: Mass and Spin

The XMM-Newton spectra of several black holes allowed the detection of a broad skewed Fe K_α line. These lines are emitted by the material of an accretion disk which often extends down to a few Schwarzschild radii. The line-profile allows the determination of the innermost orbit and sometimes a decision between a Schwarzschild black hole and Kerr black hole, which makes them one of our most powerful diagnostics of the strong gravitational field.

The observed profile of the Fe K_α line in the spectrum of XTE J1650-500, taken during its 2001 outburst, suggests that the primary in the system may be a Kerr black hole. Its steep emissivity profile, that is hard to explain in terms of a standard accretion disk model, may be explained by the extraction and dissipation of rotational energy from the black hole with nearly maximal angular momentum [6].

The XMM-Newton outburst spectrum of the galactic black hole GX 339-4 reveals an extremely skewed relativistic Fe K_α emission line and an ionised disk reflection spectrum. The observed line profile requires an inner disk radius, which is not compatible with a Schwarzschild black hole. From the profile Miller et al. could constrain the black hole angular momentum to be $a > 0.8 - 0.9$ (where $r_g = GMc^{-2}$ and $a = cJG^{-1}M^{-2}$) [7].

The AGN MCG-6-30-15 was twice targeted by XMM-Newton, both times showing broad skewed Fe K_α lines. The comparisons between the XMM-Newton EPIC spectrum of MCG-6-30-15 with the corresponding Chandra High Energy Transmission Grating Spectrometer spectrum and warm absorber models are very illustrative, because the warm absorbers fail to account for the broadening of the iron line [10]. During the first XMM-Newton observation of MCG-6-30-15, the AGN was in a deep minimum state allowing the broad line to be seen down to energies in the 2 to 3 keV region. The shift in energy and the line profile are difficult to understand in a pure accretion disc model. Very likely XMM-Newton is witnessing the extraction and dissipation of rotational energy from a spinning black hole [11].

XMM-Newton observed the brightest X-ray flare detected so far from Sgr A*, the supermassive black hole in the centre of our Galaxy [12]. The corresponding power density spectrum peaks at 5 different periods [13] showing twin high frequency quasi-periodic oscillations with a ratio of 3:2 and 3:1, similar to the micro-quasars GRO J1655-40, XTE J 1550-564 and GRS 1915-105. Studying this behaviour theoretically, Aschenbach found topology changes in the motion of particles orbiting rapidly rotating Kerr black holes [14]. Zdenêk et al. demonstrated that these changes occur also for non-geodesic circular orbits with constant specific angular momentum [15]. Combining this theoretical progress with the observational measurements, the mass and spin of the supermassive black hole in the Galactic centre can be determined with highest precision: $M = (3.28 \pm 0.13) \times 10^6 M_\odot$ and $a = 0.99616$.

Turner et al. [16] analysed the energy-time plane of XMM-Newton data for the narrow line Seyfert galaxy Mrk 766 in the 4-8 keV energy range. A component of Fe K_α emission shows a variation of photon energy with time that appears both to be statistically significant and to be consistent with sinusoidal variation. The orbital period could be determined to be $\sim$165 ks and the line-of-sight velocity $\sim$13,500 km s^{-1} [16]. These measurements imply a constraint to the black hole mass of $4.9 \times 10^5 M_\odot < M_{BH} < 4.5 \times 10^7 M_\odot$.

Iwasawa, Miniutti and Fabian tentatively detected a modulation of a transient, red-shifted Fe K_α emission feature in the X-ray spectrum of the Seyfert galaxy NGC 3516

[17]. The feature varies systematically in flux at intervals of 25 ks whereas its peak moves in energy between 5.7 and 6.5 keV. The spectral evolution of the feature agrees with Fe K emission arising from a spot on the accretion disc at a radius of (3.5-8) Schwarzschild radii modulated by Doppler and gravitational effects. The measurements allow the mass of the black hole to be estimated to be $(1\text{-}5)\times 10^{7} M_{\odot}$, which agrees well with values obtained from other techniques.

5 Resumé

With the current observations of black holes we are witnesses to a new area of astrophysics: the era of the strong, rotating gravitational field. The collected results prove its importance for several topics: galactic and extragalactic black holes, supermassive black holes, cosmology and theory. But we must keep in mind that we are taking only the first steps in this new scientific landscape.

References

1. A. Einstein: Annalen der Physik **7**, 769 (1919)
2. S. Chandrasekhar: *The Mathematical Theory of Black Holes*, Oxford University Press, Oxford 1983, pp 136 – 137
3. K. S. Thorne: *Black Holes & Time Warps*, W. W. Norton & Company, New York London 1994, pp 314 – 315
4. Y. Tanaka et al: Nature **375**, 659 (1995)
5. N. H. Reeves et al: Nature **416**, 51 (2002)
6. J. M. Miller et al: ApJ **570**, L69 (2002)
7. J. M. Miller et al: ApJ **606**, L131 (2004)
8. S. Komossa et al: ApJ **630**, L17 (2004)
9. A. Streblyanska et al: A&A **432**, 395 (2005)
10. A. J. Young et al: ApJ **631**, 733 (2005)
11. J. Wilms et al: MNRAS **328**, L27 (2001)
12. D. Porquet et al: A&A **407**, L17 (2003)
13. B. Aschenbach et al: A&A **417**, 71 (2004)
14. B. Aschenbach: A&A **425**, 1075 (2004)
15. S. Zdenêk et al: PhRvD **71**, 4037 (2005)
16. T. J. Turner: A&A **445**, 59 (2006)
17. K. Iwasawa, G. Miniutti, A. C. Fabian: MNRAS **355**, 1073 (2004)

Evolution of Supermassive Black Holes

M. Volonteri

Institute of Astronomy, University of Cambridge, Madingley Road, Cambridge CB3 0HA, UK,
marta@ast.cam.ac.uk

Summary. Supermassive black holes (SMBHs) are nowadays believed to reside in most local galaxies, and the available data show an empirical correlation between bulge luminosity - or stellar velocity dispersion - and black hole mass, suggesting a single mechanism for assembling black holes and forming spheroids in galaxy halos. The co-evolution between galaxies and quasars is indicated by the observation of quiescent SMBHs in nearby normal (i.e. not active) galactic centers. The inferred mass density of the inactive black holes is in good agreement with the estimate of the total mass accreted by quasars, through integration of their luminosity function over time. In hierarchical models of galaxy formation major mergers are responsible for forming bulges and elliptical galaxies. Galactic interactions also trigger gas inflows, and the cold gas may be eventually driven into the very inner regions, fueling an accretion episode and the growth of the nuclear SMBH. In cold dark matter cosmogonies, small-mass subgalactic systems form first to merge later into larger and larger structures. In this paradigm galaxy halos experience multiple mergers during their lifetime. If every galaxy with a bulge hosts a SMBH in its center, and a local galaxy has been made up by multiple mergers, then a BH binary is a natural evolutionary stage. The evolution of the supermassive black hole population clearly has to be investigated taking into account both the cosmological framework and the dynamical evolution of SMBHs and their hosts.

1 Introduction

It is well established observationally that the centers of most galaxies host supermassive black holes (SMBHs) with masses in the range $M_{BH} \sim 10^6 - 10^9 \, M_\odot$ (e.g., Ferrarese et al. 2001; Kormendy & Gebhardt 2001; Richstone 2004). The evidence is particularly compelling for our own galaxy, hosting a central SMBH with mass $\simeq 4 \times 10^6 \, M_\odot$ (Ghez et al. 2000, 2005; Schödel et al. 2002). Dynamical estimates indicate that, across a wide range, the central black hole (BH) mass is about 0.1% of the spheroidal component of the host galaxy (Häring & Rix 2004; Marconi & Hunt 2004; Magorrian et al. 1998). A tight correlation is also observed between the BH mass and the stellar velocity dispersion of the hot stellar component (Ferrarese & Merritt 2000; Gebhardt et al. 2000; Tremaine et al. 2002). For galaxies, these correlations may well extend down to the smallest masses. For example, the dwarf Seyfert 1 galaxy POX 52 is thought to contain a BH of mass $M_{BH} \sim 10^5 \, M_\odot$ (Barth et al. 2004). At the other end, however, the Sloan Digital Sky survey detected luminous quasars at very high redshift, $z > 6$. Follow-up observations confirmed that at least some of these quasars are powered by supermassive black holes (SMBHs) with masses $\simeq 10^9 \, M_\odot$ (Barth et al. 2003; Willott et al. 2003). We are therefore left with the task of explaining the presence of very big SMBHs when the Universe is less than 1 Gyr old, and of much smaller BHs lurking in 13 Gyr old galaxies.

2 Sowing the Seeds of Black Holes

SMBHs with $M_{BH} \sim 10^6 - 10^9 \, M_\odot$ cannot form from simple stellar evolution of comparable mass stars. They probably evolve from seeds of intermediate mass. One first possibility is the direct formation of a BH from a collapsing gas cloud (Haehnelt & Rees 1993; Loeb & Rasio 1994; Eisenstein & Loeb 1995; Bromm & Loeb 2003; Koushiappas, Bullock & Dekel 2004). The main issue for this family of models is how to get rid of the angular momentum of the gas, so that it can form a central massive object, which eventually becomes subject to post-Newtonian gravitational instability and forms the seed BH. The mass of the seeds vary according to different models, but typically are in the range $M_{BH} \sim 10^3 - 10^6 \, M_\odot$. An alternative scenario proposes the gravitational core collapse of relativistic star clusters which may have been produced in starbursts at early times (Shapiro 2004). This scenario leads to a formation of seed BHs in the mass range $M_{BH} \sim 10^2 - 10^4 \, M_\odot$. The seeds of SMBHs can also be associated with the remnants of the first generation of stars. The first stars which form at $z \sim 20$ in halos which represent high-σ peaks of the primordial density field. At zero metallicity, the gas cooling proceeds in a very different way compared to local molecular clouds. The inefficient cooling might lead to a very top-heavy initial stellar mass function, and in particular to the production of very massive stars with masses $> 100 M_\odot$ (Carr, Bond, & Arnett 1984). If very massive stars form above 260 $M_\odot$, they would rapidly collapse to massive BHs with little mass loss (Fryer, Woosley, & Heger 2001), i.e., leaving behind seed BHs with masses $M_{BH} \sim 10^2 - 10^3 \, M_\odot$ (Madau & Rees 2001).

It is not possible to distinguish among the proposed models with current observations: the initial conditions are mostly washed out when SMBHs gain most of their mass, between $z = 3$ and $z = 1$. There are two ways we can find out in which mass range the seed BHs were formed. One is to detect a subpopulation of relic seed BHs still lurking in present-day galaxy halos. The prediction of a population of massive BHs wandering in galaxy halos and the intergalactic medium at the present epoch is an inevitable outcome, if the assembly history of SMBHs goes back to early times. These intermediate mass BHs have been left over by inefficient galaxy mergers, or dynamical interactions (see §3). The number and mass function of these wandering BHs differs largely from model to model. If these relic seeds are massive, $M_{BH} \sim 10^5 - 10^6 \, M_\odot$, they can perturb the gravitational potential, and generate wave patterns in the gas within a disc close to galaxy centers. The observed density pattern can be used as a signature in detecting the most massive wandering black holes in quiescent galaxies (Etherington & Maciejewski 2006). Several attempts have been made to investigate the emission properties of these wandering BHs (Islam, Taylor & Silk, Volonteri & Perna 2005, Mii

& Totani 2005), and their possible association to the population of ultraluminous off-nuclear ('non-AGN') X-ray sources (ULXs) that have been detected in nearby galaxies (van der Marel 2003, Fabbiano 2005). No strong conclusion has been reached yet.

A second way to discern the seeds of SMBHs is through gravitational waves. Merging black holes are expected to be a strong source of gravitational waves, and the planned gravitational wave interferometer LISA is expected to detect black hole mergers out to very large redshifts, and down to very small masses, being sensitive to mergers of BHs with masses in the range $10^3 - 10^7 M_{\odot}$. The difference in prediction between models can be up to 2-3 orders of magnitude (Haehnelt 2004), and therefore LISA will be extremely helpful in solving this riddle (Figure 1).

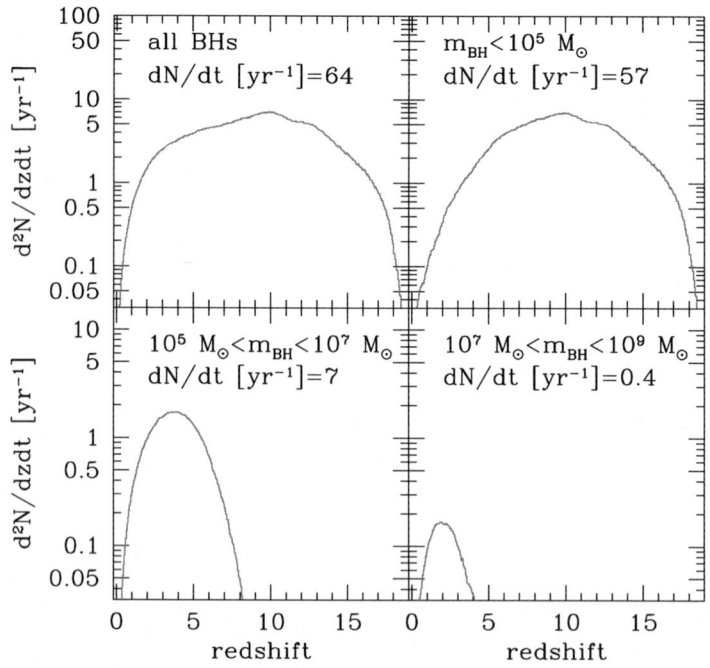

Fig. 1. Predictions on the number of MBH binary coalescences observed by LISA per year at $z = 0$, per unit redshift, in different BH mass intervals. Each panel also lists the integrated event rate, dN/dt, predicted by Sesana et al. (2004). If SMBHs form from very massive seeds at low redshift, the expected merger rate is of order a few per year (cfr. bottom panel). If SMBHs evolve from high redshift small seeds the merger rate can be up to tens or hundred per year (cfr. upper panels).

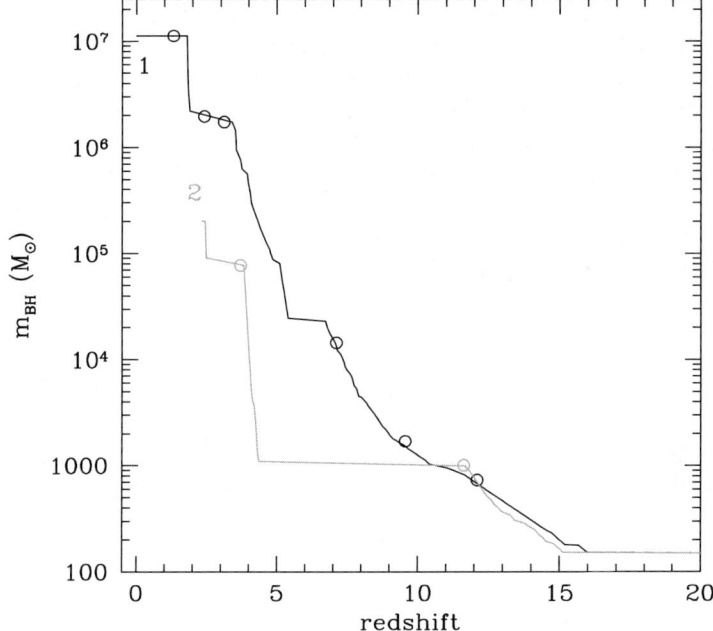

Fig. 2. Mass-growth history of two SMBHs, one ending in a massive halo ('1') at $z = 0$, and one in a satellite ('2') at $z = 2.3$. The SMBH mass grows by gas accretion after major merger events and by SMBHs mergers (*circles*). Note how most of the mass of the black holes is gained in accretion episodes and not through SMBH mergers.

3 Spinning Top Toys and Yo-Yos

Whatever your favourite choice of seed holes, several models, starting from different seeds, have proved equally successful in explaining the evolution of SMBHs and quasars in the redshift range $1 < z < 3$ (Kauffmann & Haehnelt 2000; Wyithe & Loeb 2002, 2003 Cattaneo et al. 1999). Accretion episodes are believed to be triggered by major mergers of galaxies, which can destabilize the gas in the interacting galaxies. The cold gas may be eventually driven into the very inner regions, fueling an accretion episode and the growth of the nuclear SMBH (Fig. 2).

This last year has been especially exciting, as the first high resolution simulations of galactic mergers including BHs and quasar feedback, showed that the co-evolution scenario is generally correct (Springel, di Matteo & Hernquist 2005; di Matteo, Springel & Hernquist 2005). The complete and detailed picture, in particular regarding the physics in the vicinity of the SMBH, is not clear yet, as simulations do not allow one to resolve the sphere of influence of the BHs.

The above mentioned models for the evolution of SMBHs and quasars all reproduce satisfactorily the observed luminosity function of quasars. Integration, over redshift and luminosity, of the luminosity function of quasars, gives the total light emitted by known quasars (Yu & Tremaine 2002; Elvis, Risaliti, & Zamorani 2002; Marconi et al. 2004). Using the Soltan's argument, this quantity can be converted into the total mass density accreted by black holes during the active phase, by assuming a mass-to-energy conversion efficiency, ϵ, and normalizing to the local black hole mass density (Aller & Richstone 2002; Merloni, Rudnick & Di Matteo 2004). The average radiative efficiency results somehow larger than $\epsilon = 0.06$, which is the standard value for non-spinning BHs.

The spin of a SMBH is modified by a merger with a secondary BH, or by interplay with the accretion disc and magnetic fields. If accretion proceeds from a thin disk, and magnetic processes are not important, rapidly spinning SMBHs are to be expected (Volonteri et al. 2004). BH mergers, in fact, do not lead to a systematic effect, but simply cause the BHs to random-walk around the spin parameter they had at birth. BH spins, instead, efficiently couple with the angular momentum of the accretion disk, producing Kerr holes independent of the initial spin. This is because, for a thin accretion disk, the BH aligns with the outer disk on a timescale that is much shorter than the Salpeter time, corresponding to an e-folding time for accretion at the Eddington rate, leading to most of the accretion being from prograde equatorial orbits. In any model in which SMBH growth is triggered by major mergers, every accretion episode must typically double the BH mass to account for the local SMBH mass density. As a result, most of the mass accreted by the hole acts to spin it up, even if the orientation of the spin axis changes in time. Most individual accretion episodes thus produce rapidly-spinning BHs independent of the initial spin.

Our investigation finds that the spin distribution is skewed towards rapidly spinning holes, is already in place at early epochs, and does not change significantly below redshift 5. As shown in Fig. 3, about 70% of all SMBHs have spins larger than $\hat{a}a/m_{BH} = 0.9$, corresponding to efficiencies approaching 20-30% (assuming a "standard" spin-efficiency conversion). The spin parameter of a few SMBHs in the local Universe seems indeed to be non-zero, as shown by the prominent soft X-ray excess (Crummy et al. 2006), or the profile of the Iron line (Miniutti, Fabian & Miller 2004, Streblyanska et al. 2005).

Highly spinning black holes pose a challenge at early times, however. The accretion of mass at the Eddington rate causes the BH mass to increase in time as

$$M(t) = M(0) \ \exp \left(\frac{1 - \epsilon}{\epsilon} \frac{t}{t_{\text{Salp}}} \right), \tag{1}$$

where $t_{\text{Salp}} = 0.45\,\epsilon/(1 - \epsilon)$ Gyr. Given an initial BH mass, $M(0)$, the higher the efficiency, the longer it takes for the BH to grow in mass up to its final mass (Shapiro 2005). The highest redshift quasar currently known, SDSS 1148+3251, at $z = 6.4$, has estimates of the SMBH mass in the range $(2 - 6) \times 10^9 M_\odot$ (Barth et al. 2003, Willott, McLure & Jarvis 2003). If the accretion efficiency is 40%, and the SMBH powering SDSS 1148+3251 has been accreting continuously for the whole Hubble time (about 1Gyr at $z = 6$), then it has grown by a factor $\simeq 1000$, requiring a $10^6 M_\odot$ seed well before halos of comparable mass have populated the Universe.

A way out is to consider a large contribution to the BH mass by mergers. At $z < 5$ MBH mergers do not play a fundamental role in building up the mass of SMBHs

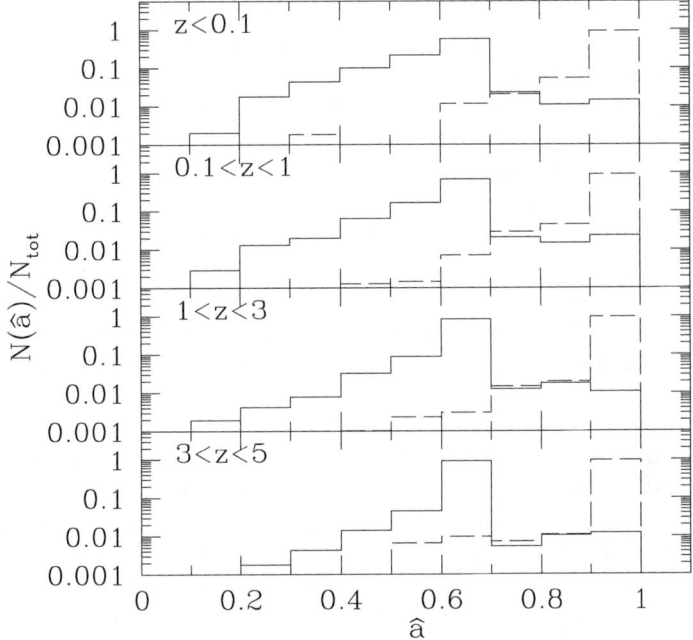

Fig. 3. Distribution of SMBH spins in different redshift intervals. "Seed" holes are born with a spin parameter $\hat{a} = 0.6$. *Solid histogram:* effect of black hole binary coalescences only. *Dashed histogram:* spin distribution from binary coalescences and gas accretion via a thin disk.

(Yu & Tremaine 2003), but they can be possibly important at $z > 5$, where we do not have constraints from a Soltan-type argument. Mergers can have a positive contribution to the build-up of the SMBH population that has been suggested could be up to $10^9 M_\odot$ (Yoo & Miralda-Escudé 2004). On the other hand, dynamical processes can disturb the growth of BHs, especially at high redshift (Haiman 2004, Yoo & Miralda-Escudé 2004), and give a negative contribution to SMBH growth.

4 Playing Pools with Black Holes

The lifetime of BH binaries can be long enough (Begelman, Blandford & Rees; Quinlan & Hernquist 1997; Milosavljevic & Merritt 2001; Yu 2002) that following another galactic merger a third BH can fall in and disturb the evolution of the central system. The three BHs are likely to undergo a complicated resonance scattering interaction, leading to the final expulsion of one of the three bodies ('gravitational slingshot') and to the recoil of the binary. Any slingshot, in addition, modifies the binding energy of the binary, typically creating more tightly bound systems.

Another interesting gravitational interaction between black holes happens during the last stage of coalescence, when the leading physical process for the binary evolution becomes the emission of gravitational waves. If the system is not symmetric (e.g. BHs have unequal masses or spins) there would be a recoil due to the non-zero net linear momentum carried away by gravitational waves in the coalescence ('gravitational rocket'). The coalesced binary would then be displaced from the galactic centre, leaving straightaway the host halo or sinking back to the centre owing to dynamical friction.

Slingshots and rockets basically give BHs a recoil velocity, that can even exceed the escape velocity from the host halo and spread BHs outside galactic nuclei (Volonteri, Haardt & Madau 2003). In the shallow potential wells of mini-halos, the growth of BHs from seeds can be halted by these ejections. Yoo & Miralda-Escudé (2004) explored the minimum conditions that would allow the growth of seed MBHs up to the limits imposed by SDSS 1148+3251 when recoils are taken into account. To meet the $z = 6$ constraint, continued Eddington-limited accretion must occur onto MBHs forming in halos with $T_{\rm vir} > 2000$K at $z \leq 40$.

One important piece of information which is still missing, is the typical timescale for a BH binary to merge. After dynamical friction ceases to be efficient, at parsec scale or so, there is still a large gap before emission of gravitational waves can be efficient in bringing the binary to coalescence in less than a Hubble time. In massive galaxies at low redshift, the subsequent evolution of the binary may be largely determined by three-body interactions with background stars (Begelman, Blandford, & Rees 1980). In gas-rich, star-poor high redshift systems, gas processes can lead the binary evolution. Dynamical friction in a gaseous environment is much more efficient than against a stellar background, and, also, if an accretion disc is surrounding the MBH binary, the secondary BH can be dragged to the primary within a viscous timescale, much shorter than any other involved timescale. Figure 4 compares the MBH merger rates in two different models accounting for the orbital evolution of MBH binaries in the phase preceding emission of gravitational waves. If gas processes (viscous drag, as shown) drive the MBH orbital decay, MBHs start merging efficiently at very early times, when host DM halos are still small. Though the absolute number of mergers is larger, more MBHs are therefore ejected from DM halos due to the rocket effect.

5 Conclusions

The last few years have seen exciting developments in our understanding of the evolution of supermassive black holes. Though we start having a coherent picture on how the population of SMBHs evolved, there are several crucial issues that remain to be clarified, both at high and low redshift.

References

1. Aller, M. C., & Richstone, D.: AJ **124**, 3035 (2002)
2. Barth, A. J., Martini, P., Nelson, C. H. & Ho, L. C.: ApJ **594**, L95 (2003)
3. Barth, A.J., Ho, L.C., Rutledge, R.E., & Sargent, W.L.W.: ApJ **607**, 90 (2004)

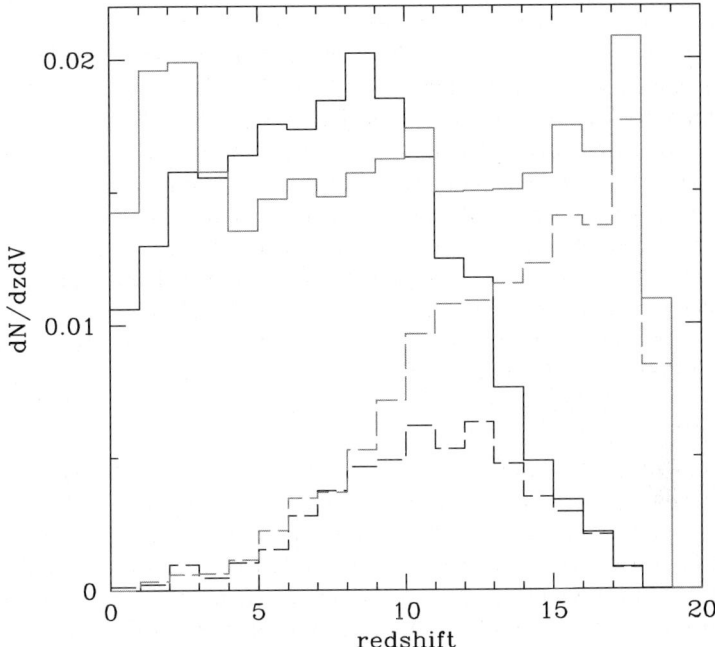

Fig. 4. Rates of binary MBH mergers (*solid line*) and ejections due to the gravitational radiation recoil (*dashed line*) per comoving Mpc^3. The subset of ejected binaries is selected requiring that the recoil velocity is larger than the escape velocity from the host halo. At redshifts $z > 10$ about 80% of the merging MBH binaries are ejected, and therefore lost, into the intergalactic medium. The black lines refer to a model with inefficient BH mergers, as binaries shrink by stellar dynamics. The red / grey lines show the case in which BH binaries evolve on the viscous timescale, if an accretion disc is present.

4. Begelman, M. C., Blandford, R. D., Rees, M. J.: Nature **287**, 307 (1980)
5. Bromm V., Loeb A.: ApJ **596**, 34 (2003)
6. Carr, B. J., Bond, J. R., & Arnett, W. D.: ApJ **277**, 445 (1984)
7. Cattaneo, A., Haehnelt, M. & Rees, M. J.: MNRAS **308**, 77 (1999)
8. Crummy, J. et al.: MNRAS **365**, 1067 (2006)
9. di Matteo, T., Springel, V., & Hernquist, L.: Nature **433**, 604 (2005)
10. Eisenstein D. J., Loeb A.: ApJ, **443**, 11 (1995)
11. Elvis, M., Risaliti, G., & Zamorani, G.: ApJ **565**, L75 (2002)
12. Etherington, J. & Maciejewski, W.: MNRAS **367**, 1003-1010 (2006)
13. Fabbiano, G.: Science **307**, 533 (2005)
14. Ferrarese, L., & Merritt, D.: ApJL, **539**, 9 (2000)
15. Ferrarese, L.: ApJ **578**, 90 (2002)
16. Fryer, C. L., Woosley, S. E., & Heger, A.: ApJ **550**, 372 (2001)
17. Gebhardt, K. et al. : ApJL **543**, 5 (2000)
18. Ghez, A.M., et al.: Nature **407**, 349 (2000)

19. Ghez, A.M., et al.: ApJ **620**, 744 (2005)
20. Häring, N., & Rix, H.-W.: ApJ, **604**, L89 (2004)
21. Haehnelt M. G., Rees M. J.: MNRAS **263**, 168 (1993)
22. Haehnelt, M.: Joint Formation of Supermassive Black Holes and Galaxies. In *Coevolution of Black Holes and Galaxies* , ed. L.C. Ho (Cambridge Univ. Press, Carnegie Obs. Ast. Ser. Vol. 1), 405
23. Haiman, z.: ApJ **613**, 36 (2004)
24. Islam R. R., Taylor J. E. & Silk, J.: MNRAS **354**, 443 (2003)
25. Kauffmann, G., & Haehnelt, M. G. MNRAS **311**, 576 (2000)
26. Kormendy, J., & Gebhardt, K.: Supermassive Black Holes in Galactic Nuclei. In *20th Texas Symposium on Relativistic Astrophysics*, ed. H. Martel & J.C. Wheeler, p. 363
27. Koushiappas S. M., Bullock J. S., Dekel A.: MNRAS **354**, 292 (2004)
28. Loeb A., Rasio F. A.: ApJ, **432**, 52 (1994)
29. Madau, P., & Rees, M. J.: ApJ **551**, L27 (2001)
30. Magorrian, J. et al.: AJ **115**, 2285 (1998)
31. Marconi A. et al.: MNRAS **351**, 169 (2004)
32. Marconi, A. & Hunt, L. K.: Apj **589**, L21 (2003)
33. Merloni, A., Rudnick, G. & di Matteo, T.: MNRAS **354**, 37 (2004)
34. Mii, H. & Totani, T.: ApJ **628**, 873 (2005)
35. Milosavljevic, M., & Merritt, D.: ApJ **563**, 34 (2001)
36. Miniutti, G., Fabian, A. & Miller, J. M.: MNRAS **351**, 466 (2004)
37. Quinlan, G. D., & Hernquist, L.: New Astronomy **2**, 533 (1997)
38. Richstone, D.: Supermassive Black Holes: Demographics and Implications. *In Coevolution of Black Holes and Galaxies*, ed. L.C. Ho (Cambridge Univ. Press, Carnegie Obs. Ast. Ser. Vol. 1), 281
39. Schödel, R., et al.: Nature **419**, 694 (2002)
40. Shapiro, S. L.: ApJ **613**, 1213 (2004)
41. Shapiro, S. L.: Apj **620**, 59 (2005)
42. Springel, V., di Matteo, T., & Hernquist, L.: ApJ **620**, L79 (2005)
43. Streblyanska, A. et al.: A&A **432**, 395 (2005)
44. Tremaine, S., et al.: ApJ **574**, 740 (2002)
45. Volonteri M. et al.: ApJ, 620 **69** (2005)
46. Volonteri, M. & Perna, R.: MNRAS **358**, 913 (2005)
47. Volonteri, M., Haardt, F. & Madau, P.: ApJ **582**, 559 (2003)
48. Willott, C. J. et al.: ApJ **626**, 657 (2005)
49. Wyithe, J. S. B. & Loeb, A.: ApJ **581**, 886 (2002)
50. Wyithe, J. S. B. & Loeb, A.: ApJ **595**, 614 (2003)
51. Yoo J. & Miralda-Escudé, J.: ApJ **614**, L25 (2004)
52. Yu, Q. & Tremaine, S.: MNRAS **335**, 965 (2002)
53. Yu, Q.: MNRAS **331**, 935 (2002)
54. van der Marel, R. P.: Intermediate-Mass Black Holes in the Universe: Constraints from Dynamics and Lensing. In *THE ASTROPHYSICS OF GRAVITATIONAL WAVE SOURCES*, AIP Conference Proceedings, Volume **686**, pp. 115-124

Part IV

Active Galactic Nuclei

AGN and XRB Variability: Propagating-Fluctuation Models

P. Arévalo[1] and P. Uttley[2]

[1] University of Southampton, Southampton So17 1BJ, UK
`patricia@astro.soton.ac.uk`
[2] NASA Goddard Space Flight Center, Greenbelt, MD 20771, USA
`pu@milkyway.gsfc.nasa.gov`

1 Introduction

X-ray light curves from active galactic nuclei (AGN) and galactic black hole candidates (GBHC) can be strongly variable over a wide range of time-scales. The X-ray emission from these systems is believed to arise from the vicinity of the central black hole so the information contained in the timing properties can be used to study the accretion and emission processes.

The variability is usually quantified through the Fourier Transform (FT) of the light curves, to disentangle their behaviour at different variability time-scales. The modulus squared of the FT measures the variability power at each Fourier frequency, or Power Spectral Density (PSD). The Cross Spectrum between two simultaneous light curves (e.g. same observation in different energy bands) measures their coherence and any lags between the fluctuations in each band.

2 Observed X-ray Timing Properties

X-ray light curves are usually variable in a broad range of time-scales. Most measured AGN PSDs have a power law shape with slopes of ∼-1, bending to steeper slopes at short time-scales. This PSD shape is also common in GBHC in the high/soft state, while in the low/hard state there is, at least, an additional cut-off in the variability power at long time-scales.

Different X-ray energy bands normally have highly coherent fluctuations but show slightly different behaviours: higher energies tend to show more short time-scale variability, i.e. flatter high-frequency PSD slopes. Secondly, fluctuations in harder energy bands tend to lag behind their counterparts at lower energies. These time lags increase both with the time-scale of the fluctuations and with the separation of the energy bands compared.

One last important characteristic of the X-ray light curves is that the amplitude of the fluctuations is larger when the average flux is higher [4]. The linear relation between rms variability and flux implies that fluctuations on different time-scale must be coupled together. This coupling implies that the variability cannot be produced by a distribution of independent emitting regions of different sizes, each producing fluctuations on a different time-scale.

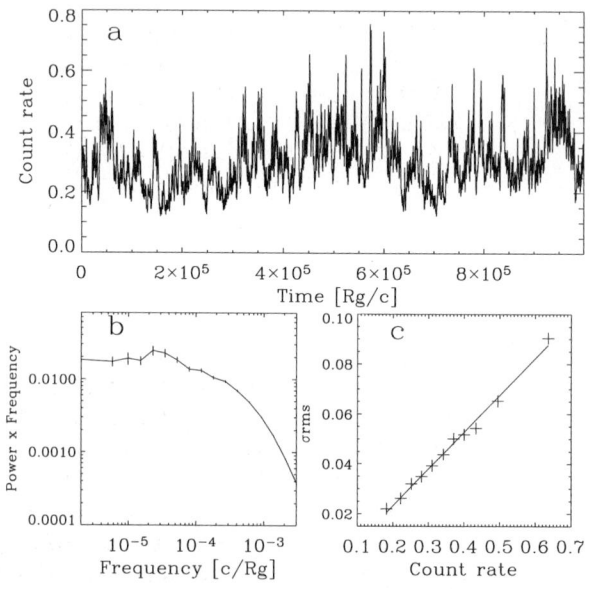

Fig. 1. Simulated light curve calculated with a propagating fluctuation routine (a). The light curves has a bending power law PSD (b, notice y axis denotes Power × frequency), and follows a linear rms-flux relation (c).

3 Fluctuating Accretion Models

Lyubarskii (1997) [3] proposed a model where accretion rate fluctuations propagate down the accretion disc towards the centre, modulating the flux emitted in the inner-most region. The characteristic time-scales of the fluctuations depend on their radius of origin, relating shorter time-scales to smaller radii. This scheme solves two prob-lems: Firstly, as the fluctuations have a wide range of radii to arise from, they can carry a wide range of variability time-scales to the emitting region, producing broad PSDs. Secondly, the fluctuations couple together as they propagate so that inner, high fre-quency, fluctuations are modulated by longer time-scale fluctuations, giving rise to the rms-flux relation.

Kotov et al. (2001) [2] suggested that if the hard X-rays are produced through a more concentrated emitting region than the soft X-rays, two more properties can be reproduced: the PSDs become energy dependent and time lags appear between X-ray energy bands.

In this work, we chose power law radial emissivity profiles with different power law indices for each energy band, giving steeper profiles to harder bands. Then, as the fluctuations propagate inward through the emitting region, they are 'seen' first in the soft band and later in the hard band. This setup also produces time-scale dependent lags: as the fluctuations are produced within the emitting region, shorter time-scale variability, that is only produced in inner regions, travel shorter distances and so their time lags are smaller.

Keeping the fluctuation frequencies tied to the propagation speed (e.g. both proportional to the viscous time-scale of the accretion disc) produces quasi power-law lag spectra with slopes of -0.5– -1 and amplitudes of 0.1–10 % of the time-scale[1]. At the same time, the extended emitting regions assumed, act as a low-pass filter on the PSDs, producing the desired higher power at short time-scales for harder energy bands.

4 Fits to Real Data

The propagating-fluctuation scheme, complemented with different emissivity profiles for each energy band can reproduce simultaneously the energy dependence of the PSDs and the shape of the time lag spectrum [1]. Figure 2 shows a fit to Cyg X-1 data in the high/soft state. The top panel shows the time lags between 2-5 keV and 8-13 keV bands, fitted with a model lag spectrum (solid line). Here, the velocity with which the fluctuations propagates needs to be twice the viscous velocity, to reproduce the right amplitude of the lags. The bottom panel shows the ratio between the PSDs of the same energy bands, fitted with the model using the same emissivity profiles required by the lag spectrum.

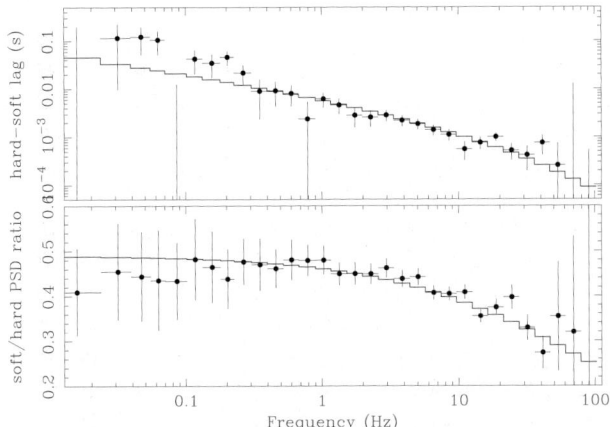

Fig. 2. Simultaneous fitting of the time lag spectrum (top) and PSD ratio (bottom) of an observation of Cyg X-1 in the high/soft state. The fit was made with a simple realisation of the propagating fluctuations model and is shown by the solid lines.

References

1. P. Arévalo, P. Uttley: MNRAS **357**, 2 (2006)
2. O. Kotov, E. Churazov, M. Gilfanov: MNRAS **327** 799 (2001)
3. Y. Lyubarskii: MNRAS **292**, 679 (1997)
4. P. Uttley, I. McHardy: MNRAS **323**, 26 (2001)

The Source of Variable Optical Emission is Localized in the Jet of the Radio Galaxy 3C 390.3

T.G. Arshakian[1], A.P. Lobanov[1], V.H. Chavushyan[2], A.I. Shapovalova[3], and J.A. Zensus[1]

[1] Max-Planck-Institut für Radioastronomie, Auf dem Hügel 69, 53121 Bonn, Germany
(tigar,alobanov,azensus)@mpifr-bonn.mpg.de
[2] Instituto Nacional de Astrofísica Óptica y Electrónica, Apartado Postal 51 y 216, 72000 Puebla, Pue, México, vahram@inaoep.mx
[3] Special Astrophysical Observatory of the Russian AS, Nizhnij Arkhyz, Karachaevo-Cherkesia 369167, Russia

1 The Link between Variable Radio Emission of the Jet and Optical Continuum Emission

The strong continuum emission is believed to originate in the central pc-scale region of active galactic nuclei (AGN) and is responsible for ionizing the cloud material in the broad-line region (BLR). Locating the source of the variable continuum emission in AGN is therefore central for understanding the mechanism for release and transport of energy in active galaxies. In radio-loud AGN, the continuum emission from the relativistic jets dominates at all energies. The presence of a positive correlation between beamed radio luminosity of the jet and optical nuclear luminosity in the sample of radio galaxies suggests that the optical emission is non-thermal and may originate from a relativistic jet [1, 2]. The detection of a correlation between the variations of radio and optical nuclear emission in a single source would be the most direct evidence of optical continuum emission coming from the jet.

To search for a relation between variability of the optical continuum flux and radio flux density in AGN on scales of $\sim 1\,\mathrm{pc}$, we combine [3] the results from monitoring of the radio-loud broad emission-line galaxy 3C 390.3 ($z = 0.0561$) in the optical and ultraviolet regimes with ten very long baseline interferometry (VLBI) observations of its radio emission at $15\,\mathrm{GHz}$ [4] made from 1992 to 2002 using the VLBA (Very Long Baseline Array).

The model fitting of a single epoch VLBA image (Fig. 1, left panel) shows five radio components on the scale of 2 mas. For ten VLBA images, we identified five moving components (C4-C8) and two stationary components (S1 and S2) separated from D by (0.28 ± 0.03) mas and (1.50 ± 0.12) mas, respectively (Fig. 1, right panel). The proper motions of moving components correspond to apparent speeds of $0.8\,c$ to $1.5\,c$. We also measured the epoch, t_{D}, at which each component was ejected from the component

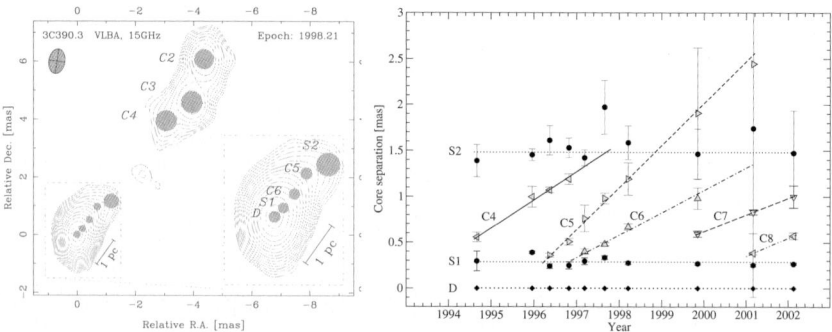

Fig. 1. *Left*: A single epoch (1998.21) radio structure of 3C 390.3 observed with VLBI at 15 GHz (1 mas = 1.09 pc). The labels in the inset mark three stationary features (D, S1 and S2) and two moving components (C5 and C6) identified in the jet. Images of later epochs showed also new components C7 and C8. *Right*: Separation of the jet components relative to the stationary feature D (filled diamonds) for ten epochs of VLBI observations. Moving components are denoted by triangles (C4-C8), and stationary components S1 and S2 are marked by filled circles.

D and the epoch, t_{S1}, when it passed through the location of the stationary feature S1 (Fig. 1, right panel).

Employing minimization methods we found significant correlations between variations in the radio emission from D and S1 and in the optical continuum emission (Fig. 2, left panel). A time delay of ~ 1 year was found between flux density variations of D and S1 components of the jet. We also showed that the radio light curves of D and S1 correlate significantly (≥ 95 % confidence level) and lead the optical light curve with time delays of ~ 1.4 yr and < 0.4 yr respectively. This finding indicates that *there is a physical link between the jet components and optical continuum: the variable optical continuum emission is located in the innermost part of the sub-pc-scale jet near to component S1.* A large fraction of the variable optical continuum emission in 3C 390.3 is non-thermal [5] and produced in a region of 0.004 pc in size [6] at a distance of > 0.4 pc from D [3].

The link between the optical continuum and the jet is also supported by the correlation between the local maxima in the optical continuum light curve and the epochs t_{S1} (Fig. 2, left panel). The null hypothesis that this happens by chance is rejected at a confidence level of 99.98%. This suggests that *radio component ejection events are coupled with the long-term variability of optical continuum.*

2 The Central Sub-Pc-Scale Region in 3C 390.3

To understand the structure of the central engine and its radiation mechanism one needs to locate the stationary features with respect to the central nucleus. The component D is identified with the base of the jet which is most likely to be located in the central engine of 3C 390.3 near the accretion disk or hot corona. This is supported by the link between the ejection epoch (t_D) of the component C5 and the dip in the X-ray emission (similar to 3C 120; [8]). The feature S1 can be associated with the stationary radio feature which

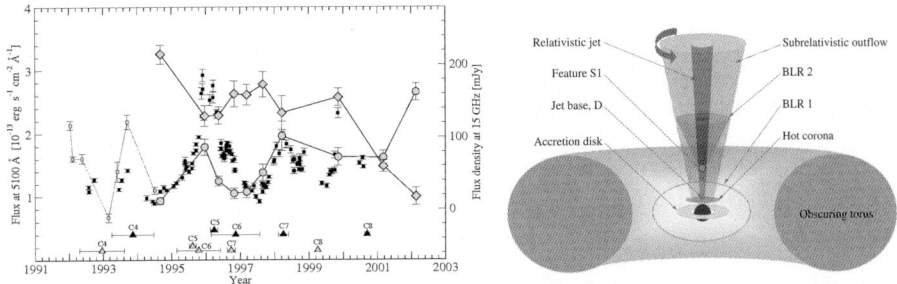

Fig. 2. *Left*: Optical continuum fluxes at 5100Å (black squares) and 1850Å (open squares) superimposed with the flux density at 15 GHz from the stationary components D (gray diamonds) and S1 (gray circles) of the jet and the epochs t_D and t_{S1} marked by open and filled triangles respectively. *Right*: A sketch of the nuclear region in 3C 390.3 which shows only the approaching jet. The broad-line emission is likely to be generated both near the central nucleus (BLR 1, ionized by the emission from a hot corona or the accretion disk) and near the S1 in a rotating sub-relativistic outflow [7] surrounding the jet (BLR 2, ionized by the emission from the relativistic plasma in the jet).

may be produced by internal oblique shock formed in the continuous relativistic flow (Fig. 2, right panel). We suggest that the beamed synchrotron radio emission from S1 produces the optical continuum emission via the Inverse Compton effect. The variable optical continuum emission associated with the jet and counterjet forms two conical shaped BLRs (BLR2; Fig. 2, right panel) at a distance of about 0.4 pc from the central nucleus.

The presence of the BLR2 at a large distance from the central nucleus challenges [3] the existing models of BLRs and the assumption of virialized motion in the BLR and hence, the BH mass estimates of radio-loud AGN.

References

1. M.J. Hardcastle, D.M. Worrall: MNRAS, 314, 359 (2000)
2. M. Chiaberge, A. Capetti, A. Celotti: AA, 394, 791 (2002)
3. T.G. Arshakian, et al: astro-ph/0512393 (2005)
4. K.I. Kellermann, et al.: ApJ, 609, 539 (2004)
5. W. Wamsteker, W. Ting-gui, N. Schartel, R. Vio: MNRAS, 288, 225 (1997)
6. M. Dietrich, et al.: ApJSS, 115, 185 (1998)
7. D. Proga, J.M. Stone, T.R. Kallman: ApJ, 543, 686 (2000)
8. A.P. Marscher, et al.: Nature, 417, 625 (2002)

XMM-Newton RGS Spectra in Type 2 Seyfert Galaxies

S. Bianchi and M. Guainazzi

European Space Astronomy Centre of ESA, Villafranca del Castillo, Spain,
Stefano.Bianchi@sciops.esa.int

We present the XMM-*Newton* RGS spectra of 8 Seyfert 2 galaxies, showing that all the spectra are dominated by emission lines with generally a very low level of continuum (see Figures 1-3). This allows us to detect strong emission lines even in very short observations of sources with relatively low fluxes, or heavily absorbed in the soft X-rays (Circinus). On the other hand, long exposure times give high S/N spectra (190 ks for Mrk 3). In this case, the use of several diagnostic tools becomes possible and useful information on the physical properties of the gas responsible for the soft X-ray emission in this class of objects can be extracted. For details see Bianchi, M. Guainazzi and, M. Chiaberge 2006, A&A 448, 499.

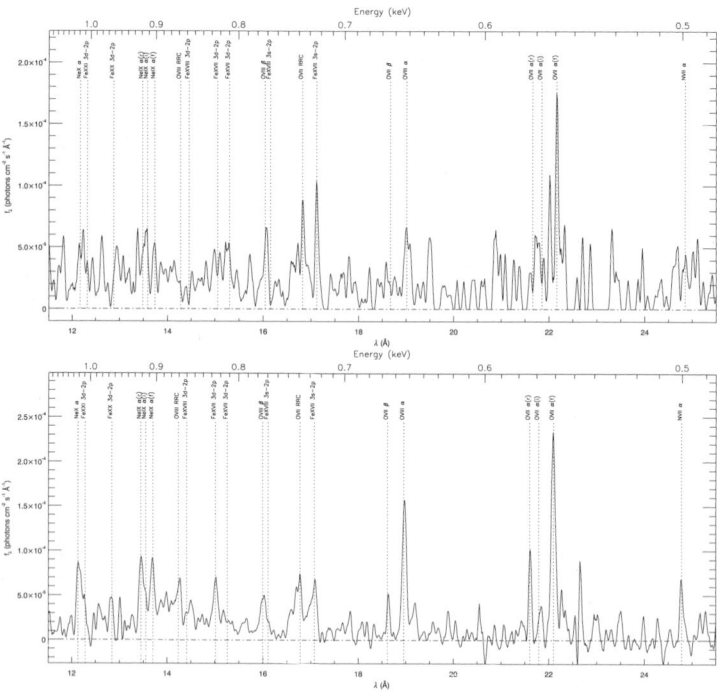

Fig. 1. Combined RGS1/RGS2 spectrum in the rest frame of the source, for NGC 1386 (17 ks, $F_{0.5-2\,keV} = 1.8 \times 10^{-13}\,\mathrm{erg\,cm}^{-2}\,\mathrm{s}^{-1}$) and Mrk 3 (190 ks, $4.7 \times 10^{-13}\,\mathrm{erg\,cm}^{-2}\,\mathrm{s}^{-1}$)

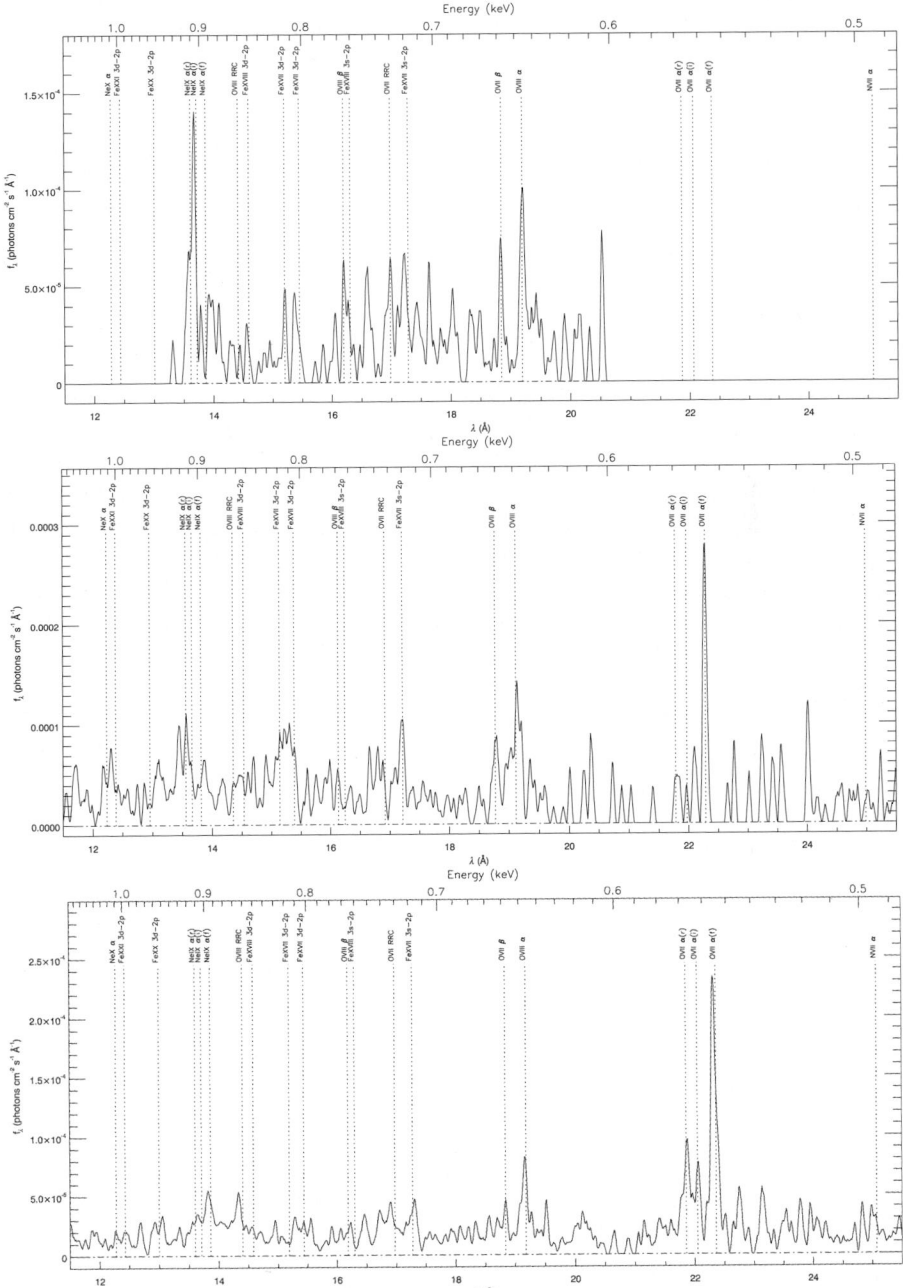

Fig. 2. Same as Fig. 1, for NGC 3393 (14 ks, 2.2×10^{-13},erg cm^{-2} s^{-1}), NGC 4388 (12 ks, 3.4×10^{-13},erg cm^{-2} s^{-1}) and NGC 4507 (46 ks, 3.3×10^{-13} cgs)

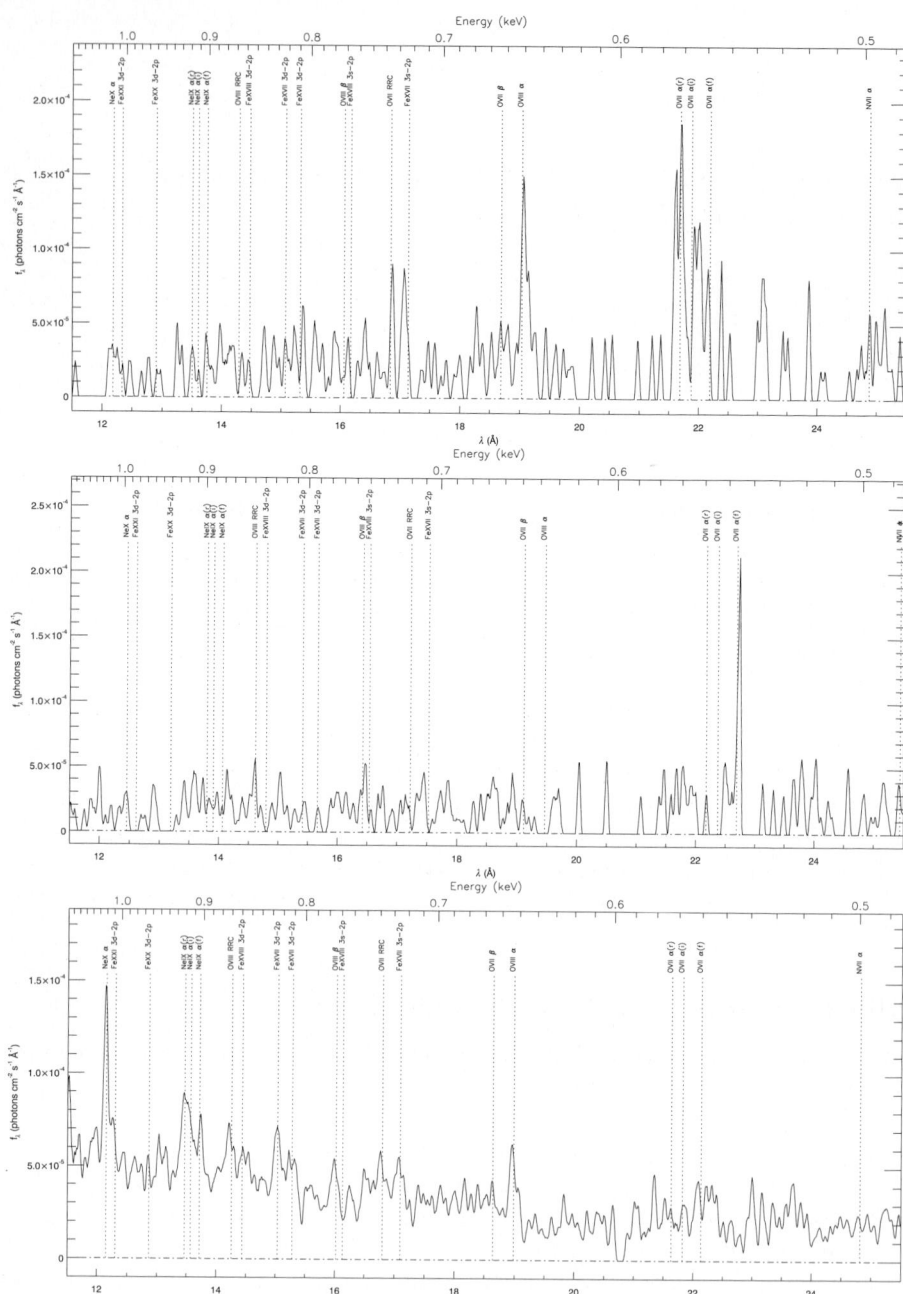

Fig. 3. Same as Fig. 1, for NGC 5643 (10 ks, 1.4×10^{-13}, erg cm^{-2} s^{-1}), NGC 7212 (14 ks, 0.8×10^{-13}, erg cm^{-2} s^{-1}) and Circinus (105 ks, 8.7×10^{-13} cgs)

First Results from the Extended Chandra Deep Field-South (E-CDF-S) Survey

W.N. Brandt[1] and the E-CDF-S Team[2]

[1] Department of Astronomy and Astrophysics, The Pennsylvania State University, 525 Davey Lab, University Park, PA 16802, USA, niel@astro.psu.edu

[2] Arcetri, CMU, ESO, GSFC, IAP, JHU, MPE, Penn State, STScI, Trieste

The E-CDF-S survey has been designed to complement the Chandra Deep Fields by significantly increasing the solid angle of sky with coverage to very sensitive X-ray flux levels. It is composed of four contiguous 250 ks ACIS-I observations flanking the original Chandra Deep Field-South (CDF-S). These observations cover a region with superb and growing multiwavelength coverage (e.g., COMBO-17, GEMS, GOODS, Spitzer, VLT), and they have the sensitivity to detect the X-ray emission from moderate-luminosity active galactic nuclei (AGNs) to $z \approx 4$–6 as well as X-ray luminous star-burst galaxies to $z \approx 1$. All of the Chandra observations have now been successfully performed and analyzed, and nearly 1000 X-ray sources, mostly AGNs, are detected in the E-CDF-S region (about 600 of these are new). Multiwavelength follow-up studies are underway to investigate topics including the cosmic evolution and luminosity dependence of AGN X-ray emission, the efficacy of X-ray versus optical AGN-selection techniques, AGN clustering, moderate-luminosity AGNs in the high-redshift universe, and off-nuclear X-ray sources.

Pointers to the Refereed Scientific Literature

The main results presented in this talk have now been delivered as three substantial articles for refereed journals. Instead of repeating this material in a cursory manner here, I refer you to these articles:

1. Lehmer et al. (2005) provide a detailed description of the Chandra observations along with X-ray source catalogs and other useful data products.
2. Lehmer et al. (2006) investigate the properties and redshift evolution of intermediate-luminosity, off-nuclear X-ray sources from $z \approx 0.05$–0.3.
3. Steffen et al. (2006) constrain the X-ray-to-optical properties of optically selected AGNs over wide ranges of luminosity and redshift.

The E-CDF-S World Wide Web site will be updated as new results are delivered; see http://www.astro.psu.edu/users/niel/ecdfs/ecdfs-chandra.html.

We gratefully thank the Chandra X-ray Center, NASA, and the NSF for supporting this work.

References

1. B.D. Lehmer, W.N. Brandt, D.M. Alexander et al: ApJS **161**, 21 (2005)
2. B.D. Lehmer, W.N. Brandt, A.E. Hornschemeier et al: ApJ **131**, 5, 2394 (2006)
3. A.T. Steffen, I. Strateva, W.N. Brandt et al: ApJ **131**, 6, 2826 (2006)

The Optical and X-ray Properties of AGN in COSMOS

M. Brusa,, V. Mainieri, and G. Hasinger on behalf of the XMM-COSMOS collaboration

Max Planck Institut für extraterrestrische Physik, Giessenbachstrasse 1, D–85748 Garching, Germany, (marcella,vmainieri,ghasinger)@mpe.mpg.de

1 The XMM-COSMOS Survey

One of the primary goals of observational cosmology is to trace star formation and nuclear activity along with the mass assembly history of galaxies as a function of redshift and environment. The detailed study of the nature of the "active" source population is being pursued by complementing deep pencil beam observations [1, 2, 3] (see [4] for a recent review; see also [5]), with shallower, larger area surveys [6, 7, 8]. The results from the identification campaigns have unambiguously unveiled a differential evolution for the low– and high–luminosity AGN population [9, 10, 11, 12]. In these studies, the evolution of the luminosity function of obscured AGN remains the key parameter still to be determined. Moreover, characterizing the N_H distribution of the AGN and the evolution of the AGN population (i.e. Type-1/Type-2 ratio) are crucial inputs in the X-ray background modeling and in the understanding of the connection between host galaxy and black hole evolution.

The best strategy to address this issues is to increase the area covered in the hard X–ray band, and the corresponding optical-NIR photometric and spectroscopic follow-up, down to $F_{2-10keV} \sim 5 \times 10^{-15}$ erg cm^{-2} s^{-1}, where the bulk of the XRB is produced. The XMM–*Newton* wide-field survey in the COSMOS field (hereinafter: XMM-COSMOS) is expected to be an important step forward in addressing the topics described above. The 2 deg^2 contiguous area of the HST/ACS COSMOS Treasury program [13] has been surveyed with XMM–*Newton* for a total of $\sim$800 ks during AO3, and additional 600 ks have been already granted in AO4 to this project. The XMM-COSMOS project is described in [14], and the X–ray source counts from the first 800 ks of the XMM–*Newton* observations obtained so far are presented in [15]. Here we present a few selected highlights on the optical and X-ray properties of X-ray selected AGN in COSMOS, drawn from Brusa et al. 2006 and Mainieri et al. 2006.

2 Multicolor Properties of hard X-ray Sources

We cross-correlated the optical counterparts of the X-ray sources detected in the first set of XMM-COSMOS observations and identified using the "likelihood ratio technique" (as described in detail in [16]) with the Subaru multicolor and ACS catalogs, and with the Magellan/IMACS, zCOSMOS and literature spectroscopic data, all available in the framework of the COSMOS consortium (see ApJ special issue dedicated to COSMOS, in press). We constructed a sample of 524 sources with multicolor and ACS information, almost half of them (248) with spectroscopic redshift and classification. Even if this sample is not "complete", it can be considered representative of the X-ray source population at fluxes $> 10^{-15}$ erg cm^{-2} s^{-1}.

Optical and near infrared color diagrams

Figure 1 (left panel) shows the U-B vs. B-V diagram for all field objects (small black points) with B>19 classified as point-like in the full ACS catalog (on the basis of the analysis of the available data; Leauthaud, private communication). The locus occupied by stars is very well defined in this diagram, with the two densely populated regions in the blue (hot subdwarf stars) and red (dwarf M stars) parts of the sequence. Most of the points at U-B<0.3 (UVX objects) are expected to be AGN at z<2.3, while the objects on the right of the stellar sequence are likely to be compact galaxies classified as point-like. Overplotted as blue symbols are the X–ray sources classified as point-like from ACS. Spectroscopically confirmed Broad Line AGN (BL AGN) and objects without broad line in the optical spectra (NOT BL AGN) are also indicated (green and yellow symbols overplotted on the blue ones, respectively). Finally, the expected track of quasars from redshift 0 to 3.5 in this color-color diagram is also reported, computed using the SDSS AGN template from [17].

Comparing the optical color-color properties of AGNs in the COSMOS field with those of field objects, we estimate that X–ray data with a flux limit of $S_{0.5-2\text{keV}} \sim 10^{-15}$ erg cm^{-2} s^{-1} recover up to 70% of the AGN candidates which would be selected on the basis of their ultraviolet excess (U-B<0.3) at B<23 (see [18] for a similar estimate at a somewhat brighter X–ray flux from *ROSAT* observations in the Marano field). On the other hand, a significant fraction ($\sim$40%) of the X-ray selected AGNs, especially those without broad lines in their optical spectra, would have not been easily selected as AGN candidates on the basis of purely optical criteria, either because of colors similar to those of normal stars or field galaxies at z$\sim$1-3, or because of morphological classification not consistent with that of point–like sources. Figure 1 (right panel) shows the R−K vs. K–band magnitude (Vega magnitudes) for the subsample of objects with ACS morphological information. In addition, sources spectroscopically identified with BL AGN are marked with green symbols, while sources identified with NOT BL AGN are marked by yellow symbols. A significant difference in the R−K distributions for point-like and extended sources is present: while the widths of the two distributions are similar ($\sigma \sim$0.8 for both of them), extended sources are significantly redder ($\langle R-K \rangle$=4.05$\pm$0.05) than point-like sources ($\langle R-K \rangle$=2.91$\pm$0.06). When the spectroscopic information is also considered, red objects are preferentially associated with NOT BL AGN (yellow circles), while blue objects are preferentially associated with BL AGN (green circles).

We have then investigated the distribution of the R-K colors as a function of the X–ray hardness ratio (HR). The hardest sources (i.e. HR> −0.3, solid histogram in the right part of Fig. 1) are mostly associated with red and very red objects (R-K>4). This

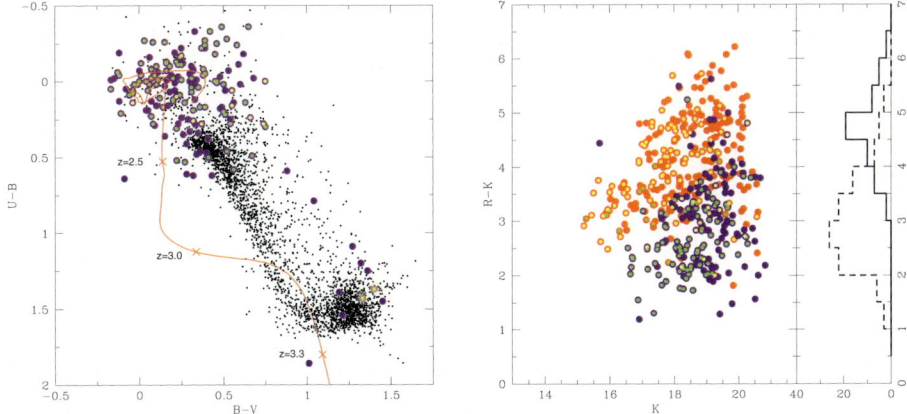

Fig. 1. Left panel: U-B vs. B-V diagram for all the fields objects in COSMOS with B> 19 (small black points) classified as point-like in the full ACS catalog. Overplotted as blue symbols are the counterparts of X–ray sources. Spectroscopically confirmed BL AGN and NOT BL AGN are also indicated (green and yellow symbols overplotted on the blue ones, respectively; see text for further details). Right panel: R-K color vs. K band magnitude (Vega system) for the X-ray sources counterparts with ACS information. Symbols as in left panel. The histogram on the right part of the figure shows the distribution of R-K colors for the sources with HR> −0.3 (i.e. candidate X–ray obscured AGN, solid histogram) and the sources detected only in the soft band (HR=−1, dashed histogram).

indicates an excellent consistency between optical obscuration of the nucleus as inferred from ACS (extended morphology), optical to near infrared colors, and the presence of X–ray obscuration as inferred from the hardness ratio and spectral analysis (see also next section; [19, 20]). Conversely, sources detected only in the soft band (i.e. HR=−1, dashed histogram) have preferentially blue R-K colors typical of those of optically selected, unobscured quasars [21].

3 X-ray Spectral Properties

We performed X–ray spectral analysis of the 133 objects with spectroscopic identification *and* more than 100 net counts in the [0.3-10] keV band. We have developed an automated procedure to extract the X-ray spectrum of a source from single exposures and combine them (details are given in [22]). We fitted all the sources with a basic model (APL), a *powerlaw* with intrinsic absorption (*zwabs*). An additional photoelectric absorption component (*wabs*) is fixed to the Galactic column density in this area ($N_H^{Gal} \sim 3 \times 10^{20}$ cm^{-2}). This fit yields the power-law photon index Γ, the intrinsic column density N_H, and the X-ray luminosity in the [0.5-2] and [2-10] keV rest-frame bands. We divide our sample in two: the 81 sources with more than 180 net counts form *sample-1*, while the 52 sources with less that 180 net counts are in *sample-2*. Performing the spectral fit we leave free both Γ and N_H to vary in sample-1, while we fix Γ to

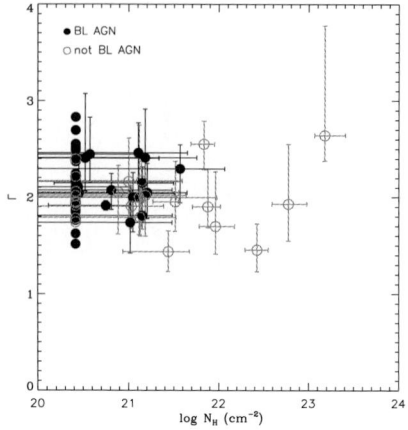 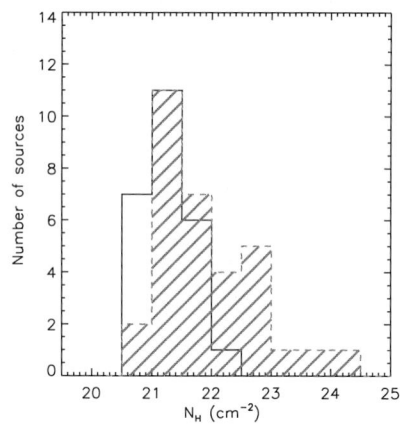

Fig. 2. Left panel: Γ versus N_H for the X-ray sources with more than 180 net counts in [0.3-10] kev (bright sample) and spectroscopically identified. Filled circles are Type-1 AGN, while empty circles are Type-2 AGN. Error bars correspond to 1σ. Right panel: Intrinsic column density (N_H) distribution for BLAGN (empty histogram) and not BLAGN (hatched histogram).

the average value obtained for sample-1 while fitting sources of sample-2. We carefully check for each source the results obtained with the basic *APL* model and, if large residuals are present, we refine the fit using more complex models.

The spectral properties of the sources in *sample-1* are summarised in Fig.2 (left panel). The average value of Γ does not change as a function of N_H as already noticed in deep surveys (e.g. [23]). For the whole sample we found $< \Gamma >= 2.09$ and the dispersion of the distribution of the best fit values is $\sigma \approx 0.27$, in agreement with the results presented by [24, 25] Adopting the classical optical classification, the mean value for the spectral slope for BL AGN (57 sources) is $< \Gamma >= 2.14$ with a dispersion of $\sigma \approx 0.26$ while for not BL AGN (24 sources) we obtain $< \Gamma >= 1.99$ and $\sigma \approx 0.29$. Furthermore, the photon index does not vary in the redshift range z=[0.0,3.0] covered by our sample.

The other physical quantity that we derive from the spectral fitting is the column density N_H. We detect an intrinsic absorption larger than the Galactic value for 57 ($\sim 43\%$) sources of our sample. Figure 2 (right panel) shows the distribution of N_H values for these sources, separately for BLAGN and NOT BLAGN. The visual impression that the two distributions are different is confirmed by a Kolmogorov-Smirnov test that gives a probability for them being different larger than 99.9%. Most (12 out 14) of the objects with $N_H > 10^{22}$ cm^{-2} are NOT BLAGN.

This work is just the first phase of COSMOS AGN studies. For the first time, on a statistical and meaningful sample of X–ray sources we were able to perform a combined multicolor, spectroscopic and morphological analysis. The results presented in this paper clearly show the need for a full multiwavelength coverage to properly study and characterize the AGN population as a whole. As more data become available, the selection of COSMOS AGNs by all available means - X-ray, UV, optical, near-IR, and radio

- will build up to give the first bolometric selected AGN sample, fulfilling the promise of many years of multi-wavelength studies of quasars [26].

References

1. D.M Alexander, F.E. Bauer, W.N. Brandt et al: AJ **126**, 539 (2003)
2. R. Giacconi, A. Zirm, J. Wang, J. et al.: ApJS **139**, 369 (2002)
3. G. Hasinger, B. Altieri, M. Arnaud et al.: A&A, **365**, L45 (2001)
4. W.N. Brandt & G. Hasinger: ARA&A **43**, 827 (2005)
5. W.N. Brandt et al.: these proceedings
6. F. Fiore, M. Brusa, F. Cocchia et al.: A&A **409**, 79 (2003)
7. P. Green, J.D. Silverman, R.A. Cameron et al.: ApJS **150**, 43 (2004)
8. A.T. Steffen, A.J. Barger, P. Capak et al.: AJ **128**, 1483 (2005)
9. A.J. Barger, L.L Cowie, R.F. Mushotzky et al: AJ **129**, 578 (2005)
10. J.D. Silverman, P.J. Green, W.A. Barkhouse et al.: ApJ **624**, 630 (2005)
11. G. Hasinger, T. Miyaji, & M. Schmidt: A&A **441**, 417 (2005)
12. F. La Franca, F. Fiore, A. Comastri et al.: ApJ **635**, 864 (2005)
13. N.Z. Scoville et al.: ApJS, in press, (2006)
14. G. Hasinger, N. Cappelluti, H. Brunner et al.: ApJS, in press, (2006)
15. N. Cappelluti, G. Hasinger, H. Boehringer et al.: ApJS, in press, (2006)
16. M. Brusa, G. Zamorani, A. Comastri et al.: ApJS, in press, (2006)
17. T. Budavari, I. Csabai, A.S. Szalay et al: AJ **121**, 3266 (2001)
18. G. Zamorani, M. Mignoli, G. Hasinger et al.: A&A **346**, 731 (1999)
19. D.M Alexander, W.N. Brandt, A.E. Hornschemeier et al: AJ **122**, 2156 (2001)
20. M. Brusa, A. Comastri, E. Daddi et al.: A&A **432**, 69 (2005)
21. W.A. Barkhouse & P.B. Hall: AJ **121**, 2843 (2001)
22. V. Mainieri, G. Hasinger et al.: ApJS, in press, (2006)
23. V. Mainieri, J. Bergeron, G. Hasinger et al.: A&A **393**, 425 (2002)
24. P. Tozzi, R. Gilli, V. Mainieri et al.: A&A, **451**, 457 (2006)
25. S. Mateos, X. Barcons, F.J. Carrera et al.: A&A **444**, 79 (2005)
26. M. Elvis, B.J. Wilkes, J.C. McDowell et al.: ApJS **95**, 1 (1994)

Relativistic Iron Lines at High Redshifts

A. Comastri[1], M. Brusa[2], and R. Gilli[1]

[1] INAF-Osservatorio Astronomico di Bologna, via Ranzani 1, I–40127 Bologna, Italy,
(andrea.comastri,roberto.gilli)@oabo.inaf.it
[2] Max Planck Institut für Extraterrestrische Physik, Giessenbachstr. 1, D–85748 Garching,
Germany, marcella@mpe.mpg.de

Summary. The shape and the intensity of the 6.4 keV iron line bring unique information on the geometrical and physical properties of the supermassive black hole and the surrounding accreting gas at the very center of Active Galactic Nuclei. While there is convincing evidence of a relativistically broadened iron line in a few nearby bright objects, their properties at larger distances are basically unknown. We have searched for the presence of iron lines by fully exploiting *Chandra* observations in the deep fields. The line is clearly detected in the average spectra of about 250 sources stacked in several redshift bins over the range z=0.5-4.0. We discuss their average properties with particular emphasis on the presence and intensity of a broad component.

1 Introduction

Most of the accretion power, which makes Active Galactic Nuclei luminous hard X–ray sources, is released in the innermost regions around the central Supermassive Black Hole (SMBH), where the relativistic effects in the strong field limit (gravitational redshift and light bending) significantly affect the emerging spectrum. In particular the iron emission line at 6–7 keV is by far the most significant feature in the AGN X–ray spectra and a detailed study of its profile provides unique information about the gas properties and the nature of the spacetime in the proximity of the SMBH (see [1] for an exhaustive review including the most recent observational results).

A key issue concerns the shape of the line profile and in particular of the red wing, which is much broader if the emitting gas is accreting over a rapidly spinning SMBH. If this is the case, the innermost stable orbit is much closer ($\simeq 1.24 r_g$ where $r_g = GM/c^2$) to the central BH than for a non rotating BH ($\simeq 6 r_g$) and the emitted photons suffer from stronger general relativistic effects which are in principle visible in the line profile provided that enough sensitivity is reached in the 2–10 keV band.

The profile of the broad emission line in the nearby Seyfert galaxy MCG–6–30–15 represents a textbook case. The presence of a broad relativistic line with a skewed profile extending down to 4 keV (and during some episodes even at lower energies [2]), was first unambiguously determined by ASCA [3], and then confirmed by essentially all the X–ray experiments (i.e. *BeppoSAX*, *XMM–Newton* and *Chandra*). While the evidence for a line profile distorted by general relativistic effects is quite solid in

MCG–6–30–15, and tested against alternative solutions such as complex absorption
[4], the observational constraints on other nearby Seyfert galaxies are not such to rule
out alternative possibilities. Moreover the search for broad iron lines beyond the nearby
Universe is strongly hampered by the sensitivity of the present instruments and only a
few tentative positive results were reported [5, 6]

Deep XMM–*Newton* and *Chandra* observations offer the possibility to search for
iron line emission at much larger distances ($z > 0.5$) at least in a statistical sense, by
stacking the X–ray counts of a large number of X–ray sources.

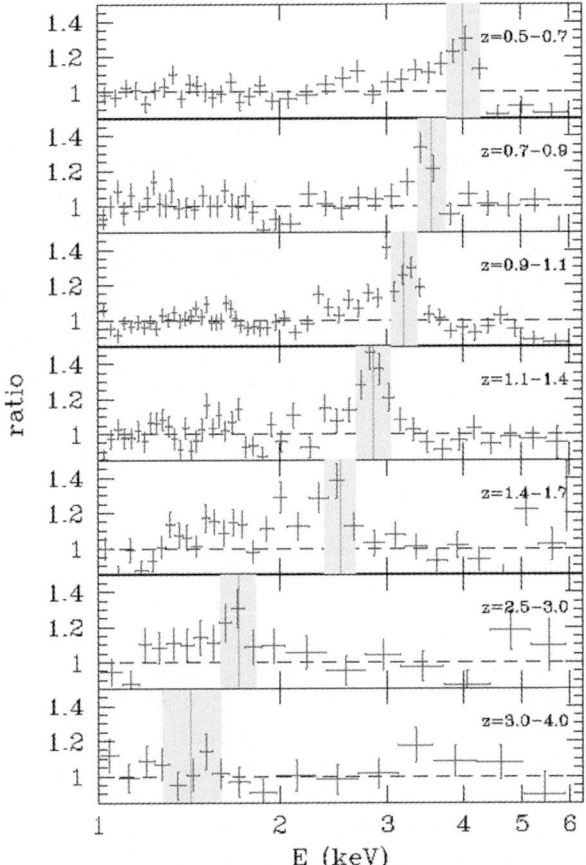

Fig. 1. Residuals of a simple power–law fit to the source spectra in seven different redshift bins
as labeled. The vertical line in each panel is at the expected position for the redshifted 6.4 keV
Fe $K\alpha$ line while the shaded region encompasses the bin width reported in Table 1 and defined
as $\Delta E = 6.4/(1+z_{max}) - 6.4/(1+z_{min})$ keV.

2 Stacking in XMM–*Newton* and *Chandra* Deep Fields

In the deep XMM–*Newton* pointing (~ 800 ksec) in the Lockman Hole the individual spectra of 53 Type 1 AGN and 43 Type 2 AGN for which a spectroscopic redshift is available, were brought to rest–frame and then summed together [7]. The residuals of a single power law fit to the stacked spectra clearly show a prominent iron line at 6.4 keV with a skewed profile extending towards lower energies. A good fit to both (type 1 and type 2) line profiles is obtained with a disc model with an innermost orbit around $\sim 3r_g$, suggesting that, on average, most of the SMBH in distant AGN are spinning. The above described approach maximizes the counting statistic in the energy band covered by the line, but at the same time loses, by definition, any information about the redshift dependence of line properties.

A complementary approach [8] has been pursued by stacking the X–ray counts, in the observed frame, of spectroscopically identified sources in the *Chandra* deep fields (CDFN & CDFS) in seven redshift slices spanning the z=0.5–4 range. The choice of bin sizes and distribution is driven by a trade–off between the number of counts in each bin and the need to sample an observed energy range narrow enough to detect the spectral feature, keeping at the same time the instrumental response as uniform as possible. The sample includes 171 sources in the CDFN and 181 in the CDFS, spanning the luminosity range $L_{2-10keV} = 10^{41} - 10^{45}$ ergs s^{-1}. With the exception of the highest redshift bin, a significant excess above a power–law continuum is present at the expected energy (see Fig. 1).

The residuals leftward of the iron line (which are more prominent in the bins with the highest number of sources and counting statistics z=0.5–0.7 and 0.9–1.1) suggest the presence of a broad redshifted component, similar to that observed by XMM [7].

3 Searching for Broad Lines in Deep Fields

The stacking results described above, motivated by two different key scientific objectives, were obtained following two different approaches in the stacking procedure. The XMM–*Newton* stacked spectra were obtained by unfolding the instrumental response with a power law model. In this way it is possible to shift and add the individual spectra and maximize the S/N ratio around the iron line energy. The drawback is a model dependent parameterization of the underlying continuum. The alternative strategy [8], following a more conventional approach, does not suffer of the model dependent unfolding of the instrumental response, but at the same time it is not well suited for a study of the line profile which is smeared in each redshift bin by the bin size itself. A third possibility which combines the pros of the two above mentioned methods would be to stack a large number of source spectra within a redshift range which is small enough that the energy (redshift) spread is negligible or at least smaller than the instrumental energy resolution.

The presence of significant spikes in the redshift distribution observed in both CDFS [9] and CDFN [10] has prompted us to further investigate such a possibility. The residuals of a single power law fit to the stacked spectra of the 37 X–ray sources in the CDFS redshift slice ($0.664 < z < 0.742$) and the 12 objects in the CDFN redshift slice

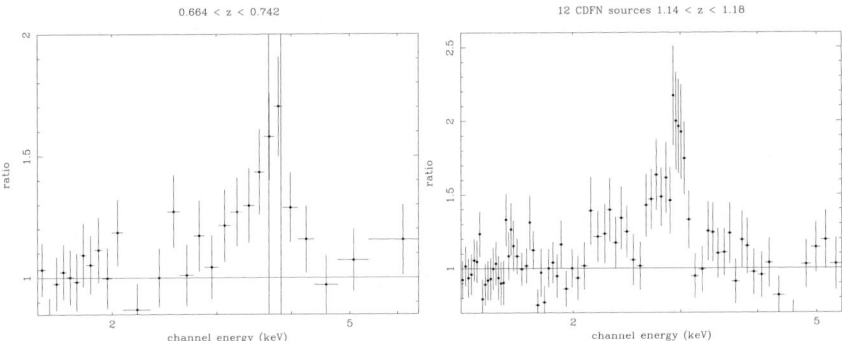

Fig. 2. *Left panel*: The residuals of a single power law fit to the stacked spectra of the 37 X–ray sources in the CDFS redshift slice $0.664 < z < 0.742$. *Right panel*: The fit and residuals of a single power law fit to the stacked spectra of the 12 objects in the CDFN redshift slice $1.14 < z < 1.18$.

$(1.14 < z < 1.18)$ are shown in Fig. 2. In both cases, a broad wing is clearly visible redward of a narrow core corresponding to the redshifted energy of the 6.4 keV line. In both samples the line shape is well fitted by a relativistic disk line model. However, the accuracy in the determination of the inner disk radius is not good enough to constrain the SMBH spin (Fig. 3). While this is not surprising given the available counting statistics, the similarity of the very shape of the observed profile to what is expected by a relativistic line is tantalizing. Although the perspectives of the detection of spinning BH at high redshift are promising a few words of caution appear to be appropriate. It is well known that the line intensity and especially the line profile are dependent on an accurate modelling of the underlying continuum and in particular are sensitive to the presence of complex absorption [1]. The latter possibility is especially relevant for the

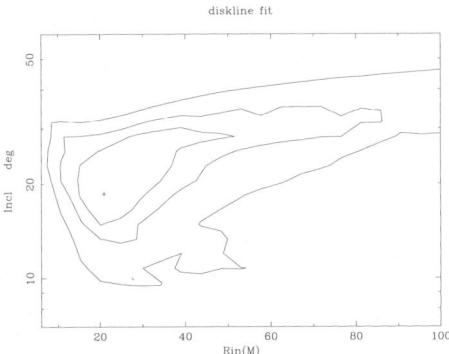

Fig. 3. 68, 90 and 99 % confidence contours of the disk inner radius (units of r_G and inclination angle (degrees)

analysis of the stacked spectrum of the deep field sources which are mostly obscured AGN. The 2–7 keV continuum resulting from the superposition of absorbed spectra with different redshifts is modified with respect to a single power law, by the absorption cut–off and most important by the 7.1 keV iron edge. The net effect of the latter can be approximated with a smeared edge which effectively enhances the X–ray continuum just below the iron line emission. The residuals of a single power law fit to the sources in the z=0.5–0.7 bin suggest the presence of an extended broad wing. (Fig. 4, left panel) which is almost completely accounted for by the continuum predicted by X–ray background synthesis models [11] in the same redshift bin (Fig. 4, right panel).

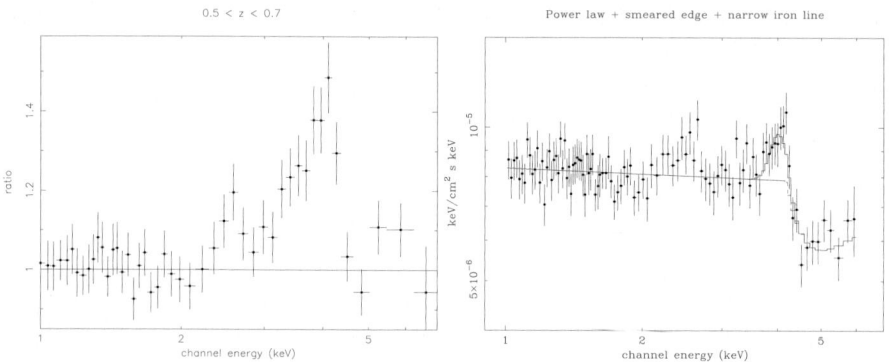

Fig. 4. *Left panel*: The residuals of a single power law fit to the *z*=0.5–0.7 bin adapted from [8]. *Right panel*: The unfolded spectrum of a complex absorption model fit to the same data set

4 Conclusions

The detection of spinning BH at high redshift is probably close to the capabilities of present instrumentation provided that deeper observations and/or larger samples of sources are collected and appropriately analyzed especially for what concern the modelling of the underlying continuum. Future missions with large collecting area will surely provide a step forward in such a direction.

References

1. A.C. Fabian, G. Miniutti: preprint (astro–ph/0507409), (2005)
2. K. Iwasawa, A.C. Fabian, Reynolds C.S. et al.: MNRAS, **282**, 1038
3. Y. Tanaka, K. Nandra, Fabian A.C. et al.: Nature, **375**, 659 (1995)
4. A.J. Young, J.C. Lee, A.C. Fabian, C.S. Reynolds, R.R. Gibson, C.R. Canizares: ApJ, **631**, 733 (2005)
5. A. Comastri, M. Brusa, F. Civano: MNRAS **351**, L9 (2004)
6. G. Miniutti, A.C. Fabian: MNRAS **366**, 115 (2006)
7. A. Streblyanska, G. Hasinger, A. Finoguenov, et al: A&A **432**, 395 (2005)
8. M. Brusa, R. Gilli, A. Comastri: ApJ **621**, L5 (2005)
9. R. Gilli, A. Cimatti, E. Daddi et al: ApJ **592**, 721 (2003)
10. A. J. Barger, L.L. Cowie, P. Capak et al: AJ **126**, 632 (2003)
11. R. Gilli, M. Salvati, & G. Hasinger: A&A **366**, 407 (2001)

An Explanation for the Soft X-Ray Excess in Active Galactic Nuclei

J. Crummy[1], A.C. Fabian[1], L.C. Gallo[2], and R.R. Ross[3]

[1] Institute of Astronomy, Madingley Road, Cambridge, CB3 0HA,UK
(jc,acf)@ast.cam.ac.uk
[2] Max-Planck-Institut für extraterrestrische Physik, Postfach 1312, 85741 Garching, Germany,
lgallo@mpe.mpg.de
[3] Physics Department, College of the Holy Cross, Worcester, MA 01610, USA,
rross@holycross.edu

Summary. We fit a large sample of type 1 AGN spectra taken with *XMM-Newton* with the relativistically blurred photoionized disc model of Ross & Fabian. This model is based on an illuminated accretion disc of fluorescing and Compton-scattering gas, and includes relativistic Doppler effects due to the rapid motion of the disc and general relativistic effects such as gravitational redshift due to presence of the black hole. The disc model successfully reproduces the X-ray continuum shape, including the soft excess, of all the sources. It provides a natural explanation for the observation that the soft excess is at a constant temperature over a wide range of AGN properties. We use the model to measure the rotation of the black holes.

1 Introduction

The soft excess is a major component of the X-ray spectra of many AGN and is present in every source in this survey. It is defined as the enhanced emission below ∼2 keV compared to a power law extrapolation of the approximately power law spectrum in the 2 − 10 keV band. This extra emission is generally approximately thermal in shape, and well fit with a black body of energy 0.1 − 0.2 keV. This result stands over several decades in AGN mass, e.g. [1, 2, 3]. This poses problems for the thermal interpretation of the data, as the temperature does not scale in the expected way with luminosity ($L \propto T^4$), as well as being too high for the standard model of AGN discs [4]. Several alternative models have been proposed to account for one or both of these results, none of which have yet been broadly accepted by the community. We investigate photoionized emission blurred relativistically by motion in an accretion disc, which has been previously studied by e.g. [5]. We use the latest models of Ross & Fabian [6], in which a semi-infinite slab of cold optically thick gas of constant density is illuminated by a power law, producing a Compton component and fluorescence lines from the ionized species in the gas. To produce the relativistically blurred photoionized disc model this reflected emission is added to the illuminating power law and the summed emission is convolved with a Laor [7] profile to simulate the blurring from an accretion disc around

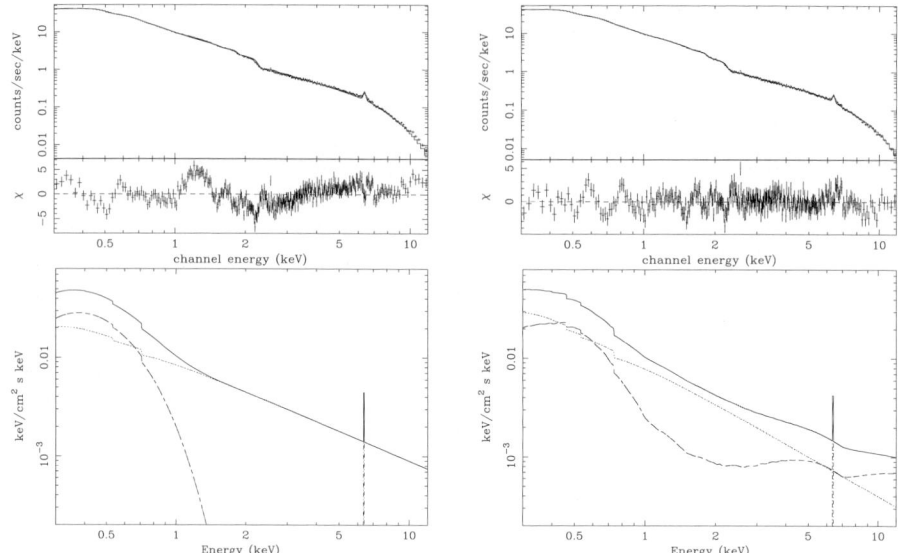

Fig. 1. This figure shows a comparison between the thermal model and the relativistically blurred photoionized disc reflection model for NGC 4051. The top panels show spectra and residuals to the thermal (left) and disc reflection (right) models, and the bottom panels show the components of the thermal (left) and disc reflection (right) models. Both include a power law (dotted line) and narrow iron emission (sharp line at ~6.4 keV), the components shown in dashed lines are a black body and the relativistically blurred disc reflection respectively. As the residuals show, the disc reflection model is a much better fit. Most of the remaining residuals to NGC 4051 in the soft band (top right panel, 0.5 – 1.0 keV) have been shown to be due to narrow line emission (Ponti et al., in prep).

a maximally rotating (Kerr) black hole. Since the model is physically motivated measuring the free parameters, such as the disc inclination, can give us information about the sources. An example of the model is shown in Figure 1, and the effect of the blurring is shown in Figure 2.

2 Data & Analysis

We used publicly available archival *XMM-Newton* data on 34 type 1 AGN with high-quality observations available. We produced spectra in the standard way (see [8] for details), using pn data exclusively.

We fit the data with two classes of models, the standard power law with a black body to model the soft excess, and the relativistically blurred photoionized disc reflection model[4]. Both models are subject to absorption from cold gas in our Galaxy, we also

---

[4] Models available from
http://www-xray.ast.cam.ac.uk/~jc/kdblur.html and
http://heasarc.gsfc.nasa.gov/docs/xanadu/xspec/models/reflion.html

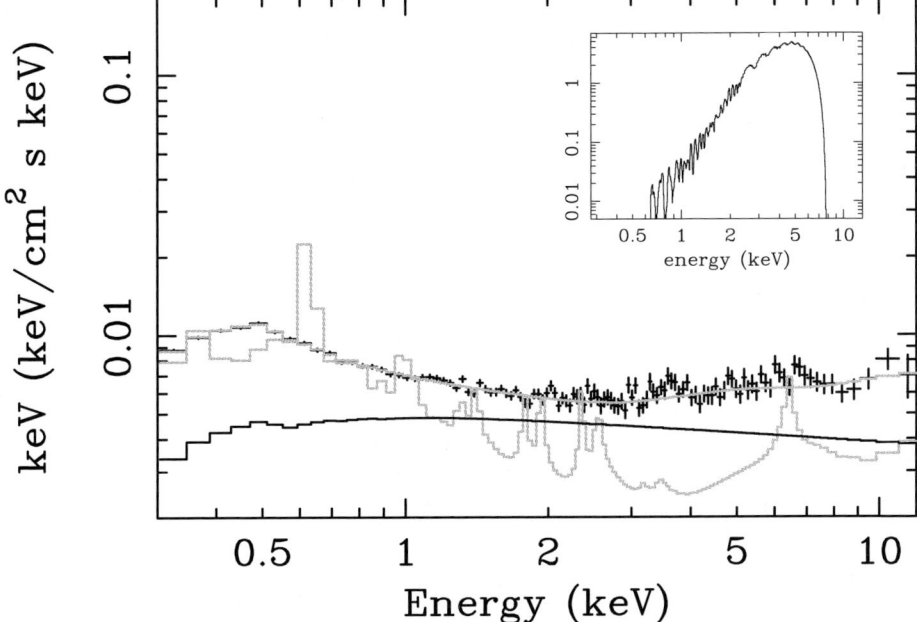

Fig. 2. An example disc reflection model fit plotted in νF_ν, showing the amount of flux emitted as a function of energy. The crosses are data, the data-following (grey) line is the fit, the smooth (black) line the power law component and the spiky (grey) line the reflection component. The insert shows the shape of the Laor line used in the convolution. The soft excess is shown to be the result of blurred line emission, and the extreme blurring effect on the iron lines around 6.4 keV is visible.

allow for absorbing matter at the AGN by fitting for extra cold absorption and up to two absorption edges in the 0.45 – 1.1 keV band. These components are redshifted such that they are local to the AGN. We finally allow for cold reflection (e.g. from distant gas such as a torus) by including a narrow iron line at 6.4 keV.

3 Results & Discussion

Our results are presented in full in [8]. One of the sources in the sample, PG 1404+226, has been investigated with the same model in a previous paper [9]. We find that the relativistically blurred photoionized disc reflection model is a better fit to the data, with 25 of the 34 sources showing an improvement in $\chi^2 > 2.7$ per degree of freedom (note that this does not correspond to 90 per cent probability, the χ^2 distribution is not calibrated across models). 6 sources show a significant worsening and 3 sources are inconclusive. In some sources the improvement is very marked, see Figure 1.

The disc reflection model reproduces the shape of the continuum well, and naturally explains the constant temperature of the soft excess. The model also reproduces many

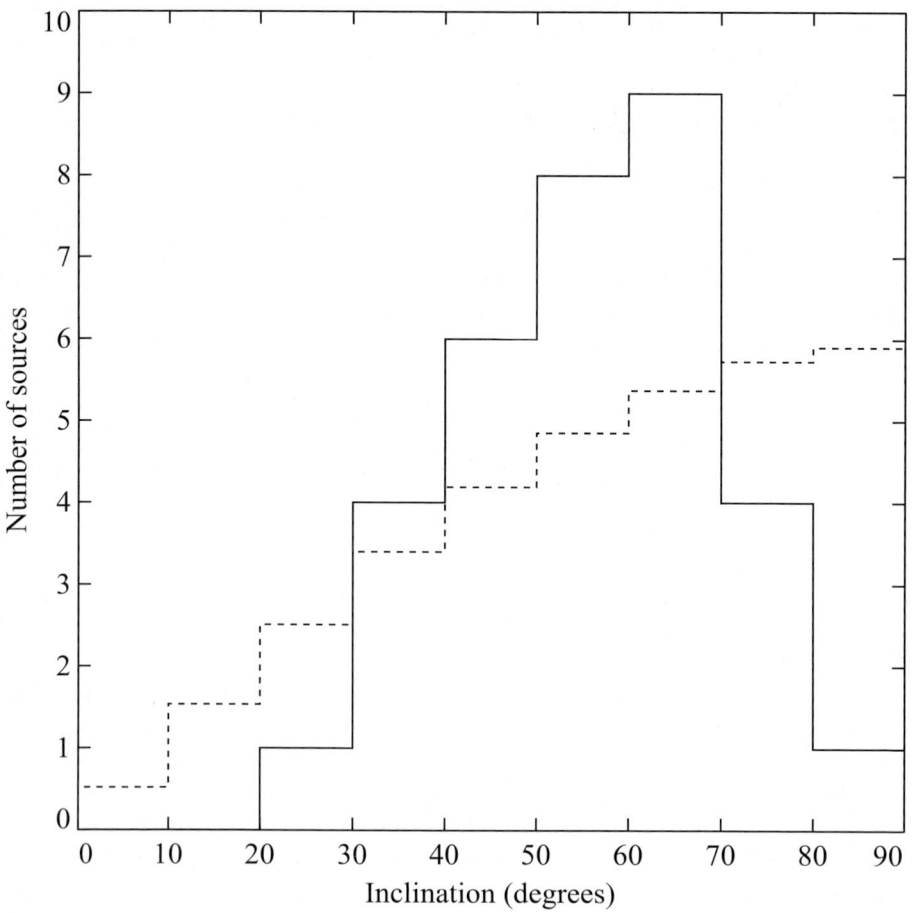

Fig. 3. This histogram shows the measured inclinations of the sources (the solid line) and the expected distribution if the inclinations were random (the dashed line). There is a deficit at high inclinations.

features that would be otherwise interpreted as absorption edges in medium resolution pn data. The disc reflection model is somewhat less smooth than the thermal model (see Figure 1), and bumps on the scale of absorption edges are possible. The model also explains the apparent lack of broad iron lines in most AGN spectra. Very few broad iron lines have been clearly detected (the most well studied being MCG –6-30-15, [10]). The results suggest that the broad iron lines do occur, but are broadened to the point of near undetectability. When fitting the soft excess and the iron line region of MRK 0586 independently with the model, consistent parameters are recovered. This illustrates how the hard and soft bands are connected, they key to the absence of obvious broad lines is in the shape of the soft excess. Conversely, it is clear that the soft excess is nothing more than a blend of many broad lines.

Our measured inclinations are plotted in Figure 3. We find that the inclinations are inconsistent with being random, with a low Kolmogorov-Smirnov (K-S) test probability. The data is somewhat consistent with being random over the range $0° - \sim 81°$. The deficit at high inclinations could be due to torus obscuration, with the sources that do have high inclination perhaps not having a torus, or with the torus unaligned with the central disc.

The rotation of the central black hole may be measured using this model. It is possible to use a model based on a non-rotating (Schwarzschild) black hole [11], instead of the maximally rotating black hole we use for other fits. We performed fits to all our sources using this non-rotating model, and found that in all cases the non-rotating model has a worse goodness of fit, with only two of the 34 sources where the fit is of comparable quality. This shows that rotating black holes dominate our sample. All the sources have inner disc radii consistent with being below 6 gravitational radii, and 29 of the 34 have inner radii consistent with being below 1.3 gravitational radii, which requires near maximal rotation. This is a strong indication that most black holes are maximally rotating, in agreement with theoretical predictions, e.g. Volonteri et al. [12] predict that 70 per cent of AGN black holes are maximally rotating.

One further interesting result from the fits is that the illuminating power-law component is often undetected. This might be due to relativistic effects hiding the illuminating continuum from us. If the source of the illuminating continuum is above the black hole, the light from it will be bent and impact on the disc, with little escaping to be observed [13]. Emission from the disc is Doppler beamed away from the black hole and will still escape. This primarily applies to sources which do not share the rotation velocity of the disc, e.g. the base of a weak jet.

Acknowledgments. ACF and RRR thank the Royal Society, and the College of the Holy Cross for support respectively. The *XMM-Newton* satellite is an ESA science mission (with instruments and contributions from NASA and ESA member states).

References

1. R. Walter, H.H. Fink, A&A, **274**, 105 (1993)
2. M. Gierliński, C. Done, MNRAS, **349**, L7 (2004)
3. D. Porquet, J.N. Reeves, P. O'Brien, W. Brinkmann, A&A, **422**, 85 (2004)
4. N.I. Shakura, R.A. Sunyaev., A&A, **24**, 337 (1973)
5. D.R. Ballantyne, K. Iwasawa, A.C. Fabian, MNRAS, **323**, 506 (2001)
6. R.R. Ross, A.C. Fabian, MNRAS, **358**, 211 (2005)
7. A. Laor, ApJ, **376**, 90 (1991)
8. J. Crummy, A.C. Fabian, L. Gallo, R.R. Ross, MNRAS, **365**,1067 (2006)
9. J. Crummy, A.C. Fabian, W.N. Brandt, Th. Boller, MNRAS, **361**, 1197 (2005)
10. Y. Tanaka, et al., Nature, **375**, 659 (1995)
11. A.C. Fabian, M.J. Rees, L. Stella, N.E. White, MNRAS, **238**, 729 (1989)
12. M. Volonteri, P. Madau, E. Quataert, M.J. Rees, ApJ, **620**, 69 (2005)
13. G. Miniutti, A.C. Fabian, MNRAS, **349**, 1435 (2004)

Extended Inverse-Compton Emission from Distant, Powerful Radio Galaxies

M.C. Erlund[1], A.C. Fabian[1], K.M. Blundell[2], A. Celotti[3], and C.S. Crawford[1]

[1] Institute of Astronomy, University of Cambridge, Madingley Road, Cambridge, CB3 0HA, UK, mce@ast.cam.ac.uk
[2] Department of Astrophysics, University of Oxford, Oxford, OX1 3RH, UK
[3] SISSA/ISAS, via Beirut 4, 34014 Trieste, Italy

1 Introduction

Inverse-Compton scattering of the cosmic microwave background (ICoCMB, CMB) in powerful high-redshift radio sources should be detectable across the universe, since the energy density of the CMB increases steeply with redshift, z, counterbalancing surface brightness dimming [1]. Most high-redshift radio galaxies should therefore have extended X-ray emission produced by ICoCMB, providing a direct tracer of lower Lorentz factor ($\gamma \sim 10^3$) electrons than those producing the radio synchrotron emission ($\gamma \sim 10^5 - 10^6$). These ICoCMB electrons will cool slower than their synchrotron-emitting counterparts, so ICoCMB emission also traces an older population of electrons. Several such sources have been detected [1, 2, 3, 4, 5, 6].

Here, *Chandra* observations of two high-redshift radio galaxies (3C 432 and 3C 191) are analysed with the aim of characterising their extended X-ray emission.

2 3C 432

A *Chandra* observation of 3C 432 ($z = 1.785$, RA 21h22m46.2s Dec +17d04m38s), which was 19.77 ks-long, was taken on 2005 January 7 in VFAINT (very faint) mode. Extended X-ray emission is clearly seen in Fig. 1 lying along the radio jet, not only in the lobes but also in the bridge. Spectra were extracted from the nucleus; the total extended X-ray emission (i.e. that within a $8'' \times 18''$ region aligned along the radio axis); the emission within the 1.54 GHz radio contours, excluding the central source, and the background. The nucleus has a photon index of $\Gamma = 1.84^{+0.09}_{-0.11}$ and intrinsic absorption $N_H = 2.1^{+0.36}_{-0.21} \times 10^{21}$ cm^{-2} assuming a Galactic absorption of 7.4×10^{20} cm^{-2}, giving a $2 - 10$ keV-band luminosity $L_x = 3.381 \times 10^{45}$ ergs^{-1}. The total extended emission has $\Gamma = 1.57^{+0.27}_{-0.36}$ and $L_x = 3.08^{+0.76}_{-0.94} \times 10^{44}$ ergs^{-1}, finally the extended emission within the 1.54 GHz contours has $\Gamma = 1.52^{+0.27}_{-0.48}$ and $L_x = 1.82^{+0.65}_{-0.60} \times 10^{44}$ ergs^{-1}. These values were calculated using C-statistics, however χ^2-statistics gives consistent values.

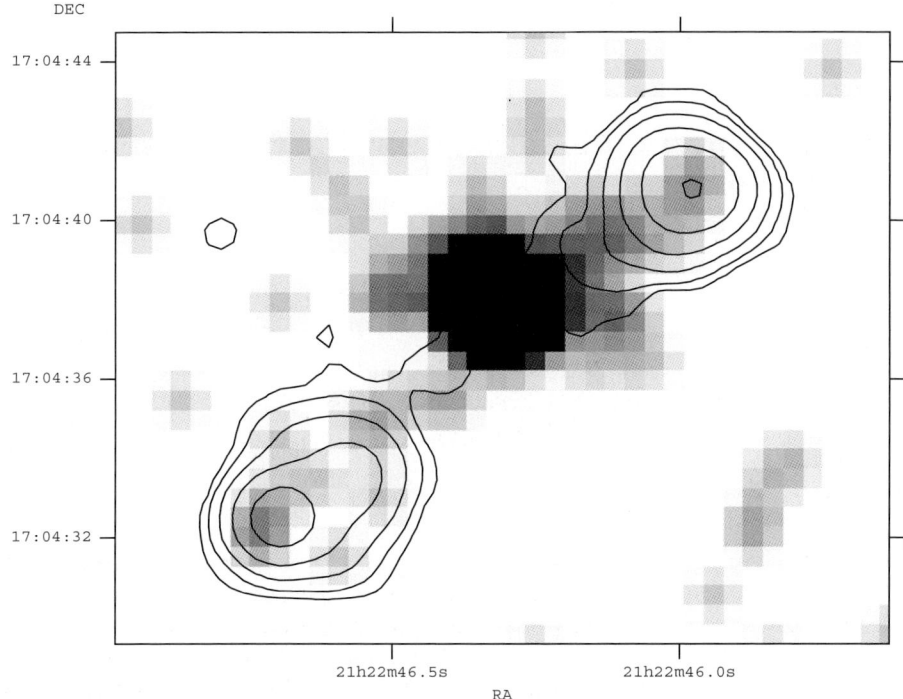

Fig. 1. Image of 3C 432, Gaussian-smoothed by 0.49", with 1.54 GHz radio contours overlaid (0.001, 0.003, 0.012, 0.042, 0.144, 0.5 Jy/beam).

3 3C 191

Two observations of 3C 191 have been analysed ($z = 1.965$, RA 08h04m47.9s Dec +10d15m23s): the first is 8.32 ks-long [5], taken with *Chandra* in FAINT mode on 2001 March 7 and the second is 19.77 ks-long, taken on 2004 December 12 in VFAINT mode. The extended X-ray emission is apparent in Fig. 2, which is the sum of the two observations, and is aligned along the direction of radio jet, continuing well beyond the radio emission to the south. The nucleus has an intrinsic absorbed power law with $N_H = 0.46^{+0.29}_{-0.35} \times 10^{22}$ cm^{-2}, $\Gamma = 1.79^{+0.12}_{-0.12}$ and $L_x = 4.04 \times 10^{45}$ ergs^{-1}. The total extended emission (6.5" × 10.5" region excluding the central source) has $\Gamma = 1.66^{+0.32}_{-0.29}$ and $L_x = 3.85^{+1.07}_{-1.30} \times 10^{44}$ ergs^{-1}, and the emission within the lowest 8.46 GHz radio contour shown in Fig. 2 has $\Gamma = 1.95^{+0.44}_{-0.20}$ and $L_x = 1.79^{+0.49}_{-0.58} \times 10^{44}$ ergs^{-1} (using C-statistics).

References

1. D. A. Schwartz. X-Ray Jets as Cosmic Beacons. ApJ **569**, L23 (2002)

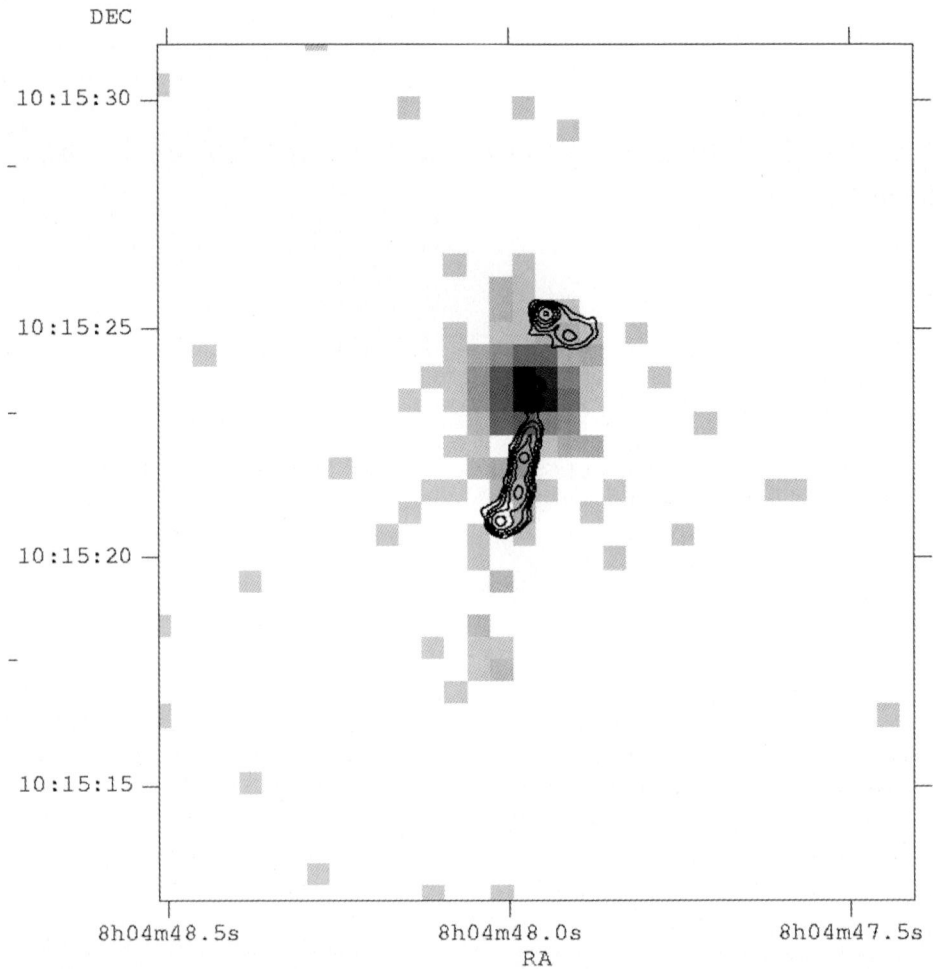

Fig. 2. X-ray image of 3C 191 with 8.46 GHz radio contours overlaid (0.0005, 0.0012, 0.0029, 0.0069, 0.0167, 0.04 Jy/beam).

2. E. Belsole, D. M. Worrall, M. J. Hardcastle, M. Birkinshaw, and C. R. Lawrence. XMM-Newton observations of three high-redshift radio galaxies. MNRAS **352**, 924 (2004)
3. R. A. Overzier, D. E. Harris, C. L. Carilli, L. Pentericci, H. J. A. Röttgering, and G. K. Miley. On the X-ray emission of $z \sim 2$ radio galaxies: IC scattering of the CMB and no evidence for fully-formed potential wells. A&A **433**, 87 (2005)
4. C. Scharf, I. Smail, R. Ivison, R. Bower, W. van Breugel, and M. Reuland. Extended X-Ray Emission around 4C 41.17 at z = 3.8. ApJ **596**, 105 (2003)
5. R. M. Sambruna, J. K. Gambill, L. Maraschi, F. Tavecchio, R. Cerutti, C. C. Cheung, C. M. Urry, and G. Chartas. A Survey of Extended Radio Jets with Chandra and the Hubble Space Telescope. ApJ **608**, 698 (2004)
6. K. M. Blundell, A. C. Fabian, C. S. Crawford, M. C. Erlund, and Celotti A. Discovery of the low-energy cutoff in a powerful radio galaxy. ApJ **644**, L13 (2006)

The Most Distant Radio Quasar as seen with the Highest Resolution

S. Frey[1], Z. Paragi[2], L. Mosoni[3], and L.I. Gurvits[2]

[1] FÖMI Satellite Geodetic Observatory, Budapest, Hungary, frey@sgo.fomi.hu
[2] Joint Institute for VLBI in Europe, Dwingeloo, The Netherlands
[3] Konkoly Observatory, Hungarian Academy of Sciences, Budapest, Hungary

Summary. We discuss our VLBI imaging observations of J0836+0054, the highest redshift ($z = 5.774$) radio-loud quasar known at present.

1 Introduction

SDSS J0836+0054 is one of the highest redshift quasars known to date discovered using the multicolor imaging data from the Sloan Digital Sky Survey (SDSS, [1]). Its redshift is $z = 5.774$ [7]. This is the only object in the sample of nineteen $z > 5.7$ SDSS quasars found to date ([2] and references therein) that is identified with a previously known radio source. J0836+0054 therefore provides a unique possibility to study a radio quasar at an extremely high redshift, and to compare its high-resolution structure with that of the quasars in lower-redshift samples. The compact radio emission in radio-loud active galactic nuclei (AGN) is known to originate from pc-scale jets in the close vicinity of the central black hole via incoherent synchrotron process.

The mass of the central black hole in J0836+0054 is estimated as $M = 4.8 \times 10^9 M_\odot$ [1]. There is growing evidence from the last couple of years that supermassive black holes have already formed at early cosmological epochs, less than 1 Gyr after the Big Bang. The apparent absence of strong gravitational lensing for $z \sim 6$ quasars suggests that their central black hole mass estimates based on the Eddington argument are not seriously biased.

2 Observations

We observed the quasar J0836+0054 with ten radio telescopes of the European Very Long Baseline Interferometry (VLBI) Network (EVN) at 1.6 GHz and 5 GHz frequency. The observations took place on 8 June 2002 and 4 November 2003. Since the source is very weak (~ 1 mJy or less), we employed the technique of phase-referencing. By regularly switching between the target and a nearby bright and compact reference source within the atmospheric coherence time, and interpolating solutions derived for the calibrator, sufficiently long integration could be achieved in order to reach a low image noise level. The total on-source times were almost 3 and 4 hours at 1.6 and 5 GHz, respectively. The radio images obtained are shown in Fig. 1. Details of the observing and the data reduction are given by [3, 4].

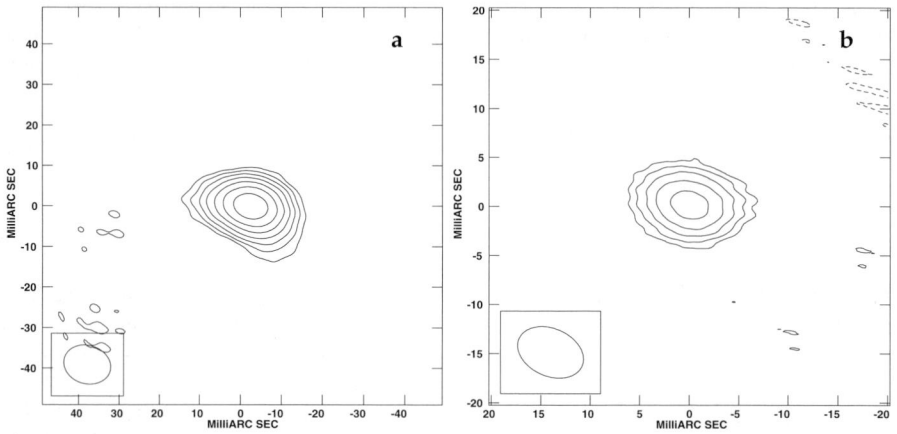

Fig. 1. Naturally weighted 1.6-GHz **(a)** and 5-GHz **(b)** EVN images of J0836+0054 [3, 4]. The positive contour levels increase by a factor of $\sqrt{2}$. **(a)** The first contours are drawn at -75 and 75 μJy/beam. The peak brightness is 770 μJy/beam. The restoring beam is 11.9 mas $\times$ 9.4 mas at PA=74°. **(b)** The first contours are drawn at -70 and 70 μJy/beam. The peak brightness is 333 μJy/beam. The restoring beam is 7.0 mas $\times$ 4.8 mas at PA=66°. The images are centered at R.A. $8^h 36^m 43.86020^s$ and Dec. $+0°54'53.2290''$ (J2000), the position determined with 2-mas uncertainty [4].

In order to assess the fraction of the flux density originating from the milli-arcsecond (mas) scale compact core, we conducted observations with the NRAO Very Large Array (VLA) on 5 November 2003. The results indicate that the quasar is indeed very compact: its total radio emission comes from within the region imaged with VLBI, i.e. the central 40 pc (assuming $\Omega_m = 0.3$, $\Omega_\Lambda = 0.7$ and $H_0 = 65$ km s^{-1} Mpc^{-1} for calculating the linear size).

3 Results and discussion

Our VLBI images indicate that J0836+0054 is compact at mas angular scales. The structure is reminiscent of a typical radio "core" often observed with VLBI in radio-loud AGN. We did not find any indication of multiple images produced by gravitational lensing. Based on the observed high-resolution radio structure and the radio luminosity, this extremely distant quasar appears similar to other powerful AGN at much later cosmological epochs. The estimated black hole mass, the radio and X-ray luminosities place J0836+0054 very close to the best-fitting "fundamental plane" of black hole activity derived by [5].

Various optical, infrared, X-ray observations, as well as our radio measurements suggest that the few $z \sim 6$ quasars known at present are basically "indistinguishable" from their lower-redshift cousins, as far as their physical parameters (luminosity, black hole mass, metallicity, etc.) are concerned. This is a challenging constraint for cosmological structure formation models. However, recent numerical simulations suggest that hierarchical structure formation in a cold dark matter dominated universe can indeed lead to the existence of massive galaxies and quasars as early as $z \sim 15$ [6]. The descendants of the first supermassive black holes are now in the middle of rich galaxy clusters. According to these simulations, quasars at $z \sim 6$ lie in the center of very massive dark matter halos, surrounded by a large number of fainter galaxies. As an observational support, most recently a factor of ~ 6 relative overdensity of associated faint galaxy candidates was found in the 5 square arcminute field around J0836+0054, using HST ACS data [8].

Acknowledgments. This research was supported by the Hungarian Scientific Research Fund (OTKA, grant T046097) and the European Commission's FP6 I3 Programme RadioNet under contract No. 505818. SF acknowledges the Bolyai Research Scholarship received from the Hungarian Academy of Sciences. The European VLBI Network is a joint facility of European, Chinese, South African and other radio astronomy institutes funded by their national research councils. The National Radio Astronomy Observatory is a facility of the National Science Foundation operated under cooperative agreement by Associated Universities, Inc.

References

1. X. Fan, V.K. Narayanan, R.H. Lupton et al: Astron. J. **122**, 2833 (2001)
2. X. Fan, M.A. Strauss, G.T. Richards et al: AJ **131**, 1203 (2006)
3. S. Frey, L. Mosoni, Z. Paragi, L.I. Gurvits: Mon. Not. R. Astron. Soc. **343**, L20 (2003)
4. S. Frey, Z. Paragi, L. Mosoni, L.I. Gurvits: Astron. Astrophys. **436**, L13 (2005)
5. A. Merloni, S. Heinz, T. Di Matteo: Mon. Not. R. Astron. Soc. **345**, 1057 (2003)
6. V. Springel, S.D.M. White, A. Jenkins et al: Nature **435**, 629 (2005)
7. D. Stern, P.B. Hall, L.F. Barrientos et al: Astrophys. J. **596**, L39 (2003)
8. W. Zheng, R. Overzier, R.J. Bouwens et al: AJ **640**, 574 (2006)

Investigating Narrow-Line Seyfert 1 with X-Ray Spectral Complexity

L. C. Gallo[1,2] and I. Balestra[1]

[1] Max-Planck-Institut für extraterrestrische Physik, Postfach 1312, 85741 Garching, Germany,
lgallo@mpe.mpg.de, balestra@mpe.mpg.de
[2] Institute of Space and Astronautical Science, Japan Aerospace Exploration Agency,
Yoshinodai 3-1-1, Sagamihara, Kanagawa 229-8510, Japan

1 The high-energy spectral complexity in narrow-line Seyfert 1 galaxies

The high sensitivity achieved with *XMM-Newton* in the $2 - 10$ keV range has revealed spectral complexity in several narrow-line Seyfert 1 galaxies (NLS1). This complexity was first realised in 1H0707–495 (Boller et al. 2002) as a sharp spectral drop at about 7 keV. Observations of NLS1 over the years have lead to the discovery of several objects, which display similar features at energies between $\sim 7 - 9$keV, or more generally, spectral curvature between $2 - 10$ keV. The exact nature of this behaviour, which seems to be a characteristic of NLS1, is uncertain. Two models which appear probable are partial-covering (e.g. Tanaka et al. 2004 and references within) and reflection (e.g. Fabian et al. 2002).

In terms of partial-covering, the ~ 7 keV drop is produced by absorption of the continuum emission by a dense material, which only partly obscures the primary emitter. This can, in principle, explain the absence of other absorption features (e.g. intrinsic cold absorption, Fe L absorption, fluorescence emission), and if the absorber is allowed to be in radial motion, it can possibly account for the various edge energies which are seen (e.g. Gallo et al. 2004).

Reflection of the power law continuum source off the cold accretion disc can also adequately describe the X-ray spectra of NLS1 (e.g. Fabian et al. 2002). In this case, the sharp drop at high energies is the blue wing of a relativistically broadened iron line.

It stands to reason that regardless of the correct model, the process may be ubiquitous in NLS1 and probably in active galactic nuclei (AGN). By varying the prevalence of the physical process (e.g. the degree of absorption or amount of reflection) one can potentially describe the different types of X-ray spectra that are observed.

In this work we examine what, if anything, is unique about the NLS1 which appear to possess high-energy complexity. Complete analysis and results are presented in Gallo (2006).

2 Sample definition

Examining all available *XMM-Newton* observations of NLS1 we compiled a sample of 28 which satisfied our criteria for high signal-to-noise and broadband coverage (see Gallo 2006). From this parent sample we identified 7 NLS1, which exhibited the X-ray complexity described above (hereafter known as the C-sample; filled circles in figures). The remaining objects formed the general or simple NLS1 (S-sample; open squares in figures).

3 Results

For the two samples described in Sect. 2 we compared various multiwavelength para-menters to determine in exactly what way they differed. In most cases, the two samples were comparable, but there were some parameters that showed discrepancies.

3.1 X-ray flux dependence on sample definition

We examined the long-term X-ray light curves of each source utilising data from past missions. We found that the S- and C-samples defined in this work are representative of NLS1 in X-ray typical- and low-flux states, respectively. It follows that objects could transit from one sample to another depending on their X-ray flux state.

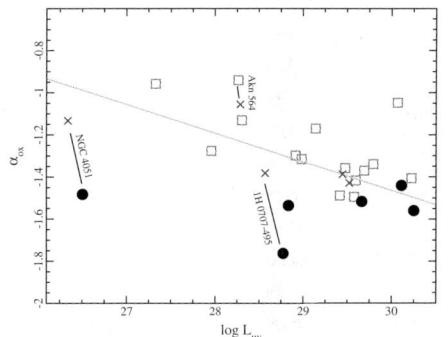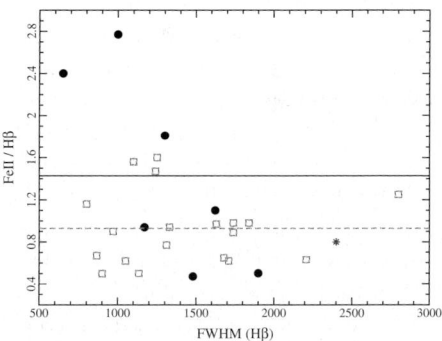

Fig. 1. In each figure the members of the S- and C-sample are shown as black dots and red squares, respectively. *Left panel*: α_{ox} as a function of 2500 Å luminosity measured from contemporaneous X-ray and UV data. For comparison, the average relation for AGN (Stateva et al. 2005) is shown as the solid line. The S-sample appears to follow the Strateva relation well, but the complex NLS1 are X-ray weak. Multiple *XMM-Newton* observations exist for some of the objects in our sample. These are shown as blue crosses. The α_{ox} variations displayed by 1H 0707–495 and NGC 4051 indicate significant changes in the X-ray flux and more moderate variations in the UV. *Right panel*: The Fe II/Hβ ratio is plotted against the FWHM(Hβ). The average ratio for the S-sample and the C-sample are marked by a dashed and solid line, respectively. Although there is considerably more scatter in the average ratio of the complex NLS1 sample, the objects appear to possess much stronger Fe II emission.

3.2 X-ray weak NLS1

By utilising contemporaneous UV and X-ray data to measure the optical-to-X-ray spectral index (α_{ox}), we determined that objects that possess complexity in their $2-10\,\mathrm{keV}$ spectrum appear to be X-ray weaker (i.e. they are found below the average $L_{uv}-\alpha_{ox}$ relation; Figure 1 left panel). At least in two of the objects (NGC 4051 and 1H 0707–495) for which multiple, simultaneous, X-ray/UV observations are available, extreme epoch-to-epoch changes are seen in α_{ox}, demonstrating that when the spectrum appears less complex, X-ray emission is stronger and α_{ox} is more typical.

3.3 Extreme Fe II emission

Not all of the NLS1 in the C-sample possess strong optical Fe II emission (specifically Fe II/Hβ), but it does appear that the most extreme Fe II emitters (e.g. Fe II/H$\beta > 1.7$) are spectrally complex (right panel Figure 1). It could be that objects with extreme Fe II emission do possess the mechanism to create high-energy complexities. A simple test would be to make X-ray observations of a few extreme Fe II emitters, say with Fe II/H$\beta > 1.7$, as suggested by Figure 1, and determine if their high-energy spectra are complex.

However, the fact that a source can transit from one sample to another depending on its X-ray flux state (Section 3.1) complicates this assertion since the variability behaviour of Fe II is not well understood. It is also possible that NLS1 with extreme Fe II emission simply show the greatest variability; therefore making it more likely of catching these objects in an extreme (low) flux state.

References

1. T. Boller, A. Fabian, R. Sunyaev et al: MNRAS **329**, 1 (2002)
2. A. Fabian, D. Ballantyne, A. Merloni et al: MNRAS **335**, 1 (2002)
3. L. C. Gallo, Y. Tanaka, T. Boller et al: MNRAS **353**, 1064 (2004)
4. L. C. Gallo: MNRAS **368**, 479 (2006)
5. I. Strateva, W. Brandt, D. Schneider et al: AJ **130**, 387 (2005)
6. Y. Tanaka, T. Boller, L. C. Gallo et al: PASJ **56**, 9 (2004)

A Survey of Gaussian Flares in AGN

M. Guainazzi[1], P. Rodriguez-Pascal[1], and F. Favata[2]

[1] European Space Astronomy Center of ESA, Apartado 50727, E-28080 Madrid, Spain,
   Matteo.Guainazzi@sciops.esa.int
[2] Astrophysics Division, ESTEC, Postbus 2200, Noordwijk, The Netherlands

1 Description of the study and motivation

We present in this paper preliminary results of a spectroscopic study of a sample of isolated X-ray flares in Active Galactic Nuclei. The motivation of this study is to provide clues on the nature of X-ray variability in AGN. Although there is nowadays general consensus that the physical processes responsible for the often extreme X-ray variability observed in AGN occurs in the accretion flow onto the supermassive black holes, there is little consensus about its mechanism and the ultimate physical driver. The now almost twenty years old controversy between X-ray disk illumination and accretion flow instabilities is still basically unresolved ([4]), despite several multiwavelength monitoring programs ([3]). The recent discovery of "narrow" blue-shifted features in the X-ray spectra of nearby Seyfert Galaxies has revived the hypothesis of localized "flares" illuminating a small portion of the accretion flow, whose combination produce the observed overall X-ray variability ([2]), possibly in analogy to the shot noise observed in Galactic X-ray binaries ([1]).

To address this issue, we have selected a sample of "isolated flares" detected in light curves of XMM-Newton observations of AGN. We have not attempted at an "objective" definition of what an isolated flare is. It suffices here to day that we consider hereafter events, where a trailing and leading edge above a constant quiescent level can be recognized.

2 Results

The preliminary results can be summarized as follows:

- we have identified 6 flares in the observations of 5 AGN: (ESO 0244-G17; Mkn 493, NGC4051, NAB0205+024, PG1211+143; see Fig. 1)) with typical duration of 10^4 s (there is an obvious selection effect favoring the detection of these timescales due to the typical length of an XMM-Newton observation)
- they cover a wide range in luminosity, from 10^{40} erg s^{-1} to 10^{43} erg s^{-1}
- they are generally softer than the quiescent emission (see Fig.2)

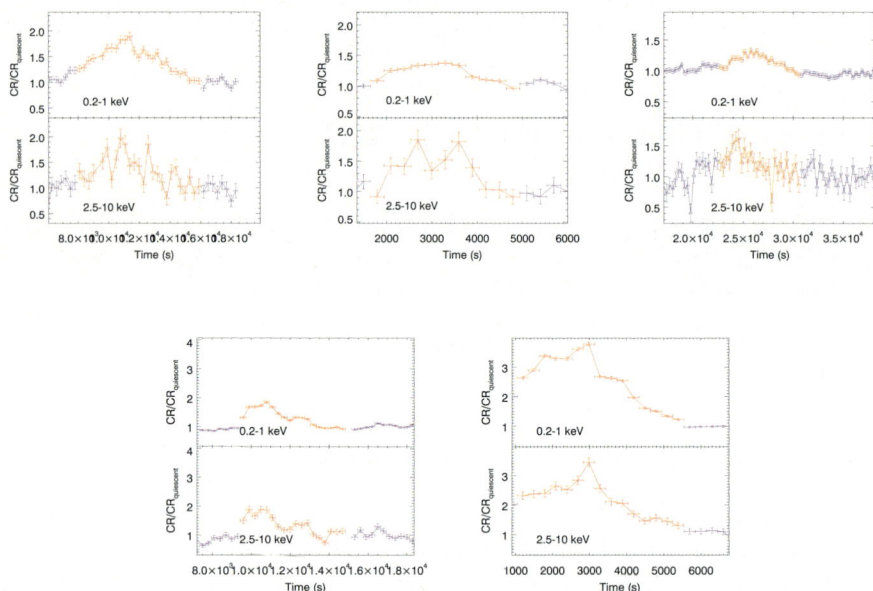

Fig. 1. Flare profiles. From *top left* to *bottom right*: ESO0244-G17, Mkn 493, PG1211+143, NGC 4051 (last two flares)

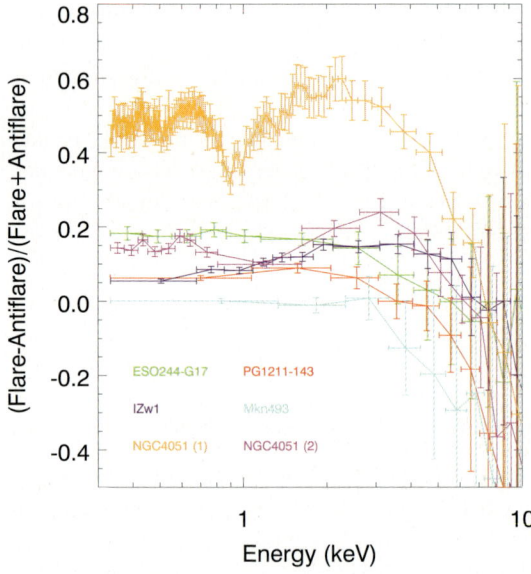

Fig. 2. Normalized ratio between the flare and the quiescent spectra

- with only one exception, there is no evidence for cross-correlation lags between the flare soft and the hard X-ray (see Fig. 3), suggesting that they represent single

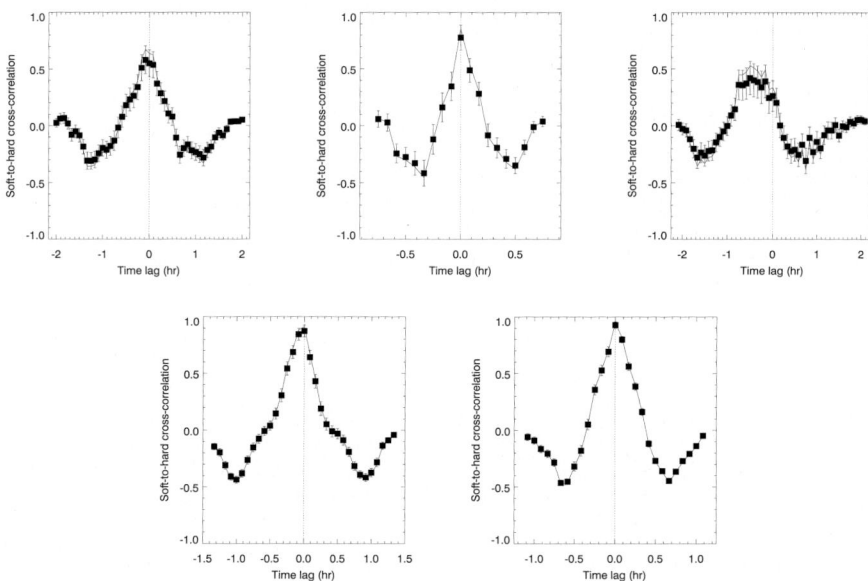

Fig. 3. Cross-correlation between the 0.2–1 keV and the 3–10 keV energy bands for the light curves of Fig. 1

explosive events, not modified by further Comptonization

References

1. L. Burderi, N.R. Robba, N. La Barbera, M. Guainazzi, ApJ, **481**, 943 (1997)
2. A. Merloni, A.C. Fabian, MNRAS, **328**, 958 (2001)
3. K. Nandra, I. Papadakis, ApJ, **554**, 710 (2001)
4. P. Uttley, et al., ApJ, **584**, L53 (2003)

A Simple Model for Quasar Density Evolution

H. Horst[1,2] and W.J. Duschl[2,3]

[1] European Southern Observatory, Casilla 19001, Santiago 19, Chile, hhorst@eso.org
[2] Institut für Theoretische Astrophysik, Albert-Ueberle-Str. 2, 69120 Heidelberg, Germany, wjd@ita.uni-heidelberg.de
[3] Steward Observatory, The University of Arizona, 933 N. Cherry Ave, Tucson, AZ 85721, USA

Summary. It is widely agreed upon that AGN and Quasars are driven by gas accretion onto a supermassive black hole. The origin of the latter however still remains an open question. In this work we present the results of an extremely simple cosmological model combined with an evolutionary scenario in which both the formation of the black hole as well as the gas accretion onto it are triggered by major mergers of gas-rich galaxies. Despite its very generous approximations our model reproduces the quasar density evolution in remarkable agreement with observations.

1 Introduction

While it is widely agreed upon that galaxy-galaxy interactions and, in particular, mergers play a crucial role for the growth of supermassive black holes in quasars and also for providing the fuel for quasar activity, there is still dispute whether these black holes are of primordial nature or not. N-body simulations (e.g. by [1]) show that the tidal forces in interacting galaxies can trigger strong gas inflows towards the center of the merger. The mass of this gas is sufficient to build a supermassive BH of $10^7 \ldots 10^{10} M_\odot$ and provide enough fuel for quasar activity. On the basis of the β-viscosity model by [3] calculations by [4] show that it takes less than 10^9 a from the merger to form a fully developed quasar even if no primordial supermassive BH was originally present. This model can be tested by comparing the resulting co-moving space density of quasars to those derived from observations (e.g. [5]).

2 Outline of the model

Based on results by [4] we assume an average time delay of $5 \cdot 10^8$a between the merger and the peak quasar activity. The individual values of this delay depend, of course, on details like the size (s_{disk}) and initial mass (M_{disk}) of the disk. Estimates of the relevant (viscous) timescale show its dependence $\propto s_{\mathrm{disk}}^{3/2} M_{\mathrm{disk,i}}^{1/2}$. This leads to the – at first glance surprising—finding of the faster formation of the more massive black holes. For our present purpose, however, an average value of this delay serves its purpose. For more details of this model, we refer the reader to the contribution by Duschl and Strittmatter in this volume and in [4].

For our purpose it is sufficient to use a "test universe" comprised of 50 000 galaxies. This test universe is expanding according to an Einstein-de Sitter-cosmology with a Hubble constant of $H_0 = 72 \frac{km}{s \cdot Mpc}$. Our galaxies are treated as particles with a finite cross section and a thermal velocity dispersion of $v_s = 300 \frac{km}{s}$. In this framework we compute the merger rate for each simulated time bin ($10 \cdot 10^6$ a) by taking into account two different processes: Direct geometrical hits and gravitationally driven mergers. For the first process we assume every galaxy to have a spherical cross section with a radius of 15 kpc. For the second process the cross section radius is $r_{cs} = 2G^* M v_{rel}^{-2}$, with M being the mass of one galaxy and v_{rel} being the relative velocity between both. We assume a delay between the merger and the onset of the quasar phase of $5 \cdot 10^8$ a for the geometrical hit and $1 \cdot 10^9$ a for the gravitationally driven merger. The quasar phase in turn is assumed to last for another $5 \cdot 10^8$ a.

In our model the Universe is treated in very simple manner: As an expanding spherical box. The size of this box is determined by deriving today's matter density from the assumed Hubble constant ($\left(\frac{8}{3}\pi G^* \rho_0\right)^{1/2} = H_0$) and then using the Friedmann-Lemaître-equation to calculate the according radius at any given time. We completely neglect structure formation and start our simulation with large galaxies ($M = 10^{11} M_\odot$) already in place $2 \cdot 10^8$a after the Big Bang. To account for the formation of galaxies we increase their number over the first $5 \cdot 10^8$ a of our simulation until the final number of 50 000 is reached (see fig. 1 for the effect of this procedure).

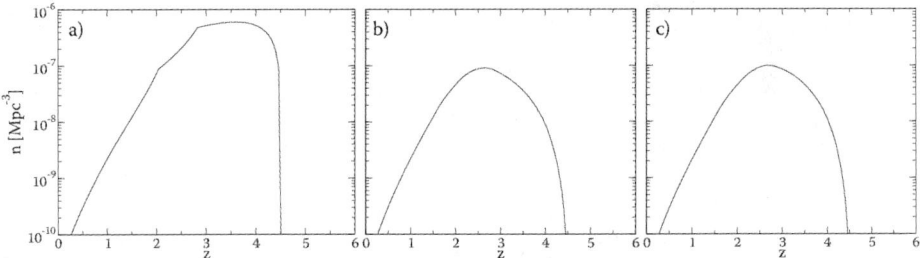

Fig. 1. Co-moving space density of quasars as computed by our simulation. In panel a) all galaxies are in place at the start of the simulation. In panels b) and c) the number of galaxies is gradually increased over the first $5 \cdot 10^8$a. In panel b) a constant galaxy formation rate and in panel c) a Gaussian formation rate was used – the resulting co-moving space density of quasars is almost exactly the same.

3 Results

The comparison between our model results and observational data compiled by [5] is shown in fig.2. The quasar density evolution at lower redshifts and the position of the peak at $z \approx 2.5$ match very well. Please note that the X-ray selected sample contains lower luminosity AGN in addition to quasars. In this respect it is natural that our results resemble the 2dF curve (from [2]) better than the data from [5]. At higher redshifts the deviation between simulated and observed co-moving spatial density is increasing. In our model we clearly miss the earliest quasars which arise from exceptionally fast evolving mergers. Despite this shortcoming of our results their overall agreement with observational data is remarkable.

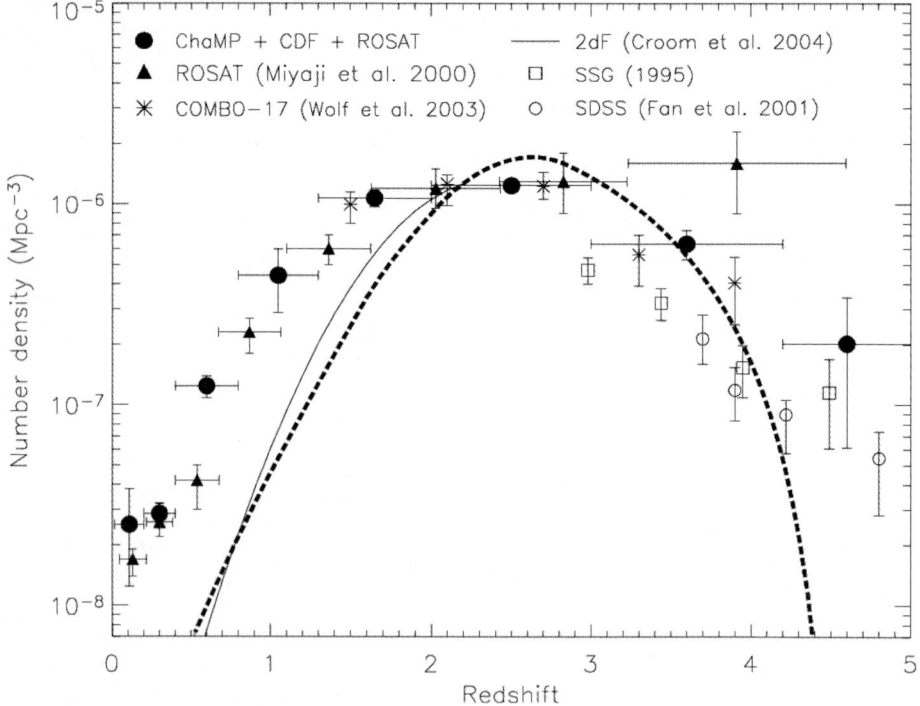

Fig. 2. Comparison of our simulated quasar density evolution with observational results. The broken line is the same as in panel b) of Fig. 1, while the underlying graph has been taken from [5]. The theoretical curve has been scaled to match the 2dF results.

References

1. J. E. Barnes, L. E. Hernquist: ApJ **471**, 115 (1996)
2. S. M. Croom, et al.: MNRAS **349**, 1397 (2004)
3. W. J. Duschl, P. A. Strittmatter, P. L. Biermann: A&A **357**, 1123
4. W. J. Duschl, P. A. Strittmatter: in prep. (2006)
5. J. D. Silverman, et al.: ApJ **624**, 630 (2005)

The Dispersion of the MIR – Hard X-ray Correlation in AGN

H. Horst[1,2], A. Smette[1,3], P. Gandhi[1,4], and W. J. Duschl[2,5]

[1] European Southern Observatory, Casilla 19001, Santiago 19, Chile,
(hhorst,asmette,pgandhi)@eso.org
[2] Institut für Theoretische Astrophysik, Albert-Ueberle-Str. 2, 69120 Heidelberg, Germany,
wjd@ita.uni-heidelberg.de
[3] F.N.R.S., Rue d'Egmont 5, 1000 Bruxelles, Belgium
[4] Institute of Astronomy, Madingley Road, Cambridge CB3 0HA, UK
[5] Steward Observatory, The University of Arizona, 933 N. Cherry Ave, Tucson, AZ 85721,
USA

Summary. We present new 12.5μm IR photometry (MIR) of 8 AGN, obtained with VISIR on the VLT and study the dispersion of the correlation between their MIR and hard X-ray luminosities. The unified scenarios predict that hard X-rays can provide a measure of the bolometric luminosity of the AGN. On the other hand, the nuclear infrared continuum is due to AGN emission reprocessed by dust in a torus-like geometry. Our results imply that the dispersion in the relation between the hard X-ray and mid infrared luminosities of AGN is much smaller than that found in earlier studies. While our sample is still too limited to yield statistically robust results, our work demonstrates the importance of high angular resolution for the investigation of the IR properties of AGN.

1 Introduction

The unification model for active galactic nuclei (AGN) attributes the different appearances of Seyfert 1 (Sy1) and Seyfert 2 (Sy2) galaxies uniquely to an orientation effect [1]. The basic idea is that the central engine is surrounded by an optically and geometrically thick molecular torus. Associated with this torus are large masses of dust which supposedly reprocess the X-ray emission from the accretion disk and re-emit it in the MIR regime.

In consequence it is very attractive to correlate the IR continuum of AGN with hard X-ray observations in order to test the unification scenarios. Recently Lutz et al. found a correlation between MIR luminosity (measured with ISOPHOT) and absorption-corrected hard X-ray luminosity for a sample of 71 AGN [3]. However, they reported two problems that the unification model faces when confronted with their observations. (i) The scatter of this relation is about an order of magnitude larger than expected from a previous study [2]. (ii) Contrary to what one expects, if an optically and geometrically thick torus dominates an AGN's MIR continuum, they did not find a significant difference between type I and II in their average MIR – hard X-ray – ratio.

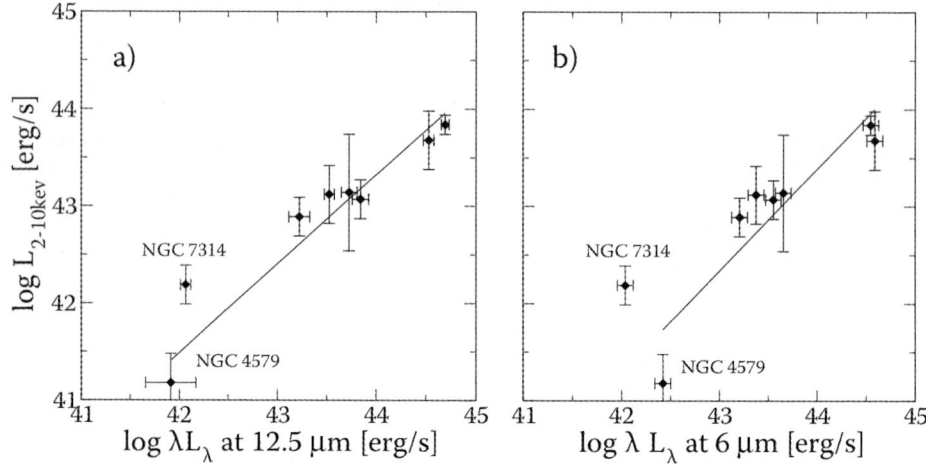

Fig. 1. Correlation of MIR – hard X-ray Luminosities for our VISIR sample. In the left hand panel, our own $12.5\,\mu$m rest frame measurements are shown while the right hand panel shows the ISOPHOT measurements published in [3] for the same objects. Note that the X-ray error bars indicate the peak-to-peak variability of the individual sources rather than measurement errors. All luminosities were calculated using $H_0 = 71\frac{\text{km}}{\text{s Mpc}}$, $\Omega_m = 0.27$ and $\Lambda = 0.73$. The solid lines show the power-law fit to the data. NGC 7314 was excluded from this fit since we did not obtain photometry redward of 11.5μ for this object.

The goal of our project is to test whether these problems arise because of the large aperture (24"×24") of their ISOPHOT spectra or whether they are intrinsic and thus require a modification of the unified scenario.

2 Our project and data

Between April and July 2005 our group obtained N-band photometry with VISIR for 8 objects, consisting of 4 early-type Seyferts (Sy type 1.0 - 1.4) and 4 late-type Seyferts (Sy type 1.5 - 2.0) taken from the sample by [3]. Most objects were observed in 3 filters between 9.82μm and 12.81μm. Here we report only the measurements obtained at wavelengths corresponding to 12.5μm in the rest frame.

Fig. 1 shows the correlation between MIR and absorption-corrected hard X-ray luminosities. The latter are the average values of different hard X-ray observations compiled from the literature. The error bars indicate the peak-to-peak variations of these different measurements. For most of our objects at least three different hard X-ray observations taken over a range of more than 10 years were available. Only for one object only one observation could be found in the literature. Here we assumed a variability by a factor of 2.

In the left hand panel, our own $12.5\,\mu$m rest frame measurements are shown while the right hand panel shows the ISOPHOT measurements published in [3] for the same objects. The black lines show the least square fit to the data. Here we excluded NGC 7314 for which we did not obtain photometry redward of $11.5\mu m$. For both data sets the slopes of the fit are unity within their respective errors. Note however that the standard deviation of the residuals is smaller by a factor of 1.7 for the VISIR data.

Also noteworthy is the fact that, if one assumes a standard AGN SED which is strongly rising from the NIR to MIR regimes, the luminosities we find are systematically smaller by a factor of ~ 2 than expected from the ISOPHOT $6\mu m$ ones. We can thus estimate the host galaxy contribution to the ISOPHOT fluxes to be of the order of 50%. The only exception to this is NGC 4579, the object with the lowest hard X-ray luminosity in our sample. Here the contribution from the host galaxy clearly dominates.

3 Conclusion

The surprisingly tight correlation we find between the MIR and X-ray luminosities for our sample of AGN (excluding NGC 7314) in combination with the lower luminosities indicates that the larger scatter found in earlier ISOPHOT data is due to contamination from the AGNs' host galaxies. However our results confirm the earlier finding of Sy1s and Sy2s having the same MIR – hard X-ray -ratio. This as well as the low intrinsic scatter exhibits a challenge to current clumpy tori models (e.g. [4]). More observational data as well as more detailed modeling of the radiation transfer in such tori are required to test the latter's validity.

References

1. R. Antonucci: ARA&A **31**, 473 (1993)
2. A. Krabbe, T. Böker, R. Maiolino: APJ **557**, 626 (1001)
3. D. Lutz, R. Maiolino, H. W. W. Spoon, A. F M. Moorwood: A&A **418**, 465 (2004)
4. M. Nenkova, Z. Ivezic, M. Elitzur: ApJ **570**, L9 (2002)

Gamma-Ray Probe of the QSO's Obscured Evolution

A. Iyudin[1], J. Greiner[2], G. Di Cocco[3], and S. Larsson[4]

[1] SINP, Moscow State University, Moscow, Russia, aiyudin@srd.sinp.msu.ru
[2] MPI für extraterrestrische Physik, Garching, Germany, jcg@mpe.mpg.de
[3] IASF/INAF-Bologna, Bologna, Italy, dicocco@bo.iasf.cnr.it
[4] AlbaNova Center, Stockholm University, Sweden, stefan@astro.su.se

Summary. The obscured phase of QSOs, as well as their accretion history, can be best followed by observing QSOs bright in >10 MeV gamma-rays. By analysing the resonant absorption troughs in spectral energy distribution of flaring QSOs one can measure the (baryonic) absorbing column and baryonic content of the QSO host galaxy, while the flare strength will give information on the accretion rate of the QSO powering supermassive black hole. By measuring the baryonic absorbing column for QSOs at different redshifts one can follow the early obscured evolution of AGN at redshifts up to z~6.

1 Obscured AGNs

The idea that substantial absorption in AGN could be a definite characteristic of the early phases of QSO evolution [2] was invoked to explain the submillimeter observations of X-ray absorbed AGN at z~1 – 3, that have shown strong emission at 850 μm [10]. This is a signature of copious star formation. In the AGN evolution model, the main obscured growth phase of the QSO coincides with the formation of the host galaxy spheroid, the completion of which indicates the beginning of the luminous, unobscured phase of the QSO's evolution [12]. The verification of this model at redshifts up to z~6 appears possible via application of the γ-ray resonant absorption method.

1.1 Gamma-ray absorption

The recently introduced absorption method based on the detection of resonant absorption troughs in the gamma regime [5] has all the qualities to become a practical tool to measure baryonic absorption columns along the line-of-sight from the point source towards the observer. The pencil-like γ-ray beam from the QSO (or GRB) probes all absorbers along the line-of-sight providing their appropriate columns and redshifts, including the QSO's (GRB) host galaxy, as well as the matter in the Milky Way halo [6]. This γ-ray absorption method relies on the resonance-like photoabsorption by atomic nuclei. The photoabsorption on nuclei has three peaks in the cross section, at energies of ~7 MeV, in the region of the "pygmy" dipole resonance (PDR), at 20-30 MeV (giant dipole resonance (GDR)), and at ~325 MeV (Δ-resonance), see references to the relevant processes in [5].

Best studied of the above three processes are the GDR, and Δ-isobar resonances, with the typical absorption cross sections per nucleon of $\sigma_{GDR} \sim 1.1$ mb, and $\sigma_\Delta \sim 0.5$ mb. The Δ-isobar resonance has the energy of the absorption peak in the cross section, and the cross section peak value per nucleon, which is the same for all nuclei. The GDR cross section peak energy varies with different nuclei, but for the solar mixture of elements it is primarily defined by the properties of GDR on ^{4}He [5]. Either of GDR or Δ-isobar resonances, can be used to derive the absorbing column value.

In the γ-ray absorption experiment we have a rather simple geometry where a pencil-like beam of γ-ray photons has to pass through one or many absorbers on the way from a point source (QSO) towards an observer. Therefore the differential photon flux observed at the Earth can be written as a function of the photon energy and of redshift:

$$\frac{dN}{dE} = \left(\frac{dN}{dE}\right)_{unabsorbed} \cdot e^{-\tau(E,z)} \quad . \tag{1}$$

The dependence on E and z are quite complex, to simplify it we assume that we are dealing with two absorbers, one in the QSO host galaxy, and the second absorber in the Milky Way. To derive the absorbing column(s) one has to fit the QSO's spectral energy distribution, including troughs.

Absorption columns of a given quasar derived from the UV or optical observations are always $\sim$100 times smaller than the X-ray derived columns (see [1] and Fig. 1 (left)). Similarly, from the same kind of relation, absorbing columns derived in the γ-ray regime are larger than those derived via X-ray measurements (Fig. 1(right)).

The absorption columns of γ-ray bright quasars found with the γ-ray absorption method [5, 6] are of the order of 10^{26} cm^{-2}. With the superior sensitivity of *GLAST* absorbing columns as low as $\sim 10^{25}$ cm^{-2} will become measurable and thus close the gap (Fig. 1) between the X-ray and gamma-ray column distributions. The absorption columns and redshifts of γ-ray bright quasars detected by *GLAST* will also enable us to follow the evolution of the halo mass of QSO hosting galaxies up to redshifts of z$\approx$6.

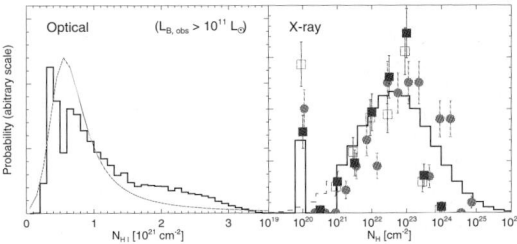 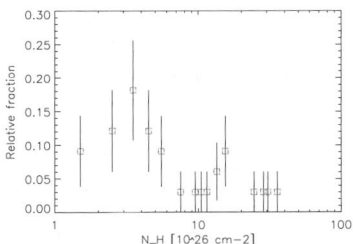

Fig. 1. Left: Typical distributions of the absorbing column densities derived towards luminous quasars in optical samples (left), and from the hard X-ray sample (right). Figure is adopted from Hopkins et al. (2005). X-ray data shown are from Treister et al. (2004) (blue / dark squares) and from Mainieri et al. (2005) (red / grey circles). Right: Column densities of absorbers in QSO hosting galaxies and in the Milky Way halo derived by the gamma-ray absorption method from SEDs of QSOs and EUIDs detected by EGRET.

2 Conclusions

Assuming that gravitational coupling of the baryonic and the Dark Matter holds [7], it is possible to use the measured absorption columns to probe the evolution of the baryonic matter content in the halo of QSO host galaxies. Different constituents of the Dark Matter halo can be traced via the gamma-ray absorption method: (1)- ordinary baryonic matter, like hydrogen, helium etc..., that constitutes the cold globules of Pfenniger (2004) [11]; (2)- dense color superconducting clumps [9]; (3)- dark baryons, relatives of X-particle [14]; or (4)- mixture of heavy and light eigenstates.

The γ-ray absorption method can provide an info on the evolution of Ω_b starting from the beginning of the reionization epoch. The sensitivity and an energy range of *GLAST* will allow to follow evolution of the baryonic halo of QSO host galaxies up to z$\approx$6. To probe the proto-galactic haloes at z$\approx$15 one will need the $\sim$1 MeV to 500 MeV range telescope with the sensitivity comparable or exceeding that of the *GLAST*. This energy range is suitable to cover both resonance-like photoabsorption troughs at $\sim$25/(1+z) MeV (GDR) and 325/(1+z) MeV (Δ-isobar), and to have a few energy bins outside of the absorption troughs to constrain the continuum shape.

References

1. N. Arav, J. Kaastra, K. Steenbrugge, et al: ApJ **590**, 174 (2003)
2. A.C. Fabian, MNRAS, **308**, L39 (1999)
3. A.C. Fabian, **astro-ph/0304122** (2004)
4. P.F. Hopkins, L. Hernquist, Th.J. Cox, et al., ApJ, **632**, 81 (2005)
5. A.F. Iyudin, O. Reimer, V. Burwitz, et al., A&A, **436** 763 (2005a)
6. A.F. Iyudin, V. Burwitz, J. Greiner, et al., **astro-ph/0501039** (2005b)
7. D.B. Kaplan, Phys.Rev.Let, **68**, 741 (1992
8. V. Mainieri, P. Rosati, P. Tozzi, et al., A&A, **437**, 805 (2005)
9. D.H. Oaknin, and A. Zhitnitsky, Phys.Rev., **D71**, 3519 (2005)
10. M.J. Page, J.A. Stevens, R.J. Ivison, F.J. Carrera, ApJ **611**, L85 (2004)
11. D. Pfenniger: Dark Molecular Hydrogen. In: *PoS: Baryons in Dark Matter Halos*, eds. R. Dettmar, U. Klein, P. Salucci, **087** (SISSA publ.) (2004)
12. J. Silk, & M. Rees, A&A, **331**, L1 (1998)
13. E. Treister, C.M. Urry, E. Chatzichristou, et al., ApJ, **616**, 123 (2004)
14. G. Zaharijas, G.R. Farrar, Phys.Rev., **D72**, 3502 (2005)

Optical Observations of SBS1520+530 at TUG

I. Khamitov[1], I. Bikmaev[2,3], Z. Aslan[1,4], N. Sakhibullin[2,3], V. Vlasyuk[5], and A. Zheleznyak[6]

[1] TÜBİTAK National Observatory of Turkey, irekk@tug.tug.tubitak.gov.tr
[2] Kazan State University, Russia
[3] Academy of Sciences of Tatarstan, Kazan, Russia
[4] Akdeniz University, Antalya, Turkey
[5] Special Astrophysical Observatory of Russian Academy of Sciences, Russia
[6] Institute of Astronomy of Kharkov National University, Ukraine

Summary. We present the light curves of the components of gravitationally lensed source SBS1520+530 in R_c band, obtained during the interval 2001-2005 with the 1.5 m Russian-Turkish Telescope RTT150 at TÜBİTAK National Observatory of Turkey. The time delay of brightness fluctuations between the two components of the gravitationally double-imaged quasar corresponding to 2001-2002 period of observations is determined. Using all the available data, at least two microlensing events, one of them with long-time linear behavior and second one with duration of a few hundred days, are detected.

1 Introduction

A legacy of the Einstein's, the deflection of light rays by gravity, has been used successively in the study of the Universe at cosmological distances by Gravitational Lensing methods. An example of a gravitational lens system, among many others, is the gravitationally lensed quasar SBS1520+530. This source was discovered by [3] as a double quasar, separated by 1."56, with spectrally identical components at a redshift z=1.855. The system was optically monitored by several groups of observers ([2, 4, 5, 8]). These observations indicate that the system shows measurable variations of brightness on time scales of about 100 days in both quasar images. This made it possible to determine a time delay between the components of 130±3 days from the light curves ([2], and confirmed by [5]). This parameter is highly important in studying the mass distribution in a lensing galaxy and in estimating H_0 by Refsdal's method [7]. The results of time delay measurement and some signs of gravitational microlensing, as a linear trend of the difference between the light curves [2], were reported at GLITP-2001 workshop by I. Burud. Prompted by these results, we decided to observe the SBS1520+530 at TÜBİTAK National Observatory of Turkey for possible microlensing events, which in principle allows one to scan the continuum emission regions of the quasar on microarc-second scale.

2 Observations and photometry

We observed the gravitationally lensed system (GLS) SBS1520+530 with 1.5m Russian-Turkish Telescope RTT150 [1] in R_c-band in every clear night between 14 June 2001 – 29 August 2002 with seeing better than 1".8 using 1k×1.5k ST-8E CCD (binning 2×2, 0."32/pixel). Because of a detected non-linearity of the CCD at signals higher than about 15000ADU, we made a series of short 60sec exposures with total integration time 10-15 minutes. In V-band, images were taken only monthly. Observations were continued in R_c, and V bands using the 2k×2k Andor CCD (0."24/pixel) from 2003 up to now with sampling about 3–4 observations per year. A series of 5–10 exposures with duration of 300 seconds were made.

All data were pre-processed (BIAS, DARK subtraction, flat-field correction) using IRAF.NOAO.CCDRED facilities. The SBS1520+530 is a close system. The images with FWHM better than 1."3 were selected for photometry. We finally used 64 nights in the R_c-band and 13 in the V-band. In photometry, we used co-added images of a given night. The photometry was performed using the IDLPHOT package from the IDL Astronomy User's Library. The algorithm of Iterative PSF Subtraction was used for brightness estimation of SBS1520+530 components. The application of the algorithm to this GLS is described in [8]. Detailed photometry is given in [6].

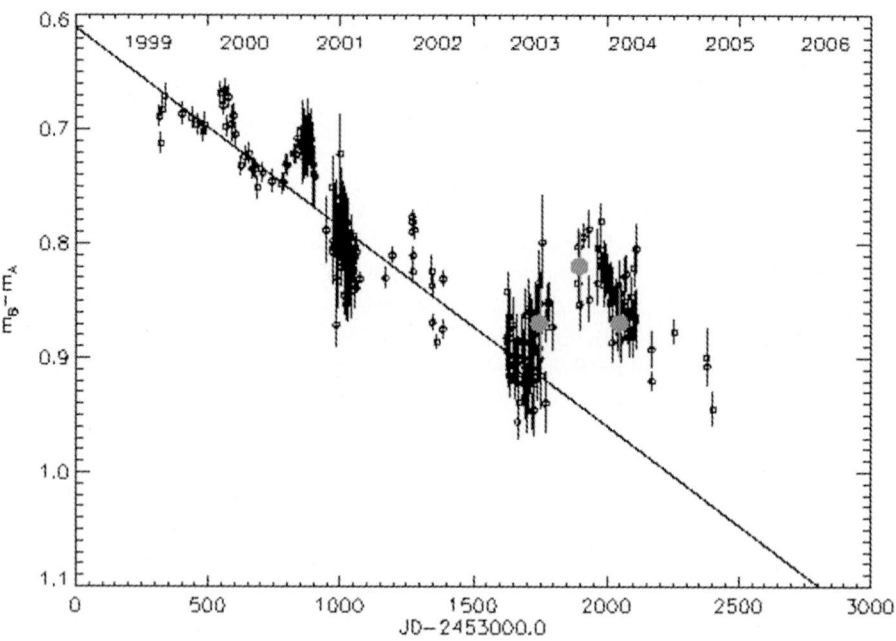

Fig. 1. The magnitude difference m_B-m_A of light curves, B curve is shifted by the time delay 128^d. A fit of linear drift of 0.064 mag/year from 1999 to the beginning of 2003 is shown.

3 Results

The light curves of the components of gravitationally lensed source SBS1520+530 in VR_c band were obtained during the interval 2001-2005. We did not find any noticeable changes in the V-R color. The average values are the same as found by [5]: $(V-R)_A = 0^m.15$ and $(V-R)_B = 0^m.18$.

To estimate the time delay, we used the most sampled period of observations in 2001-2002, which is about 400 days long. The time delay was determined as the minimum of function of goodness of polynomial fit to the composite light curve A and B, with B light curve shifted by time delay Δt and with Δm_B magnitude shift of B relative to A light curve. Detailed algorithm is given in [6]. We found the value of time delay in 128^d which is in good agreement with early determinations $130^d \pm 3$ [2] and $130^d.5 \pm 2.9$ [5], determined on the different periods of observations.

Both light curves, shifting B light curve by this time delay, are aligned reasonably well except the period between the end of 2003 and 2005. Detected non-correlated variations might be a sign of microlensing. To study this behavior, we used all the published data - electronically available CDS catalogues J/A+A/391/481/table2, J/A+A/440/53/table1 and the table of photometry from [8]. Our data set falls in the gap of published data between 2001-2002, allowing us to have an almost uninterrupted light curve between 1999 and 2005. As a result of merging all the published photometry, the long term drift of B light curve relative to A light curve is finally confirmed (Fig.1). Another noticeable detail on the difference light curve is a re-brightening between 2003-2005 . It seems real because the data satisfy the direct differences m_B-m_A obtained from the overlapped A and B light curves (large grey filled circles in Fig.1) and can be interpreted as microlensing event.

References

1. Z. Aslan et al: AstrL **27**, 398 (2001)
2. I. Burud et al: A&A **391**, 481 (2002)
3. V.H. Chavushyan et al: A&A **318**, L67 (1997)
4. V. Dudinov et al. In: Proc. of Internat. Conf. *"Kinemat. And phys. Of Celest. Bodies"*, Spec. Issue, 170 (2001)
5. E.R. Gaynullina et al: A&A **440**, 53 (2005)
6. I.M. Khamitov et al: AstrL, *32*, 8, 514-519 (2006)
7. S. Refsdal: MNRAS **128**, 307 (1964)
8. A.P. Zheleznyak, A.V. Sergeev, O.A. Burkhonov: Astronomy Reports **47**, 717 (2003)

Less is More? Are Radiogalaxies Below the Fanaroff-Riley Break More Polarised on Pc-Scales?

P. Kharb[1], P. Shastri[2], and D.C. Gabuzda[3]

[1] Rochester Institute of Technology, Rochester, USA, kharb@cis.rit.edu
[2] Indian Institute of Astrophysics, Bangalore, India, pshastri@iiap.res.in
[3] University College Cork, Cork, Ireland

1 Introduction

Magnetic (B-) fields are believed to be instrumental in the formation and collimation of relativistic jets from radio-powerful active galaxies (Blandford & Payne 1982[1], Lovelace et al. 1987[6]). Ordered parsec-scale B-fields have been detected in the nuclei of several of the Doppler-beamed BL Lacs and quasars via VLBI polarimetry (Gabuzda et al. 1994[3]). However, such evidence is meagre in radio-loud galaxies whose jets are purportedly closer to the sky plane and therefore not heavily Doppler beamed. (Taylor et al. 2001[8], Middelberg et al. 2003[7]). In particular, there have been no detections of parsec-scale nuclear polarisation for any galaxy below the Fanaroff-Riley (**FR**) luminosity break (Fanaroff & Riley 1974[2]) that broadly separates radio galaxies with plume-like structures from those with well-collimated jets terminating in hot spots. Here we present our detections of nuclear parsec-scale polarisation in four FRI galaxies, and discuss their implications in the framework of Unification.

2 Observations and Results

We observed the four FRI radio galaxies $viz.$, 3C 66B, 3C 78, 3C 264 and 3C 270, at 8.4 GHz (λ 3.6cm) on 1 March 2002 with a global VLBI array consisting of the Very Large Baseline Array (VLBA)[4] and five stations of the European VLBI Network.[4] The calibration and imaging were done using standard techniques.

The spatial resolution is ~0.5 mas, corresponding to ~0.2–0.4 pc. Polarisation of 0.4–1% was detected from the inner parsec of all the galaxies (Fig. 1). In each case, the brightest component has a flat or inverted spectral index as inferred from comparison with images at other frequencies (e.g., Giovannini et al. 2001[4], Jones & Wehrle

[4] The NRAO is operated for the National Science Foundation (NSF) by Associated Universities, Inc. (AUI), under a cooperative agreement. The European VLBI Network (EVN) is a joint facility of European, Chinese, South African and other radio astronomy institutes funded by their national research councils.

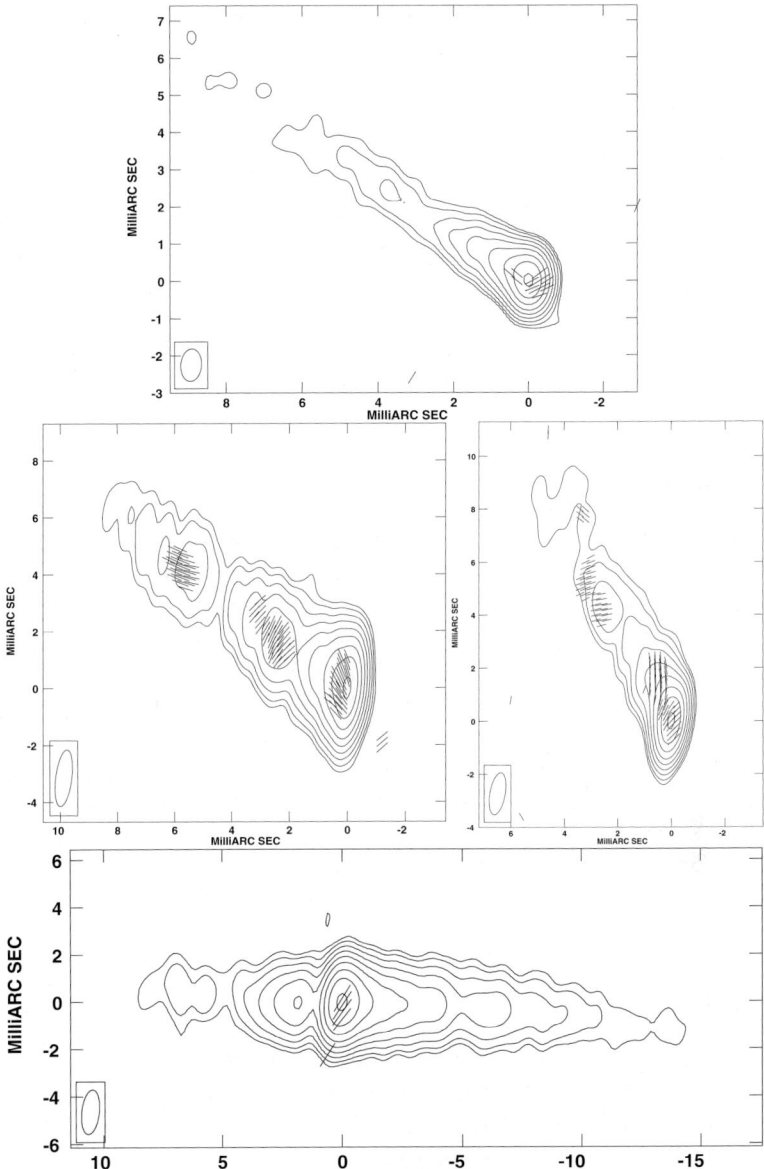

Fig. 1. Total intensity radio maps of the galaxies with polarisation electric vectors superimposed. Top: 3C 66B; Middle: 3C 78 & 3C 264; Bottom: 3C 270. The surface brightness peaks in are 118.2, 285.9, 165.3 & 136.3 mJy/beam respectively. The lowest contour is $\pm 0.35\%$ of the peak surface brightness, and the contour levels are in percentage of the peak, increasing in steps of two. The polarisation measured is $\simeq 0.4 - 1\%$, in the cores, rising to $\simeq 5 - 10\%$ in the inner jet (≤ 1 pc from the core), and reaching 60% in a knot about 1.5 pc from the core of 3C 264.

1997[5]), and can therefore be identified as the VLBI "core." 3C 270 has no detected polarisation beyond its core.

3 Conclusions

1. The fractional polarisation is comparable to the corresponding values in BL Lacs and quasars, indicating the presence of appreciably ordered B-fields in the inner parsec of FRI galaxies and little depolarisation.
2. Albeit for a small number, our 100% detection rate is higher than that for narrow-lined FRII galaxies (only 1/4 detected: Taylor et al. 2001[8], Middelberg et al. 2003[7], Zavala & Taylor 2002[9]). If this trend is due to depolarisation by the inner ionized edge of an obscuring torus, it would imply that the FRIs lack such ionized material in their inner pc.
3. The inner-jet polarisation is parallel to the local jet in at least 3C 66B and 3C 78, implying a transverse B-field as is typical of BL Lacs. In 3C 264 and 3C 270, the nuclear B-field orientation cannot be unambigously inferred due to possible resolution and/or Faraday rotation effects. A longitudinal B-component develops further along the jet in 3C 78 and 3C 264, and there are regions of alternating B-field orientation in 3C 78, similar to BL Lacs. On the whole, the qualitative and quantitative similarities between these polarisation structures and those of BL Lacs are striking, and our images are thus consistent with the prediction of the simple unification in which FRIs are the parent population of BL Lacs.
4. The two galaxies with the highest jet-to-counterjet ratios (*i.e.*, the highest implied Doppler-beaming), *viz.*, 3C 78 and 3C 264, have more detected polarisation and structure in their jets.
5. The ordered nuclear B-field in 3C 66B is on a scale ~ 0.3 pc, and therefore within ~ 1650 Schwarzschild radii of the putative black hole, based on its estimated black-hole mass of $\sim 2 \times 10^9 M_\odot$ (Noel-Storr et al.'05, preprint).

References

1. Blandford, R. D. & Payne, D. G.: MNRAS **199** 883 (1982)
2. Fanaroff, B. L. & Riley, J. M.: MNRAS **167** 31P (1974)
3. Gabuzda, D. C., et al.: ApJ **435** 140 (1994)
4. Giovannini, G., et al.: ApJ **552**, 508 (2001)
5. Jones, D. L. & Wehrle, A. E.:ApJ **484** 186 (1997)
6. Lovelace, R. V. E., et al.: ApJ **315** 504 (1987)
7. Middelberg, E., Roy, A. L., Bach, U., Gabuzda, D. C., & Beckert, T. 2003, in Proceedings of conference "Future Directions in High Resolution Astronomy", ASP Conf. Series, (astro–ph/0309385) (2003)
8. Taylor, G. B., et al.:: ApJ **559** 703 (2001)
9. Zavala, R. T. & Taylor, G. B.: ApJ **566** L9 (2002)

Relativistic Effects on the Observed AGN Luminosity Distribution and Spectral Shape of Seyfert Galaxies

Y. Liu[1], S.N. Zhang[1,2,3,4], and X.L. Zhang[3]

[1] Physics Department and Center for Astrophysics, Tsinghua University, Beijing, 100084, China, zhangsn@tsinghua.edu.cn, yuan-liu@mails.tsinghua.edu.cn

[2] Key Laboratory of Particle Astrophysics, Institute of High Energy Physics, Chinese Academy of Sciences, Beijing, China

[3] Physics Department, University of Alabama in Huntsville, Huntsville, AL35899, USA, zhangsn@uah.edu, zhangx@email.uah.edu

[4] Space Science Laboratory, NASA Marshall Space Flight Center, SD50, Huntsville, AL 35812, USA

1 Introduction

In the unified model of Seyfert galaxies (Antonucci 1993), it is expected that the observed luminosity distribution and spectral shapes of Seyfert I and Seyfert II should be identical except for the effects of obscuration by the torus material. However in several recent X-ray surveys of AGN (Ueda et al. 2003; Hasinger 2004) it has been found that the fraction of type-II AGNs is anti-correlated with the observed X-ray luminosity, such that the luminosity dependent dusty torus model may be required. However this observed anti-correlation could be explained in the framework of the unified model if the inclination angle effects of X-ray radiation are taken into account (Zhang 2005, hereafter Z05). For simplicity, relativistic effects produced when the X-rays are emitted in the accretion disk are ignored in Z05. We will consider relativistic corrections to the observed luminosity distributions of AGNs in §2 of this paper. Moreover some observations have found that the spectra of Seyfert II objects are systematically harder than those of Seyfert I objects (Zdziarski, Poutanen, & Johnson 2000; Malizia et al. 2003, hereafter M03), which lacks satisfactory explanations within the unified model of AGNs. In §3 we also investigate the relativistic effects on the observed spectral shapes of type I and type II systems. In §4 we show our conclusions.

2 AGN Luminosity Distribution with Relativistic Corrections

In the non-relativistic case when we view an AGN system with an inclination angle θ , defined as the angle between the normal direction of the accretion disk (also the spin axis of the black hole) and the line of sight, the observed luminosity will be reduced by a factor of $\cos\theta$ compared with the face-on case, i.e., $\theta = 0°$. This projection factor will become complicated after considering the relativity correction, as shown in Figure 1 (left). If the limb-darkening effect is considered, an additional factor of $(1 + 2\cos\theta)/3$ to the observed luminosity should be included (Netzer 1987).

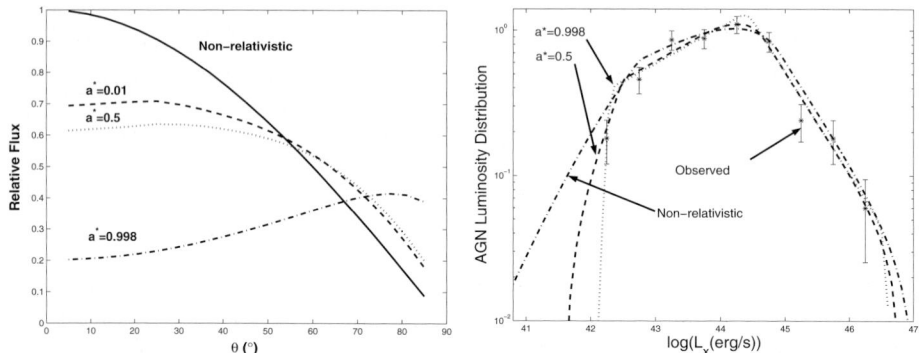

Fig. 1. *Left:* The observed X-ray flux and inclination angle dependence for non-relativistic and relativistic cases in 2-10 keV. The relativistic effects include the Doppler shift and boosting, the gravitational redshift and light bending. The correction is calculated by the ray-tracing method (Fanton et al. 1997). *Right:* Observed and intrinsic AGN luminosity distributions with the same assumptions in Z05.

Detailed comparisons with data for the relativistically corrected luminosity distribution and type II fraction is shown in Figure 1 (right) and Figure 2 (left). In summary, the observed luminosity distribution and type II fraction prefer the mildly spinning black hole case for high luminosity ($L_x \geq 10^{43.5}$ erg/s) AGNs (but without any constraint for low luminosity AGNs), and the dividing inclination angle between type II and type I AGNs is between 60 and 70 degrees, slightly different from that of 68 and 76 degrees in Z05.

3 X-Ray Spectra from Seyfert I and Seyfert II AGNs with Relativistic Corrections

In this section, we investigate the difference caused by the relativistic corrections of spectral indices of Seyfert I and Seyfert II galaxies. With the same assumption of §2, in Figure 2 (right) we show the dependence of the spectral index α as a function of inclination angle calculated only with relativistic corrections for $a^* = 0.998$, compared with the observed spectral indices for the average Seyfert I and II AGNs from M03, and together with those calculated for the anisotropic Comptonization by several authors (Haardt & Maraschi 1993, hereafter HM93; Poutanen & Svensson 1996, hereafter PS96; Maraschi & Haardt 1997, hereafter MH97). Clearly only the relativistically corrected spectral shape is consistent with data.

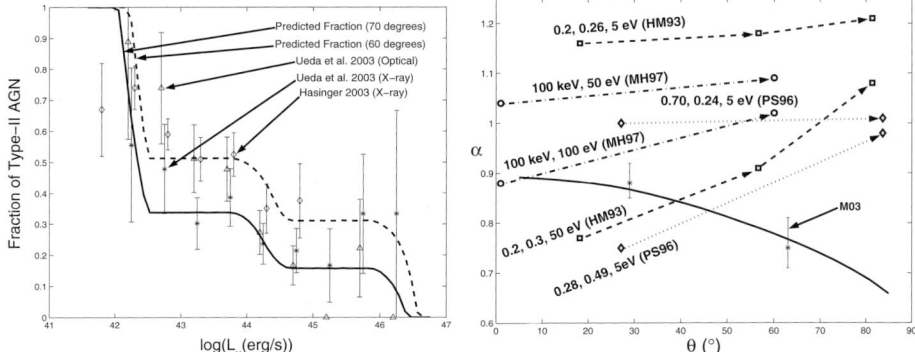

Fig. 2. *Left:* Type-II AGN fraction as the function of the observed apparent X-ray luminosity (after absorption corrections) for $a^* = 0.5$. *Right:* The relationship between spectral index and inclination angle: the solid curve is the result of relativistic corrections with angular momentum $a^* = 0.998$. Other symbols are for the results of the anisotropic Comptonization. The numbers on the figure are τ (the optical depth), $\Theta \equiv kT_e/m_e c^2$ (the temperature of electrons) and kT_{bb} (the temperature of blackbody spectra), or kT_e and kT_{bb}. The observed results of Seyfert galaxies in M03 are also shown in the figure.

4 Conclusion and Discussion

The main results of this paper are: (1) The relativistic corrections improve the luminosity distribution and require that high luminosity AGNs contain black holes spinning only mildly; (2) The relativistic corrections could also explain the observed spectral difference between Seyfert I and II AGNs and require Seyfert AGNs to contain black holes rapidly spinning, whereas the anisotropic Comptonization gives the opposite trends to the observed results. These results suggest that distant high luminosity AGNs (quasars) are powered by the post-merger accretion which has not spun-up the black hole which is on the average only spinning mildly after a major merger event. However nearby Seyfert galaxies should have been in an extended accretion state which has spun-up the black hole significantly.

References

1. R. Antonucci: ARA&A **31**, 473 (1993)
2. C. Fanton et al: PASJ **49**, 159 (1997)
3. F. Haardt, L. Maraschi: ApJ **413**, 507 (1993)
4. G. Hasinger: Nucl. Phys. B Proc. Suppl. **132**, 86 (2004)
5. A. Malizia et al: ApJ **589**, L17 (2003; M03)
6. L. Maraschi, F. Haardt: 1997, ASPC **121**, 101M (1997)
7. H. Netzer: MNRAS **225**, 55 (1987)
8. J. Poutanen, R. Svensson: ApJ **470**, 249 (1996)
9. Y. Ueda, M. Akiyama, K. Ohta, T. Miyaji: ApJ **598**, 886 (2003)
10. A.A. Zdziarski, J. Poutanen, W.N. Johnson: ApJ **542**, 703 (2000)
11. S.N. Zhang: ApJ **618**, L79 (2005; Z05)

Nuclear Activity in Galaxies driven by Binary Supermassive Black Holes

A.P. Lobanov

Max-Planck-Institut für Radioastronomie, Auf dem Hügel 69, 53121 Bonn, Germany,
alobanov@mpifr-bonn.mpg.de

Summary. Nuclear activity in galaxies is closely connected to galactic mergers and supermassive black holes (SBH). Galactic mergers perturb substantially the dynamics of gas and stellar population in the merging galaxies, and they are expected to lead to formation of supermassive binary black holes (BBH) in the center of mass of the galaxies merged. A scheme is proposed here that connects the peak magnitude of the nuclear activity with evolution of a BBH system. The scheme predicts correctly the relative fractions of different types of active galactic nuclei (AGN) and explains the connection between the galactic type and the strength of the nuclear activity. It shows that most powerful AGN should result from mergers with small mass ratios, while weaker activity is produced in unequal mergers. The scheme explains also the observed lack of galaxies with two active nuclei, which is attributed to effective disruption of accretion disks around the secondary in BBH systems with masses of the primary smaller than $\sim 10^{10} \, M_\odot$.

1 Binary Black Holes and Nuclear Activity in Galaxies

Important roles played by galactic mergers and binary black holes in galaxy evolution was first recognized several decades ago [1, 22]. A large number of subsequent studies have addressed the problems of evolution of binary black holes in post-merger galaxies (see [19] for a recent review) and connection between mergers and nuclear activity in galaxies [3, 8, 25]. The correlations observed between the masses, M_{BH}, of nuclear black holes in galaxies, the and masses, $M_\star$, [18] and velocity dispersions, $\sigma_\star$, of the host stellar bulges [6, 7, 26] suggest a connection between the formation and evolution of the black holes and galaxies. Growth of black holes in galactic centers is self-regulated by outflows generated during periods of supercritical accretion [5, 9, 24]. This mechanism offers a plausible explanation for the observed M_{BH}–σ relation [11].

Supermassive black holes are expected to form in the early Universe, with multiple SBH likely to be common in galaxies [8]. However, the detailed connection between the SBH evolution and the nuclear activity is somewhat elusive. Observational evidence for binary SBH is largely indirect (see [17, 19] and references therein), with only two double galactic nuclei (NGC 6240 [12] and 3C 75 [23]) observed directly in early merger systems, at large separations. There are no convincing cases for secondary black holes within active galaxies, although some of them may be hiding among the extranuclear

X-ray point sources detected by ROSAT and Chandra [2, 20]. This implies that the activity of secondary companions is quenched at early stages of the merger, possibly due to disruption of the accretion disk.

The nuclear activity depends strongly on the availability of accreting material in the immediate vicinity of a black hole, and AGN episodes are believed to last for $\sim 10^7$–10^8 yr. This is likely to be smaller than typical lifetimes of nuclear binary black holes in galaxies [19]. This suggests that nuclear binary black holes systems may provide a mechanism necessary for instilling and supporting high accretion rates over timescales implied by large-scale relativistic outflows produced in AGN [4, 21]. Evolutionary stages of BBH systems can also be connected phenomenologically to different types of AGN [16]. Here, an analytical model is proposed that connects the evolution of central SBH in to the nuclear activity in galaxies.

1.1 BBH Evolution

The main constituents of the model are: 1) binary system of supermassive black holes, 2) accretion disk, 3) central stellar bulge. The BBH is described by the masses M_1, M_2 ($M_1 \geq M_2$) of the two black hole and their separation r. The accretion disk is assumed to be a viscous Shakura-Sunyaev disk, with a mass M_d. The disk extends from $\rho_{in} R_g$ to $\rho_{out} R_g$, where R_g is the gravitational radius and $\rho_{in} \approx 6$ and $\rho_{out} \approx 10^4$ [10]. The central bulge extends over a region of radius $r_\star$ and has a mass $M_\star$ ($M_\star > M_{12} = (M_1 + M_2)$) and a velocity dispersion $\sigma_\star$.

The evolution of the BBH is described in terms of *reduced mass*, $\tilde{M}$, and *reduced separation*, $\tilde{r}$ of the binary. The reduced mass is defined as $\tilde{M} = 2 M_2 / M_{12}$. This definition implies $\tilde{M} = 0$ for $M_2 = 0$ and $\tilde{M} = 1$ for $M_2 = M_1$. If $q = M_2 / M_1$ is the mass ratio in the system, then $\tilde{M} = 2 q / (1 + q)$. The reduced separation is given by $\tilde{r} = r / (r + r_c)$, where r_c is the separation at which the two black holes become gravitationally bound (this happens at $\tilde{r} = 1/2$). Binary systems have $\tilde{r} \leq 1/2$, while unbound pairs of SBH have $\tilde{r} > 1/2$.

1.2 Accretion Disk Disruption in Binary Black Holes

Two SBH in a merger galaxy are expected to form a binary system at a separation $r_c = r_\star (M_{12} / M_\star)^{1/3}$, with an initial orbital speed $v_{init} = \sigma_\star (M_{12} / M_\star)^{1/3}$ [1]. Assuming that the relative speed of the two black holes reaches asymptotically its Keplerian value, the *approach speed* can be defined as $v_{appr}(r) = \sigma_\star (r_\star / r)^{1/2} (M_{12} / M_\star)$. Both black holes are assumed to have active accretion disks at early stages of the merger. The separations r_d at which the accretion disks are disrupted and eventually destroyed can be estimated for each of the two black holes by equating the approach speed to the Keplerian velocity at the outer edge of the disk: $v_{k,out} = c/\sqrt{\rho_{out}}$. This yields $r_d = \rho_{out} r_c (\sigma_\star / c)^2 (M_{12} / M_\star)^2$. The bulge mass and velocity dispersion must satisfy the M_{BH}–$\sigma_\star$ and M_{BH}–$M_\star$ relations [11]. The resulting reduced separation is $\tilde{r}_d = 1/(1 + \xi)$, where $\xi = M_1/[1.86 \times 10^7 \, M_\odot \, \rho_{out} \, \phi^2 \, (2 - \tilde{M})^3]$ and ϕ is the collimation angle of the outflow carrying the excess energy and angular momentum from the immediate vicinity of the black hole. This corresponds to a critical mass

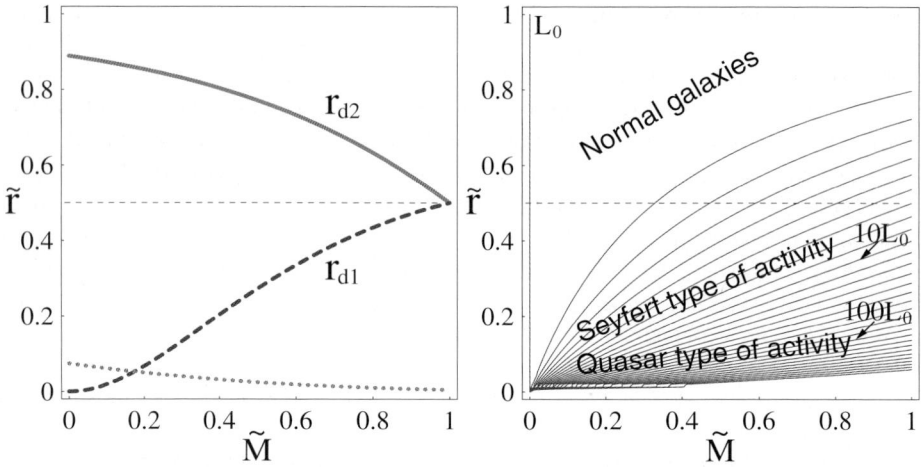

Fig. 1. Properties of binary black holes in the $\tilde{M}$–$\tilde{r}$ plane ($\tilde{r} = 1/2$ signifies the capture distance at which a pair of black holes becomes gravitationally bound). **Left:** Reduced separations for the disruption distance of the accretion disks around the secondary (solid line) and the primary (dashed line) black holes. The separations are calculated for $M_1 = M_{eq}$. Above the $\tilde{r}_{d2}$ line, both black holes retain accretion disks, while only one accretion disk (around the primary) exists. The dotted line shows the limiting distance below which a circumbinary disk may exist. This becomes feasible at $\tilde{M} \lesssim 0.2$. **Right:** Peak luminosities of AGN calculated for a range of values of the reduced mass $\tilde{M}$ and reduced distance $\tilde{r}$ in binary systems of SBH in the centers of galaxies. Equal luminosity contours are drawn at a logarithmic step of 0.1, starting from a unit luminosity L_0 marked by the vertical line at $\tilde{M} = 0$.

$M_{eq} = 1.86 \times 10^7 \, M_\odot \, \rho_{out} \, \phi^2$ for which an equal mass binary system will undergo disk destruction at the time of gravitational binding (at $\tilde{r}_d = \tilde{r}_c = 1/2$. In systems with $M_1 < M_{eq}$ the destruction of accretion disk around the secondary will occur before the formation of a gravitationally bound system. For typical values of $\rho_{out} \approx 10^4$ and $\phi = 0.1$–0.3, M_{eq} reaches 10^9–$10^{10} \, M_\odot$. It implies that most of active galaxies formed by galactic mergers should undergo destruction of the disk around the secondary BH before or during the formation of a gravitationally bound systems. Since masses of the nuclear black holes in galaxies rarely exceed $10^{10} \, M_\odot$, this offers a natural explanation for the observed lack of active galaxies with double nuclei, since it predicts that *in most galaxies with binary black hole systems the secondary companion will be inactive*.

Denoting $\epsilon_1 = M_1/M_{eq}$, the disruption distances are

$$\tilde{r}_{d1} = \left(1 + \frac{\epsilon_1}{\tilde{M}^2(2 - \tilde{M})}\right)^{-1} \quad \text{and} \quad \tilde{r}_{d2} = \left(1 + \frac{\epsilon_1}{(2 - \tilde{M})^3}\right)^{-1},$$

for the primary and secondary black hole, respectively. A circumbinary disk can exist at orbital separations smaller than $\sim G \, M_1 \rho_{out}^{1/2} c^{-2}$. These three characteristic distances are shown in left panel of Fig. 1, for $M_1 = M_{eq}$.

1.3 Peak Luminosity of AGN

The peak magnitude of the nuclear activity in a galaxy hosting a binary black hole system can also be connected with the reduced mass and orbital separation of the two black holes. Assuming that the accretion rate increases proportionally to the tidal forces acting on stars and gas on scales comparable to the accretion radius, $2\,G\,M_{\mathrm{bh}}/\sigma_{\star}^2$, the peak luminosity from an AGN can be crudely estimated from

$$L_{\mathrm{peak}} = L_0 \left(1 + \frac{\tilde{M}}{2 - \tilde{M}} \frac{\tilde{M}}{\tilde{r}^2} \right) ,$$

where L_0 is the "unit" luminosity of a typical single, inactive galactic nuclei. The peak luminosities calculated in this fashion are plotted in the right panel of Figure 1 for the entire range of $\tilde{M}$ and $\tilde{r}$.

The peak luminosity increases rapidly with increasing $\tilde{M}$ and decreasing $\tilde{r}$, and it reaches $L_{\mathrm{peak}} = 1000\,L_0$ for an equal mass binary SBH at $r \approx 0.03\,r_{\mathrm{c}}$. This corresponds most likely to powerful quasars residing in elliptical galaxies. At the same $\tilde{r}_{\mathrm{c}}$, an unequal mass binary, with $\tilde{M} = 0.15$, will only produce $L_{\mathrm{peak}} \approx 10\,L_0$, which would correspond to a weak, Seyfert-type of active nucleus. Assuming that galaxies are distributed homogeneously in the $\tilde{M}$–$\tilde{r}$ diagram, this scheme implies that about 70% of all galaxies should be classified as inactive, while the Seyfert-type of galaxies, with $L_{\mathrm{peak}} = 10$–$100\,L_0$, should constitute 25% of the galaxy population, and the most powerful AGN, with $L_{\mathrm{peak}} > 100\,L_0$, should take the remaining 5%. It shows that the most powerful AGN with $L_{\mathrm{peak}} > 1000\,L_0$ should be found in binary SBH with nearly equal masses of the primary and secondary black holes. Binary SBH with smaller secondary companions should produce (at the peak of their nuclear activity) weaker, Seyfert-type AGN. Evidence exists in the recent works [13], [15] that the nuclear luminosity does indeed increase with the progression of the merger, but more systematic and detailed studies are required.

2 Conclusion

The model described above can be applied effectively to high-resolution optical studies and data from large surveys that can be used to obtain estimates of the nuclear luminosities and black hole masses in active galaxies. The most challenging task is to assess the state of the putative binary, since the secondary black holes are very difficult to detect. For wide binaries, Direct evidence may be sought in galaxies with double nuclei and extranuclear compact sources. Close binaries can probably be identified only indirectly, through periodic perturbations caused by the secondary companion. Other indicators, such as flattening of the galactic nuclear density profile due to BBH [19], can also be considered. Once the binary separations have been estimated, it would be possible to populate the $\tilde{M}$–$\tilde{r}$ diagram and study whether different galactic and AGN types occupy distinctively different areas in the diagram.

References

1. M.C. Begelman, R.D. Blandford, M.J. Rees: Nature **28**, 307 (1980)
2. E.J.M. Colbert, A.F. Ptak: AJSS **143**, 25 (2002)
3. T. Di Matteo, R.A.C. Croft, V. Springel, et al.: ApJ **593**, 56 (2003)
4. V.I. Dokuchaev: MNRAS **251**, 564 (1991)
5. A.C. Fabian: MNRAS **308**, L39 (1999)
6. L. Ferrarese, D. Merritt: ApJ **539**, L9 (2000)
7. K. Gebhardt, et al: ApJ **539**, L13 (2000)
8. M.G. Haehnelt, G. Kauffmann: MNRAS **336**, 61 (2002)
9. M.G. Haehnelt, P. Natarajan, M.J. Rees: MNRAS **300**, 817 (1998)
10. P.B. Ivanov, I.V. Igumenshchev, I.D. Novikov: ApJ **507**, 131 (1998)
11. A. King: ApJ **596**, L27 (2003)
12. S. Komossa, V. Burwitz, G. Hasinger, et al.: ApJ **582**, L15 (2003)
13. S. Laine, R.P. van der Marel, J. Rossa, et al.: AJ **126**, 2717 (2003)
14. B.D. Lehmer, W.N. Brandt, A.E. Hornschemeier, et al.: ApJ **131**, 2394 (2006)
15. G. Letawe, P. Magain, F. Courbin, et al.: *(in prep.)*
16. A.P. Lobanov: MemSAIt **76**, 164 (2005)
17. A.P. Lobanov, J. Roland: A&A **431**, 831 (2005)
18. J. Magorrian, S. Tremaine, D. Richstone, et al.: AJ **115**, 2285 (1998)
19. D. Merritt, M. Milosavljević: Living Reviews in Relativity **8**, (2005), http://www.livingreviews.org/lrr-2005-8
20. M.C. Miller, E.J.M. Colbert: IJMPD **13**, 1 (2004)
21. A.G. Polnarev, M.J. Rees: A&A **283**, 301 (1994)
22. N. Roos: A&A **104**, 218 (1981)
23. F.N. Owen, C.P. Odea, M. Inoue, J.A Eilek: ApJL **294**, L85 (1985)
24. J. Silk, M.J. Rees: A&A **331**, L1 (1998)
25. V. Springel, T. Di Matteo, L. Hernquist: ApJ **620**, 79 (2005)
26. S. Tremaine, K. Gebhardt, R. Bender, et al.: ApJ **574**, 740 (2002)

Extragalactic Photon Background above GeV Energies: High Peaked BL Lacertae Objects or Dark Matter?

K. Mannheim[1], D. Elsässer[1], and T. Kneiske[2]

[1] Institut für Theoretische Physik und Astrophysik, Universität Würzburg
[2] Department of Physics, University of Adelaide, Australia

Summary. Blazars produce a large fraction of the observed extragalactic gamma ray background between 100 MeV and 100 GeV. The average spectrum of the resolved blazars in the 3rd EGRET catalogue is too steep to be consistent with the shape of the background spectrum above 1 GeV. The most proliferant emitters above 100 GeV, the high-peaked blazars, which are too faint to have been detected with EGRET, but of which an increasing number is now being detected with imaging air Cherenkov telescopes such as MAGIC, H.E.S.S., and VERITAS, could supply a substantial contribution to the background above GeV energies. We show that dark matter annihilation in clumpy halos could naturally produce an emission component peaking at ~ 20 GeV. Adding this to the blazar-origin component, a best-fit model for the extragalactic gamma ray background is obtained.

1 Introduction

From the all-sky survey performed with the EGRET spark chamber onboard the Compton Gamma Ray Observatory (CGRO) a catalogue of point sources has been obtained, the 3rd EGRET catalogue [1], of which a sizeable number were identified with flat-spectrum radio quasars (FSRQ), low-peaked BL Lac objects (LBL), and high peaked BL Lac objects (HBL). The flux-limited survey also lead to the determination of an isotropic, diffuse gamma ray background, the extragalactic gamma ray background (EGRB), after proper subtraction of the galactic foreground due to cosmic ray interactions [2, 3]. Presumably, the EGRB is comprised of photons from sources too faint to be resolved by EGRET [4]. Modeling the luminosity function and redshift distribution of the resolved gamma ray sources, however, showed that the total contribution of the gamma ray blazars to the EGRB seems not to exceed $\sim 50\%$ [5]. Moreover, the average spectrum of the resolved sources is too steep ($\alpha = 2.37 \pm 0.04$) to explain the spectrum of the EGRB which is overall flatter ($\alpha = 2.10 \pm 0.03$). In the recent re-analysis of the EGRB [3], the EGRB spectrum below 1 GeV is consistent with the average spectrum of resolved blazars, but significantly hardens above (10σ deviation from the best-fit power law below 1 GeV with a slope of $\alpha = 2.33$, see. Fig. 2).

Whereas the effective area of EGRET above 10 GeV may be subject to debate, the arrival directions of the highest energy photons above 10 GeV from the EGRET survey can be determined quite accurately. In general, they can not be traced back to the positions of known blazars [6], although some photons correlate with known BL Lacs [7]. This may either reflect the fact that many of the blazars contributing to the EGRB are still unknown or the emission is of a more diffuse nature. Extremely high-peaked blazars could have escaped a radio, optical or soft X-ray (ROSAT) identification so far, but would show up above GeV energies (GLAST) and in the hard X-ray band (eROSITA). Gamma rays in the EGRET band can also be of secondary origin, resulting from pair cascades initiated from gamma rays of higher energies interacting with low energy photons from the metagalactic background radiation field [8].

Several possibilities for the origin of the EGRB above GeV energies have been pointed out in the literature, of which the most appealing are (i) cascading of higher energy photons originating from gamma ray bursts [9] or from blazars [10, 11, 12], (ii) gamma radiation from clusters of galaxies storing cosmic rays escaping from AGN, starburst galaxies or mergers [13] and (iii) secondary gamma rays from the fragmenting heavy mesons produced in neutralino annihilation events occuring in clumpy dark matter halos [14].

In this contribution, we highlight the implications of the increasing number of blazars detected with MAGIC and H.E.S.S. on their role in producing part of the EGRB.

2 Blazars detected with imaging air Cherenkov Telescopes

Table 1 summarizes the high peaked blazars unambiguously detected so far with imaging air Cherenkov telescopes (for references, see ref. [15] and recent astro-ph entries by H.E.S.S., MAGIC, and VERITAS Collaborations). In addition, tentative detections of BL Lacertae (z=0.069) and 3C66A (z=0.44) have been reported by the Crimean Cherenkov Telescope group. In spite of the large amplitude variability, a baseline emission with a gamma ray luminosity comparable to the X-ray luminosity seems to be a bareable assumption for HBLs, which is also in line with the expectation from simple one-zone SSC models [17, 18].

Table 1. High peaked blazars unambiguously detected with imaging air Cherenkov telescopes (as of 2/2006)

#	Name	Redshift	Recent Reference
1	Mkn421	0.030	astro-ph/0508244
2	1ES1101-232	0.186	astro-ph/0508073
3	1ES1218+304	0.182	astro-ph/0508244
4	1H1426+428	0.129	astro-ph/0305579
5	PG1553+113	0.36(?)	astro-ph/0601545
6	Mkn501	0.033	astro-ph/0508244
7	1ES1959+650	0.051	astro-ph/0508543
8	PKS2155-304	0.116	astro-ph/0506593
9	1ES2344+514	0.044	astro-ph/0508499
10	H2356-309	0.165	astro-ph/0602435

The contribution of HBLs to the EGRB can be coarsely estimated by adopting a space density of

$$\Phi_{\mathrm{HBL}}(L) = 10^{-7}(L/10^{44}\ \mathrm{erg/s})^{-1}\ \mathrm{Mpc}^{-3} \tag{1}$$

for gamma ray luminosities $L \geq 10^{44}$ erg s^{-1} from [16]. Owing to pair creation, HBLs beyond a luminosity distance of 1 Gpc will not contribute directly to the EGRB at 100 GeV [8]. However, photons at higher energies can induce electromagnetic cascades due to interactions with the metagalactic radiation field contributing to the EGRB below 100 GeV. Due to the absence of positive evolution of HBLs, the contribution due to cascading from HBLs at higher redshifts can be largely ignored (thinning out of this radiation component due to cosmic expansion). There may be a significant cascading contribution from the nearby HBLs, if they possess a generic bump with $L_\gamma > L_X$ well above 100 GeV before absorption [12]. Fig. 1 shows the possible contribution of HBLs with a peak at multi-TeV energies adopting an HBL template spectrum taking into account pair absorption and subsequent cascading. Presumably, the cascade emission is isotropized by weak magnetic fields, and would be received from the entire (unbeamed) host population of HBLs.

A simple estimate of the total intensity of HBLs (without metagalactic cascading) yields

$$I_{\mathrm{HBL}} = \frac{1}{4\pi} \int_{\mathrm{Gpc}} \int_{10^{44}\ \mathrm{erg/s}}^{10^{47}\ \mathrm{erg/s}} \Phi_{\mathrm{HBL}}(L) \frac{dLdV}{4\pi D_L^2} \simeq 3 \times 10^{-7}\ \mathrm{GeV\ cm^{-2}\ s^{-1}\ sr^{-1}} \tag{2}$$

in agreement with the more detailed calculations shown in Fig. 1.

This is at the level of an extrapolation of the spectrum of the EGRB with a slope of $\alpha = 2.33$ to the energy of 100 GeV. The number of these nearby ($z < 0.27$) HBLs is not large, i.e. $N_{\mathrm{HBL}} = \int_{\mathrm{Gpc}} \int \Phi_{\mathrm{HBL}} L^{-1} dLdV \sim 100$, and their average energy flux level $\bar{F}_\gamma = 4 \times 10^{-8}$ GeV cm^{-2} s^{-1} large enough for detection. The minimum flux $F_X = F_\gamma = L_{\mathrm{min}}/4\pi D_L^2 = 5 \times 10^{-10}$ GeV cm^{-2} s^{-1} (corresponding to 0.3 μJy at 1 keV) is detectable by current telescope facilities if a large amount of observation time will be invested. A total of 10 blazars have been clearly detected by searching candidates from HBL compilations, as can be seen from Tab. 1. In the case of the well-known sources, the observed gamma ray fluxes correspond to low state fluxes of previous detections, and no significant variability has been detected, suggesting that a steady baseline emission has been seen (e.g., [17]). The total census will very likely approach the number expected from the above estimate in the coming years.

In a similar way, the evolving blazars (FSRQ, LBL) could produce a convex EGRB spectrum from their superimposed quiescent (steep) and flaring (hard) spectrum components [4]. As a toy model we evaluate

$$\Phi_\gamma^{\mathrm{AGN}} \propto \int d\mathrm{V} \mathrm{n(z)} (1+\mathrm{z})^2 [(\mathrm{E}(1+\mathrm{z})/\mathrm{E_b})^{-2.33} + (\mathrm{E}(1+\mathrm{z})/\mathrm{E_b})^{-1.5}] \cdot \kappa(\mathrm{E, z})$$

with the source density in the comoving frame $\mathrm{n(z)} \propto (1+\mathrm{z})^{3.4}$ in the redshift range $0.03 \leq \mathrm{z} \leq 1.5$ using an attenuation factor κ [12] to account for pair production. This spectrum can be considered as an upper limit to the blazar contribution. Together with the steep power law component $\propto E^{-2.33}$ (the upper end of which we have shown to be consistent with being produced by the HBLs), this brackets the expected contribution from unresolved blazars to the EGRB (Fig. 2).

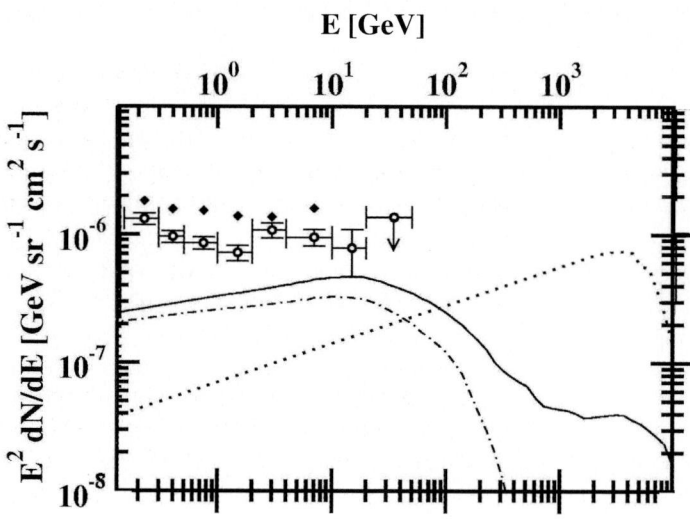

Fig. 1. Contribution of HBLs with multi-TeV peak to the EGRB (dotted line), after pair attenuation (dot-dashed line), and including the effect of subsequent inverse-Compton scattering off the microwave photons (solid line) (for details see [12]).

3 Gamma Rays due to annihilating SUSY Dark Matter

Given that gamma ray bursts remain well behind blazars when comparing their output in terms of calorimetry, the only other conceivable high energy source in the Universe which could be important for the EGRB are dark matter particles with a mass that exceeds the energies of the photons comprising the EGRB. Weakly interacting dark matter particles have long been suspected to exist, since they are independently predicted by the theory of supersymmetry. Supersymmetry could have a conserved quantum number and thus a lightest, stable particle, usually the neutralino [19]. Self-annihilation of the neutralinos produces gamma rays, starting from the time of thermal freeze out in the hot Big Bang and extending until the present day, thereby producing diffuse, isotropic background radiation. Since the neutralinos do not carry electrical charge, $\chi\chi \to \gamma\gamma$ and $\chi\chi \to Z^0\gamma$ is loop-suppressed. For an observed halo at redshift z and energy E, annihilation will instead dominantly lead to jet fragmentation, dumping much of the energy into continuum gamma rays with differential energy distribution $df\,(E(1+z))$. The gamma ray intensity due to "cosmological" neutralino annihilation is then given by

$$I_\gamma\,(E) = c\,(8\pi H_0)^{-1}\,\langle\sigma v\rangle_\chi\,\Omega_{DM}^2\,\rho_{crit}^2 m_\chi^{-2}$$

$$\int_0^{z_{max}} [(1+z)^3\,\kappa(E,z)\,\Gamma\,(z)\,\Psi df_{E(1+z)}]\xi^{-1}\,(z)\,dz \quad (3)$$

[19, 20, 21], where ρ_{crit} is the critical density. Gamma ray attenuation is included via the attenuation function $\kappa(E,z)$ (for $0 < z \le 5$ we use the attenuation derived from the star formation history [12] and for z > 5 the absorption derived from interactions with the cosmological relic radiation field [22]). $\Gamma(z)$ denotes the intensity enhancement due to structure formation [23]. The range of integration is limited to $0 \leq z \leq 20$; gamma rays from higher redshifts are negligible. The parameter $\xi(z)$ is given by

$$\xi(z)^2 = \Omega_M (1+z)^3 + \Omega_K (1+z)^2 + \Omega_\Lambda .$$

In this work, we employ the cosmological "concordance model" of a flat, dark energy and dark matter dominated Universe with the parameters $(\Omega_\chi, \Omega_{DM}, \Omega_M, \Omega_K, \Omega_\Lambda) = (0.23, 0.23, 0.27, 0, 0.73)$. For the dimensionless Hubble-Parameter h we use the value 0.71. The annihilation induced intensity scales quadratically with the local overdensity and thus strongly depends on the amount of structure present in the dark matter. The intensity enhancement due to structure formation is sensitive to the predominant density profile of the dark matter halos. We employ the NFW halo profile [24], and a lower mass cutoff for the dark matter halos of 10^5 solar masses. Additional clumping from smaller scales (1...10) is accounted for by the factor Ψ. Adding the annihilation spectrum to the steep power law, and fitting this combined model to the measured EGRB, the best fit values for the neutralino mass is $m_\chi = 515^{+110}_{-75}$ GeV, which is well in line with what can be expected within the minimal supersymmetric standard model [14] where the mass of the neutralino is close to the electroweak symmetry breaking scale $m_\chi \sim (\sqrt{2}G_F)^{-0.5} = 246$ GeV.

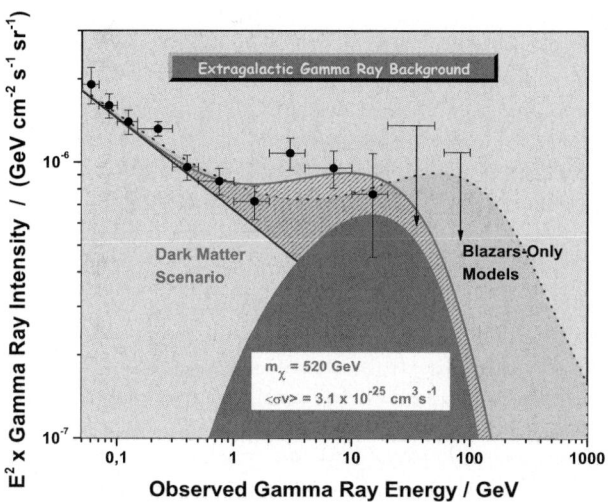

Fig. 2. Extragalactic gamma-ray background: spectrum as determined from EGRET data by Strong et al. (data points); the upper limit in the (60–100) GeV range is from Sreekumar et al.; the model curves are explained in the text

4 Comparison of astrophysical and combined EGRB Models

The maximum astrophysical model implies that the fraction of "hard-spectrum" sources or source states at an energy of 1 GeV is 20% - considerably more than the fraction of the hard-spectrum sources in the EGRET catalogue. It may be questioned, if flares with a hard spectrum contribute significantly to the EGRB due to their low duty cycle. Also, the origin of the highest energy photons detected by EGRET has been shown to be generally inconsistent with known blazar positions [6, 7], perhaps hinting that blazars may in fact not be the dominating contributors to the EGRB at energies beyond 10 GeV. The number of sources discovered above ~ 100 GeV by imaging air Cherenkov telescopes seems to be in line with what is expected for HBLs, but even including the effect of metagalactic casacading on a generic multi-TeV bump does not seem to fully explain the excess in the newly determined spectrum of the EGRB. However, only the details of the gamma ray spectra unfolded from the effects of pair attenuation will tell the story at completeness. There is an urgent need of increasing the number counts of HBLs and measuring their spectra at highest energies. The imaging air Cherenkov telescopes (e.g., H.E.S.S., MAGIC, VERITAS) play the key role in the latter respect, whereas the survey capability of GLAST will increase the overall census by a large margin.

If the sharpness of the multi-GeV excess is not a spurious feature due to an erroneous foreground subtraction, the neutralino model provides a natural explanation. The decay spectrum of heavy dark matter particles is sharply peaked at an energy of $m_\chi c^2/20 \sim 20$ GeV, and the flux is set by the amount of clumping in dark matter halos (structure formation simulations) and the dark matter density (freeze-out and cosmological precision measurements). Taking seriously the neutralino hypothesis, one may ask about complimentary astrophysical means of detection. The gamma luminosity of the Galactic Center (GC) itself strongly depends on the existence of an inner cusp in the galactic dark matter halo, which has been questioned in the recent literature. Here, we settle for the case of a moderately cusped NFW profile, a scenario in which the Galactic Center would exhibit a luminosity due to annihilations of our neutralino candidate of $5 \cdot 10^{35}$ ergs/s above 1 GeV from within 10^{-3} sr. We note that even a very shallow profile in the case of the Milky Way would not be in contradiction with the assumption of a high extragalactic intensity due to dominantly cusped profiles, since most of the contribution to the EGRB is due to halos on a different mass scale than the Milky Way halo. Gamma ray emission from the GC due to dark matter annihilation could be searched for in future ground based and satellite experiments. However, the already detected strong TeV-emission from the GC (probably due to supernova remnants) will make background suppression difficult [25, 26, 27]. There should also be a faint excess from the galactic halo which is, however, washed out by the inverse-Compton radiation component. An excess of positrons above the expectations from cosmic ray transport models corresponding to a similar neutralino mass has been reported [28]. External galaxies like M 87 should also be interesting targets for IACT observations [29]. With an energy threshold of 50 GeV and the halo profile from [30], in case of the presented dark matter scenario M87 would be detectable with 5σ in 100 hours of observation time using the MAGIC-II stereo telescope.

Acknowledgments. Support by the BMBF Verbundforschung (O5CM0MG1), the Helmholtz Gemeinschaft (VIHKOS), and the Bavarian Network of Excellence (ENB) is gratefully acknowledged.

References

1. R. C. Hartman et al.: ApJS **123**, 79 (1999).
2. P. Sreekumar et al.: Astrophys. J. **494**, 523 (1998).
3. A. W. Strong, I. V. Moskalenko and O. Reimer: Astrophys. J. **613**, 956 (2004).
4. F. W. Stecker and M. H. Salamon: Astrophys. J. **464**, 600 (1996).
5. J. Chiang and R. Mukherjee: Astrophys. J. **496**, 752 (1998).
6. D.J. Thompson, D.L. Bertsch and R.H.Jr. O´ Neall: Astrophys. J.Suppl. **157**, 324 (2005).
7. D.S. Gorbunov, P.G. Tinyakov, I.I. Tkachev and S.V. Troitsky: MNRAS **362**, L30 (2005).
8. T. Kneiske et al.: Astron. Astrophys. **413**, 807 (2004).
9. T. Totani: Astropart. Phys. **11**, 451 (1999).
10. R. Protheroe and T. Stanev: MNRAS **264**, 191 (1993).
11. Z.G. Dai, B. Zhang, L. Gou, P. Mészaros and E. Waxman: Astrophys.J.Lett. **580**, L7 (2002).
12. T.M. Kneiske and K. Mannheim: AIPC **745**, 578 (2005).
13. C.A. Scharf and R. Mukherjee: Astrophys.J. **580**, 154 (2002).
14. D. Elsaesser and K. Mannheim: Phys.Rev.Lett. **94**, 171302 (2005).
15. D. Horan and T. C. Weekes: New Astron. Rev. **48**, 527 (2004).
16. V. Beckmann, D. Engels, N. Bade and I. Wucknitz: Astron. Astrophys. **401**, 927 (2003).
17. J. Albert i Fort et al.: (MAGIC Collaboration) Astrophys.J. **639**, 761 (2006).
18. J. Albert i Fort et al.: (MAGIC Collaboration) Astrophys.J. **submitted**, (astro-ph/0603) (2006).
19. L. Bergstrom, J. Edsjo and P. Ullio: Phys. Rev. Lett. **87**, 251301 (2001).
20. P. Ullio, L. Bergstrom, J. Edsjo and C. Lacey: Phys. Rev. D **66**, 123502 (2002).
21. D. Elsaesser and K. Mannheim: Astropart. Phys. **22**, 65 (2004).
22. A. A. Zdziarski and R. Svensson: Astrophys. J. **344**, 551 (1989).
23. J. Taylor and J. Silk: MNRAS **339**, 505 (2003).
24. J. F. Navarro, C. S. Frenk and S. D. M. White: Astrophys. J. **462**, 563 (1996).
25. K. Kosack et al. (VERITAS Collaboration): Astrophys.J. **608**, L97 (2004).
26. F. Aharonian et al. (H.E.S.S. Collaboration): Astron. Astrophys. **425**, L13 (2004).
27. J. Albert i Fort et al. (MAGIC Collaboration): Astrophys.J. Lett. **638**, L110 (2005).
28. M. Duvernois et al. (HEAT Collaboration): APS Meeting, Albuquerque, New Mexico, April 20-23 , 11006 (2002).
29. E. A. Baltz et al.: Phys. Rev. D **61**, 023514 (1999).
30. J. C. Tsai: Astrophys. J. **413**, L59 (1993).

Resolving the Dust Tori in AGN with the VLT Interferometer

K. Meisenheimer[1], K. Tristram[1], and W. Jaffe[2]

[1] Max-Planck-Institut für Astronomie, Königstuhl 17, D–69117 Heidelberg, Germany,
    meise@mpia.de
[2] Sterrewacht Leiden, 2333 CA Leiden, The Netherlands

Summary. The MID-infrared Interferometric instrument (MIDI) at the VLT Interferometer (VLTI) is the ideal instrument to resolve the emission of the putative dust tori in nearby AGN. Indeed, the first successful MIDI observations of the brightest and nearest Seyfert II galaxies – NGC 1068 and the Circinus galaxy – already provided us with an unexpected wealth of information about the dust distribution and properties in the inner few parsec. Observations of the closest radio galaxy – Centaurus A – reveal a unresolved $10\,\mu$m source ($<\ 0.2\,$pc) which presumably represents the base of the radio jet. About a dozen extragalactic sources will be accessible to MIDI, and will allow us to tackle two important issues of AGN physics: (1) are torus-shaped dust structures and our various viewing angles into them indeed responsible for the apparent differences in Seyfert I and II galaxies, and (2) how does the gas reservoir in these tori feed the accretion disks around the black holes?

1 Dust Tori in AGN: what do we expect ?

There are several lines of indirect evidence that in many Active Galactic Nuclei (AGN) the central engine – a super-massive black hole fed by an surrounding accretion disk – is embedded into a axi-symmetric, geometrically thick structure of gas and dust: the so-called torus. This dusty torus is held responsible for the observed dichotomy between Seyfert galaxies of type I and II, respectively: in type I, an unobscured view along the torus axis allows us to look directly into the central engine, while in type II objects the torus is seen edge-on and thus blocks our view towards the very center.

There exist many different models for the geometry and density distribution of those tori in the literature. They mostly try to explain the overall SEDs of Type I and type II AGN as well as their relative abundance by radiative transfer in disk- or torus-shaped dust distributions. In order to understand how these dust tori might look at various mid-infrared wavelengths and spatial resolutions, we have carried out three-dimensional radiative transfer calculations for various dust geometries, wich predict SEDs and images at various wavelengths [6]. Examples can be found in the contribution by Schartmann et al. (these proceedings). However, experience shows that it is impossible to decide between different models, as long as the emission of the dust structure remains unresolved. Most models predict, that the AGN-heated dust (which radiates at mid-infrared wavelengths) is confined to the innermost few parsecs. Thus, even for the closest AGN, a resolution of $\lesssim 20$ mas is mandatory. This can only be achieved by interferometry.

2 The Seyfert II Case: Resolving the Torus

2.1 First Case Study: NGC 1068

First MIDI observations of the proto-typical Seyfert II galaxy, NGC 1068, have been obtained during VLTI *Science Demonstration Time* (SDT).[3]

Two independent baselines were observed in 2003: a longer (78 m) and a shorter baseline (42 m). As the results of these observations have been published already [3], we highlight only our essential conclusions: NGC 1068 contains a well resolved warm dust component of T = 320 ± 20 K and size $d \times h = 49 \times 30$ mas^2 (i.e. 3.4×2.1 parsec at the distance of NGC 1068). Embedded into it is a hot component of T> 800 K, which is marginally resolved ($h = 0.7 \pm 0.2$ pc) in North-South direction but poorly constrained ($d \lesssim 1$ pc) along East-West. We interpret the 300 K component as the emission of a geometrically thick dust structure ("torus"). The hot component can naturally be explained by the emission of dust near sublimation temperature, expected at the inner walls of an axial funnel which allows the radio jet and ionized gas to emerge along the source axis. New insight into the central parsec of NGC 1068 has been gained by comparing the model inferred from the MIDI observations with recent VLBA maps in the radio continuum (5 GHz) by Gallimore et al. [1]: the radio continuum (most likely due to free-free emission of ionized gas) displays a disk-like structure with some indication of flaring towards its outer rim. This disk has the size of the hot component, suggesting that we witness the evaporation process at the inner wall of the funnel.

2.2 Second Case: the Circinus Galaxy

The Circinus galaxy is the closest Seyfert II galaxy (at 4 Mpc distance), and has been observed as part of the MIDI guaranteed time program on nearby AGN. This list contains 14 targets for which preparatory observations with TIMMI2 and NACO showed, that their sub-arcsec core fluxes exceed MIDI's sensitivity limit: $S_N \simeq 400$ mJy at 10 μm.

Archival HST observations of Circinus show a wide ionization cone c.f. Fig. 2) but the very core remains hidden behind the patchy dust structure in the disk of the galaxy. Thus, only our recent NACO observations [5] at $\lambda > 2\,\mu$m allow a first glimpse of the core. As it seems marginally resolved already with NACO, we chose the shortest baselines of the VLTI for the MIDI observations. The left panel of Fig. 1 shows the results obtained with the telescope combination UT2–UT3 during two runs in February and June 2004. As MIDI disperses the N-band by a prism, both the total flux F_{tot} (from single telescopes) and the interferometric, correlated flux F_{corr} are spectrally

---

[3] The Science Demonstration Time was introduced to demonstrate the capabilities of VLTI in different areas of astronomy. It is coordinated by Francesco Paresce. The AGN observations are coordinated by Huub Röttgering.

resolved ($\lambda/\Delta\lambda \simeq 30$). The first thing to notice is that very low levels of correlated flux $\langle F_{\mathrm{corr}}\rangle \simeq 400$ mJy are sufficient to track the interferometric fringes with MIDI. Second, the N-band spectrum is dominated by the broad the silicate absorption feature. When inspecting the *visibility* $V := F_{\mathrm{corr}}/F_{\mathrm{tot}}$ (not shown) the silicate feature is absent, indicating that the SiO absorption is a global feature (due to foreground material), and does not deepen when picking out the core emission by interferometry.

Due to its southern declination ($\delta = -65°3$), Circinus allows us to obtain a very nice coverage of the $uv-$plane (see left panel in Fig. 1). Although we are still far from understanding the wealth of information contained in these 12 independent, spectrally resolved visibility measurements, one can derive a first impression by assuming that the brightness distribution can be approximated by a Gaussian shape. Then each visibility point (in the Fourier plane) corresponds to a Gaussian HWHM (in the image plane). The result of this approach is shown as points in Fig. 2. The HWHM points roughly follow a box-like contour with largest diameter 1.2 pc and a shorter width of 0.75 pc. Allowing for projection effects (the axis of Circinus is believed to be inclined by about 20°w.r.t. the plane of sky, see [8]), we derive a minimum width of about 0.35 pc ($h/d = 0.29$). Obviously the dust torus of Circinus is geometrically much thinner than that in NGC 1068 – in agreement with the fact that its ionization cone is much less collimated than that of NGC 1068. This finding and the perfect alignment with the ionization cone is the most striking observational evidence that dust tori indeed shape the ionization cones of Seyfert II galaxies.

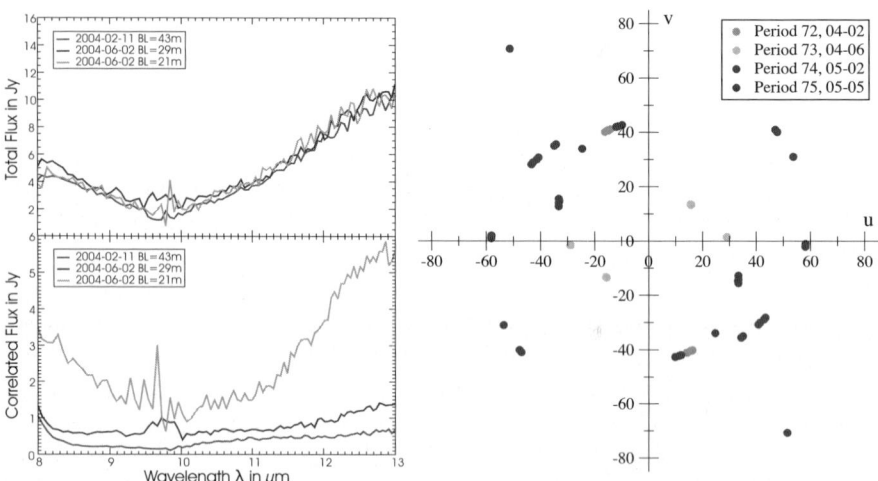

Fig. 1. Results of the first MIDI observations of the Circinus galaxy.*Left:* Total flux and correlated flux for three observations in 2004. Note that the region $9.5 < \lambda < 10.1\,\mu$m is strongly affected by atmospheric ozone absorption and should be discarded. *Right:* Current UV coverage of Circinus with MIDI observations given in units of m for the projected baseline.

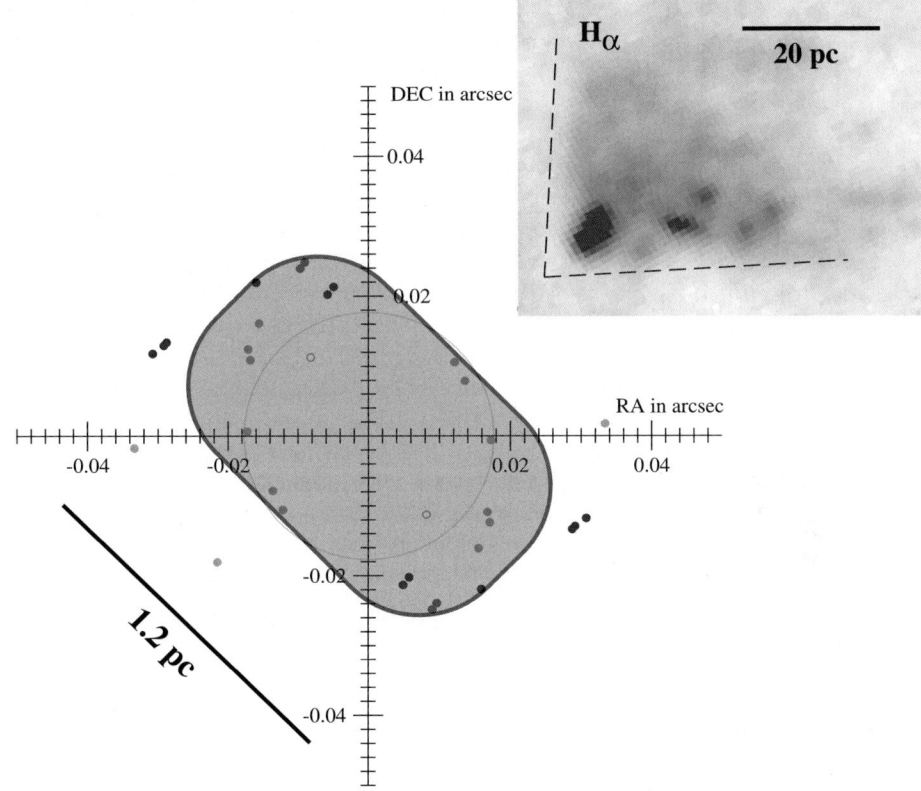

Fig. 2. The dust torus in the Circinus galaxy. The Gaussian half-width at half maximum HWHM (•) is derived from the observed visibility at 12 μm at each of the uv-points shown in Fig. 1. They can be represented by a boxy structure. The insert shows the ionization cone observed in H_α by [8]. Note the different scales.

3 A Special Case: the Radio Galaxy Centaurus A

The faintest extragalactic source which MIDI has measured so far is the famous radio galaxy Centaurus A (average core flux in the N-band < 1 Jy). Two projected baselines of about 60 m were observed with UT2–UT4 in February 2005 and another two with the almost orthogonal baseline UT2–UT3 in May 2005. As demonstrated in Fig. 3, the correlated flux can well be measured and leads to consistent results. Due to the low contrast between the total flux in a 0.″6 aperture and the bright background, the total flux measurements are much more uncertain. Nevertheless, it is clear that the visibility of Centaurus A is on the level of 80% at $\lambda < 9\mu$m, with some indication of a decrease towards the longest wavelengths. We interpret the measurements in terms of an unresolved source ($d < 0.2$ pc) which might be surrounded by a faint and well resolved component of cool dust (T < 300 K). The limited wavelength coverage of the MIDI observations alone would not allow to decide between a thermal or non-thermal origin

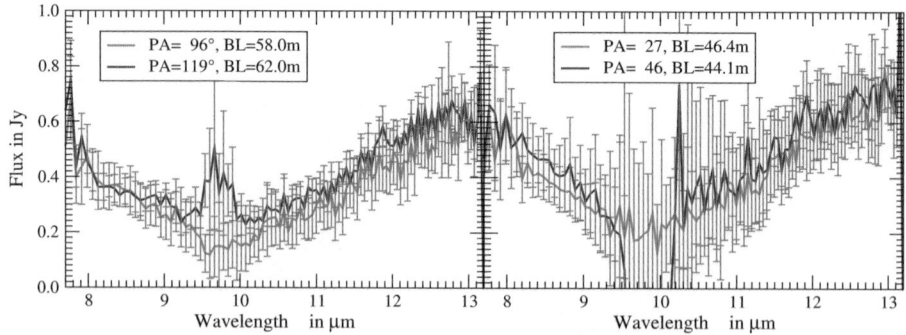

Fig. 3. Spectrum of the correlated flux F_{corr} from Centaurus A as observed on Febuary 28 (left) and on May 26, 2005 (right). Wavelengths $9.5 < \lambda < 10.1\,\mu$m are affected by the atmospheric ozone band. F_{corr} is almost constant, despite the three months interval between the observations and orthogonal baseline orientations. As in Circinus the spectral shape at $8.5 < \lambda < 12\,\mu$m is dominated by SiO absorption.

of the unresolved radiation. However, when considering the overall spectrum from the radio to the near-IR at sub-arcsec resolution (Fig. 4), it is likely that the unresolved mid-infrared core of Centaurus A is a synchrotron source, presumably the base of the radio jet seen on VLBI maps. We find no hint of a compact AGN-heated dust structure.

4 First Conclusions

From these first interferometric observations of the three closest AGN observable from Cerro Paranal we can already draw some rather general conclusions:

- The sensitivity and resolution of the VLT Interferometer using the unit telescopes at $\lambda \simeq 10\,\mu$m is well adapted to study the dusty tori in nearby AGN. The low-dispersion mode of MIDI gives us completely new insights into the dust (and hence gas) distribution within the central few parsecs.
- Our first case studies of two nearby Seyfert II galaxies, NGC 1068 and the Circinus galaxy, reveal that compact AGN-heated dust structures of radius $r_{300\mathrm{K}} \leq 2\mathrm{pc} \simeq 4r_{\mathrm{sub}}$ do indeed exist. In NGC 1068 it is geometrically thick (torus-like: $h/r > 0.5$), in Circinus the dust distribution is thinner (disk-like: $h/r < 0.3$) which naturally accounts for the large opening angle of its ionization cone. In NGC 1068 we find a very hot compact component $r \simeq r_{\mathrm{sub}}$, which might be the emission from the hot inner walls of the axial funnel within the torus.
- In the radio galaxy Centaurus A, we find no hints of a compact nuclear dust distribution. So even in cases where a high dust column obscures our view into the core region on the hundred pc scale, there might exist pretty "naked" (almost gas and dust free) AGN cores.

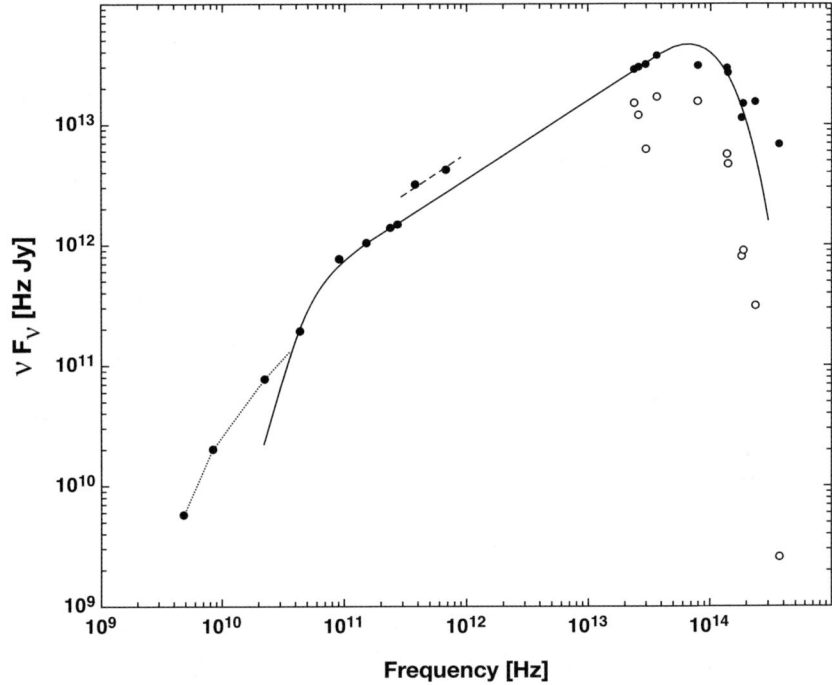

Fig. 4. Overall spectrum of the core of Centaurus A. Our MIDI results have been supplemented by own mm photometry and NACO results, plus literature values from [7],[2],and [4]. Open circles show observed flux, filled dots are corrected for an extinction of $A_V = 15.1$. The synchrotron spectrum (solid line) shows a power-law $F_\nu \sim \nu^{-0.35}$ which cuts off above $\nu_c = 8 \times 10^{13}$ Hz, and is self-absorbed below $\nu_1 = 5.5 \times 10^{10}$ Hz. Note the apparent variability around 3×10^{11} Hz and above ν_c.

Acknowledgments. Nothing of this would have been possible without the MIDI team, of whom we mention Ch. Leinert, R. Waters, U. Graser, B. Lopez, G. Perrin, O. Chesneau and R. Köhler and the VLTI team: A. Glindemann, A. Richichi, S. Morel, M. Schöller, M. Wittkowski, and many others.

References

1. J. Gallimore, S. A. Baum and C.P. O'Dea: ApJ **613**, 794 (2004)
2. T.G. Hawarden, G. Sandel, H.E. Matthews et al.: MNRAS **260**, 844 (1993)
3. W. Jaffe, K. Meisenheimer, H. Röttgering et al.: Nature **429**, 47 (2004)
4. A. Marconi, E.J. Schreier, A. Koekemoer at al.: ApJ **528**, 276 (2000)
5. A. Prieto, K. Meisenheimer, O. Marco et al.: ApJ **614**, 135 (2004)

6. M. Schartmann, K. Meisenheimer, M. Camenzind, S. Wolf and Th. Henning: A&A **437**, 861 (2005)
7. S.J. Tingay, D.L. Jauncey, J.E. Reynolds et al.: AJ **115**, 960 (1998)
8. A. S. Wilson, P. L. Shopbell, C. Simpson at al.: AJ **120**, 1325 (2000)

Iron K Lines of AGN in the X–Ray Background

A. Müller and G. Hasinger

Max–Planck–Institute for Extraterrestrial Physics, Garching, Germany,
(amueller,ghasinger)@mpe.mpg.de

Summary. X–ray deep field observations of the Lockman hole reveal a wealth of active super-massive black holes at large distances. All these AGN together contribute to a significant and strong relativistic iron K feature around 6.5 keV rest frame energy. This XMM–observation is modeled by superimposing line emission of standard disks around Kerr black holes. The simulation is based on relativistic ray tracing techniques accounting for radial drift motion of the infalling plasma in the plunging region. This work shows that it is in principle possible to explain the observed feature by a superposition of line emission from a number of disks. Thereby, low inclined disks seem to dominate the feature – a diagnosis that is in contradiction to simple AGN unification schemes based on orientation effects.

1 X–Ray Deep Field Observations of the Lockman Hole

From the 770 ks XMM–Newton deep field observation of the Lockman hole 53 type–1 AGN, 41 type–2 AGN (AGN1s, AGN2s hereafter) and 10 galaxies are extracted that cover $0 < z < 4.5$ in redshift space. Spectral analysis is done for each X–ray source by fitting a single power law modified by intrinsical absorption. Finally, each spectrum is shifted to the rest frame. After this procedure a significant and strong feature remains around 6.5 keV for both, AGN1s and AGN2s [1]. The morphology of the feature is triangular and strongly peaked in each case. Analysing the ratio plots provides a blue line edge at $\simeq 6.7$ keV for AGN1s and at $\simeq 6.8$ keV for AGN2s. FWHM amounts $\simeq 1.2$ keV for AGN1s and $\simeq 0.8$ keV for AGN2s. The similarity of the AGN1 and the AGN2 feature is unexpected. A tiny difference is that the AGN2 feature exhibits a faint blue shoulder around 7 keV rest frame energy.

2 Model of the Iron K Line Features

The simulations of relativistic emission lines are carried out by using ray tracing techniques on the Kerr geometry as presented in earlier work [2, 3]. We only consider the strongest X–ray fluorescence line i.e. iron $K\alpha$ at 6.4 keV rest frame energy. Depending on the specific velocity field of the plasma, the emissivity profile, disk size and disk inclination angle to the observer very different line shapes can result. The whole observed feature is build up by AGN in the Lockman hole field that are captured in different accretion stages, distances, luminosities and orientations. We aim to reproduce that with a suitable AGN population getting an insight into the properties of the black hole–disk systems. The emitting regions are modelled as standard accretion disks including radial drift motion of the plasma as first presented in [2]. Radial motion starts at some radius R_t. The radial emissivity profile is modelled by a cut–power law, $\epsilon(r) \propto \exp\left(-\alpha R_t/r\right) r^{-\beta}$ where $\alpha = \beta = 3.0$ is adopted. The inclination angle has the strongest impact onto line profiles: higher inclinations shift the blue line wing to higher energies and change line morphology from triangular to double–horned. In the following, a set of differently inclined disks is tested. In a first simple approach all other parameters are fixed: Kerr parameter $a = 0.8M$ as suggested by recent black hole growth studies [4], inner and outer disk edge $r_{in} = r_H = 1.6M$, $r_{out} = 30.0M$, and drift radius $R_t = 3.0M$. Fig. 1 displays the results of the X-ray feature simula-

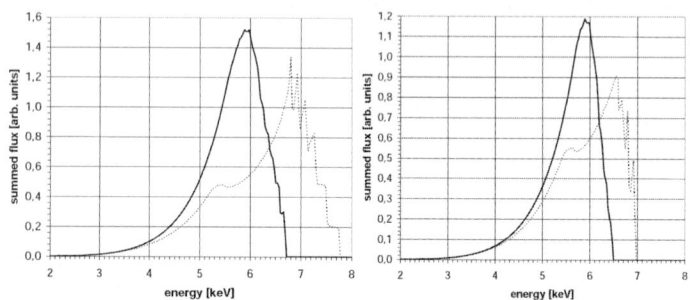

Fig. 1. Simulated Iron K features for AGN1s and AGN2s, details see text.

tions. On the basis of a simple geometrical AGN unified picture [5] the AGN1 sample is composed of low inclined standard disks whereas moderately to high inclined disks contribute to the AGN2 feature. We test different break inclination that divide into AGN1 and AGN2. The feature in the *left panel* is produced by an inclination sample that consists of $i = 5, 10, 15, 20, 25, 30°$ disks to simulate the AGN1s (*solid line*). The blue edge cuts at $\simeq 6.35$ keV in average and FWHM amounts to $\simeq 1.14$ keV. In contrast, the AGN2 feature is composed of standard disks with $i = 35, 40, 45, 50, 55, 60°$ (*dashed line*) endowed with a blue edge at $\simeq 7.22$ keV and a FWHM of $\simeq 1.05$ keV. Due to a high maximum inclination angle, $i = 30°$ respectively $i = 60°$, both features are relatively broad. Comparing to observation, FWHM for the AGN1 feature is in correspondance but *not* the blue edge. The simulated AGN2 feature does *not* at all resemble to the narrow observed feature – either in width nor in edge. Therefore, we adjust a lower break inclination, $i = 20°$, and additionally reduce the maximum inclination to $i = 40°$, see *right panel*. The summed feature for AGN1s orginates from disks with $i = 5, 10, 15, 20°$ (*solid line*). The blue edge cuts at $\simeq 6.23$ keV and FWHM amounts only to $\simeq 0.95$ keV for AGN1s. AGN2s are simulated with disks obeying $i = 25, 30, 35, 40°$ that result in a blue edge at $\simeq 6.75$ keV and FWHM of $\simeq 1.57$ keV (*dashed line*). Comparing again to the observation, both, FWHM and blue edge energy for AGN1 feature are too low in the simulation. In contrast, the blue edge of the AGN2s is compatible with observation but the FWHM is far too broad.

3 Conclusions

In this contribution a strong X–ray feature observed with XMM–Newton in the Lockman hole is simulated by superimposing single iron $K\alpha$ emission lines originating from differently inclined accretion disks. The observed narrow and triangular X–ray feature can in principle be reproduced in that way. However, in this first analysis it was not possible to exactly reproduce the observed FWHM *and* the position of the blue line edge simultaneously. It was shown that a consideration of high inclined disks shifts the feature to far into the blue. Obviously, low inclination disks dominate the feature, surprisingly for both, AGN1s *and* AGN2s. But simple AGN unification models that involve only orientation effects from a dusty torus [5] demand that AGN2s exhibit moderately to high inclined accretion disks. Therefore, we conclude that the distinction of AGN1 vs. AGN2 is more than a one–parameter problem governed by inclination. This hypothesis gets independent support from hard X–ray observations that showed a decreasing fraction of AGN2s with increasing luminosity [6]. The observed weak blue shoulder at $\simeq 7$ keV for AGN2s may indicate that some higher inclined disks with $i > 40°$ contribute. However, as been shown here, a strong contribution ends up with a line composit that is too broad. Alternatively other fluorescing elements such as Nickel may play a role to generate the blue shoulder. The peaky AGN2 feature may have underlying narrow iron lines that are reflections at the dust torus. Further investigations are needed to get support to these preliminary results.

References

1. A. Streblyanska, G. Hasinger, A. Finoguenov, et al: A&A **432**, 395 (2005)
2. A. Müller, M. Camenzind: A&A **413**, 861 (2004)
3. A. Müller: Black Hole Astrophysics: Magnetohydrodynamics on the Kerr geometry. PhD Thesis, University of Heidelberg, Germany (2004)
4. S. L. Shapiro: ApJ **620**, 59 (2005)
5. R. R. J. Antonucci & J. S. Miller: ApJ **297**, 621 (1985)
6. Y. Ueda, M. Akiyama, K. Ohta & T. Miyaji: ApJ **598**, 886 (2003)

Black Hole Mass and Growth Rate and Metal Enrichment at Low and High Redshift

H. Netzer

School of Physics and Astronomy and the Wise Observatory, Israel,
netzer@wise.tau.ac.il

Summary. Large spectroscopic samples of active galactic nuclei (AGNs), combined with improved methods for calculating black hole mass and accretion rate, enable a meaningful statistical investigation of the mass, accretion rate and metallicity of such sources as a function of redshift. The results shown here are based on the study of about 10,000 $z \leq 0.75$ radio-quiet SDSS AGNs and about 30 $z = 2.3 - 3.4$ high redshift, very high luminosity AGNs. Their combination suggests that: 1. The fraction of more massive active black holes is larger at earlier times. 2. The normalized accretion rate (L/L_{Edd}) is an increasing function of redshift for black holes of all masses. 3. The slope of the accretion rate vs. redshift correlation is similar to the slope of the cosmic star formation rate over the redshift interval 0–0.75. 4. Metallicity as measured by the NVλ1240/CIVλ1549 line ratio at high redshifts, and by FeII/H_β at low redshifts, is an increasing function of L/L_{Edd}. 5. Most AGNs do not have enough time to grow to their present size given their present (observed) accretion rate. 6. There must have been several episodes of increased broad line gas metallicity, as well as of low metallicity, in the history of most active black holes. Thus, the BLR metallicity goes through cycles and is likely correlated with the large scale galactic star formation history.

1 Introduction

The evolution of massive black holes (BHs), their growth rate and the metallicity of the gas in their immediate vicinity can now be studied in a quantitative way in big active galactic nuclei (AGNs) samples. The most important development in this area is the improvement of reverberation mapping techniques for type-I AGNs which provides a reliable method for deriving the size of the broad line region (BLR) as a function of the optical continuum luminosity, the H_β line luminosity, the ultraviolet continuum luminosity and the X-ray (2–10 keV) luminosity (Kaspi et al. 2005 and references therein). This has been used in numerous papers to obtain a "single epoch" estimate of the BH mass by combining the derived BLR size with a measure of the gas velocity, e.g. FWHM(H_β) (see Netzer 2003; Vestergaard 2004; Shemmer et al. 2004; Baskin and Laor 2005 for more details as well as for discussion of the various limitations of this method). The uncertainty in the method, using the optical continuum luminosity as the BLR-size indicator, is about a factor of 2–3.

This review illustrates the application of such methods to two types of AGNs: a small number of very high luminosity AGNs at high redshifts and a large number of AGNs at $z \leq 0.75$. The first of those is based on Shemmer et al. (2004) and the second on Netzer and Trakhtenbrot (2006). The Netzer and Trakhtenbrot paper makes use of the Sloan Digital Sky Survey (SDSS) spectroscopic archive and follows and extends the earlier work of Dunlope and Jarvis (2004) and others.

In what follows I shall use the following expressions to convert optical continuum luminosity (λL_λ measured at 5100Å) and FWHM(H_β) to black hole mass and accretion rate:

$$M(BH) = 1.05 \times 10^8 \left[\frac{L_{5100}}{10^{46}\, \text{erg sec}^{-1}} \right]^{0.65} \left[\frac{FWHM(H_\beta)}{1000\, km/sec} \right]^2 M_\odot, \quad (1)$$

$$L/L_{Edd} = \frac{f_L L_{5100}}{M(BH)}, \quad (2)$$

where the bolometric correction factor, f_L, is a function of L(5100Å). The exact luminosity dependence of f_L is not known but it is suggested that lower luminosity AGNs have, on the average, larger f_L. The value adapted here is $f_L = 7$.

2 Mass Accretion Rate and Metallicity at High Redshift

Mass and accretion rates can be measured in those high redshift sources whose H_β line falls in one of the infrared bands (J, H or K). This restricts the redshift bands to around 1.6, 2.3 and 3.4. Work of this type has been done by McIntosh et al. (1999), Yuan and Wills (2003), Shemmer et al. (2004) and others.

Shemmer et al. (2004) used their IR spectra to derive black hole masses and accretion rates for a sample of about 30 very luminous AGNs. Some of the masses found in this work approach $10^{10}\, M_\odot$. Most of these high luminosity sources were found to have extremely large accretion rates, with L/L_{Edd} in the range of 0.3–1.0.

Shemmer et al. also used published spectroscopy to estimate the metallicity of the gas in the BLRs of the sources. The method used is based on the Hamann and Ferland (1993) suggestion that the NVλ1240/CIVλ1549 line ratio is a good metallicity indicator (see also Hamann et al. 2002; Dietrich et al. 2002; Dietrich et al. 2003). The sample was supplemented by a group of low luminosity narrow line Seyfert 1 galaxies (NLS1s) observed from the ground and from space. The luminosity and the black hole masses for those low redshift sources are 2–3 orders of magnitudes smaller than those of the high-z AGNs. Using this enlarged sample they were able to show that the prime metallicity driver is the mass accretion rate. This result, which is shown in Fig. 1, is in contradiction to the original Hamann and Ferland (1993) suggestion that the broad line region (BLR) metallicity is driven by the source luminosity.

3 Mass Accretion Rate and Metallicity at Low Redshift

The availability of large AGN samples, like the SDSS and the 2dF (e.g. Adelman-McCarthy et al. 2006; Croom et al. 2004), makes statistical analysis of the luminosity,

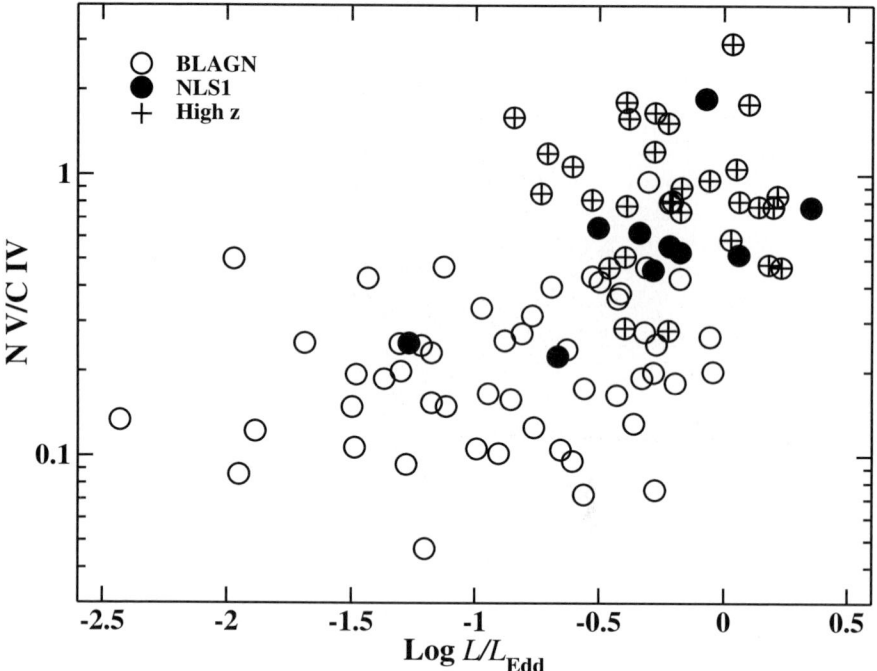

Fig. 1. Metallicity as a function of normalized accretion rate, for a sample of high luminosity high redshift AGNs as well as low luminosity, low redshift NLS1s (adapted from Shemmer et al. 2004)

mass and accretion rate of thousands of active BHs feasible for the first time. In a recent work (Netzer and Trakhtenbrot 2006) we have calculated those properties for about 10,000 radio quiet type-I AGNs with $z \leq 0.75$. This redshift limit was chosen to allow a proper fitting of the H_β line profile, thus obtaining the most reliable mass estimate. The large number of sources allow us to split the sample into several mass and accretion rate sub-groups and to study their properties as a function of redshift. One such correlation is shown in Fig. 2 where we plot L/L_{Edd} as a function of redshift for such sources. The diagram shows all radio-quiet sources as well as three mass groups (see figure caption for details). We also show, for comparison, the high redshift Shemmer et al. (2004) AGNs that are characterized by much larger masses and very large accretion rates.

Fig. 2, as well as other correlations given in Netzer and Trakhtenbrot (2006), clearly illustrates the increasing fraction of more massive BHs at larger redshifts as well as the fact that the accretion rate of all sources, regardless of their mass group, is an increasing function of redshift. This last correlation can be expressed as

$$L/L_{Edd} \propto z^{\gamma(M)} \tag{3}$$

where $\gamma(M)$ is varying from about 0.8 for M(BH)=10^7 $M_\odot$ to about 2.2 for M(BH)=$10^{8.6}$ $M_\odot$. The actual value of $\gamma(M)$ must be treated with care since it is influenced by the flux limit of the SDSS sample (see those limits in Fig. 2 and further discussion in Netzer

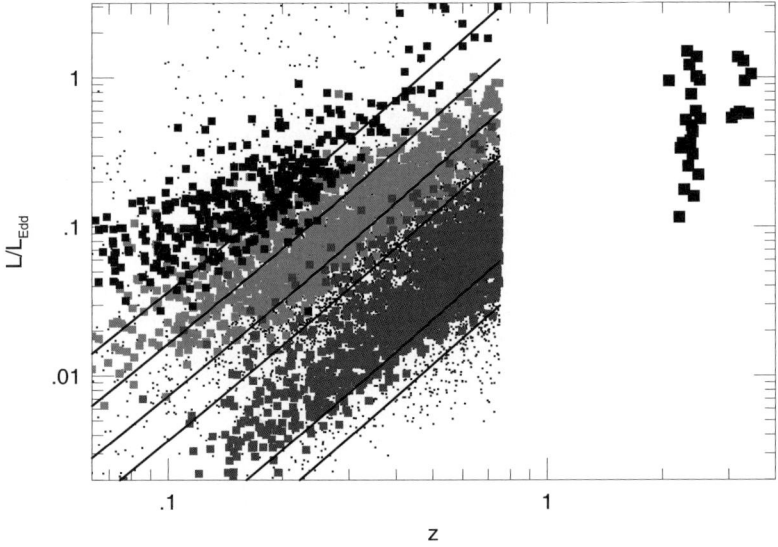

Fig. 2. L/L_{Edd} vs. redshift for radio-quiet type-I SDSS AGNs (small black points) and for high redshift high luminosity AGNs (large black squares). The three mass groups shown on the diagram are M(BH)=$10^{6.8-7.2}$ $M_\odot$ (large black points), M(BH)=$10^{7.5-7.8}$ $M_\odot$ (large red / light grey points), M(BH)=$10^{8.5-8.8}$ $M_\odot$ (large blue / dark grey points). The solid lines are the SDSS flux limits for the various groups (two per mass group, for the lower and upper mass limits).

and Trakhtenbrot 2006). It is interesting to note that the slope ($\gamma(M)$) is similar to the slope of the star formation rate (ρ_*) over the same redshift interval.

The SDSS sample allows us to study also the FeII/H_β line ratio as a function of redshift and other parameters. Here FeII stands for the combined intensity of several FeII blends in the wavelength interval 4400-4700Å(e.g. Boroson and Green 1992). We find a strong correlation of this line ratio with the normalized accretion rate, L/L_{Edd}, a trend that has been suggested by the study of several previous small samples (e.g. Boroson and Green 1992; Grupe et al. 2004). This correlation is shown in Fig. 3.

The understanding of the FeII line emission in type-I AGNs has been an outstanding issue for many years. The hundreds of energy levels, thousands of emission lines, inaccurate atomic data and complicated radiative transfer have severely limited progress in this area. Early detailed works by Netzer and Wills (1983), and by Wills, Netzer and Wills (1985), reached the conclusion that the total FeII line intensities can be understood to within a factor of ~ 3 and the predicted relative strength of some easy to measure features roughly agree with the observations. However, the theory is too uncertain to deduce the iron abundance by comparing theory to observed line ratios such

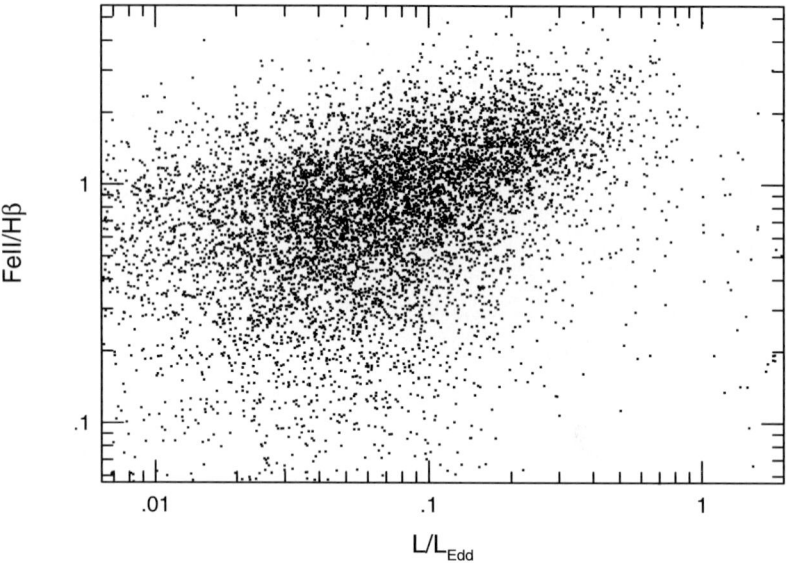

Fig. 3. FeII/H_β vs. L/L_{Edd} for radio-quiet type-I AGNs

FeII/H_β. More detailed work by Baldwin et al. (2004), based on improved atomic data and sophisticated calculations, reached a similar conclusion.

The correlations shown in Figs. 1 and 3 here suggest an alternative solution to this problem. Since L/L_{Edd} is strongly correlated with metallicity, in particular with the N/C abundance ratio, and since FeII/H_β is strongly correlated with L/L_{Edd}, it is suggested that FeII/H_β is indeed an iron abundance indicator. Thus, observed correlations in SDSS objects provide an answer to the iron abundance problem that could not have been obtained by the most advanced emission line models.

4 Growth Time of Massive Black Holes and Time Dependent Metallicity

The growth time of the BHs in the SDSS and Shemmer et al. (2004) samples can be calculated using

$$t_{\text{grow}} = 4 \times 10^8 \frac{\eta/(1-\eta)}{L/L_{Edd}} \log \frac{M}{M_{\text{seed}}} \frac{1}{f_{\text{active}}} \text{ yr}, \qquad (4)$$

where f_{active} is the fraction of time the BH is active and the efficiency of accretion is taken to be $\eta = 0.2$. For the mass of the seed BH, M_{seed}, I assumed 10^4 $M_\odot$ but the calculations are insensitive to the exact value.

Given the growth time, one can calculate the time required to grow to the observed size relative to the age of the universe at the observed redshift, $t_{grow}/t_{universe}$, assuming $f_{\text{active}} = 1$. This is shown in Fig. 4 for all SDSS and for the Shemmer et al. (2004) sources as functions of L/L_{Edd}. It is clear from the diagrams that most sources do not have enough time to grow to their observed mass given their observed accretion rate. This is true for all redshifts and all mass groups. Thus, in the majority of cases there must have been one or more past episodes where the accretion rate was higher than the value found here.

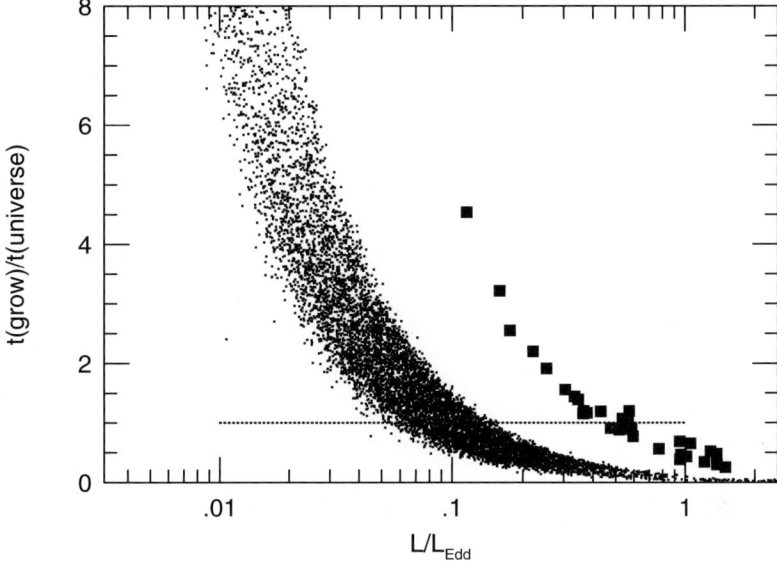

Fig. 4. Growth time in units of the age of the universe as a function of the normalized accretion rate. Small points are SDSS AGNs with $z \leq 0.75$ and large symbols are high redshift, high luminosity objects from Shemmer et al. 2004.

The results obtained for the iron abundance and the growth time, can be combined to infer the changes in the abundance of the BLR gas. Since FeII/H_β is an iron abundance indicator which depends on L/L_{Edd}, and since most type-I AGNs had higher accretion rate episodes in their past, one cannot avoid the conclusion that those past episodes were also characterized by enhanced metallicity relative to the observed value. Thus, BLR metallicity goes through cycles and it is not monotonously increasing with decreasing redshift. One possibility is an association of fast accretion phases with enhanced star formation rate that leads to increased metallicity in star forming regions throughout the galaxy. Such enriched material must find its way to the center of the system where it is enriching the BLR gas and perhaps also the accretion disk. Such a scenario must also involves ejection and/or accretion of the enriched material such that the metal content of the BLR at later times, when the accretion rate is lower, is reduced relative to the abundance at the peak activity phase.

5 Conclusions

The availability of large spectroscopic type-I AGN samples, combined with improved methods for calculating black hole mass and accretion rate, allow a meaningful statistical investigation of several of the more important properties of such sources as a function of redshifts. The results shown here are based on the study of about 10,000 $z \leq 0.75$ radio-quiet SDSS AGNs and some 30 high luminosity high redshift sources. They allow us to draw the following conclusions:

- The fraction of more massive active BHs is larger at earlier times.
- The normalized accretion rate is an increasing function of redshift for all mass BHs.
- The slope of the accretion rate vs. redshift correlation is similar to the slope of the star formation rate over the redshift interval $z \leq 0.75$.
- Metallicity as measured by the NVλ1240/CIVλ1549 line ratio at high redshift, and by FeII/H_β at low redshift, is an increasing function of L/L_{Edd}.
- Most AGNs do not have enough time to grow to their observed size given their present (observed) accretion rate.
- Given those dependences, there must have been several episodes of enhanced metallicity, as well as of low metallicity, in the history of most active BHs. metallicity in the innermost part of AGNs goes through cycles and is different from the monotonously increasing metallicity of the host galaxy.

Acknowledgments. I am grateful to my collaborators Ohad Shemmer, Paulina Lira and Benny Trakhtenbrot whose work and advise contributed a lot to this work. Funding for this work has been provided by Israel Science Foundation grant 232/03 and by the Jack Adler chair of Extragalactic astronomy at Tel Aviv University. I thank the Humboldt foundation for their prize and the host institution, MPE Garching, where most of this work has been conducted.

References

1. Adelman-McCarthy, J. K., et al.: ApJSupp **162**, 38 (2006)
2. Baldwin, J. A., Ferland, G. J., Korista, K. T., Hamann, F., & LaCluyzé, A.: ApJ **615**, 610 (2004)
3. Baskin, A., & Laor, A.: MNRAS **356**, 1029 (2005)
4. Boroson, T. A. & Green, R. F.: ApJSupp **80**, 109 (1992)
5. Croom, S. M., Smith, R. J., Boyle, B. J., Shanks, T., Miller, L., Outram, P. J., & Loaring, N.S.: MNRAS **349**, 1397 (2004)
6. Dietrich, M., Appenzeller, I., Vestergaard, M., & Wagner, S. J.: ApJ **564**, 581 (2002)
7. Dietrich, M., Hamann, F., Shields, J. C., Constantin, A., Heidt, J., Jäger, K., Vestergaard, M., & Wagner, S. J.: ApJ **589**, 722 (2003)

8. Grupe, D.: AJ **127**, 1799 (2004)
9. Hamann, F., Korista, K. T., Ferland, G. J., Warner, C., & Baldwin, J.: ApJ **564**, 592 (2002)
10. Hamann, F. & Ferland, G.: ApJ **418**, 11 (1993)
11. Kaspi, S., Maoz, D., Netzer, H., Peterson, B. M., Vestergaard, M., & Jannuzi, B. T.: ApJ **629**, 61 (2005)
12. McIntosh, D. H., Rieke, M. J., Rix, H.-W., Foltz, C. B., & Weymann, R. J.: ApJ **514**, 40 (1999)
13. McLure, R. J., & Jarvis, M. J.: MNRAS **353**, L45 (2004)
14. Netzer, H., & Wills, B. J.: ApJ **275**, 445 (1983)
15. Netzer, H.: ApJLett **583**, L5 (2003)
16. Netzer, H., & Trakhtenbrot, B.: ApJ **654**, 2, 754-763 (2007)
17. Vestergaard, M.: ApJ **601**, 676 (2004)
18. Wills, B. J., Netzer, H., & Wills, D.: ApJ **288**, 94 (1985)
19. Yuan, M. J. & Wills, B. J.: ApJLett **593**, L11 (2003)

Is the Light Bending Effect at Work in the Core of NGC 4051?

G. Ponti[1,2,3], G. Miniutti[2], M. Cappi[3], A.C. Fabian[2], L. Maraschi[4], and K. Iwasawa[2]

[1] Dipartimento di Astronomia, Università di Bologna, Via Ranzani 1, I–40127, Bologna, Italy,
`ponti@iasfbo.inaf.it`
[2] Institute of Astronomy, Madingley Road, Cambridge CB3 0HA, UK
[3] INAF–IASF Sezione di Bologna, Via Gobetti 101, I–40129, Bologna, Italy
[4] Osservatorio Astronomico di Brera, via Brera 28, 20121 Milan, Italy

Summary. A new interpretation of the X-ray spectral variability of the Narrow Line Seyfert 1 galaxy NGC 4051 is presented. The source shows a high degree of flux and spectral variability. The data show the presence of two constant components; one due to photoionized plasma emission and the other due to neutral reflection from distant material. The nuclear emission, lightly absorbed by warm material ($N_H \sim$few$\times 10^{21}$ cm^{-2}), is decomposed into two variable components: direct power-law emission and relativistic ionized reflection from the accretion disc. The reflected component seems to correlate with the direct component only at low flux, being almost constant elsewhere. This behaviour is predicted by models in which the light bending effect is dominant, thus, these data are consistent with the nuclear emission coming from a few gravitational radii from the central black hole.

We present here the results of a detailed spectral variability analysis of two XMM-Newton observations of the Narrow Line Seyfert 1 galaxy NGC 4051 (z=0.00233).

Figure 1 shows the Root Mean Square (RMS) variability spectra (Vaughan et al. 2003; Ponti et al. 2004) during the medium and low flux observations, left and right panel respectively. When the flux is low a big drop of variability at 0.8-0.9 keV, due to the presence of constant emission from a photo-ionized plasma, is evident. A second drop is present at 6.4 keV indicating the presence of a second constant component, associated with the narrow core of the Fe Kα emission line, coming from material distant from the variable illuminating source. During the high flux observation (rev. 263, left panel of Fig. 1) these narrow features are weaker and the broad band RMS spectrum has the typical shape observed in type 1 objects, with the peak of variability around 1 keV and lower variability at higher energy. According to the RMS spectra, the Γ–flux relation and the time resolved spectral variability, the source emission could be decomposed into a direct power law component with constant spectral index ($\Gamma \sim 2.2$) and an ionized reflection component coming from the inner part of the accretion disc (see Fig. 2). In this model the soft excess is due to the ionized reflection, as in Crummy et al. (2006), and the large variations in the spectral index above 2 keV (see left panel of Fig. 3) are due to the changes in flux between the two components (i.e. changes in the reflection fraction). The flux of the reflected component is correlated with that of the

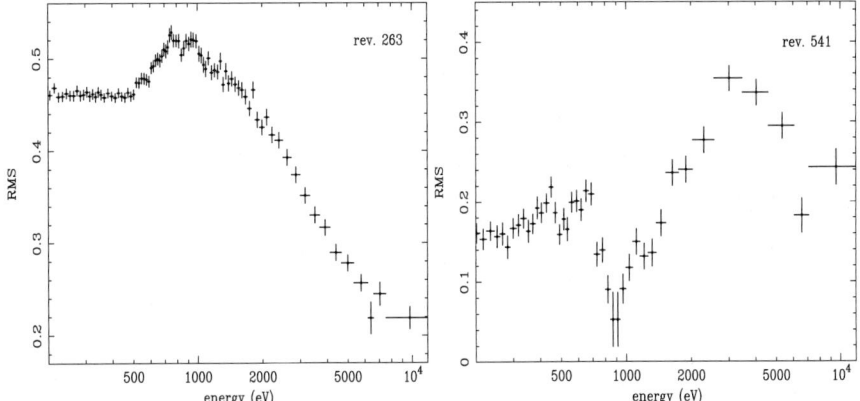

Fig. 1. RMS spectra of the medium (**left panel**) and low (**right panel**) flux observations. The RMS spectrum highlights the presence of at least two constant or weakly variable components: 1) X-ray reflection from distant material (the lower variability at 6.4 keV); 2) photo-ionized plasma emission (that generates the big drop around 0.9 keV in the low flux observation).

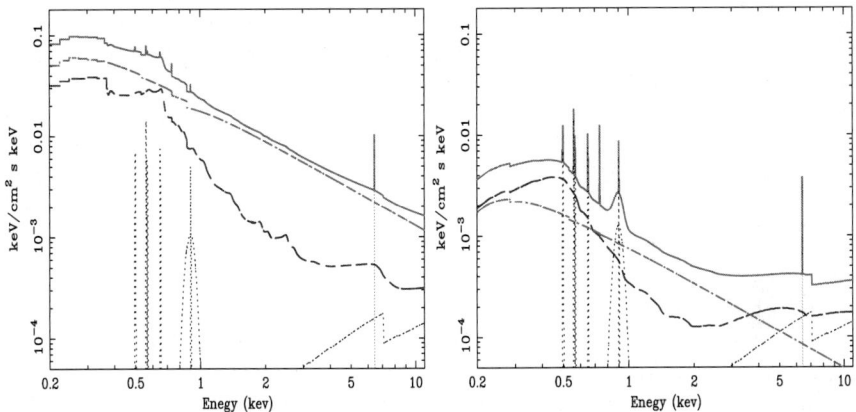

Fig. 2. Typical best fit high flux (**left panel**) and low flux (**right panel**) spectra of NGC 4051. In red (dashed-dotted line) the power law (Γ=2.2) component; in blue (dashed line) the blurred ionized reflection component (Ross & Fabian 2005); in grey (dotted line) the constant reflection from distant material and the emission from photo-ionized gas and in magenta (grey solid line) the total emission is shown.

power law component at low fluxes and it is almost constant at medium/high fluxes (see right panel of Fig. 3). A similar behaviour has been observed in MCG-6-30-15 (Vaughan & Fabian 2004) and has been interpreted in terms of a strong gravitational light bending by Miniutti et al (2003) and Miniutti and Fabian (2004). In this interpretation the source emission has to come from a few (less than 10-20) gravitational radii from the central black hole. For a more extensive and detailed description of the analysis see Ponti et al. (2006) and Ponti et al. (2005).

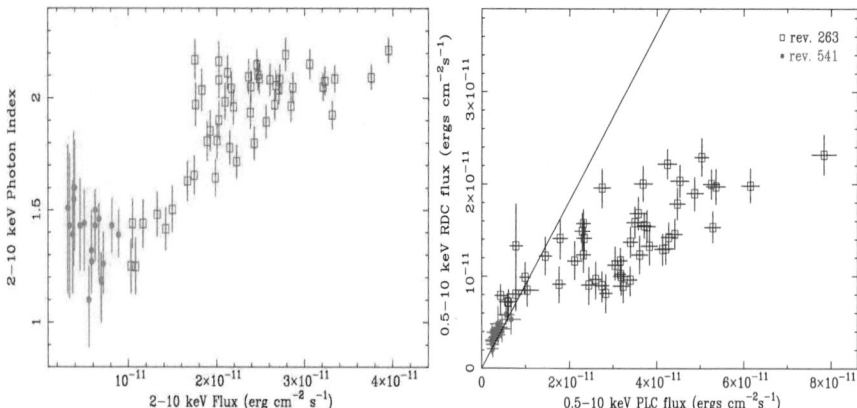

Fig. 3. (Left panel) Best fit spectral index vs. flux in the 2-10 keV band. The photon index increases with flux and seems to saturate at high flux ($\Gamma \sim 2.2$) and low fluxes to $\Gamma \sim 1.3$-1.4. This relationship could be due to the inter-combination of a steep ($\Gamma=2.2$) power law component and an almost stable flatter reflection component that is more important at low fluxes. **(Right panel)** 0.5-10 keV flux of the relativistic ionized reflection dominated component (RDC) vs. the flux of the direct power law component (PLC). The line shows the expected correlation if the reflected components is not affected by relativistic effect. The data seem, instead, to show a linear correlation at low fluxes and an almost constant reflection component at medium/high fluxes as expected if the light bending is at work.

Other interpretations of the data are possible with the major spectral variations due to a pivoting spectral power law (Uttley et al. 2003) or to variations in the opacity of an intervening absorber (Pounds et al. 2004; Elvis et al. 2005). The model proposed here seems however to require the minimum number of free parameters and catches the bulk of the spectral and temporal properties. Similar models have been successfully applied to MCG-6-30-15, 1H0707-495, 1H0429-577 and many other NLSy1s (Fabian & Vaughan 2003; Fabian et al. 2004, 2005; Foschini et al. 2004).

References

1. Crummy, J., Fabian, A. C., Gallo, L., & Ross, R. R.: MNRAS **365**, 1067 (2006)
2. Elvis et al.: Proceedings of "The X-Ray Universe 2005" conference, 2005 Sept 26-30, El Escorial, Madrid, Spain (2005)
3. Fabian, A. C., & Vaughan, S.: MNRAS **340**, L28 (2003)
4. Fabian, A. C., Miniutti, G., Gallo, L., Boller, T., Tanaka, Y., Vaughan, S., & Ross, R. R.: MNRAS **353**, 1071 (2004)
5. Fabian, A. C., Miniutti, G., Iwasawa, K., & Ross, R. R.: MNRAS **361**, 795 (2005)
6. Foschini, L., et al.: A&A **428**, 51 (2004)
7. Miniutti, G., Fabian, A. C., Goyder, R., & Lasenby, A. N.: MNRAS **344**, L22 (2003)
8. Miniutti, G., & Fabian, A. C.: MNRAS **349**, 1435 (2004)
9. Ponti, G., Cappi, M., Dadina, M., & Malaguti, G.: A&A **417**, 451 (2004)
10. Ponti, G., Miniutti, G., Cappi, M., Maraschi, L., Fabian, A.C. & Iwasawa, K.: Proceedings of "The X-Ray Universe 2005" conference, 2005 Sept 26-30, El Escorial, Madrid, Spain (2005)

11. Ponti, G., Miniutti, G., Cappi, M., Maraschi, L., Fabian, A.C. & Iwasawa, K.: MNRAS **368**, 903 (2006)
12. Pounds K.A., Reeves J.N., King A.R., Page K.L.: MNRAS **350**, 10 (2004)
13. Ross, R. R., & Fabian, A. C.: MNRAS **358**, 211 (2005)
14. Uttley P., Fruscione A., McHardy I.M., Lamer G.: ApJ **595**, 656 (2003)
15. Vaughan, S., Edelson, R., Warwick, R. S., & Uttley, P.: MNRAS **345**, 1271 (2003)
16. Vaughan, S., & Fabian, A. C.: MNRAS **348**, 1415 (2004)

Jet Activity in Supermassive Binary Black Holes

F.M. Rieger

UCD School of Mathematical Sciences, University College Dublin, Belfield, Dublin 4, Ireland,
`frank.rieger@ucd.ie`

Summary. Supermassive binary black holes (SBBHs) are a natural outcome of galaxy mergers. Here we show that low-frequency ($f \leq 10^{-6}$ Hz) quasi-periodic variability observed from radioloud, jet-emitting AGNs provides substantial inductive support for the presence of close ($d \leq 0.1$ pc) SBBHs in their centers. We estimate the properties for a sample of SBBH candidates and discuss the results within the context of jet activity and binary evolution.

1 Introduction

Galaxy mergers are known to play a vital role in the cosmological evolution of galaxies and the growth of supermassive black holes (BHs). Since almost every galaxy seems to contain a black hole (BH), frequent formation of supermassive binary black holes (SBBHs) is naturally expected to happen during cosmic time, e.g., [2, 4, 9]. In this context, elliptical galaxies such as the host galaxies of radio-loud AGNs are usually thought to form when two spiral galaxies collide and merge. Interacting galaxies are indeed observationally well-known today and the observation that the relative number of spiral to elliptical galaxies tends to increase in distant clusters, e.g., [3], gives additional credit to the underlying evolutionary picture. Direct observational evidence for the formation of a wide SBBH system has been recently established by the Chandra discovery of two activity centers (separation $d \sim 1$ kpc) in the merging ultraluminous infrared galaxy NGC 6240 [8]. While the existence of wide SBBHs seems thus well grounded, the possible existence of close ($d \sim 0.01 - 0.1$ pc) SBBHs appears much more ambiguous. Dynamical friction and slingshot interactions with stars normally ensure that a binary system gets quickly closer, e.g., [2]. However, due to the formation of a loss cone in the stellar distribution around the binary and the possible ejection of a BH in a subsequent merger event, it is still theoretically unclear today whether under normal circumstances a substantial fraction of SBBHs can really coalesce within at most a Hubble time.

2 Periodic Variability in Close SBBHs

The presence of close SBBH systems has been repeatedly invoked as plausible source for a number of observational findings in blazar-type AGN sources ranging from the misalignment and precession of jets to helical trajectories and quasi-periodic variability, cf. [12] for a review. In particular, interactions of the companion with the accretion disk around the primary BH naturally provide an explanation for the longterm optical periodicity with periods of the order $P_{\mathrm{obs}}^{\mathrm{opt}} \sim$ several years as observed in a number of blazars, cf. [13]. Such interactions can significantly accelerate the evolution of the binary and are also likely to induce some eccentricity $e \sim 0.1$ [1, 7]. In any case, one may derive an upper limit for the intrinsic Keplerian orbital period $P_k \leq \frac{2}{(1+z)} P_{\mathrm{obs}}^{\mathrm{opt}}$, by assuming that the observed periodicity is caused by the secondary BH crossing the disk around the primary on a slightly non-coplanar orbit. Note, however, that the occurrence of Newtonian-driven disk precession naturally introduces some deviation from strict periodicity.

The observations of helical jet paths in many blazar sources suggest that periodic variability (especially in the energy range dominated by the jet, e.g., radio, X- and γ-ray) can also arise due to differential Doppler boosting effects for a periodically changing viewing angle, e.g. [10]. For non-ballistic helical motion, classical travel time effects result in a reduction of the observable period P_{obs} with respect to the real driving period P such that $P_{\mathrm{obs}} \simeq (1+z) \frac{P}{\gamma_b^2}$, where $\gamma_b \sim (5-15)$ is the typical bulk flow Lorentz factor [11]. Orbital motion and (Newtonian) precession generally belong to the most obvious driving sources for helical jet paths. If, as it is usually believed, the high energy emission is produced on smaller jet scales, the emission may be primarily modulated by the orbital SBBH motion resulting in observable periods $P_{\mathrm{obs}} \sim 30 \, (P_{\mathrm{obs}}^{\mathrm{opt}}/10\,\mathrm{yr}) \, (15/\gamma_b)^2$ d. On the other hand, the main part of the radio emission usually originates from larger jet scales where Newtonian jet precession is likely to dominate the jet motion. As the associated precessional period is (at least) an order of magnitude higher than P_k, i.e., $P = P_p \geq 10 P_k$ [11], observable radio periods are expected to satisfy $P_{\mathrm{obs}}^{\mathrm{radio}} \geq 20 P_{\mathrm{obs}}^{\mathrm{opt}}/\gamma_b^2$. Note, that if P_p is rather small (say $P_p \sim 10 P_k$), moderate bulk Lorentz factors are still sufficient to account for $P_{\mathrm{obs}}^{\mathrm{radio}} < P_{\mathrm{obs}}^{\mathrm{opt}}$ as sometimes observed. *The detection of quasi-periodic variability on different timescales in different energy bands can thus add strong support to close binary scenarios*, cf. Table 1.

Table 1. Properties of a sample of blazar SBBH candidates, cf. [12] for details, with masses in $10^8 \, M_\odot$, and upper limits for the associated binary lifetime τ_{grav} due to gravitational radiation in units of 10^8 yrs for typical mass ratios ≥ 0.01.

name	redshift z	periods P_{obs}	$(m+M)$	P_k [yr]	$d/10^{16}$cm	τ_{grav}
Mkn 501	0.034	23.6 d (X-ray) ~ 23 d (TeV)	(2-7)	(6-14)	(2.5-6)	5.50
BL Lac	0.069	13.97 yr (optical) ~ 4 yr (radio)	(2-4)	(13-26.1)	(4.8-9.7)	28.9
3C 273	0.158	13.65 yr (optical) 8.55 yr (radio)	(6-10)	(11.8-23.5)	(6.5-12)	3.51
OJ 287	0.306	11.86 yr (optical) ~ 12 yr (infrared) ~ 1.66 yr (radio)	6.2	(9.1-18.2)	(5.5-8.8)	1.68
0235+16	0.940	2.95 yr (optical)? 5.7 yr (radio)	≥ 1	(1.5-3.1)	≥ 0.95	0.31

3 Jet Activity and Evolution of SBBHs

Table 1 shows properties derived for a sample of blazar SBBH candidates, where the observed periodicities have been used to estimate the last three columns. In all cases the gravitational lifetime of the binary is (a) significantly smaller than the Hubble time for the most likely range of mass ratios, a result that gives strong phenomenological support to the notion that SBBHs can really coalesce, and (b) still large enough to satisfy the minimum source lifetime required for jet fuelling (e.g., $\tau \sim 10^6$ yr for 3C 273). The derived separations are of order of the maximum size $r_d \sim 1000\, r_g$ for a simple standard Shakura & Sunyaev disk (as implied by the Toomre stability parameter $Q \simeq 1$) around the primary BH. Note that more complex models that take magnetic torques from the disk wind into account also suggest that r_d can be as high as ~ 1 pc, e.g. [5].

The interaction of the secondary with the disk around the primary provides a natural trigger for enhanced jet activity and may even lead to recurrent ejection of superluminal jet components. Moreover, such interactions may also result in the formation of two-sided (hot) outflows of gas from the disk, with maximum luminosities corresponding to the Eddington limit of the secondary, cf. [6].

Acknowledgments. Discussions with G. Romero are gratefully acknowledged.

References

1. P.J. Armitage, P. Natarajan: ApJ **634**, 921 (2005)
2. M.C. Begelman, R.D. Blandford, M.J. Rees: Nature **287**, 307 (1980)
3. A. Dressler et al.: ApJ **490**, 577 (1997)
4. L. Ferrarese, H. Ford: Space Science Reviews **116**, 523 (2005)
5. J. Goodman: MNRAS **339**, 937 (2003)
6. P.B. Ivanov, I.V. Igumenshchev, I.D. Novikov: ApJ **507**, 131 (1998)
7. P.B. Ivanov, J.C.B. Papaloizou, A.G. Polnarev: MNRAS **307**, 79 (1999)
8. S. Komossa et al.: ApJL **582**, L15 (2003)
9. A.P. Lobanov: Mem. S.A.It. **76**, 164 (2005)
10. F.M. Rieger, K. Mannheim: A&A **359**, 948 (2000)
11. F.M. Rieger: ApJL **615**, L5 (2004)
12. F.M. Rieger: in Proc. 22nd Texas Symposium on Relativistic Astrophysics (Stanford 2004), eds. P. Chen et al. (eConf:C041213), 1601 (2005)
13. E. Valtaoja et al.: ApJ **531**, 744 (2000)

Statistics of Local Hard X-Ray Selected AGN: Contribution of Obscured Accretion Onto Supermassive Black Holes

S. Sazonov[1,2], M. Revnivtsev[1,2], R. Krivonos[2,1], E. Churazov[1,2], and R. Sunyaev[1,2]

[1] Max-Planck-Institut für Astrophysik, Karl-Schwarzschild-Str. 1, D-85740 Garching, Germany, sazonov@mpa-garching.mpg.de
[2] Space Research Institute, Russian Academy of Sciences, Profsoyuznaya 84/32, 117997 Moscow, Russia

1 Introduction

X-ray surveys play a key role in studying the demography and cosmological evolution of active galactic nuclei. Deep surveys performed in the standard X-ray band (2–10 keV) have greatly improved our knowledge of AGN at $z > 0.3$, but due to the small area covered they are not well suited for studying the low-z AGN population. Moreover, there is evidence that most AGN are strongly intrinsically obscured at optical and soft X-ray wavelengths. To obtain an unbiased picture of the local AGN population there is a need in very large area surveys in hard X-rays. We have recently made a survey of the high Galactic latitude sky in the 3–20 keV band with RXTE and are now conducting an all-sky survey at energies above 20 keV with INTEGRAL. The main results of the former survey and progress with the latter are summarized below.

2 RXTE Slew Survey (XSS) at 3–20 keV

We used RXTE/PCA slew observations to perform a serendipitous survey of the high Galactic latitude ($|b| > 10°$) sky in the 3–20 keV band, in which a total of ~300 sources with fluxes down to ~0.5–1 mCrab were detected, including 100 identified AGN [7]. Most of the detected AGN are located at $z < 0.1$. We carried out a statistical analysis of this sample to investigate several key properties of the local AGN population, in particular the absorption column distribution and X-ray luminosity function [9].

Rather surprisingly, the N_H distribution proved very different for low-luminosity ($L_X < 10^{43.5}$ erg/s) and high-luminosity ($L_X > 10^{43.5}$ erg/s) AGN: while two thirds of the former are X-ray absorbed ($22 < \log N_H < 24$), the corresponding fraction is $< 20\%$ for the latter. This does not take into account Compton thick AGN ($\log N_H > 24$), to which the XSS is not sufficiently sensitive. Interestingly, the 3–20 keV AGN luminosity function flattens below $L_X \sim 10^{43.5}$ erg/s, approximately where obscured AGN start to dominate (see Fig. 1). Comparison with high-z surveys indicates that the AGN population continues to experience rapid downsizing at very low redshifts (approaching $z = 0$) [1].

In the published XSS catalog 35 sources remained unidentified, presenting a significant uncertainty for ensuing statistical studies of AGN. We have thus undertaken a number of steps to improve the localization of unidentified sources (from the original $\sim 1°$ error boxes) and to complete the identification of the XSS catalog. In particular, using RXTE scans and observations with SWIFT/XRT and INTEGRAL, we identified 4 XSS sources with nearby (z =0.017–0.098) luminous ($L_X \sim 5 \times 10^{42}–10^{44}$ erg/s) AGN, two of them being heavily absorbed ($N_H \sim 10^{23}$ cm^{-2}) [8]. We hope to shortly fill the remaining gaps in the XSS catalog. The updated catalog will consist of up to $\sim$130 X-ray bright (> 0.5–1 mCrab), mostly nearby ($z < 0.1$) AGN and will serve as a valuable reference sample for AGN studies.

3 INTEGRAL All-Sky Survey above 20 keV

A comparable or even better (with respect to heavily absorbed AGN and/or in crowded regions) capability for studying bright (> 1 mCrab) hard X-ray sources is now provided [4, 2] by the IBIS telescope on INTEGRAL, owing to its sensitivity peaking above 20 keV, large field of view and good ($\sim 10'$) angular resolution. Public and our proprietary INTEGRAL/IBIS observations cover more than half of the sky as of Spring 2005. We recently started a campaign to observe with INTEGRAL the missing sky regions, which is to be completed in 2006. Once the survey of the whole sky is done, it will be used for statistical analysis of AGN, similarly to the XSS.

So far more than 70 INTEGRAL sources have been identified as AGN. However, there remains a significant number of unidentifed sources. In a pilot program, we observed with Chandra 8 new INTEGRAL sources and identified 5 nearby ($z = 0.025$–0.055 in the 4 cases with measured redshifts) AGN [10]. Their X-ray spectra reveal the presence of significant amounts of absorbing gas ($\log N_H$ in the range 22–24), demonstrating that INTEGRAL is starting to fill in the sample of nearby obscured AGN. Optical follow-up studies of INTEGRAL sources are underway, and a dosen or so newly discovered sources have already been classified as Seyfert 1 or Seyfert 2 galaxies, using the 1.5-meter Russian-Turkish Telescope [3] and larger telescopes [6].

4 Conclusion

The RXTE, INTEGRAL and SWIFT [5] extragalactic surveys nicely complement each other in that RXTE is the most sensitive with respect to moderately obscured ($\log N_H < 23$) AGN, INTEGRAL provides a good coverage of low Galactic-latitude regions (including the Great Attractor), and SWIFT is scanning the whole sky uniformly with high sensitivity. A combination of these three surveys will provide an accurate census of local accreting supermassive black holes.

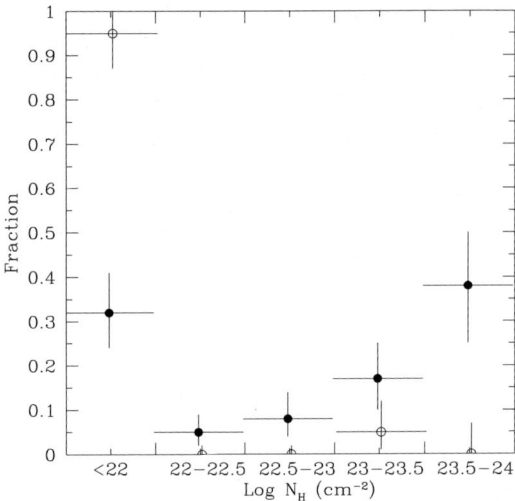

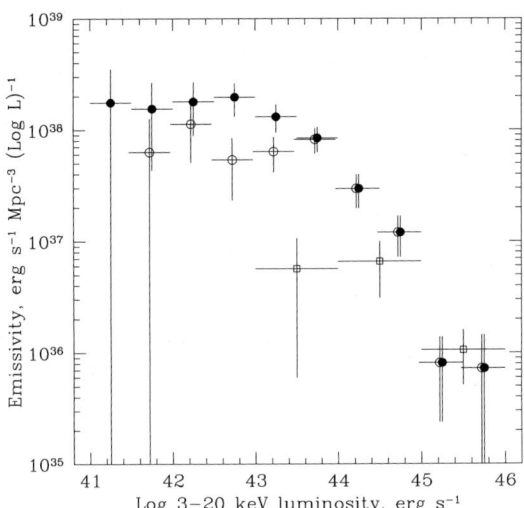

Fig. 1. *Top*: Absorption column density distribution of non-blazar AGN with $L_X < 10^{43.5}$ erg/s (filled circles) and $L_X > 10^{43.5}$ erg/s (open circles). *Bottom*: contribution of AGN with various luminosities to the local 3–20 keV volume emissivity of: Compton-thin non-blazar AGN (filled circles), unabsorbed non-blazar AGN (open circles), and blazars (open squares). Based on the RXTE Slew Survey.

References

1. A.J. Barger, L.L. Cowie, R.F. Mushotzky et al.: AJ **129**, 578 (2005)
2. L. Bassani, M. Molina, A. Malizia et al.: ApJ **636**, L65 (2006)
3. I.F. Bikmaev, R.A. Sunyaev, M.G. Revnivtsev, R.A. Burenin: Astron. Lett. **32**, 250 (2006)
4. R. Krivonos, A. Vikhlinin, E. Churazov et al.: ApJ **625**, 89 (2005)
5. C.B. Markwardt, J. Tueller, G.K. Skinner et al.: ApJ **633**, L77 (2006)
6. N. Masetti, E. Mason, L. Bassani et al.: A&A **448**, 547-556 (2006)
7. M. Revnivtsev, S. Sazonov, K. Jahoda, M. Gilfanov: A&A **418**, 927 (2004)
8. M. Revnivtsev, et al.: A&A **448**, L49-L52 (2006)
9. S. Sazonov, M. Revnivtsev: A&A **423**, 469 (2004)
10. S. Sazonov, E. Churazov, M. Revnivtsev et al.: A&A **444**, L37 (2005)

3D-Models of Clumpy Tori in Seyfert Galaxies

M. Schartmann[1], K. Meisenheimer[1], M. Camenzind[2], S. Wolf[1], and Th. Henning[1]

[1] Max-Planck-Institut für Astronomie (MPIA), Königstuhl 17, D-69117 Heidelberg, Germany,
    schartmann@mpia.de
[2] ZAH, Landessternwarte Heidelberg, Königstuhl 12, D-69117 Heidelberg, Germany

1 Introduction

Active Galactic Nuclei consist of a central supermassive black hole ($10^6 - 10^{10}\,\mathrm{M_\odot}$), surrounded by an accretion disc, which illuminates a larger dust reservoir further out. Assuming a toroidal shape of this dust distribution (in agreement with the so-called *Unified Scheme* [1]), the view towards the centre is unobscured, when looking face-on to the torus (the Seyfert I case) and blocked when looking edge-on (the Seyfert II case). An important spectral feature of those tori in the mid-IR is due to silicate grains and has now been observed with Spitzer in a wide variety between emission and absorption (e. g. [6, 7]). We employ radiative transfer simulations to explain these different characteristics of the IR-emission. Earlier attempts of radiative transfer modelling showed that in the case of continuous dust distributions, only heavily fine-tuned models succeed in suppressing the strong silicate emission in the Seyfert I case. The proposition that a clumpy torus modelling might produce the observed variety of features [4] was recently doubted by a 2D approach [2].

2 Our Model

The establishment of powerful radiative transfer tools like MC3D [9, 8] now allows to probe real three-dimensional dust distributions. For the models presented here, we assume the shape of our continuous Turbulent Torus Models [5] and fill the volume randomly with clouds, uniformly distributed in radial direction as well as on concentric shells. The size distribution of the clumps grows proportionally to the square root of r (the radial direction) and the density of the individual clumps scales with r^{-1}. This typically leads to a distribution shown in 3D in Fig. 1 for a model with 14500 clumps - with one quarter of the torus cut out in order to be able to look into the central part. The resulting mid-infrared appearance is shown in Fig. 2.

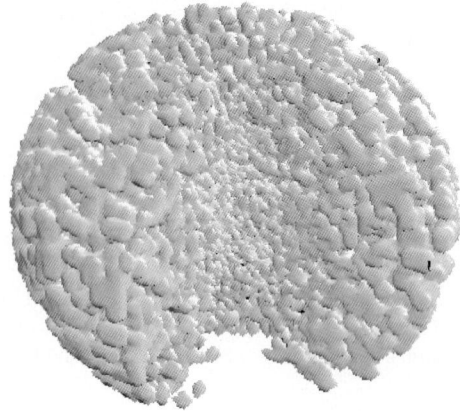

Fig. 1. 3D model density distribution of our clumpy model.

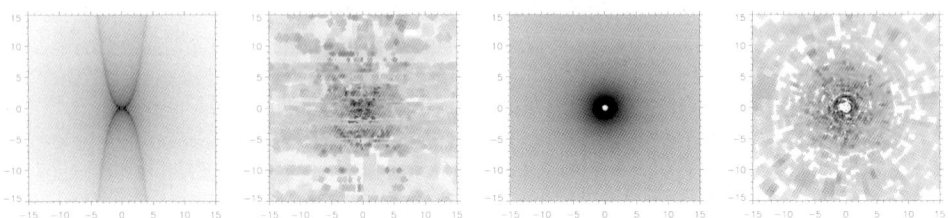

Fig. 2. Comparison of images at $\lambda = 13.2\,\mu$m for $i = 90°$ and $0°$ of the continuous and the clumpy model (axis labels are given in pc). The x-shaped emission traces the directly illuminated funnel walls and is very pronounced in the continuous case – as density drops with $1/r$ – but barely visible for the randomly distributed clumps. Looking face-on, the expected ring-like structure emerges in both models.

3 Dust: Mass Study

The leading free parameter in our model is dust mass (or the optical depth τ). Fig. 3 compares a dust mass study of our continuous torus model with our clumpy torus model. Only the face-on case is displayed here, as in the edge-on case, both models show the same behaviour. The optical depths within the equatorial plane averaged over all angles φ range from $\tau = 1$ at 9.7μm for the solid curves up to $\tau = 8$ for the dash-dotted curves. While for the SEDs of the continuum model, the relative height of the silicate emission feature changes only slightly, in the clumpy case, the absolute height of the feature differs. With increasing dust mass, it changes from emission to partial absorption to absorption. Obviously, in a clumpy model, the optical depth and the distribution of the clouds in the central region of the torus enables us to control the strength of the emission feature.

Additionally, we find a difference at short wavelengths in a sense that with increasing dust mass, continuous models show a change from a volume emission to an emission of more or less only the funnel walls, resulting in higher fluxes at short wavelengths.

Clumpy models behave in the opposite way: higher dust masses mean less flux at short wavelengths.

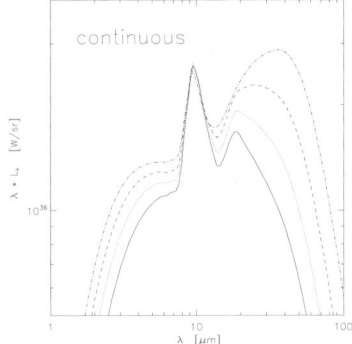 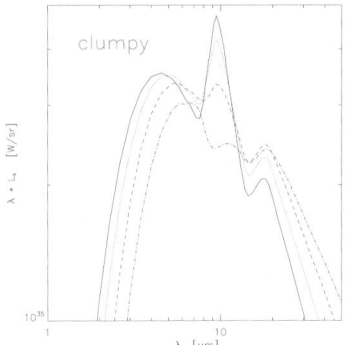

Fig. 3. Face-on case of a dust mass study for our clumpy and continuous model.

These results can be understood as follows: For the case of an optically thick model, the directly illuminated clumps are responsible for the production of the silicate feature in emission. When looking edge-on to the torus, lots of low temperature dust clouds are visible. Therefore, we expect qualitatively the same behaviour as in the continuous case. As already pointed out by [4], the situation changes for the face-on case. Many of the directly illuminated parts of the clumps are now hidden behind others and their relative contribution compared to lines-of-sight with cold dust in front (producing silicate features in absorption) decreases. Depending on the shadowing by cold clouds, the appearance of the silicate feature varies between emission, flat and absorption.

4 Conclusions

We calculated SEDs and surface brightness distributions for a 3D clumpy dust distribution with the help of the radiative transfer code MC3D in order to compare the behaviour of the silicate feature with our previous continuous torus modelling effort [5]. According to our simulations, the strength of the silicate feature – when looking face-on to the torus – depends mainly on the optical depth, the geometrical distribution and number density of the innermost clumps. Thus, many model setups will succeed in producing the observed strength of the feature. Further constraints are achieved by interferometric observations [3] in the range of 2 to 13 μm, which will allow for resolving the inner morphology of such tori. We are currently working on physically justified 3D models of clumpy tori in Seyfert galaxies by combining hydrodynamic simulations with radiative transfer calculations, in order to tackle the ambiguity problem.

References

1. R. Antonucci: ARAA, **31**, 473-521 (1993)

2. C.P. Dullemond, I.M. van Bemmel: A&A, **436**, 47-56 (2005)
3. K. Meisenheimer, K.R. Tristram, W. Jaffe: these proceedings (2007)
4. M. Nenkova, Ž. Ivezić, M. Elitzur: ApJL, **570**, L9-L12 (2002)
5. M. Schartmann, K. Meisenheimer, M. Camenzind, S. Wolf, Th. Henning: A&A, **437**, 861-881 (2005)
6. R. Siebenmorgen, M. Haas, E. Krügel, B. Schulz: A&A, **436**, L5-L8 (2005)
7. D.W. Weedman, L. Hao, S.J.U. Higdon, D. Devost, Y. Wu, V. Charmandaris, B. Brandl, E. Bass, J.R. Houck: ApJ, **633**, 706-716 (2005)
8. S. Wolf: CoPhC, **150**, 99-115 (2003)
9. S. Wolf, Th. Henning, B. Stecklum: A&A, **349**, 839-850 (1999)

Part V

Clusters of Galaxies

Cosmological Tests with Galaxy Clusters

H. Böhringer

Max-Planck-Institut für extraterrestrische Physik, Garching, Germany, hxb@mpe.mpg.de

1 Introduction

Clusters of galaxies as the largest well defined objects in our Universe form an integral part of the cosmic large-scale structure, the seeds of which have been set in the early Universe. Therefore the evolution of the galaxy cluster population is tightly connected to the evolution of the large-scale structure and the Universe as a whole. It is for this reason that observations of galaxy clusters can be used to trace the evolution of the Universe and to test cosmological models.

Galaxy clusters are composed of mostly dark matter, hot thermal plasma, and galaxies, roughly in proportions of 87%, 11%, and 2%, respectively.[1] These forms of matter are bound in the gravitational potential well of the cluster, the depth of which is characterized by the velocity dispersions of the galaxies ranging from about 400 to 1500 km s^{-1}. In total, rich galaxy clusters with masses above 10^{14} M$_\odot$ contribute about 3% to the total matter density and including galaxy groups down to masses of 10^{13} M$_\odot$ increases this fraction by about a factor of three (see Fig. 1, left).

X-ray observations play currently a prime role in the cosmological applications of galaxy clusters. The X-ray emission is due to the gaseous intracluster medium (ICM), a hot, highly ionized thermal plasma with a temperature of ten to hundred million degrees that fills the whole cluster volume and thus provides a contiguous picture of the cluster structure. The X-ray imaging and spectroscopic data provide a measure of the ICM density and temperature structure. Assuming that the sound crossing time in the ICM is shorter than the dynamical age of the cluster and thus that hydrostatic equilibrium is established in the ICM, the gravitational cluster binding mass can be calculated from the density and temperature distribution [1]. This assumption is in general only approximately fulfilled and can be locally, coarsely violated in clusters which are found in a stage of the merging of two major subunits. Fig. 1 (right) shows a comparison of the so determined masses and the X-ray luminosities of the 106 brightest galaxy clusters in the sky [2]. There is a tight correlation of the X-ray luminosity and cluster mass,

---

[1] Here and throughout the paper we adopt a Hubble constant of $H_0 = 70$ km s^{-1} Mpc^{-1} and a concordance cosmological model with a normalized matter density parameter, $\Omega_m = 0.3$, and a normalized cosmological constant parameter, $\Omega_\Lambda = 0.7$, if not stated otherwise.

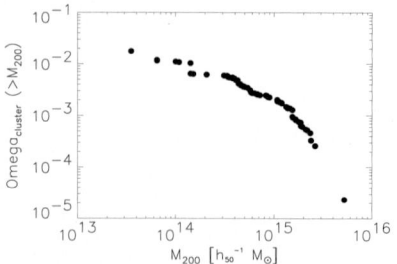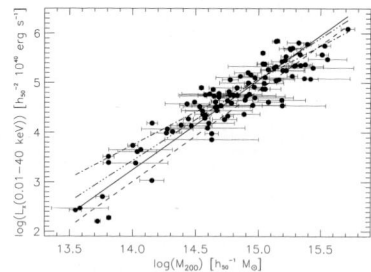

Fig. 1. Left: Mass density of galaxy clusters in the Universe as a function of the lower mass cut in units of the critical density of the Universe [2]. **Right:** Mass-X-ray luminosity relation determined from the 106 brightest galaxy clusters found in the *ROSAT* All-Sky Survey by Reiprich & Böhringer [2].

which allows to predict the cluster mass from the observed luminosity with an accuracy of about 60% (of which probably more than half is due to the uncertainty of the mass measurement). This relation is currently improved with XMM-Newton and Chandra.

2 Cosmological Tests with the Abundance and Spatial Distribution of Clusters

To trace the large-scale structure with galaxy clusters we need a statistically complete cluster sample for a large enough volume. For the comparison of the observed cluster abundance and spatial distribution with theoretical model predictions the masses of the clusters involved have to be known, at least approximately. X-ray surveys of clusters provide several advantages for the construction of cluster samples to be used for cosmological studies. First of all the X-ray luminosity is tightly correlated to the cluster mass as shown above, the presence of X-ray emission is a signature of a truly bound structure, and due to the centrally peaked surface brightness profile projection effects on the sky are minimized.

For the present illustration we use the REFLEX cluster survey, the currently largest, most homogeneous, and most complete X-ray cluster survey [3, 4]. This survey is based on the ROSAT All-Sky Survey [5] and covers an area of 4.24 sr in the southern sky up to declination $\delta = +2.5^{o}$ comprising 447 objects. The cluster catalogue is estimated to be better than 90% complete with a contamination from X-ray luminous AGN less than 9%.

The basic census of the galaxy cluster abundance is the density of clusters of a given mass, the mass function, which tells us much about the fluctuation amplitude of the dark matter density distribution, σ_8, from which the clusters form (e.g. [6]). A binned representation of the luminosity function (closely related to the mass function) for the REFLEX cluster survey is shown in Fig. 2 (left). A best fit to all data and the prediction of the concordance cosmological model are shown in this figure. The latter provides a very good fit to the luminosity function at higher luminosity, while at lower

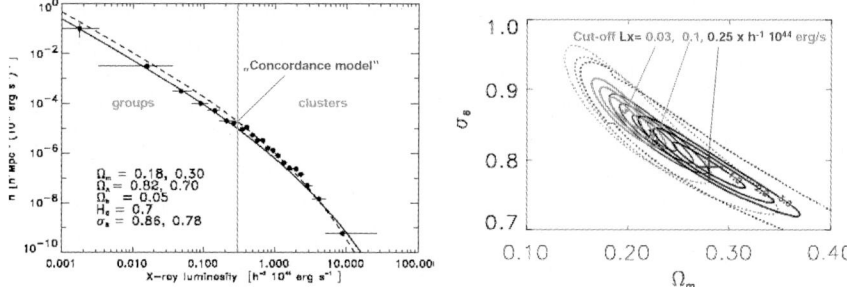

Fig. 2. Left: X-ray luminosity function of the REFLEX cluster survey. The two curves show the best fitting model to all the data (solid line) and the concordance cosmological model (dashed line) with parameters for both models given in the figure [7]. **Right:** Constraints on the cosmological parameters, σ_8, and Ω_m from the best fit to the X-ray luminosity function for various mass ranges. For the fit of the mass range above $2.5 \cdot 10^{43} h^{-1}$ erg s^{-1} we obtain values of $\sigma_8 = 0.74 - 0.85$ and $\Omega_m = 0.21 - 0.35$ (2σ limits).

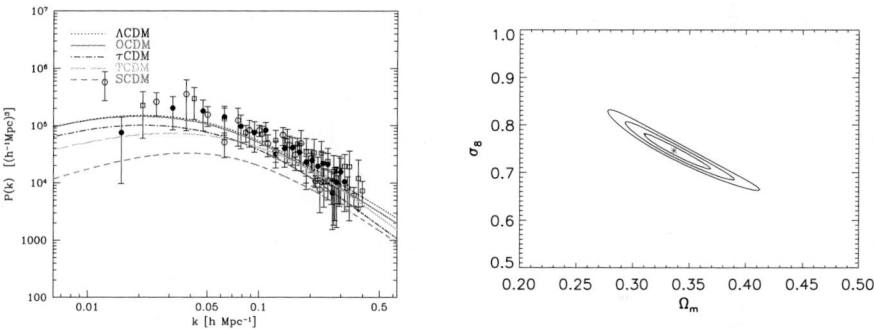

Fig. 3. Left: Power spectra of the density fluctuations in the REFLEX cluster sample together with predictions from various popular cosmological models taken from the literature. The shape of the power spectrum is best represented by the ΛCDM and OCDM models with a low matter density parameter ($\Omega_m \sim 0.3$). For details see Schuecker et al. [10]. **Right:** Constraints on the cosmological density parameter, Ω_m, and the amplitude of the matter density fluctuations on a scale of $8h^{-1}$ Mpc, σ_8, obtained from the comparison of the density fluctuation power spectrum and the cluster abundance as a function of redshift in a Karhunen-Loéve statistical analysis of the REFLEX survey data (Schuecker et al. [11]). The likelihood contours give the 1,2, and 3σ limits.

luminosity the number of galaxy groups is smaller than the prediction. This is most probably due to the fact that we know still too little about the group properties and the group masses are therefore not predicted correctly. If we restrict the cosmological test to the objects with luminosities above $2.5 \cdot 10^{43} h^{-1}$ erg s^{-1} we get reliable constraints for the parameters σ_8 and Ω_m as shown in the right panel of Fig. 2.

The spatial distribution of galaxy clusters is following the overall matter distribution in the Universe, the seeds for which were set in the early Universe. The relation of the amplitude of the cluster density and dark matter density fluctuations can be derived from statistical considerations and simulations e.g. [8]. The amplitude of the cluster density fluctuations is thus amplified compared to that of the matter density, an effect called "biasing". The most fundamental statistical description of the spatial structure is based on the second moments of the distribution, characterized by the two-point-correlation function or the density fluctuation power spectrum. The two-point correlation function of REFLEX shows a power law shaped function with a slope of 1.83, a correlation length of $18.8h_{100}^{-1}$ Mpc, and a possible zero crossing at $\sim 45h_{100}^{-1}$ Mpc [9]. The density fluctuation power spectrum (Fig. 3, [10]) is characterized by a power law at large values of the wave vector, k, with a slope of $\propto k^{-2}$ for $k \leq 0.1h$ Mpc^{-1} and a maximum around $k \sim 0.03h$ Mpc^{-1} (corresponding to a wavelength of about $200h^{-1}$ Mpc). This maximum reflects the size of the horizon when the Universe featured equal energy density in radiation and matter and is a sensitive measure of the mean density of the Universe, Ω_m, [10]. The shape of the power spectrum therefore provides strong constraints on the matter density. Combined with the cluster abundance (in the frame of a Karhunen-Loéve eigenfunction decomposition analysis [11]) we find tight constraints on the matter density and amplitude of the matter density fluctuations as shown in the right panel of Fig. 3.

The cluster population in the nearby Universe provides no constraints of a possible cosmological constant, Λ, or parameters describing the behavior of Dark Energy. Due to the dependence of the structure growth function on these parameters the observation of the evolution of the cluster population provides access to these parameters. The best high redshift cluster samples comprise the *ROSAT* Distant Cluster Survey (RDCS) by Rosati et al. [12] and the 160 deg^2 survey lead by Vikhlinin [13]. The evolutionary effect in the luminosity function and gas mass functions found in these surveys are roughly consistent with a concordance cosmological model. For a detailed study of Dark Energy parameters more detailed data are required. The high sensitivity of XMM-Newton is now providing a deeper look and the recent detection of the most distant X-ray cluster at a redshift of 1.39 [14] illustrates the prospects of ongoing and upcoming surveys.

3 Complementarity to other Cosmological Tests and Conclusions

Testing cosmological models with galaxy clusters is complementary to other important cosmological tests. Fig. 4 shows the constraints on the cosmological parameters Ω_m and Ω_Λ for three different types of cosmological tests: (i) the study of the fluctuations in the cosmic microwave background with the WMAP satellite [15], (ii) the measurements on the geometry of the Universe using supernovae type Ia as standard candles [16, 17], and the statistics of the large scale structure from X-ray cluster observations [11, 18]. All three constraints meet at the same small parameter region around the concordance cosmological model with parameters of $\Omega_m \sim 0.3$ and $\Omega_\Lambda \sim 0.7$. Other measures of the large-scale structure based on the galaxy distribution provide similar constraints as the clusters [19, 20]. The combination of several cosmological tests in one statistical parameter constraint approach was for example done for the combination of the REFLEX cluster survey and the supernova studies by Schuecker et al. [18]. The results for constraints on the parameter Ω_m and the equation of state parameter, w, of the Dark Energy shown in Fig. 4 (right) favor a value of $w = -1$, a Dark Energy model equivalent to the classical Λ cosmological model.

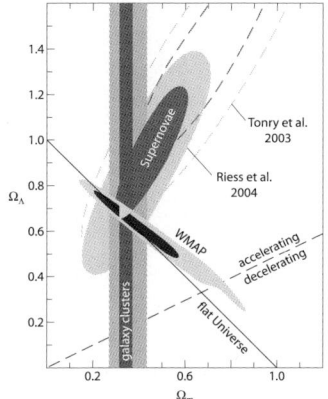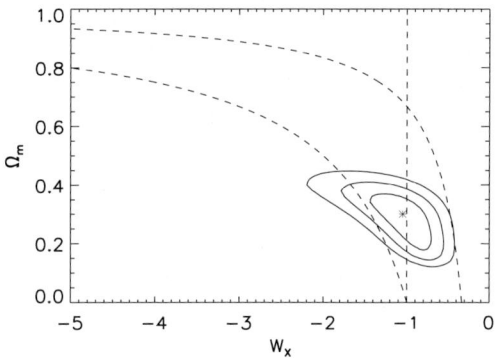

Fig. 4. Left: Constraints on the cosmological parameters Ω_m and Ω_Λ from three different cosmological tests: supernovae Ia as standard candles [16, 17], analysis of the fluctuation spectrum of the cosmic microwave background [15], and abundance and spatial clustering observations of X-ray clusters of galaxies [11]. All the tests meet at the same parameter region encircling the concordance model. **Right:** Constraints on the cosmological density parameter, Ω_m, and the equation of state parameter of the dark energy, w_x, for the REFLEX cluster data combined with literature results from distant supernovae obtained by Schuecker et al. [18]. The likelihood contours give the 1,2, and 3σ limits.

The results presented here, based on ROSAT and ASCA, are almost outdated and will soon be superseded by Chandra and XMM-Newton results. A major break-through for the study of the nature of Dark Energy with galaxy clusters will then be provided by means of an improved large sky area or all-sky X-ray survey. Several proposals of such a survey have been made e.g. to the Dark Energy task force in the US and with the eROSITA mission from a German team ([21] which was also presented at this conference), which when granted will provide a competitive cosmological test on an equal level with the most important other attempts.

References

1. Sarazin, C.L.: Rev. Mod. Phys. **58**, 1 (1986)
2. Reiprich, T.H. & Böhringer, H.: ApJ **567**, 716 (2002)
3. Böhringer, H., Schuecker, P., Guzzo, L., et al.: A& A **369**, 826 (2001)
4. Böhringer, H., et. al.: A&A **425**, 367 (2004)
5. Trümper, J.: Science **260**, 1769 (1993)
6. White, S.D.M., Efstathiou, G., Frenk, C.S.: MNRAS **262**, 1023 (1993)
7. Böhringer, H., et al.: ApJ **566**, 93 (2002)
8. Mo, H.J. & White, S.D.M.: MNRAS **282**, 347 (1986)
9. Collins, C.A., et al.: MNRAS **319**, 939 (2000)
10. Schuecker, P., et al.: A&A **368**, 86 (2001)
11. Schuecker, P., et al.: A&A **398**, 867 (2003)
12. Rosati, P., et al.: ApJ **492**, L21 (1998)
13. Vikhlinin, A., et al.: ApJ **502**, 558 (1998)

14. Mullis, C.R., et al.: ApJ **623**, L85 (2005)
15. Spergel, D.N., et al.: ApJ **148**, 175 (2003)
16. Tonry, J.L., et al.: ApJ **594**, 1 (2003)
17. Riess, A.G., et al.: ApJ **607**, 665 (2004)
18. Schuecker, P., et al.: A&A **402**, 53 (2003)
19. Peacock, J.A., et al.: Nature **410**, 169 (2001)
20. Tegmark, M., et al.: ApJ **606**, 702 (2004)
21. Predehl, P., et al.: SPIE **6266**, 19 (2006)

Supermassive Black Holes in Elliptical Galaxies: Switching from Very Bright to Very Dim

E. Churazov[1,2], S. Sazonov[1,2], R. Sunyaev[1,2], W. Forman[3], C. Jones[3], and H. Böhringer[4]

[1] MPI für Astrophysik, Karl-Schwarzschild-Strasse 1, 85741 Garching, Germany
[2] Space Research Institute (IKI), Profsoyuznaya 84/32, Moscow 117810, Russia
[3] Harvard-Smithsonian Center for Astrophysics, 60 Garden St., Cambridge, MA 02138, USA
[4] MPI für Extraterrestrische Physik, Giessenbachstrasse, 85740 Garching, Germany

Summary. Relativistic outflows (mainly observed in the radio) are a characteristic feature of both Galactic stellar mass black holes and supermassive black holes (SMBHs). Simultaneous radio and X-ray observations of Galactic sources have shown that the outflow is strong at low accretion rates, but it weakens dramatically or disappears completely at high accretion rates, manifesting structural changes in the accretion flow. If the outflow is quenched in SMBHs at high accretion rates similarly to the behavior of galactic sources, then rapidly accreting SMBHs may grow without very strong impact on the gas in the host galaxy. At much lower accretion rates the system switches to a stable state corresponding to passively evolving ellipticals, when the power of the outflow is high and it keeps the gas hot.

1 Introduction

It is now widely accepted that virtually every elliptical galaxy hosts a supermassive black hole (SMBH) at its center with its mass tightly correlated with the mass and stellar velocity dispersion of the galaxy itself (e.g [14, 26]). During the early stages of galaxy evolution, when large quantities of cold gas are presumably present, these black holes accrete matter at high rates and are observed as bright QSOs once they have grown to a considerable mass. In contrast the central SMBHs of local elliptical galaxies usually have very small bolometric luminosities. Despite low observed luminosities, these SMBHs in local ellipticals drive strong outflows which manifest themselves as bubbles of radio emitting plasma coincident with depressions in the X-ray images (e.g. [4, 10, 18]). At least in several well studied objects the power of the outflow is sufficient to offset the gas cooling losses (e.g. [3, 7, 20]).

Radio emitting flows are also observed in X-ray binaries containing stellar mass black holes and the physics there is believed to be similar to that in SMBHs. For individual binaries the radio emission has been observed to vary depending on the X-ray flux and on the spectral state of the source (e.g. [13]). In particular it was found that an outflow ceases above a certain accretion rate (e.g. [12]). There are indications that active galactic nuclei (AGNs) may also exhibit a similar kind of radio power behavior [17].

2 Black Hole Energy Release in X-ray Binaries and AGNs

Simultaneous observations of several X-ray binaries, in particular black hole candidates, in radio and X-ray bands have recently led to the following broad picture (e.g.
[13, 15]). In the so-called "low/hard" source state X-ray and radio fluxes are well corre-
lated. In this state the accretion rate is low - $\dot{M} \ll \dot{M}_{\mathrm{Edd}}$, the accretion flow is optically
thin (geometrically thick) and radiative efficiency (luminosity/$\dot{M}c^2$) is much lower than
the canonical 10%. For individual sources (a notable example is GX339-4) the F_r/F_X
correlation extends over at least three orders of magnitude in X-ray luminosity.

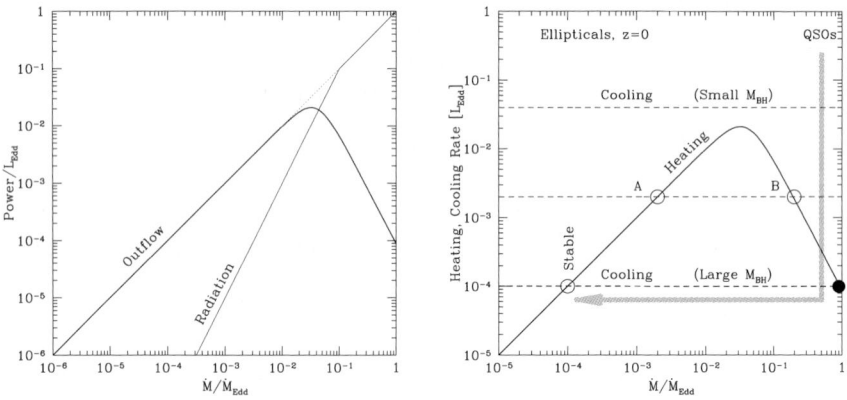

Fig. 1. a: Sketch of black hole energy release as a function of mass accretion rate. At low accretion rates a significant fraction of accretion power goes into an outflow (thick solid line). Above
a certain accretion rate, corresponding to the transition to the standard accretion disk mode, the
outflow power decreases. The radiative power (thin solid line) is instead very low at low accretion
rates and reaches a fixed fraction ($\sim$10%) of $\dot{M}c^2$ at accretion rates above 0.01-0.1 of Eddington. **b:** Illustration of gas heating and cooling in elliptical galaxies. The thick solid line shows,
as a function of the SMBH accretion rate, the heating rate due to outflow, which is comple-
mented/dominated by radiative heating near the Eddington limit. Horizontal dashed lines show
the gas cooling rate. The upper cooling line represents a young galaxy in which a large amount of
gas is present and/or the black hole is small. Feedback from the black hole is not able to compensate for gas cooling losses and the black hole is in the QSO stage with a near-critical accretion
rate, high radiative efficiency and weak feedback. As the black hole grows it moves down in this
plot. The black solid dot marks the termination of this stage, when the black hole is first able
to offset gas cooling, despite the low gas heating efficiency. The lower cooling line illustrates
present day ellipticals: a stable solution exists at low accretion rates when mechanical feedback
from the black hole compensates gas cooling losses. The radiative efficiency of accretion is very
low and black hole growth is very slow.

For binaries the F_r/F_X correlation breaks down when higher X-ray fluxes are considered [12, 15]. This break is associated with the increased mass accretion rate and a
transition of the source to the so-called "high/soft" state. In this state the bulk of the

observed X-ray emission comes from a standard, optically thick and geometrically thin accretion disk, and the spectra are reasonably well described by the multicolor black body model. In this state the radio emission drops by a large factor (or completely disappears) suggesting quenching of the outflow.

It seems reasonable to assume that the outflow behavior is similar for galactic binaries and SMBHs and consider the most straightforward implications for elliptical galaxies and their central SMBHs [8]. Previously the role of feedback from growing SMBHs on their host galaxies has been discussed by many authors using different assumptions (e.g. [1, 2, 5, 9, 16, 19, 22, 24, 25, 28]).

Guided by the observations of X-ray binaries we adopt a simple prescription for AGN energy release (shown in Fig.1a) according to which at low rates there is a strong outflow whose power correlates with the accretion rate. The outflow power drops by a large factor when the accretion rate rises from $\sim 10^{-2} \dot{M}_{\mathrm{Edd}}$ to $\sim \dot{M}_{\mathrm{Edd}}$.

We further assume that in elliptical galaxies the efficiency of converting the outflow energy into the gas heating can be high, up to 100% (e.g. [6]). For simplicity we assume that the gas heating rate due to SMBH feedback is equal to the outflow power. This leads to a dependence of gas heating rate on the accretion rate (Fig.1b) that mirrors the outflow power curve (Fig.1a).

Apart from the mechanical feedback, radiative gas heating, via photoionization and compton scattering of radiation from the SMBH, should be at work [9, 21]. Because of low efficiency of converting luminosity into heat the radiative feedback can be safely neglected at low accretion rates, but it can however be important at near-Eddington luminosities. It may therefore contribute significantly (or dominate) the high-$\dot{M}$ part of the heating curve in Fig.1b far from its maximum. This insures that total feedback power (mechanical plus radiative) does not go to zero near the Eddington rate even if the outflow quenches completely.

It is convenient to characterize the gas heating power F through the efficiency δ of converting accreted mass into heat: $F = \delta c^2 \dot{M}$. For subsequent analysis we assume that at very low accretion rates (outflow regime), the efficiency is high, $\delta_O \approx const \sim 0.1$, while close to the Eddington limit, the efficiency is much lower, $\delta_E \ll \delta_O$.

2.1 A Toy Model

We now consider the implications of our model for the evolution of elliptical galaxies. We completely neglect stellar feedback, which can be especially important for low mass systems. Thus the discussion below mainly refers to massive systems with velocity dispersions larger than $\sim$200 km/s. For simplicity we assume that mass accretion rate onto the black hole is described by the Bondi formula:

$$\dot{M}_{Bondi} \approx \pi (GM_{\mathrm{BH}})^2 c_s^{-3} \rho \propto s^{-3/2},$$

where c_s is the gas sound speed, ρ is the gas mass density, and $s \propto T_e / \rho^{\gamma - 1}$ is the gas entropy index. Then it is the gas entropy near the black hole which regulates the amount of heat supplied by the SMBH into the gas.

We follow the picture of a quasi-steady cooling flow in an isothermal halo (e.g. [27]) with the velocity dispersion σ and the gas mass fraction f_{gas}. The angular momentum of the gas is neglected. The initial mass of the black hole is assumed to be small enough that its feedback can be completely neglected (situation corresponding to the upper cooling line in Fig.1b). We assume that in such a situation cold gas will be present in large amounts in the system, stars will be actively forming, and the black hole will accrete at the Eddington rate, independently of the details. The black hole mass then grows exponentially with the Salpeter e-folding time $t_s \approx 4 \times 10^7$ yr: $M_{\rm BH} \propto e^{t/t_s}$. The mass cooling rate of the gas is instead a decreasing function of time:

$$\dot{M}_{\rm cool} = \frac{2}{\sqrt{3\pi}} \frac{f_{gas}^{3/2}\sigma^2 \Lambda(T)^{1/2}}{m_p G^{3/2} t^{1/2}} \approx 180\ \sigma_{200}^2 t_9^{-1/2} \Lambda_{23}^{1/2}\ M_\odot/{\rm yr}.$$

Therefore with time cooling losses will decrease relative to the (growing) heating power of the black hole accreting at the Eddington rate. With our definition of the feedback amplitude, the black hole accreting at the Eddington rate will be able to offset gas cooling provided that:

$$\delta_E c^2 \dot{M}_{\rm Edd} \approx C \approx 3/2kT \frac{\dot{M}_{\rm cool}}{\mu m_p}.$$

Substituting the expression for the Eddington accretion rate ($\propto M_{\rm BH}$) and solving for $M_{\rm BH}$ we find:

$$M_{\rm BH} \approx \frac{\eta}{\delta_E} \frac{\sigma^4 f_{gas}^{3/2} \Lambda(T)^{1/2} \sigma_T}{2\sqrt{3}\pi^{3/2} c G^{5/2} m_p^2 t^{1/2}} \approx 4 \times 10^8 \left(\frac{\delta_E}{10^{-5}}\right)^{-1} \sigma_{200}^4 t_9^{-1/2} \Lambda_{23}^{1/2}\ M_\odot$$

for an accretion radiative efficiency $\eta = 0.1$. Once the black hole has grown to this size it will be able to prevent gas cooling, cold gas will disappear and star formation will cease. The entropy of the gas increases dramatically and the system evolves toward a low accretion rate solution. In this regime the accretion rate required to produce the same heating power as before is substantially sub-Eddington: $\dot{M} \sim \frac{\delta_O}{\delta_E} \dot{M}_{\rm Edd}$.

The resulting system evolution is shown in Fig.1b as the gray track. The black hole first accretes at the Eddington level (vertical track) and then switches (horizontal track) to the low accretion rate mode. The vertical track corresponds to a very high radiation efficiency and active star formation (cold gas is continuously deposited), whereas in the low accretion regime the radiative efficiency is low and little star formation takes place in the galaxy.

Importantly, for the black hole mass for which the thermal equilibrium is reached, the maximum possible feedback power (maximum of the function F, see Fig.1) significantly exceeds the cooling losses of the gas at the moment of the transition. This ensures that the system will be able to support an even larger atmosphere of hot gas. The latter case is probably relevant to cD galaxies in rich clusters. Since the gas heating efficiency in the low accretion rate state is high, the total mass accumulated by the SMBH over ~ 10 Gyr of passive evolution will not be significant: $M_{\rm BH} \approx 2 \times 10^7 L_{43}\ (0.1/\delta_O)\ M_\odot$, where L_{43} is the gas cooling losses in units of 10^{43} erg s^{-1}. Only for rich clusters can the black hole significantly increase its mass in this mode (e.g. [11]).

3 Conclusions

Guided by the analogy of SMBHs and the stellar mass black holes in X-ray binaries, we postulated a non-monotonic dependence of the SMBH feedback power on the mass accretion rate associated with a transition from geometrically thick accretion flow to a geometrically thin disk. The gas heating rate, associated with the outflow of relativistic plasma, is maximal at substantially sub-Eddington accretion rates and drops strongly for near-Eddington rates. Such a scenario naturally leads to two stages in the evolution of a massive elliptical galaxy: i) QSO-like nucleus and active star formation at early epochs and ii) weak nucleus and passive evolution at the present time. The same type of scenario is possible for various functional forms of the feedback power dependence on the mass accretion rate, if a non-monotonic region is present. In particular even at super-Eddington accretion rates the feedback power is very strong the same two stages of evolution (very rapid black hole growth followed by a very slow growth) will naturally appear.

References

1. Begelman M. C., Nath B. B.: MNRAS **661**, 1387 (2005)
2. Binney J., Tabor G.: MNRAS **276** (1995) 663
3. Bîrzan L. et al.: ApJ **607**, 800 (2004)
4. Böhringer H. et al.: MNRAS **264**, L25 (1993)
5. Cavaliere A., Lapi A., Menci N.: ApJ **581**, L1 (2002)
6. Churazov E., Sunyaev R. A., Forman W., Böhringer H.: MNRAS **332**, 729 (2002)
7. Churazov E. et al.: ApJ **554**, 261 (2001)
8. Churazov E. et al.: MNRAS **363**, L91 (2005)
9. Ciotti L., Ostriker J. P.: ApJ **551**, 131 (2001)
10. Fabian A. C., et al.: MNRAS **318**, L65 (2000)
11. Fabian A. C., Voigt L. M., Morris R. G.: MNRAS **335**, L71 (2002)
12. Fender R., et al.: ApJ **519**, L165 (1999)
13. Fender R., Belloni T.: ARA&A **42**, 317 (2004)
14. Ferrarese L., Merritt D.: ApJ **539**, L9 (2000)
15. Gallo E., Fender R. P., Pooley G. G.: MNRAS **344**, 60 (2003)
16. Granato G. L. et al.: ApJ **600**, 580 (2004)
17. Maccarone T. J., Gallo E., Fender R.: MNRAS **345**, L19 (2003)
18. McNamara B. R., et al.: ApJ **534**, L135 (2000)
19. Murray N., Quataert E., Thompson T. A.: ApJ **618**, 569 (2005)
20. Owen F. N., Eilek J. A., Kassim N. E.: ApJ **543**, 611 (2000)
21. Sazonov S., Ostriker J. P., Sunyaev R. A.: MNRAS **347**, 144 (2004)
22. Sazonov S., Ostriker J. P., Ciotti L., Sunyaev R. A.: MNRAS **358**, 168 (2005)
23. Shakura N. I., Sunyaev R. A.: A&A **24**, 337 (1973)
24. Silk J., Rees M. J.: A&A **331**, L1 (1998)
25. Springel V., Di Matteo T., Hernquist L.: ApJ **620**, L79 (2005)
26. Tremaine S., et al.: ApJ **574**, 740 (2002)
27. White S. D. M., Frenk C. S.: ApJ **379**, 52 (1991)
28. Wyithe J. S. B., Loeb A.: ApJ **590**, 691 (2003)

Metal Enrichment of the ICM due to Ram-Pressure Stripping of Cluster Galaxies

W. Domainko[1], W. Kapferer[1], M. Mair[1], S. Schindler[1], E. v. Kampen[1], T. Kronberger[1,2], R. Moll[1], S. Kimeswenger[1], M. Ruffert[3], and O.E. Mangete[3]

[1] Institut für Astro- Teilchenphysik, Leopold - Franzens Universität Innsbruck, Austria,
wilfried.domainko@uibk.ac.at
[2] Institut für Astrophysik, Universität Göttingen, Germany
[3] School of Mathematics, University of Edinburgh, UK

Summary. The Intra-Cluster Medium (ICM) contains a significant amount of metals. As heavy elements can not be processed in the ICM itself, enriched material has to be produced by the stellar population of the cluster galaxies and is subsequently expelled from the host galaxy into the ambient medium. Here we present hydrodynamic simulations investigating the effect of ram-pressure stripping of cluster galaxies. The efficiency, resulting spatial distribution of the metals and the time dependency of this enrichment process on galaxy cluster scale will be shown in this presentation.

1 Introduction

Clusters of galaxies are the largest gravitationally bound systems in the universe. The bulk of baryonic matter in galaxy clusters is distributed diffusely throughout the whole system as the intra-cluster medium (ICM). This ICM is a hot, thin plasma which is observed in X-rays. Surprisingly X-ray spectra revealed that this plasma is enriched with iron to about 0.3 solar abundance [5]. With the recent X-ray satellites *Chandra* and *XMM* it is even possible to study the 2D distribution of the heavy elements (e.g.[8]). Therefore the ICM can not only be of primordial origin but part of it has to be processed in stars of galaxies. Several mechanisms of galactic mass loss can lead to an enrichment of the surrounding medium. These processes are ram-pressure stripping [7],[4], galactic winds and starbursts [2],[10], galaxy - galaxy interaction [6], [9] and jets from active galaxies. Alternatively also intra-cluster stars and supernovae can enrich the ICM (e.g. [3]). In this paper we investigate the effect of ram-pressure stripping.

2 Numerical Method

As ram-pressure stripping is an environmental dependent process, accurate treatment of the ICM is essential for this problem. We perform simulations on galaxy cluster scale to investigate the impact of the material stripped off the galaxies on the properties of the ICM. We use a combined tree N-body and grid based hydrodynamic simulation with an additional semi-numerical model for galaxy formation [13]. Hydrodynamics is treated with a PPM (Piecewise Parabolic Method: [1]) which allows to resolve shocks very well. This is crucial for the efficiency of the ram-pressure stripping which we derive from the properties of the ICM at the galaxies position and the velocity of the galaxy relatively to the ICM. If the force due to ram-pressure on the galactic gas disk exceeds the restoring gravitational force this part of the Intra-Stellar Medium (ISM) of the galaxy will be lost by the galaxy [7]. The simulation is obtained on four nested grids [11]. This technique allows to cover the cluster center where most of the stripping is expected to happen with high spatial resolution. Metallicity is used as a tracer to follow the enriched material.

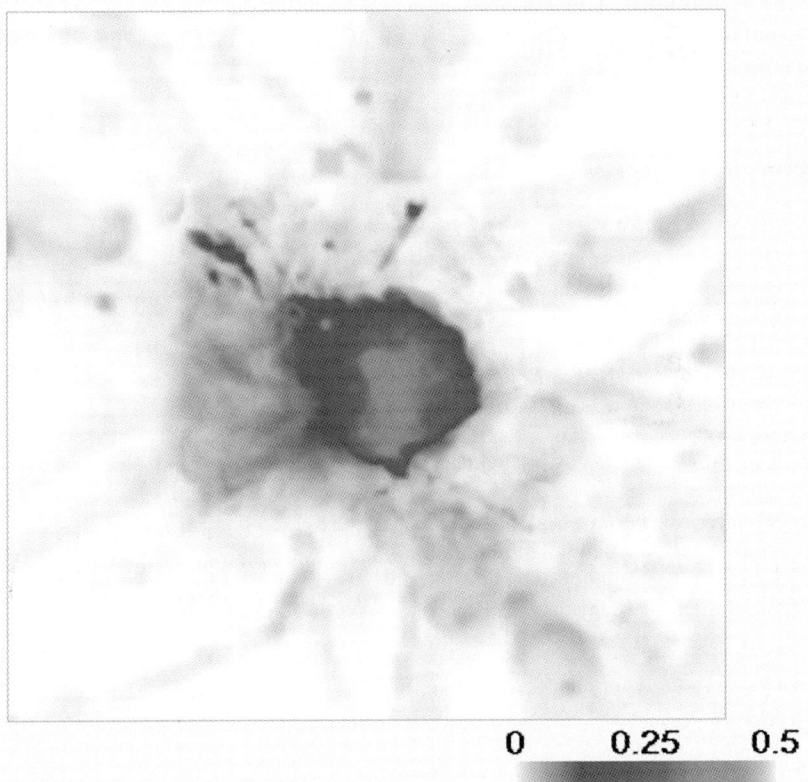

Fig. 1. Metallicity distribution of a model cluster after a simulation time of 8 Gyr. Metallicity is given in solar abundances. The size of the box is 5 Mpc on a side.

3 Results

From our simulations we find that ram-pressure stripping of cluster galaxies can account for $\sim 10\%$ of the observed level of enrichment within a radius of 1.3 Mpc in our model clusters. Additionally we show that this process is most efficient in the center of the galaxy cluster and the level of enrichment drops quite fast at larger radii. This is consistent with the fact that ram-pressure stripping is most efficient at the center of galaxy clusters due to the high ICM densities and the high velocities of the galaxies there. Enrichment by ram-pressure stripping persists over the entire simulation time and is significantly increased during subcluster mergers. Our computed 2D metallicity maps clearly show a complex pattern with stripes and plumes of metal rich material. Hence metallicity is a good tracer of the present and past interaction between the ICM and cluster galaxies. For more informations see Domainko et al. (2006 [4], Schindler et al. 2005 [12]).

Acknowledgments. The work has been performed under the FWF-Project P15868, the Project HPC-EUROPA (RII3-CT- 2003-506079) and a Doktoratsstipendium of the LFU Innsbruck.

References

1. Colella, P., Woodward, P. R.: J. Comput. Phys. **54**, 174 (1984)
2. De Young, D.S.: ApJ **223**, 47 (1978)
3. Domainko W., Gitti M., Schindler, S. & Kapferer, W.: A&A **425**, L21 (2004)
4. Domainko, W., Mair, M., Kapferer, W. et al.: A&A **452**, 795 (2006) astro-ph/0507605
5. Fukazawa, Y. Makishima, K., Tamura, T., et al.: PASJ **50**, 187 (1998)
6. Gnedin, N. Y.: MNRAS **294**, 407 (1998)
7. Gunn J.E., & Gott J.R.III: ApJ **176**, 1 (1972)
8. Hayakawa A., Furusho T., Yamasaki N.Y., Ishida M. & Ohashi T.: PASJ **56**, 743 (2004)
9. Kapferer, W., Knapp, A., Schindler, S., Kimeswenger, S. & van Kampen, E.: A&A **438**, 87 (2005a)
10. Kapferer, W., Ferrari, C., Domainko, W. et al.: A&A **447**, 827 (2006)
11. Ruffert, M.: A&A **265**, 82 (1992)
12. Schindler, S., Kapferer, W., Domainko, W. et al.: A&A **435**, L25 (2005)
13. van Kampen, E., Jimenez, R., Peacock, J. A.: MNRAS **310**, 43 (1999)

Radio Bubbles in Clusters: Relativistic Particle Content

R.J.H. Dunn[1], A.C. Fabian[1], and G.B. Taylor[2,3]

[1] Institute of Astronomy, Madingley Road, Cambridge, UK, rjhd2@ast.cam.ac.uk
[2] University of New Mexico, Albuquerque, NM, USA
[3] National Radio Astronomy Observatory, Socorro, NM USA

1 Introduction

Radio bubbles in clusters of galaxies contain magnetic fields and charged particles which emit synchrotron radiation. In the past equipartition has been assumed to infer the pressure from each component [1]. However, when the lobes are embedded in the thermal Intra-Cluster Medium (ICM) this degeneracy can be lifted with the assumption of pressure balance between the ICM and the lobes.

2 Particle Energies

Extending the work of [2] we add more active bubbles (those with GHz radio emission) into the sample. We determine k, the ratio of total particle energy to that of electrons emitting synchrotron radiation between 10 MHz and 10 GHz, assuming pressure balance between the synchrotron plasma and the ICM. Constraining the ages of the bubbles from the sound speed timescale, as no strong shocks are observed around the bubbles, we calculate the ratio of k to f, the volume filling factor. If the bubbles are completely filled by an electron-positron plasma emitting only between 10 MHz and 10 GHz then $k/f = 1$, however if it is filled by an electron-proton plasma it is expected that $k/f \sim 2000$.

The constraints on the magnetic field obtained by comparing the synchrotron cooling time to the bubble age show that none of the bubbles in the sample are in equipartition. In fact, if the bubbles were in equipartition, then they would be under pressured by factors of 1.5 to 160.

Across the whole sample the ratio lies within the range $1 \lesssim k/f \lesssim 1000$. There appears to be no correlation with any physical parameter of the bubble or its host cluster. There is no clear choice for the underlying form of the distribution of k/f, including the uncertainties on k/f by Monte Carlo simulations(Fig. 1).

The large spread in k/f may be the result of a significant population of non-relativistic particles present in some of the bubbles. Then the variation of k between clusters would indicate that the entrainment process is stochastic, and highly dependent on the environment of the jet in the first few kpc around the radio source.

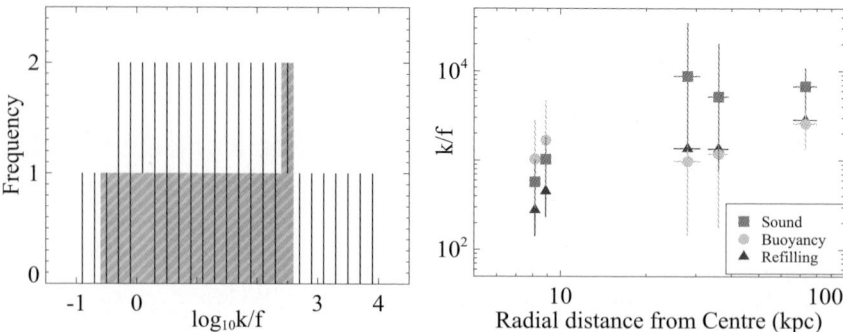

Fig. 1. LEFT: The frequency distribution of k/f from Monte-Carlo simulations of the data (grey bars). The vertical lines are the errors on the bin values. RIGHT: The evolution of k/f with bubble age in the Perseus Cluster. The different points correspond to different estimates for the timescales of the bubbles.

2.1 Ghost Bubbles

In the extended sample we also analysed "ghost bubbles," which are bubbles without GHz radio emission. For these the timescale most appropriate for the bubbles is the buoyancy timescale as these bubbles are rising up through the ICM. The values obtained for k/f are on the whole larger than those for the active bubbles. The Perseus and Centaurus Clusters contain both ghost and active bubbles, and so the evolution of k/f can be traced (Fig. 1). k/f rises as the bubbles age; the energy inferred from synchrotron emission falls as the bubble ages, and so k would be expected to rise [3].

3 Jet Matter Content

The extra particles required by the bubbles in order that they are in pressure equilibrium with their surroundings could come from two sources. Either they are inherent to the jet and are supplied by the central AGN or they are entrained by the jet as it passes up through the ICM.

To place limits on how much of the material has its origins in the jet we revisit the Synchrotron-Self-Compton (SSC) model from [4]. The emission region is assumed to be spherical, with a randomly orientated magnetic field and relativistic particles with a power-law energy distribution. We assume $\gamma_{\mathrm{min,\,e}} = 1$; for other values see the forthcoming paper (Dunn et al. in prep). The synchrotron flux of the optically thick (self-absorbed) region is independent of the particle density, placing upper limits on the magnetic field (Line 2, Fig. 2). Further limits on the $n - B$ plane arise from the optical depth at the self-absorption frequency (Line 1, Fig. 2), constraining which number densities and magnetic fields are allowed.

The kinetic luminosity of the jet is estimated from the bubble power. The energy within the radio bubble can be calculated from the pV work done in creating it. The sound speed timescale is used as these bubbles are young. The kinetic luminosity of the

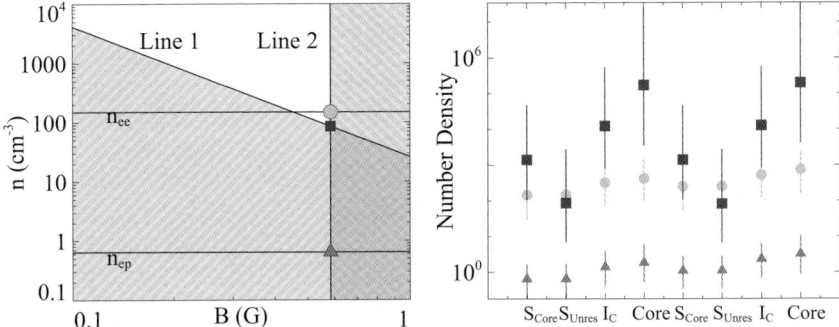

Fig. 2. LEFT: The limits on the $n - B$ plane from SSC arguments for M87 S_{Unres}. The excluded regions are shaded grey. In this case the data exclude the possibility of a heavy jet (electron-proton). RIGHT: The results for M87; the red triangles / green circles points are the expected number densities for heavy/light jets respectively. The blue squares are the minimum allowed number density. The repetition arises from the Jet/Counter-Jet cavity used for the energies.

jet depends on the matter content and so predictions on the expected number densities in the jet can be derived for the two types of jet (lines n_{ee} and n_{ep}, Fig. 2).

The clusters which have been analysed in detail so far are M87 and the Perseus Cluster. All the data for M87 are inconsistent with heavy jets, whereas in Perseus no definite conclusions can be drawn. Other clusters were too distant for the VLBI radio data to be of sufficient resolution. This causes the angular size of the radio core to be overestimated, and hence also the magnetic field and with the result that no constraints can be placed on the matter content of jets.

4 Conclusions

For radio bubbles to be in pressure balance with the ICM, they must contain a significant fraction of non-relativistic particles. The fraction varies between clusters and covers a range $1 \sim 1000$.

This work implies that matter content of the M87 jet is electron-positron, and so these extra particles are not protons from the jet, but may be entrained as it travels out through the ICM. The range in k implies that the process is stochastic, depending on the environment of the jet in the first few kiloparsec.

References

1. G. R. Burbidge. Estimates of the Total Energy in Particles and Magnetic Field in the Non-Thermal Radio Sources: ApJ **129**, 849 (1959)
2. R. J. H. Dunn and A. C. Fabian. Particle energies and filling fractions of radio bubbles in cluster cores: MNRAS **351** 862 (2004)
3. R. J. H. Dunn, A. C. Fabian, and G. B. Taylor. Radio Bubbles in Clusters: MNRAS **364**, 1343 (2005)

4. C. S. Reynolds, A. C. Fabian, A. Celotti, and M. J. Rees. The matter content of the jet in M87: evidence for an electron-positron jet: MNRAS **283**, 873 (1996)

The XMM-Newton Distant Cluster Project

R. Fassbender[1], H. Böhringer[1], J. Santos[1], G. Lamer[2], C. Mullis[3], P. Schuecker[1], A. Schwope[2], and P. Rosati[4]

[1] Max-Planck Institut für Extraterrestrische Physik, Garching, Germany,
   rfassben@mpe.mpg.de
[2] Astrophysikalisches Institut Potsdam, Germany
[3] University of Michigan, Ann Arbor, USA
[4] European Southern Observatory, Garching, Germany

1 Introduction

Distant clusters of galaxies are ideal laboratories for cosmic evolution studies and — as key tracers of large-scale structure with a high redshift leverage — they are also sensitive cosmological test objects [1]. We have initiated the XMM-Newton Distant Cluster Project (XDCP), a serendipitous survey for X-ray luminous z>1 clusters using archival XMM-Newton data, in order to provide a statistically complete sample for evolutionary and cosmological studies. The interim goal of the XDCP is the construction of an X-ray flux limited sample of ∼30 galaxy clusters beyond redshift one; the estimated XMM-Newton final archive potential is to provide more than 100 such objects. A comprehensive XDCP project overview can be found in [2]. Here we present selected results on the statistics of the X-ray data and the efficacy of the distant cluster candidate selection.

2 The XMM-Newton Distant Cluster Project

2.1 X-ray Data Analysis

For the ongoing project phase, we are focussing on VLT-accessible XMM-Newton fields outside the galactic plane with a minimum EPIC exposure time of 10ksec. The exposure time statistics for the 299 fields with the highest data quality are shown in Fig. 1(a). This data set consists of a total nominal exposure time of 9.6Msec or 6.6Msec of effective cleaned time, i.e., on average about 30% of the observing time is lost due to flares and detector overheads. The individual fields contain on average 22ksec of clean data (32ksec nominal).

Figure 1(b) illustrates the number count results of these data. About 45,000 X-ray sources are detected, of which all extended sources are carefully screened and checked for optical counterparts using the Digital Sky Survey (DSS), as illustrated in Fig. 2. In total, 450 extended X-ray sources with an optical galaxy cluster or group signature have been identified (Fig. 2(a)). 170 extended sources do not show an optical counterpart (Fig. 2(b)) and enter our list of distant cluster candidates for follow-up imaging.

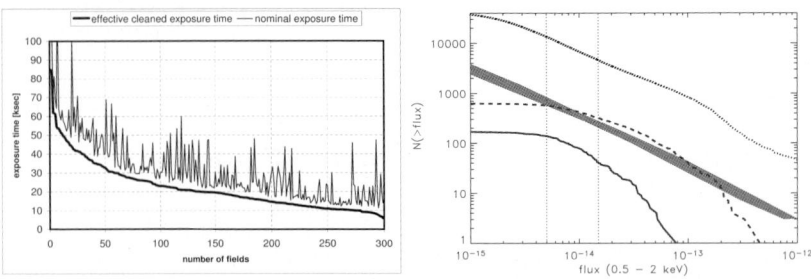

Fig. 1. X-ray data statistics of the 299 best XMM fields: **(a)** Cleaned exposure times of fields sorted in descending order *(thick line)*; the *upper thin line* indicates the raw exposure time, i.e., the difference yields the data loss due to flares and overheads. **(b)** Number counts of the same fields. *Dotted line*: all sources; *dashed line*: all identified clusters and candidates; *solid line*: distant cluster candidates. The *shaded band* indicates the RDCS cluster logNlogS as determined in [3]

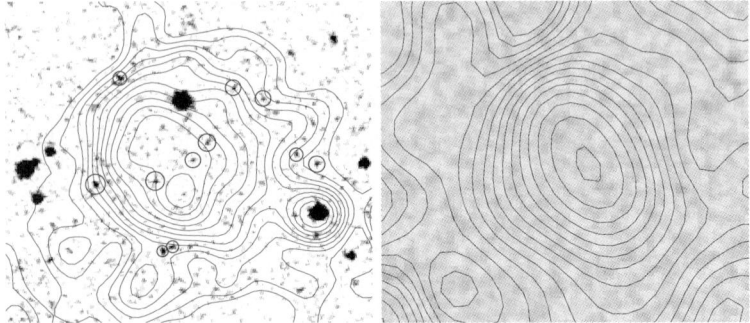

Fig. 2. Candidate selection using DSS-X-ray overlays: **(a)** DSS-identified cluster with spectroscopic redshift of z=0.73 [7]; image 3x3′. *Small circles* indicate 11 probable cluster members visible in the high contrast image. **(b)** XDCP distant cluster candidate with no optical counterpart at the DSS limit; image 1.5x1.5′

2.2 Diagnostic Tests in the COSMOS Field

We have tested our XDCP source detection and candidate selection strategy on 20 XMM-Newton pointings of the COSMOS field [6], which provides additional multi-wavelength coverage. With an average cleaned exposure time of 24.9ksec, the COSMOS fields have almost the same depth as typical XDCP survey fields and are thus ideal for internal diagnostic tests with full observational feedback on all X-ray sources.

Figure 3(a) illustrates the expected cluster redshift distribution for the XDCP survey, whereas Fig. 3(b) depicts the binned distribution as measured in 1.8 deg^2 of the COSMOS field. The dotted line in Fig. 3(b) represents the DSS-identified clusters and groups, i.e., objects with a low redshift flag; the dashed line indicates the photometric redshifts of distant cluster candidates in the COSMOS field. As shown, 6 objects with $z_{phot} > 1$ could be identified, which is more than expected (Fig. 3(a)) but could be due to low number statistics. Very interesting in this respect are results on the actual redshift limit of the DSS, i.e., the redshift past which follow-up imaging is needed. As indicated

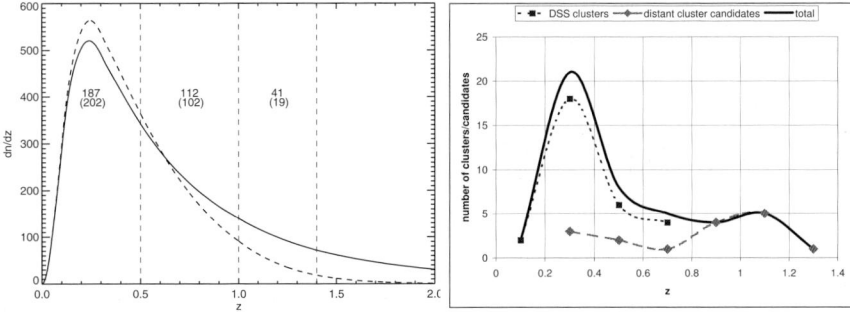

Fig. 3. Cluster redshift distributions: **(a)** Expected distribution of a $14 \, \text{deg}^2$ XMM survey based on [3] & [4] without *(solid line)* and including cluster evolution *(dashed line)*. Values indicate the total cluster number per redshift bin *(vertical lines)* for the two models. **(b)** Measured photometric redshift distribution in $1.8 \, \text{deg}^2$ of the COSMOS field [6]. *Dotted line*: DSS identified groups & clusters; *dashed line*: distant cluster candidates; *solid line*: sum of both. Data points are binned with $\Delta z = 0.2$

by the dotted line, we identified 4 optical counterparts of clusters with $0.6 < z_{phot} \leq 0.8$ on DSS images (e.g. Fig. 2(a)), all shown to be massive systems [6]. On the other hand, the five candidates at $z_{phot} \leq 0.6$ have been revealed as low mass groups, making them harder to distinguish from possible foreground objects. In summary, the COSMOS diagnostic test has shown that we can efficiently identify and flag low redshift clusters, on average out to $z \sim 0.5$–0.6, while retaining the very distant clusters in our candidate sample.

2.3 First Results and Outlook

Using an efficient R-z snapshot strategy for follow-up imaging with VLT-FORS2, the truly distant galaxy clusters at $z > 1$ can be identified from the X-ray selected candidates by means of the color of the cluster red sequence [2]. This technique resulted in the discovery of XMMUJ2235.3-2557, the most distant, X-ray luminous cluster presently known at $z = 1.393$ [5]. Spectroscopic confirmation is also pending for another 10 candidates from the pilot study with high photometric redshifts. In combination with the newly identified candidates in Sect. 2.1, the goal of a sample of ~ 30 galaxy clusters beyond redshift unity will be well achievable as soon as sufficient telescope time is available for optical follow-up observations.

References

1. Holder, G., Haiman, Z., & Mohr, J. J.: ApJL **560**, L111 (2001)
2. Böhringer, et al.: ESO Messenger **120**, 33 (2005)
3. Rosati, P., Borgani, S., & Norman, C.: ARA&A **40**, 539 (2002)
4. Mullis, C. R., et al.: ApJ **607**, 175 (2004)
5. Mullis, C. R., et al.: ApJL **623**, L85 (2005)
6. Finoguenov, A., et al., ApJS, in prep.
7. Guzzo, L., et al., ApJS, in prep.

Tracing the Mass–Assembly History of Galaxies with Deep Surveys

G. Feulner[1,2], A. Gabasch[1,2], Y. Goranova[1,2], U. Hopp[1,2], and R. Bender[1,2]

[1] Universitäts–Sternwarte München, Scheinerstraße 1, D–81679 München, Germany,
feulner@usm.lmu.de
[2] Max–Planck–Institut für extraterrestrische Physik, Giessenbachstraße 1, D–85748 Garching, Germany

Summary. We use the optical and near-infrared galaxy samples from the Munich Near-Infrared Cluster Survey (MUNICS), the FORS Deep Field (FDF) and GOODS-S to probe the stellar mass assembly history of field galaxies out to $z \sim 5$. Combining information on the galaxies' stellar mass with their star-formation rate and the age of the stellar population, we can draw important conclusions on the assembly of the most massive galaxies in the universe: These objects contain the oldest stellar populations at all redshifts probed. Furthermore, we show that with increasing redshift the contribution of star-formation to the mass assembly for massive galaxies increases dramatically, reaching the era of their formation at $z \sim 2$ and beyond. These findings can be interpreted as evidence for an early epoch of star formation in the most massive galaxies in the universe.

1 Introduction

In recent years, there has been considerable interest in the relation of the stellar mass in galaxies and their star-formation rate (SFR), since this allows to quantify the contribution of the recent star formation to the build up of stellar mass for different galaxy masses. Cowie et al. 1996 [4] were the first to investigate this connection for a K-selected sample of ~ 400 galaxies at $z < 1.5$ and noted an emerging population of massive, heavily star forming galaxies at higher redshifts, a phenomenon they termed "down-sizing". Later on, the specific star-formation rate (SSFR), defined as the SFR per unit stellar mass, was used to study this relation.

2 Connecting Star Formation and Stellar Mass

We have analysed the SSFR as a function of stellar mass and redshift z out to $z = 1.2$ [11] using a large sample of more than 6000 I-band selected galaxies from MUNICS [6, 9]. The SSFR decreases with mass at all redshifts, although we might not detect highly obscured galaxies. The low values of the SSFR of the most massive galaxies suggests that most of these massive systems formed the bulk of their stars at earlier epochs. Furthermore, stellar population synthesis models show that the most massive systems contain the oldest stellar populations at all redshifts. This is in agreement with the detection of old, massive galaxies at redshifts $1 < z < 2$ [3, 15, 14] and beyond [2]. In Fig. 1, where we have used the FDF [12, 13] and GOODS-S samples, we show that this trend continues to even higher redshifts [10].

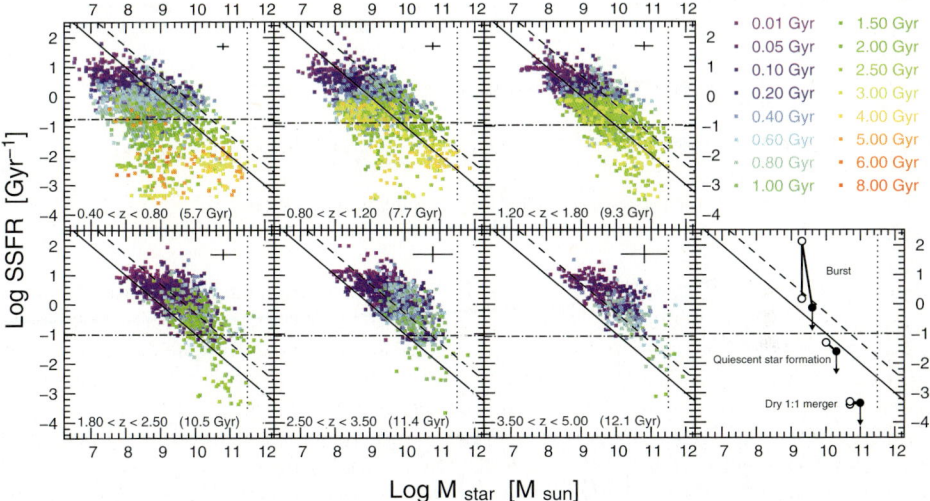

Fig. 1. *Six panels to the left:* The SSFR as a function of stellar mass and redshift. The solid and dashed lines correspond to constant SFRs. Objects are coloured according to the age of their stellar population. The dot-dashed line is the SSFR required to double a galaxy's mass between each redshift epoch and today (assuming constant SFR); the corresponding look-back time is indicated as well. The error cross in each panel gives an idea of the typical errors, while the dotted line roughly represents the high-mass cut-off of the local stellar mass function ([5, 7]). *Lower right-hand panel:* Examples for evolutionary paths yielding a doubling of a galaxy's mass. Open symbols denote the starting point, filled symbols the final state; the arrows indicated the influence of gas consumption or loss.

3 The Build-up of the Most Massive Galaxies

It is extremely interesting to investigate the average SSFR of galaxies with different masses as a function of redshift shown in Fig. 2 [10]. While at redshifts $z < 2$ the most massive galaxies are in a quiescent state, at redshifts $z > 2$ the SSFR for massive galaxies increases by a factor of ~ 10 reaching the epoch of their formation at $z \sim 2$ and beyond. While there is evidence for dry merging (i.e. without interaction-induced star formation) in the field galaxy population [1, 8], this strong increase in the SSFR of the most massive galaxies suggests that at least part of this population was formed in an early period of efficient star formation in massive haloes.

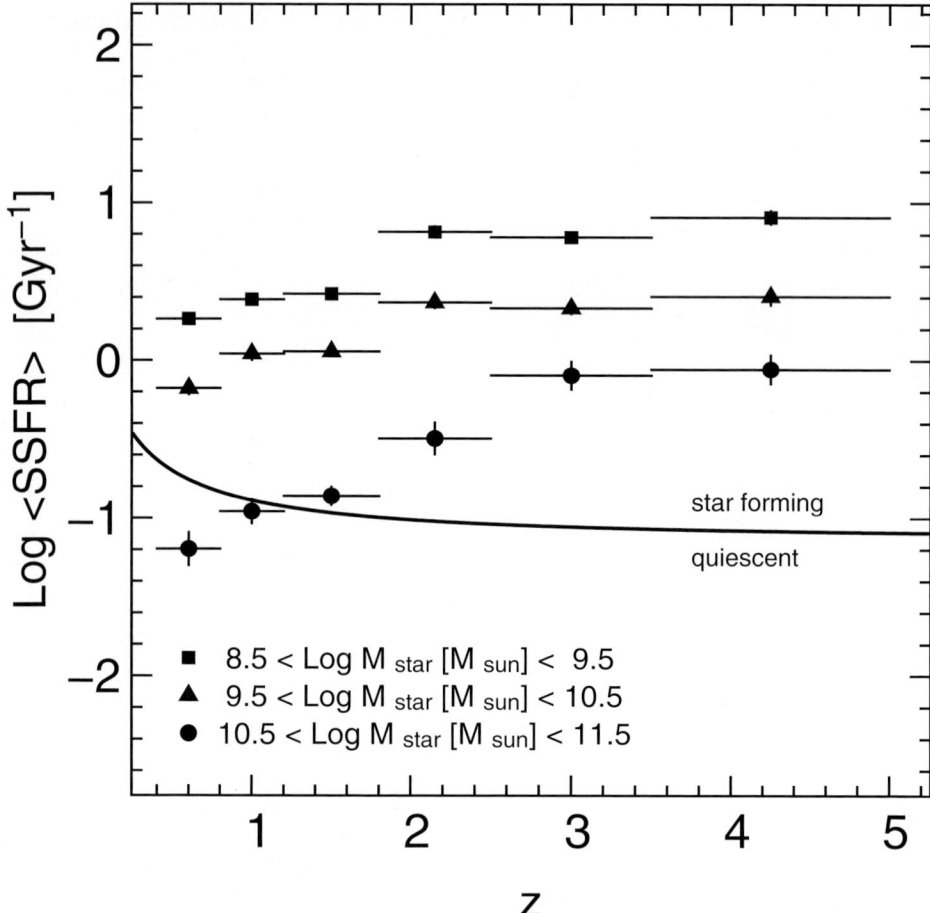

Fig. 2. Average SSFRs for galaxies with stellar masses of $\log M_\star/M_\odot \in [8.5, 9.5]$ (squares), $[9.5, 10.5]$ (triangles) and $[10.5, 11.5]$ (circles) and SFRs larger than $1\ M_\odot\ \mathrm{yr}^{-1}$ as a function of z. The error bar represents the error of the mean. The solid line indicates the doubling line of Fig. 1 which can be used to discriminate quiescent and heavily star forming galaxies.

References

1. Bell, E. et al.: ApJ **640**, 241 (2005)
2. Chen, H.-W. & Marzke, R. O.: ApJ **615**, 603 (2004)
3. Cimatti, A., Daddi, E., Renzini, A., et al.: Nature **430**, 184 (2004)
4. Cowie, L. L., Songaila, A., Hu, E. M., & Cohen, J. G.: AJ **112**, 839 (1996)
5. Drory, N., Bender, R., Feulner, G., et al.: ApJ **608**, 742 (2004)
6. Drory, N., Feulner, G., Bender, R., et al.: MNRAS **325**, 550 (2001)
7. Drory, N., Salvato, M., Gabasch, A., et al.: ApJL **619**, L131 (2005)
8. Faber, S. M. et al.: ApJ, submitted, astro-ph/0506044 (2005)
9. Feulner, G., Bender, R., Drory, N., et al.: MNRAS **342**, 605 (2003)
10. Feulner, G., Gabasch, A., Salvato, M., et al.: ApJL **633**, L9 (2005a)

11. Feulner, G.,Goranova, Y., Drory, N., Hopp, U., & Bender, R.: MNRAS **358**, L1 (2005b)
12. Gabasch, A., et al.: A&A **421**, 41 (2004a)
13. Heidt, J., Appenzeller, I., Gabasch, A., et al.: A&A **398**, 49 (2003)
14. Saracco, P., Longhetti, M., Severgnini, P., et al.: MNRAS **357**, L40 (2005)
15. Saracco, P., Longhetti, M., Severgnini, P., et al.: A&A **398**, 127 (2003)

Outbursts from Supermassive Black Holes and their Impacts on the Hot Gas in Early-Type Galaxies, Groups and Clusters

W. Forman[1], C. Jones[1], and E. Churazov[2,3]

[1] Smithsonian Astrophysical Observatory, USA (wrf,cjones)@cfa.harvard.edu
[2] Max Planck Institute for Astrophysics, Garching, Germany
[3] Space Research Institute, Moscow, Russia, churazov@mpa-Garching.mpg.de

Summary. We discuss the effects of outbursts from supermassive black holes (SMBH) in luminous galaxies on surrounding gaseous atmospheres with emphasis on the Chandra observations of M87. The Chandra observations of M87 show both a system of buoyant bubbles as well as weak shock waves. In addition, we compare SMBH outbursts in systems from early type galaxies to rich clusters.

1 Hot Gas in Early Type Galaxies, Groups, and Clusters

One of the remarkable discoveries of X-ray astronomy is the gas rich nature of early type galaxies, elliptical-dominated groups, and clusters (e.g., [13, 16, 17, 18]. The very first X-ray observations from UHURU showed that galaxy clusters are remarkably luminous X-ray sources. Subsequently, X-ray observations showed that the X-rays were produced by thermal emission from hot ($kT \sim 10^8$K) gas with a mass that exceeds the baryon content of the luminous cluster member galaxies. With the increased sensitivity of the Einstein Observatory, luminous early type galaxies, previously believed to be devoid of gas, were found to be as gas rich as their spiral counterparts, but with gas temperatures of $\sim 10^7$K, visible only in X-rays.[13]

With the launch of the Chandra observatory, we have been able to view the sky with an angular resolution comparable to that of ground-based optical telescopes. Results from Chandra showed that the central regions of hot gaseous atmospheres in clusters, groups, and early type galaxies were morphologically very complex and that this complexity was more the rule rather than the exception (e.g., [11, 12, 14]). Most importantly, the X-ray and radio observations of the hot gas and relativistic plasma provide a fossil record of the mass and energy outbursts of central AGN.

In a large fraction of clusters, the X-ray emission is strongly peaked at the cluster center on a bright cD galaxy. From the earliest X-ray observations, it was realized that high density gas with necessarily short cooling times should accrete onto the galaxies that lie at the centers of X-ray luminous gaseous atmospheres.[10, 7] However, the repository of the expected large amounts of cooled gas in the form of HI or recent star formation remained elusive. Observations with XMM-Newton finally resolved the 30 year old problem by demonstrating that mass deposition rates in "cooling flow" clusters are at least five times smaller than expected in the standard model (e.g., [24] and references therein). The XMM-Newton limits on gas cooling imply considerable energy input to compensate for the observed radiative losses.

Suggestions for providing energy input to hot gas in cluster (and group and galaxy) cores include mergers, thermal conduction (e.g. [2, 9, 15, 27, 30]) and AGN activity (e.g., [1, 3, 25, 28]). Mergers occur at irregular times and would not provide a steady energy source. While conduction may be able to provide energy to the outer regions of the cores, in the most centrally peaked cooling flow clusters, conduction is not universally effective in transporting energy to the cluster center. With the increasingly detailed X-ray observations, it has become clear that jets, bubbles and shocks, produced by periodic outbursts from the supermassive black hole in the central galaxy, have a fundamental role in (re)heating the gas in cooling atmospheres (e.g., [6], [8]).

2 M87 - A CANONICAL CLUSTER COOLING CORE

At a distance of only 16 Mpc, M87 (NGC4486), the giant elliptical at the center of the Virgo cluster, provides a unique laboratory to study the interactions of the hot intracluster gas with the energy generated by the 3×10^9 $M_\odot$ black hole at its core. M87 has long been considered a classic example of a "cooling flow" system (e.g., [5, 26]). Using radio observations, Owen et al. developed the idea that the mechanical energy produced by the SMBH in M87 was more than sufficient to compensate for the energy radiated in X-rays.[23] As we summarize below, X-ray observations of M87 show a spectacular object with shocks, filaments, buoyant bubbles, and cavities. [14, 29]

The Chandra image (see Fig. 1) shows a bright central region, X-ray arms that correspond to those seen in the radio image, and outer rings or shells that we interpret as shocks. On the largest scales, the VLA map (see [23]) shows outer lobes or "pancakes" which formed 10^8 years ago and although these are the oldest visible radio structures, they require the continual injection of energy. M87 also has bright radio arms to the East and Southwest, and in the core lies the famous jet.

The central region containing the jet and inner radio lobes (the cocoon region) originated in an episode of recent AGN activity. As shown in Fig. 2, the central core is very complex. At the center lies the jet which is surrounded by an X-ray cavity. An X-ray cavity also defines the region of the counter jet. There is radio emission and an X-ray cavity associated with another bubble southeast of the nucleus that is not coaligned with the jet (labeled "bud"). Several additional bubbles can be seen in the Eastern arm. Each of these bubbles is about 1 kpc in radius. The X-ray cavities surrounding the jet and counterjet are filled with radio emitting plasma. Assuming subsonic expansion for the radio plasma, the age of these lobes is $\sim 1.7 \times 10^6$ years. The amount of energy required to evacuate the gas from the cavities is about 10^{56} ergs.

Outside the core, (see Fig. 1) the prominent features are the X-ray arms and two shells or rings of emission. The inner shell at 14 kpc (3′) can be seen nearly all the way around M87, while the 17 kpc shell is most prominent to the northwest. To better examine the structure in the gas, we modelled M87's overall halo of X-ray emission

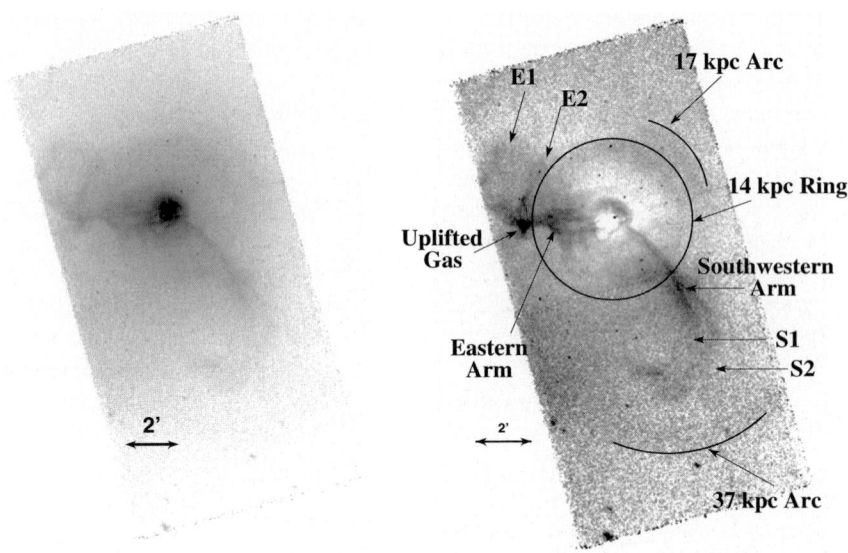

Fig. 1. The flat-fielded Chandra image in the energy band 0.5-2.5 keV (left panel) with a King model subtracted to remove the large scale radial surface brightness gradient (right panel). Many faint features are seen including 1) the prominent eastern and southwestern arms and their bifurcation (**E1, E2** and **S1, S2** identify the extensions of the eastern and southwestern arms), 2) the 14 kpc (3′) ring, 3) the 17 kpc (3.75′) arc, and 4) the faint southern 37 kpc (8′) arc.

with a smooth model and subtracted that from the Chandra image. Fig. 1 (right panel) shows the deviations from the smooth model – the arms and arcs are prominent. The sharpness of the shells as well as the completeness of the 14 kpc ring suggests they are due to shocks likely driven by the initial rapid expansion of the core radio lobes. Both X-ray arms brighten at about the radii of the shocks, suggesting that they lie nearly in the plane of the sky. A simple shock model of the radial surface brightness and temperature profiles across the 14 kpc X-ray arc requires an outburst with an energy of nearly 10^{58} ergs occurring 10^7 years ago. The shock is mildly supersonic. The outer partial ring, 17 kpc radius, would have been caused by an explosion of similar energy occurring about 4 million years earlier. Table 1 gives the mass that would need to be accreted by the SMBH to produce the outburst assuming that the mass to energy conversion efficiency is 0.1. The large accreted masses required to produce the giant outbursts in clusters like MSO735 imply a significant growth in black hole masses during the current epoch.[20]

While the rings are very prominent in the X-ray images, both the radio and X-ray images show the effects of large buoyant bubbles.[6, 23]. Churazov et al. modelled bouyant bubbles in the atmosphere of M87. The present structures, especially clear in the eastern arm with the toroidal radio structures seen in projection, correspond to bubbles produced by AGN outbursts that occurred 10^7 years ago.[6, 23] Initially a hot bubble is created by the AGN. Plasma bubbles generated in the core, rise rapidly and lose about half their energy through adiabatic expansion. As the bubble rises buoyantly

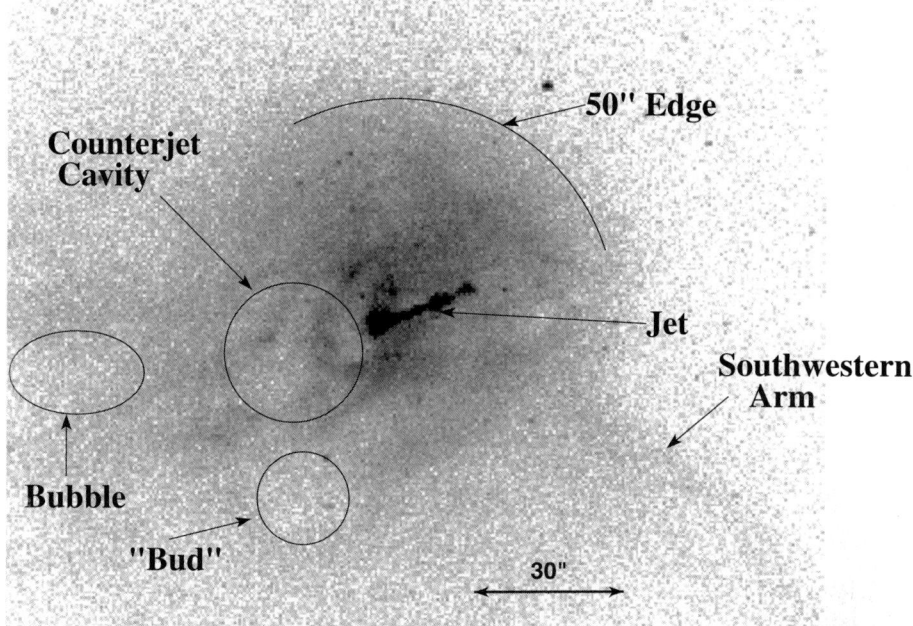

Fig. 2. The central region of M87 as seen by the Chandra ACIS-S detector in the energy band 0.5 to 2.5 keV. The core region shows the jet, the counterjet cavity, a "bud" emanating from the southeastern edge of the counterjet cavity, and the beginning of the southwestern arm.

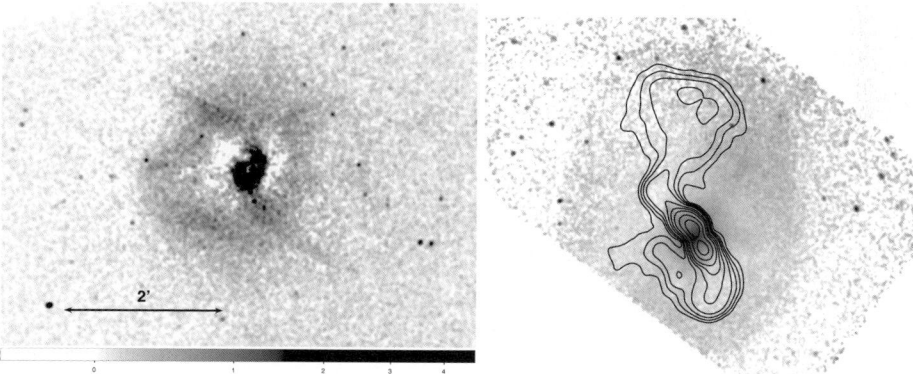

Fig. 3. (left) The Chandra X-ray image of the optically normal galaxy NGC4636 showing the distorted X-ray emission from thermal gas. The figure shows the observation after subtraction of an azimuthally symmetric profile (see [19] for details). (right) The Chandra X-ray image of Hydra A with 330 MHz radio contours showing the outer shock driven by the radio plasma (see [21] for details).

through the atmosphere, it entrains cool gas behind it. This rising, expanding gas produces the arms seen in the radio and X-ray observations. As predicted by this model, the M87 arms are observed to be cooler than the surrounding gas ([4]; see also Fig. 6 in [14]).

3 Hot Gas in Early Type Galaxies, Groups, and Clusters

It is illustrative to compare the energy and timescales of the shocks that have been seen in systems ranging from the elliptical galaxy NGC4636 to large scale energetic shocks in clusters like Hydra A (see Fig. 3 for examples of disturbed gaseous atmospheres in systems smaller and larger than M87; [19, 21]). Table 1 lists the radius of the shock, the time since the AGN outburst, the initial outburst energy and the mean power of the outburst for systems from galaxies to rich clusters.[14, 19, 20, 21, 22] As long as the energy from the outburst can be transferred to the gas, the amount of energy in the outbursts is adequate to replenish the energy lost by the gas through radiative cooling. The last column in Table 1 gives the mass that would need to be accreted by the SMBH to produce the outburst assuming that the mass to energy conversion efficiency is 0.1. The large accreted masses required to produce the giant outbursts in clusters like MS0735.6+7421 imply a significant growth in black hole masses during the current epoch.[20]

Table 1. SMBH Outburst Parameters: from Galaxies to Clusters

Source	Shock radius (kpc)	Age (My)	Energy (10^{61} erg)	Mean power (10^{46} erg s^{-1})	Mass swallowed ($10^8 M_\odot$)
NGC4636	~5	3	0.00006	0.0007	0.00003
M87	14	10.6	0.0008	0.0024	0.0005
Hydra A	210	136	0.9	0.2	0.5
Hercules A	160	59	3.0	1.6	1.7
MS0735.6+7421	240	104	5.7	1.7	3

4 Conclusion

The absence of a detected repository of cooling gas in gas rich systems has been a puzzle for almost 30 years. With the high angular resolution of Chandra applied to gas rich systems from early-type galaxies to clusters, we see clear evidence that outbursts from central supermassive black holes heat the cooling cores of X-ray luminous gaseous atmospheres. The interplay between cooling gas and SMBH's provides a remarkable laboratory for understanding not only the physics of the interactions between plasma ejected from AGN and surrounding atmospheres, but also for understanding the outburst history of AGN.

Acknowledgments. We would like to acknowledge our collaborators M. Markevitch, A. Vikhlinin, P. Nulsen, S. Heinz, H. Böhringer, R. Kraft, M. Begelman, F. Owen, and J. Eilek.

References

1. Binney, J. & Tabor, G.: MNRAS **276**, 663 (1995)
2. Bertschinger, E. & Meiksin, A.: ApJL **306**, L1 (1986)
3. Böhringer, H. & Morfill, G.: ApJ **330**, 609 (1988)
4. Belsole, E. et al.: A&A **365**, L188 (2001)
5. Böhringer, H. et al.: A&A **365**, L181 (2001)
6. Churazov, E., Bruggen, M., Kaiser, C., Böhringer, H. & Forman, W.: ApJ **554**, 261 (2001)
7. Cowie, L. & Binney, J.: ApJ **215**, 723 (1978)
8. Churazov, E., Sunyaev, R., Forman & Böhringer, H.: MNRAS **332**, 729 (2002)
9. David, L., Hughes, J. & Tucker, W.: ApJ **394**, 452 (1992)
10. Fabian, A. & Nulsen, P.: MNRAS **180**, 479 (1977)
11. Fabian, A. et al.: MNRAS **344**, L43 (2003)
12. Finoguenov, A. & Jones, C.:ApJL **547** L107 (2001)
13. Forman, W., Jones, C., Tucker, W.: ApJ **61**, 33 (1985)
14. Forman, W. et al.: ApJ **635**, 894 (2005)
15. Gaetz, T.: ApJ **345**, 666 (1989)
16. Gursky, H. et al.: ApJL **173** L99 (1972)
17. Helsdon, S., Ponman, T., O'Sullivan, E. & Forbes, D.: MNRAS, **325**, 693 (2001)
18. Jones, C. and Forman, W.: ApJ **224**, 1 (1978)
19. Jones et al.: ApJL **567**, L115 (2002)
20. McNamara, B. et al.: Nature **433**, 45 (2005)
21. Nulsen et al.: ApJ **628**, 629 (2005)
22. Nulsen et al.: ApJ **625**, L9 (2005)
23. Owen, F., Eilek, J. & Kassim, N.: ApJ **543**, 611 (2000)
24. Peterson, J. et al.: Proceedings of "The Riddle of Cooling Flows in Clusters of Galaxies" eds. Reiprich, Kempner, Soker, astro-ph/0310008 (2003)
25. Rosner, R. & Tucker, W.: ApJ **338**, 761 (1989)
26. Stewart, G. et al.: ApJ **285**, 1 (1984)
27. Tucker, W. & Rosner, R.: ApJ **267**, 547 (1983)
28. Tucker, W. & David, L.: ApJ **484**, 602 (1997)
29. Young, A., Wilson, A., Mundell, C.: ApJ **579**, 560 (2002)
30. Zakamska, N. & Narayan, R.: ApJ **582**, 162 (2003)

Tracing Gas Motions in the Centaurus Cluster

J. Graham[1], A.C. Fabian[1], J.S. Sanders[1], and R.G. Morris[2]

[1] Institute of Astronomy, Madingley Road, Cambridge, UK, jgraham@ast.cam.ac.uk
[2] Kavli Institute for Particle Astrophysics and Cosmology, Stanford Linear Accelerator Center, Stanford

1 Introduction

Galaxy clusters have an average metallicity of about a third solar, with the metals believed to originate in early type II supernovae. X-ray studies of cool core clusters have revealed an additional central peak in the metallicity, which has been interpreted as metal injection by the brightest cluster galaxy (BCG). However the observed metal distributions are significantly less peaked than the surface brightness profile of the BCG. This discrepancy suggests the possibility of using the metal distribution as a tracer for motions of the underlying cluster gas. The Centaurus cluster is particularly interesting in this regard as it has a super-solar central abundance and a very steep abundance edge, making it a limiting case for the model.

2 Model

We employed a model for iron transport pioneered in an investigation of the Perseus cluster by Rebusco et. al. (2005) [1]. The iron motion is modelled by the diffusion-advection equation:

$$\frac{\partial na}{\partial t} = \nabla \cdot (nD\nabla(a)) + S, \tag{1}$$

where a is the iron abundance, n is the hydrogen abundance (assumed to be independent of time), D is the diffusion coefficient and S represents the sources of iron. We assume that Fe injection is through stellar mass loss and type Ia supernovae [2].

The abundance profile was determined from recent *Chandra* and *XMM-Newton* observations and archive *ROSAT* data.

3 Results

3.1 Static Diffusion Coefficient

The iron abundance distribution for three models with a diffusion coefficient that is static in space and time is shown in the left hand panel of Figure 1. The observed

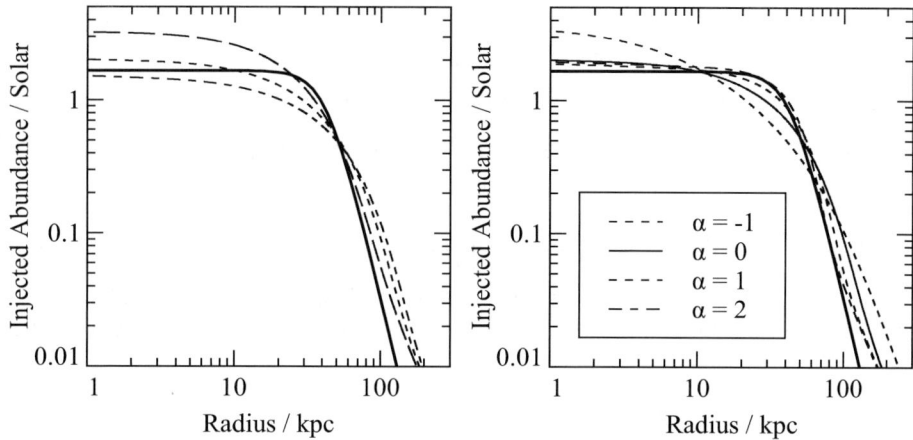

Fig. 1. LEFT: Results for uniform diffusion coefficient models with $D = 2 \times 10^{28}\,\mathrm{cm^2 s^{-1}}$ (dash-dot line), $4 \times 10^{28}\,\mathrm{cm^2 s^{-1}}$ (dotted line) and $6 \times 10^{28}\,\mathrm{cm^2 s^{-1}}$ (dashed line). The solid line represents the Centaurus model profile. None of the models fit well over the full radial range. RIGHT: Results for a variable diffusion coefficient model with $D_0 = 4 \times 10^{28}\mathrm{cm^2 s^{-1}}$, $r_0 = 25\mathrm{kpc}$ and $\alpha = -1, 0, 1, 2$. Higher values of alpha — representing more centrally concentrated diffusion profiles — fit the measured abundance distribution better.

abundance peak is best matched with $D = 4 \times 10^{28}\,\mathrm{cm^2 s^{-1}}$, about a factor of 5 smaller than the value for Perseus [1]. However, these models all produce abundance profiles that are more rounded than the observed profile, meaning it is impossible to match both the flat central abundance and the sharp drop around $30 - 70\,\mathrm{kpc}$ seen in the data.

3.2 Variable Diffusion Coefficient

To improve the agreement between model and data, we consider a situation in which the diffusion coefficient varies with radius according to a prescription from [1]:

$$D(r) = D_0 \left(\frac{n(r)}{n(r_0)} \right)^{\alpha}. \tag{2}$$

r_0 is a scale radius and α is a parameter that controls the degree of central concentration of the diffusion coefficient; larger α implies a more centrally concentrated profile. The right hand panel of Figure 1 shows the results of several simulations with $D_0 = 4 \times 10^{28}\,\mathrm{cm^2 s^{-1}}$, $r_0 = 30\,\mathrm{kpc}$ and values of α from -1 to 2. Larger values of α provide a better fit to the data, suggesting that the gas motion responsible for the spreading of the iron is concentrated in the central region of the cluster, consistent with a picture in which turbulence is created by the central AGN.

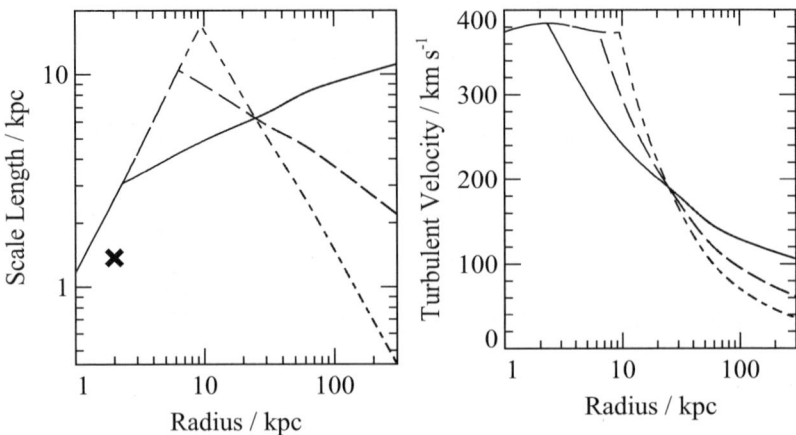

Fig. 2. Length and velocity scales needed for turbulent heating to match diffusion in the Centaurus cluster. The models are those of Figure 1a with $\alpha = 0$ (solid line), $\alpha = 1$ (dashed line) and $\alpha = 2$ (dot-dashed line). The cross indicates the western bubble position and the mean bubble radius in the cluster.

4 Turbulent Heating

Heating by turbulent motions of cluster gas has been suggested as a possible solution for the "cooling flow problem" – the observation that there is much less cool gas observed in the centres of clusters than would be expected given their age and luminosity [3].

Figure 2 shows the length and velocity scales of turbulent energy injection required to balance cooling over the inner region of Centaurus for models with values of $\alpha \geq 0$, calculated using the turbulence model from [4]. This shows that the length scales of the turbulence are larger than the size of the observed radio bubbles, implying that turbulence induced by bubble motion alone does not account for the observed abundance distribution. Other processes that may contribute to the spreading of metals include mergers and radial flow motions generated in the wake of rising bubbles. This latter idea has some support from the observed radial Hα filaments around the cluster central galaxy.

References

1. P. Rebusco, et al.: MNRAS, **359**, 1041–1048 (2005)
2. H. Böhringer, et al.: A&A, **416**, L21–L25 (2004)
3. J. R. Peterson, et al.: ApJ, **590**, 207–224 (2003)
4. T. J. Dennis and B. D. G. Chandran: ApJ, **622**, 205–216 (2005)

Simulations of Galactic Winds and Starbursts in Galaxy Clusters

W. Kapferer[1], W. Domainko[1], M. Mair[1], S. Schindler[1], E. v. Kampen[1], T. Kronberger[1,2], S. Kimeswenger[1], M. Ruffert[3], and D. Breitschwerdt[4]

[1] Institut für Astro- Teilchenphysik, Leopold - Franzens Universität Innsbruck, Austria, wolfgang.e.kapferer@uibk.ac.at
[2] Institut für Astrophysik, Universität Göttingen, Germany
[3] School of Mathematics, University of Edinburgh, UK
[4] Institut für Astronomie, Universität Wien, Austria

1 Introduction

Modern X-ray observations of galaxy clusters show clearly a non-uniform, non-spherical distribution of metals in the ICM (e.g. [1], [2]). As heavy elements are only produced in stars the processed material must have been ejected into the intra-cluster medium (ICM) by cluster galaxies. The first suggested transfer-processes were galactic winds ([3], [4]) and ram-pressure stripping ([5]). Other processes like kinetic mass redistribution due to galaxy-galaxy interactions ([6], [7]), intra-cluster supernovae ([8]) or jets of AGNs are the latest suggestions for enriching the ICM. In order to distinguish between the efficiency of the enrichment processes several approaches for simulations were carried out. De Lucia et al. (2004) used combined N-body and semi-analytical techniques to model the intergalactic and intracluster chemical enrichment due to galactic winds. Another approach are Tree+SPH simulations of galaxy clusters ([9]) including galactic winds. A comparison of the efficiency between ram-pressure stripping and quiet galactic winds was recently done by [10].
We model galactic winds and starbursts due to galaxy mergers as enrichment processes of the ICM. The resulting X-ray surface brightness, temperature and metal maps provide the key to understand the dynamics of merging and relaxed galaxy clusters.

2 Numerical methods

We use different numerical techniques to calculate the cluster components appropriately. The non-baryonic component is calculated using an N-body tree code ([11]) with constrained random fields as initial conditions, implemented by [12]. The N-body tree code provides the underlying evolution of the dark matter (DM) potential for the hydrodynamic code and the orbits of the model cluster galaxies. The properties of the galaxies are calculated by an improved version of the galaxy formation code of [13].

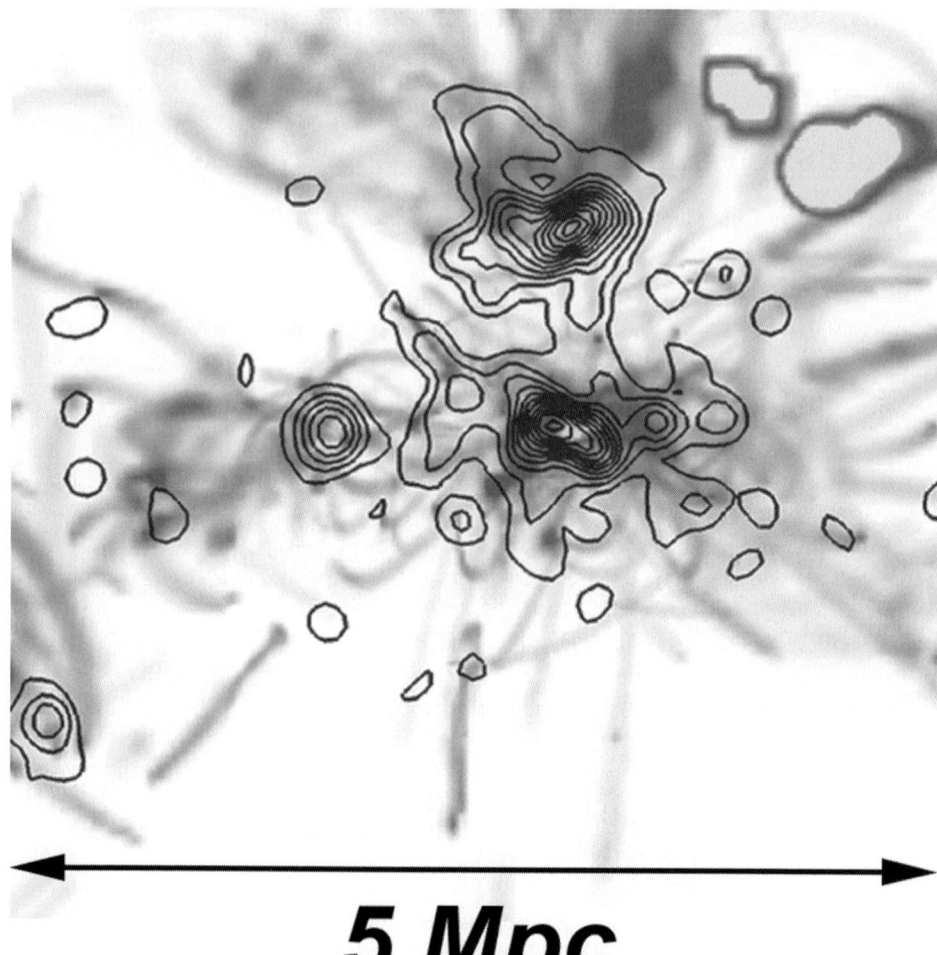

5 Mpc

Fig. 1. X-ray weighted metallicity maps (colour map) of a model cluster with galaxy-density isolines. The metallicity-gap in front (relative to the direction of motion) of the sub-structures are clearly visible. High metallicity is shown in yellow - low metallicity is shown in blue.

The adopted ΛCDM cosmology is characterized by Ω_Λ=0.7, Ω_m=0.3, σ_8=0.93 and h=0.7. For each galaxy its position and velocity are known (and stored) at all times: their orbits, luminosities, colours, etc. are a product of the N-body simulation in combination with the assumptions of the galaxy formation model. For the treatment of the ICM we use a hydrodynamic code with shock capturing scheme (PPM, [14]), with a fixed mesh refinement scheme ([15]) on four levels and radiative cooling. As the N-body tree code, the semi-numerical galaxy formation and evolution and the mass-loss rates due to galactic winds are calculated from the beginning, the hydrodynamics covers the redshift interval from z=1 to z=0, i.e. $\sim$ 57% of the whole simulation time. The mass-loss rates by quiet galactic winds for spiral galaxies are calculated with a code developed by [16]. The terminus mass loss rate in our simulations means galactic matter, which is not gravitationally bound to the originating galaxy anymore. As this enriched matter is then part of the ICM, we include it into our hydrodynamic ICM simulations.

3 Summary and Conclusions

We find that

- there is a difference between the strength of the enrichment by galactic winds and starbursts in merging and in non-merging model clusters. Merging systems show more galaxy-galaxy interactions, which cause more starbursts. On the other hand galactic winds can be suppressed by the pressure of the ICM onto the galaxies in relaxed clusters.
- especially the metal maps can help to distinguish between the pre- or post merger state of a galaxy cluster. Pre-mergers have a metallicity gap between the subclusters, post-mergers have a high metallicity between subclusters. In Fig. 1 a model cluster in the pre-merger phase is shown. The isolines indicate the galaxy density and the colours show the metallicity of the ICM. The metallicity gap in the pre-merger phase is clearly visible.
- the inhomogeneities which are introduced by galactic winds and starbursts into the ICM can survive for a long time. In the outskirts of a galaxy cluster, where the variation in the density and the velocity field of the ICM is less than in the inner parts, we find that inhomogeneities can survive up to several Gyr.
- metal enrichment of the ICM by galactic winds before z=1 is not as strong as after z=1 in the cluster centre ($(5 \text{ Mpc})^3 h^{-1}$ volume), whereas for merger-driven starbursts there is no noticeable difference.

References

1. Furusho, T., et al.: ApJ **596**, 181 (2003)
2. Fukazawa, Y., et al.: ApJ **606**, L109 (2004)
3. De Young, D.S.: ApJ **223**, 47 (1978)
4. Kapferer, W., et al., astro-ph/0508107, 2005
5. Gunn, J.E., & Gott, J.R.III: ApJ **176**, 1 (1972)
6. Kapferer, W., et al.: A&A **438**, 87 (2005)
7. Gnedin, N.Y.: MNRAS **249** 407 (1998)
8. Domainko, W., et al.: A&A **425**, L21 (2004)
9. Tornatore, L., et al.: MNRAS **349** L19 (2004)
10. Schindler, S., et al.: A&A **435** L25 (2005)
11. Barnes, J., & Hut, P.: Nature **324** 446 (1986)
12. van de Weygaert, R., & Bertschinger, E.: MNRAS **281** 84 (1996)
13. van Kampen, E., et al.: MNRAS **310** 43 (1999)
14. Colella, P., & Woodward, P.R.: J. Comp. Phys. **54** 174 (1984)
15. Ruffert M.: A&A **265** 82 (1992)
16. Breitschwerdt, D., et al.: A&A **245** 79 (1991)

The ARCRAIDER Project: A Unique Sample of X-Ray Bright, Massive Gravitational Lensing Galaxy Clusters

W. Kausch[1], M. Gitti[1,2], T. Erben[3], and S. Schindler[1]

[1] Institut für Astro- und Teilchenphysik, Universität Innsbruck, Austria,
(wolfgang.kausch,sabine.schindler)@uibk.ac.at
[2] Dept. of Physics & Astronomy, Ohio University, gitti@helios.phy.ohiou.edu
[3] Institut für Astrophysik und extraterrestrische Forschung (IAEF), Universität Bonn, Germany,
terben@astro.uni-bonn.de

Summary. We present a sample of the 22 most X-ray luminous galaxy clusters based on the ROSAT Bright Survey [15]. As all clusters are medium redshift systems ($0.1 \leq z \leq 0.52$) their lensing probability is very high. We found gravitational arc candidates in $\sim 70\%$ of the clusters. In particular we focus on a combined lensing/X-ray analysis of one of the sample clusters, Z3146, based on deep ground based wide-field (WFI@ESO2.2m, B, V and R band), archival WFPC2@HST (F606W) and XMM/Newton observations.

1 Introduction

Lensing surveys of clusters of galaxies open doors to several cosmological and statistical studies (e.g. [2, 3, 6, 16]). In particular, the combination of different techniques like lensing and X-ray offers excellent possibilities to study cluster properties in various and independent ways.

The ARCRAIDER project is aimed at these combinations of X-ray/lensing studies and consists of a sample of galaxy clusters selected from the ROSAT Bright Survey (henceforth RBS, [15]) with the following major criterions: (a) X-ray luminosity $0.5 \geq \times 10^{45}$ erg/sec (0.5-2 keV band), (b) redshift regime $0.1 \leq z \leq 0.52$, (c) classified as galaxy cluster and (d) $\delta \leq 20°$ (visibility from La Silla/Paranal). Due to their redshift and the tight $L_{\mathrm{X}} - M$ relation [12, 13] all systems have a very high probability to act as gravitational lens. The resulting 22 clusters were observed at least in the R and V band using either SUSI2@ESONTT, WFI@ESO2.2m or FORS1@ESOVLT.

In these proceedings we concentrate on an analysis of Z3146 (more details in Kausch et al. 2006, subm. to A&A). Throughout this paper we use a standard ΛCDM cosmology with $H_0 = 70\,h_{70}\,\mathrm{kms}^{-1}\mathrm{Mpc}^{-1}$ and $\Omega_M = 1 - \Omega_\Lambda = 0.3$.

2 Optical Observations of Z3146

The ground based observations were obtained during two observing programs (68.A-0255(A) and 072.A-0061(A)) with the Wide-Field-Imager WFI@ESO2.2m telescope in La Silla using the B, V and R filters. The exposure times were ~ 0.42 h in B, ~ 2.23 h in V, and ~ 6.9 h in R, respectively. For the data reduction procedure we used the GaBoDS pipeline [4]. In addition to our deep ground based observations we have used calibrated archival WFPC2@HST data obtained in a snapshot programme (ID: 8301) using the broad band V filter F606W with an exposure time of 1000 s.

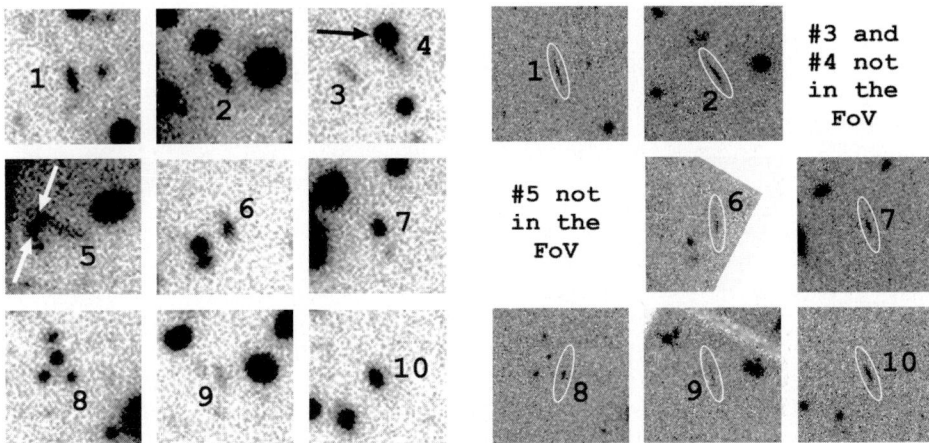

Fig. 1. Detailed comparison of arc candidates in RBS-0864 (Z3146). The left set is taken from a deep WFI@ESO2.2m image (R-band, exposure time 6.9h), the right one is an archival F606W image obtained with WFPC2. The Field-of-View in all images is 15" × 15", north is up, east to the left. Due to the better resolution and the better seeing conditions of the HST several arc candidates could be identified even on this shallow space based observation.

3 Lensing Analysis

3.1 Strong Lensing Features

We used the deep WFI frames and the shallow WFPC2 images for the identification of the arc candidates. Fig. 1 shows a morphological comparison of the 10 arc candidates in the two different instruments (see Tab. 1). The WFPC2 frame shows $\sim$ several strong lensing features which were not recognizable in the deep ground based images. In addition, the use of ARC DETECTOR, a software especially designed for an automated search for arcs [10], has confirmed their arc-like structure.

3.2 Weak Lensing Mass Reconstruction

We also performed a weak lensing mass reconstruction based on the deep R-band frame. We used a standard KS93 [7] algorithm to fix the centre of mass (Fig. 2, left

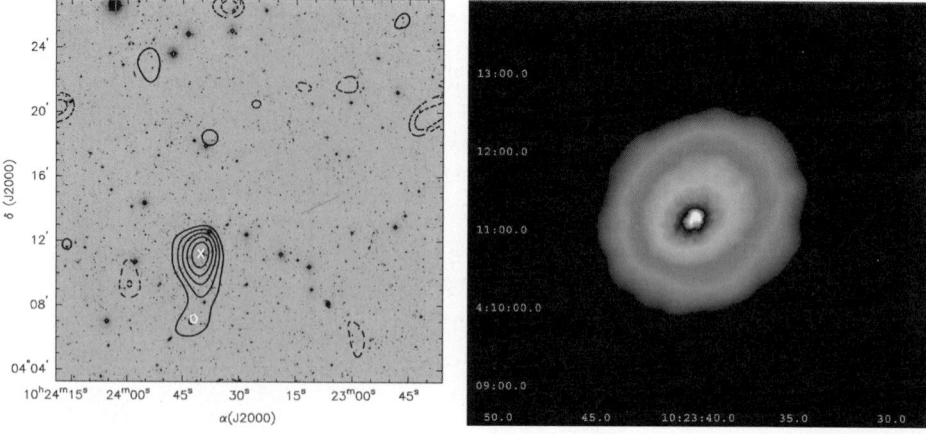

Fig. 2. Left: Shown is a KS93 weak lensing cluster mass reconstruction of Z3146. The shear field was smoothed with a Gaussian of width $1'.4$. The κ map was normalised so that the mean of κ in the entire field is zero. As we use this map only to define the centre of mass for our model fits the actual choice for this normalisation is not important. Solid contours increase in steps of $\Delta\kappa = 0.01$ ($\sim 3.3 \times 10^{13} M_\odot/\text{Mpc}^2$), assuming a source galaxy redshift of 0.75. The cluster clearly shows up as the highest peak in this reconstruction. Dashed contours represent the same levels of negative κ. The white cross marks the cluster centre of our X-ray analysis (see Sect. 4) which is in excellent agreement with the lensing centre. The peak at the cluster position is detected with a significance of more than 4.1σ, the extension to the South (marked with a white circle) with 2.7σ. Hence, this extension of the κ map is not considered a significant lensing feature. Right panel: MOS1 image of ZW3146 in the [0.3-10] keV energy band. The image is corrected for vignetting and exposure and is adaptively smoothed (signal-to-noise ratio = 20);

panel).

We estimated the shear with a KSB algorithm [8] and fitted these data to an NFW profile [11] considering c and r_{200} as free parameter. At a 90%confidence level the best fit reveals to be $c = 3.2$ and $r_{200} = 1474^{+247}_{-293} h_{70}^{-1}$ kpc (see Fig. 3).

4 X-ray Analysis

Z3146 was observed with XMM/Newton in December 2000 during rev. 182 with the MOS and pn detectors in Full Frame Mode with THIN filter, for an exposure time of 53.1 ks for MOS and 46.1 ks for pn, using SASv6.0.0 software for the calibration.

The X-ray analysis reveals a relaxed cluster without ongoing major merger indicated by the absence of distinct substructure (see Fig. 2), also the temperature map shows a rather uniform shape with a sharp central surface brightness peak at a position $10^h 23^m 40^s.01 + 04°11'09''.45$ (J2000), in very good agreement with the optical position of the central dominant cluster galaxy [15]. In addition the cluster shows a massive cooling core with a nominal mass deposition rate of ~ 1600 $M_\odot$ yr^{-1}.

Table 1. Properties of the arc candidates, 'cc' denotes the cluster center

# of arc candidate	angular distance to cc ["]	projected dist. to cc [kpc]	length l ["]	l/w WFPC2	l/w WFI
1	27	117.6	3	5.1	1.9
2	20	87.1	2.9	5.6	1.7
3	62	270.1	3.5	$--$	2.4
4	67	287.5	$4^\dagger$	$--$	1.7
5	67	287.5	$6^\dagger$	$--$	2.1
6	46	200.4	2.1	2.9	1.9
7	23	100.2	1.5	2.1	1.4
8	26	113.3	1.2	1.9	1.2
9	52	226.6	2.6	5	2.2
10	64	278.8	2.1	2.8	1.5

5 Conclusions

Using the assumption of hydrostatic equilibrium and spherical symmetry we calculate the radial mass profile derived from the X-ray observations as the projected mass along the line-of sight within a cylinder of projected radius r for a better comparison with the weak lensing mass distribution (see Fig. 3). It can be seen that the weak lensing mass is roughly half the mass derived from the XMM observations, which is an unusual large discrepancy. The reasons for that are not yet clear as the lensing mass is expected to be higher due to the assumptions of the lensing sensitivity to the accumulated mass along the line of sight (see e.g. [9], [14], or [5] for some few examples). Especially for cooling core clusters mass estimates usually seems to be in good agreement [1].

Acknowledgments. This work is supported by the Austrian Science Foundation (Fonds zur Förderung der wissenschaftlichen Forschung, FWF, Project 15868) and the "ESO-Mobilitätsstipendium" of the Austrian Ministry of Science (Österreichischen Bundesministerium für Bildung, Wissenschaft und Kunst (bm:bwk)). The authors also want to thank Joachim Wambsganss and Axel Schwope for invaluable comments and help.

References

1. S. W. Allen: MNRAS, **296**, 392, (1998)
2. M. Bartelmann, A. Huss, J. M.Colberg et al: A&A, **330**, 1 (1998)
3. N. Dalal, G. Holder, J. F. Hennawi: ApJ, **609**, 50 (2004)
4. T. Erben, M. Schirmer, J. P. Dietrich et al: AN, **326**, 432-464 (2005)
5. I.M. Gioia, E.J. Shaya, O. Le Fèvre, et al.: ApJ, **497**, 573 (1998)
6. S. Ho, M. White: APh, **24**, 257 (2005)
7. N. Kaiser, G. Squires: ApJ, **404**, 441, (1993)

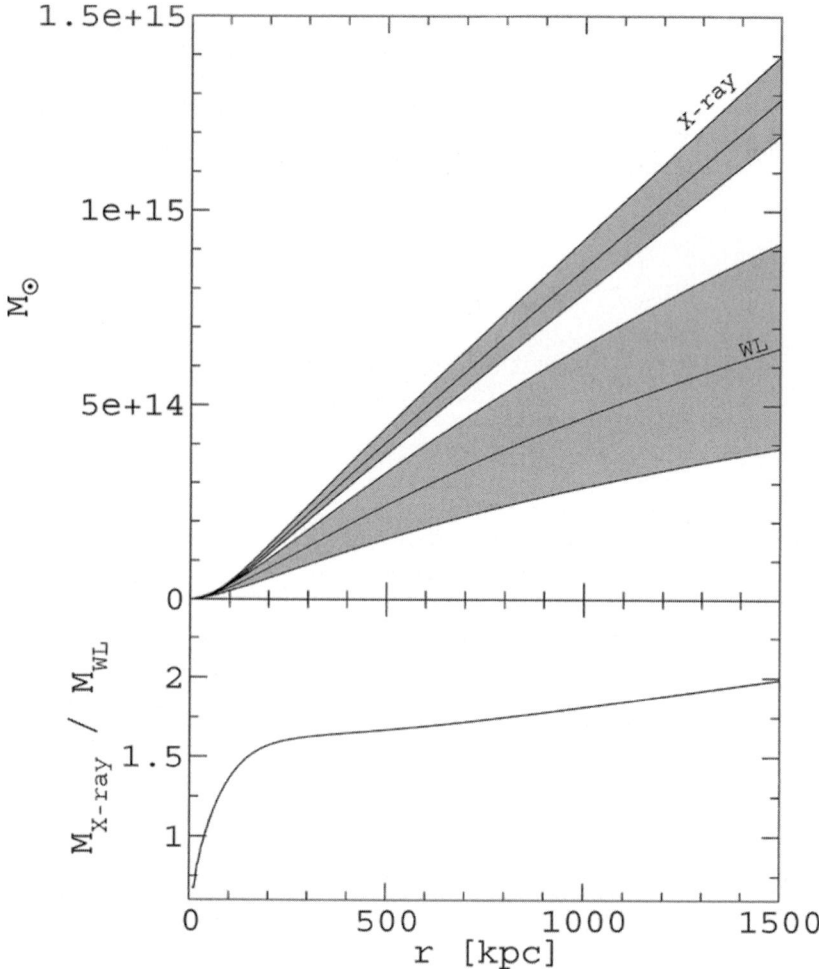

Fig. 3. Comparison of X-ray and weak lensing mass profiles. Surprisingly the mass derived from X-ray observations is roughly twice the mass obtained from the weak lensing analysis.

8. N. Kaiser, G. Squires, T. Broadhurst: ApJ, **449**, 460 (1995)
9. J. P. Kneib, Y. Mellier, R. Pellò: A&A, **303**, 27, 1995
10. F. Lenzen, O. Scherzer, S. Schindler: A&A, **416**, 391 (2004)
11. J. Navarro, C. Frenk, S.D.M. White: ApJ, **490**, 493, (1997)
12. T. H. Reiprich, H. Böhringer: AN, **320**, 296, (1999)
13. S. Schindler: A&A, **349**, 435 (1999)
14. S. Schindler, J. Wambsganss: A&A , **322**, 66 (1997)
15. A. Schwope, G. Hasinger, I. Lehmann, et al: AN, **321**, 1-52, (2000)
16. J. Wambsganss, P. Bode, J. Ostriker: ApJL, **606**, L93-L96 (2004)

APEX-SZ: A Sunyaev-Zel'dovich Galaxy Cluster Survey

R. Kneissl and the APEX-SZ collaboration

Department of Physics, 366 LeConte Hall, University of California, Berkeley, CA 94720-7300, United States, rkneissl@berkeley.edu

1 Motivation for SZ cluster surveys

The abundance and spatial distribution of massive galaxy clusters is a sensitive probe of cosmological models. In particular dark matter and dark energy have a strong effect on the observed evolution of the cluster population. To use this information to its full extent it is necessary to detect all massive clusters in a large volume to high redshift ($z > 1$). Also the total masses, which can be predicted within the cosmological model, have to be infered with high accuracy from the observables. Cluster surveys via the Sunyaev-Zeldovich (SZ) effect (inverse Compton scattering of cosmic microwave background (CMB) photons by cluster electrons) are promising in these respects. The photons interacting with the more energetic electrons gain on average a small frequency shift relative to the huge CMB background, which remains a detectable signal at the observer regardless of redshift. The intensity depends linearly on electron density and temperature. Hence the integrated cluster signal is proportional to the total thermal energy, a relatively simple quantity to relate to the total mass through gas dynamical simulations or phenomenology.

2 An SZ receiver for the APEX telescope

Advanced detector technology is required to reach the sensitivities necessary for cluster survey observations with the SZ effect and to minimise at the same time the unwanted signals. The SZ receiver built at U.C. Berkeley comprises 330 transition edge sensor bolometers, monolithically fabricated in 6 wedges, a novel pulse-tube cooling system and frequency multiplexed SQUID readout electronics. The Atacama Pathfinder Experiment telescope is a 12 m, sub-mm antenna on an extremely dry, high-altitude (5100 m) site in the Atacama desert in Chile. This modified ALMA proto-type antenna was built and is operated by the Max-Planck-Institute for Radioastronomy in Bonn, the European Southern Observatory and the Swedish Onsala Observatory. APEX-SZ [1] is the first of a new generation of large bolometer arrays for SZ surveys, such as ACT and SPT, which are under construction and will carry out an order of magnitude larger surveys in the next few years.

3 Expected galaxy cluster sample

This combination of telescope and receiver allows a 100-200 deg^2 cluster survey to a depth of 10 μK rms per 1 arcmin pixel (0.4 mJy/beam) with a well-matched resolution of 60 arcsec (FWHM) at 150 GHz. Surveys at 90 and 220 GHz are also under consideration. We expect to find a total of 700 clusters in 100 deg^2 including a wide tail extending to high redshifts (Fig. 1) with over 80 clusters being at $z > 1$. This assumes a concordance cosmology ($\Omega_\Lambda = 0.7, \Omega_M = 0.3, \sigma_8 = 0.9, h = 0.7$) and an X-ray observed gas fraction and M-T relation (e.g. [2, 3, 4]).

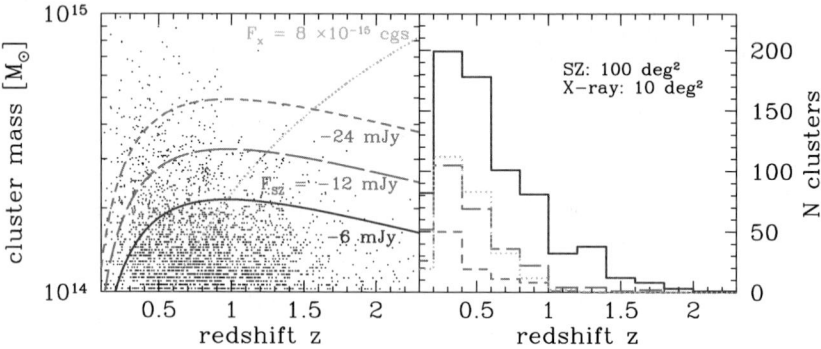

Fig. 1. Cluster selection for APEX-SZ. In the left panel we plot the density of clusters (from 'Virgo HVLC' simulations) for 100 deg^2. The limiting lines of constant, integrated SZ decrement, $Y = \int \int \sigma_T n_e \; kT/m_e c^2 \; dl \; d\Omega$, converted to flux density, correspond to APEX-SZ observations of 10, 40 and 160 days. For the low redshift end ($z < 0.4$) we assume that CMB fluctuations can be removed effectively by combining observing frequencies. For comparison we also show a medium-deep X-ray survey of 10 deg^2. In the right panel we show the corresponding number of clusters as a function of redshift in bins of $\Delta z = 0.2$.

4 Scientific goals of the APEX-SZ survey

Our aim is to study and quantify, the evolution of the cluster mass function, cluster correlation function, evolution of the gas fraction and M-T relation, dark matter density and distribution, dark energy equation of state, population of flat and inverted spectrum radio sources, high redshift dusty, star-forming galaxies, correlations of density and velocity fields, and lensing of the cosmic microwave background.

5 First light observations

The SZ receiver saw first light with the APEX telescope on Dec 23, 2005. In January 2006, despite poor weather during the Bolivian winter, the major goals of this engineering run with a single 55 bolometer wedge could be achieved: interfacing receiver and telescope, optical verification (Fig. 2 shows instrument beams), implementation of reduction software and test scans for noise characterisation.

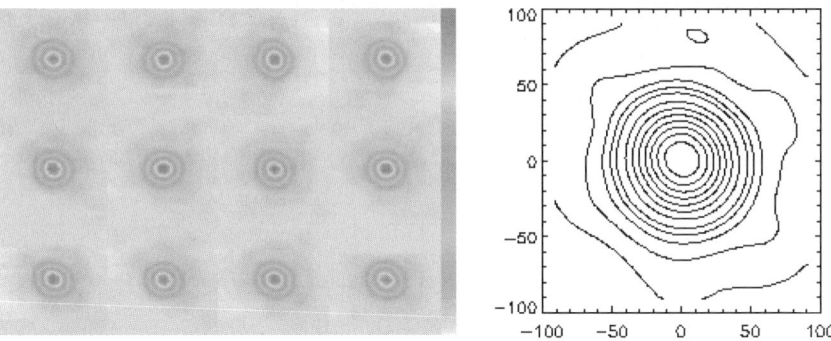

Fig. 2. Left: Examples of individual beam maps (Saturn - Jan 26, 2006); Right: Average beam, contour levels are at 2,5,10,15,20,30,...,90% of the peak value. The best fit Gaussian beam width (FWHM) is 58.4 ± 0.3 arcsec (planet size de-convolved).

6 Conclusions

APEX-SZ has demonstrated new bolometer technology, the working of a TES spider-web bolometer array with pulse-tube cooler and SQUID readout, which will enable large, sensitive sky surveys with low systematics at mm wavelengths. We expect to detect hundreds of clusters, tens of high redshift ($z > 1$) clusters, in the next two years with the APEX-SZ instrument.

References

1. D. Schwan, et al.: NewAR **47**, 933 (2003)
2. S.W. Allen, et al.: MNRAS **353**, 457 (2004)
3. A. Finoguenov, et al.: A&A **368**, 749 (2001)
4. S. Ettori, et al.: A&A **417**, 13 (2004)

Detecting Virialization Shocks Around Galaxy Clusters Through the SZ Effect

B. Kocsis[1], Z. Haiman[2], and Z. Frei[1]

[1] Institute of Physics, Eötvös University, Pázmány P. s. 1/A, 1117 Budapest, Hungary,
bkocsis@complex.elte.hu
[2] Department of Astronomy, Columbia University, 550 West 120th Street, New York, NY
10027

1 Overview

In cosmological structure formation models, massive non–linear objects in the process of formation, such as galaxy clusters, are surrounded by large-scale shocks at or around the expected virial radius [1]. Direct observational evidence for such virial shocks is currently lacking, but we show here that their presence can be inferred from future, high resolution, high-sensitivity observations of the Sunyaev-Zeldovich (SZ) effect in clusters.

We study the detectability of virial shocks in mock SZ maps, using simple models of cluster structure (gas density and temperature distributions) and noise (background and foreground galaxy clusters projected along the line of sight, as well as the cosmic microwave background anisotropies). We find that at an angular resolution of 2 arcsec and sensitivity of 10μK, expected to be reached at around 100 GHz frequencies in about 20 hr integration with the forthcoming Atacama Large Millimeter Array (ALMA) instrument (www.alma.nrao.edu), virial shocks associated with massive (10^{15} M$_\odot$) clusters will stand out from the noise, and can be detected at high significance. More generally, our results imply that the projected SZ surface brightness profile in future, high-resolution experiments will provide sensitive constraints on the density profile of cluster gas.

2 Cluster profiles

We follow [3] to obtain self-similar gas density and temperature profiles. These profiles are then truncated and normalized according to the assumed location and sharpness of a virial shock. The left panel of Fig. 1 depicts the predicted SZ surface brightness profiles with (dotted line) and without (solid line) a cutoff. Without the cutoff, the SZ profile has a nonzero limit for large radii. Various choices of the cutoff position can be compared by analyzing the difference between the associated profiles.

The right panel of Fig. 1 shows the difference between pairs of profiles with and without a cutoff, using either $D = 0, 0.2c$, or $0.4c$ (c is the concentration parameter,

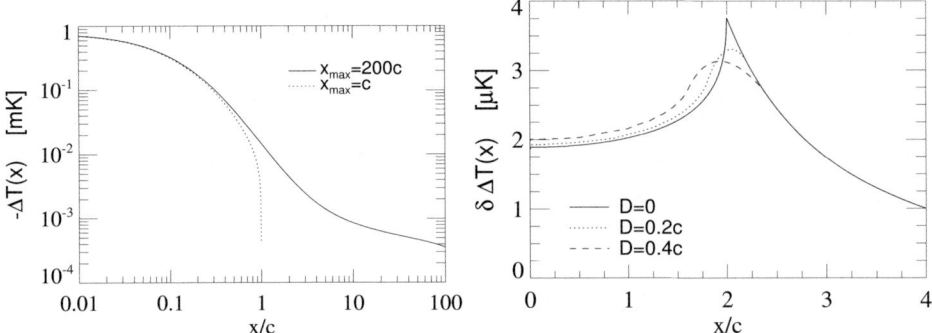

Fig. 1. Predicted SZ surface brightness profiles (left) and the difference between the normalized profiles with and without a cutoff, for various D values (right).

$x = c$ at the virial radius) for the truncated profile (we assume that the density has a linear cutoff between radii $x_{max} - D$ and $x_{max} + D$; therefore $D = 0$ means a sharp cutoff).

Considering all possible sources of contamination (detector noise, atmospheric fluctuations, etc) we conclude that these shocks are detectable.

3 Significance of detecting shocks

We quantify the significance at which virial shocks may be detected by comparing pairs of radial surface brightness profiles - one with, and one without a virial shock. The signal-to-noise ratio (S/N) is a strong function of the size of the cluster (S/N about 500 for $M_{vir} = 10^{15}M_\odot$, but S/N only about 10 for $M_{vir} = 10^{14}M_\odot$). The fractional deviation of S/N - as compared to the original value of a test model for various fiducial c (density concentration in the cluster) and ΔT_s (SZ temperature decrement scale) parameters - is plotted in Fig. 2. The relatively minor decrease in the S/N indicates that variations in c and ΔT_s cannot mimic the presence of a virial shock, i.e. that the shock remains a distinctive feature.

4 Parameter estimation

A more ambitious problem is to examine the precision with which the location and sharpness of the edge can be measured. The two panels of Fig. 3 depict contour diagrams at fixed values of this uncertainty. c, ΔT_s, and x_{max} will be precisely obtained with ALMA for the majority of the clusters, while similar precision for D is possible for only massive clusters. With ALMA's relatively small field of view, the detection of the shock proposed here will require several pointings to map out the location of the

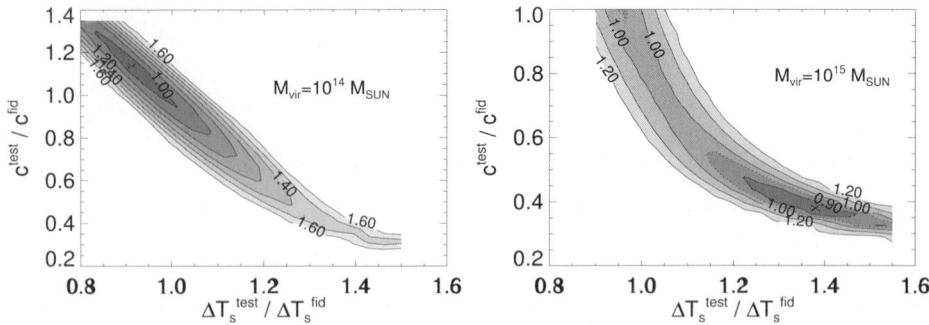

Fig. 2. The $p^{\text{test}} = (\Delta T_s^{\text{test}}, c^{\text{test}})$ parameter dependence of the S/N ratio for detecting the virial shock of a fiducial cluster with parameters $p^{\text{fid}} = (z, x_{\max}, D) = (0.3, 2c, 0.5c)$ and $M_{\text{vir}} = 10^{14} \text{M}_\odot$ (left) or $10^{15} \text{M}_\odot$ (right).

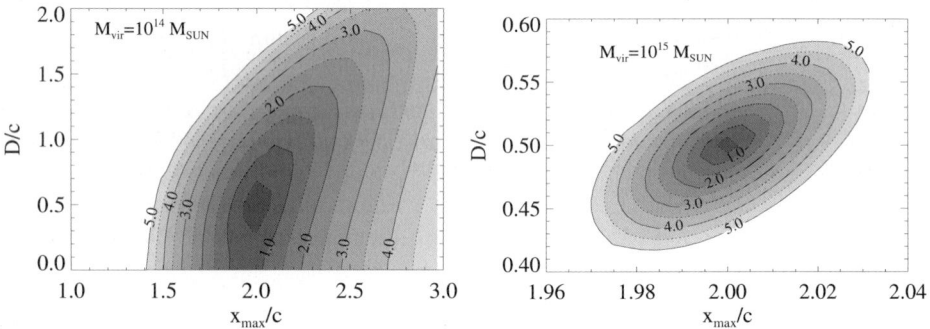

Fig. 3. The $(x_{\max}^{\text{test}}, D^{\text{test}})$ parameter dependence χ^2-contours with $(z, x_{\max}, D) = (0.3, 2c, 0.5c)$, using $M_{\text{vir}} = 10^{14} \text{M}_\odot$ (left) and $M_{\text{vir}} = 10^{15} \text{M}_\odot$ (right). The contours increase linearly in steps of $0.5\sigma_N$.

shock - but this should be feasible in a total integration time of less than 10^6 seconds for bright clusters. For detailed discussions see [2].

Acknowledgments. We acknowledge support from OTKA through grant nos. T37548 and T47042 and from EU through grant no. MRTN-CT-2004-503929. ZH thanks the US National Science Foundation and the Hungarian Ministry of Education (through a Bekesy Gyorgy Fellowship) for financial support.

References

1. Bertschinger, E.: ApJS **58** 39 (1985)
2. Kocsis, B., Z. Haiman, & Z. Frei: ApJ **623** 632 (2005)
3. Komatsu & U. Seljak: MNRAS **327** 1353 (2001)

Numerical Simulations of Metal Enrichment and Mergers in Clusters of Galaxies

M. Mair[1], W. Domainko[1], W. Kapferer[1], T. Kronberger[2], R. Moll[1], S. Schindler[1], E. van Kampen[1], S. Kimeswenger[1], C. Ferrari[1], M. Ruffert[3], and O.E. Mangete[3]

[1] Institut für Astro- und Teilchenphysik, Universität Innsbruck, Technikerstr. 25, A-6020 Innsbruck, Austria, magdalena.mair@uibk.ac.at
[2] Institut für Astrophysik, Universität Göttingen, Friedrich-Hund-Platz 1, D-37077 Göttingen, Germany.
[3] School of Mathematics and Statistics, University of Edinburgh, Edinburgh EH9 3JZ, UK.

Summary. We present combined N-body and hydrodynamic simulations of galaxy clusters to study different metal enrichment processes especially with respect to galaxy cluster mergers. We also trace the gas initially situated within subclusters to investigate the dynamics of cluster mergers.

1 Introduction

From X-ray spectra it is evident that the ICM contains heavy elements (e.g. [1]). Beside the possibility that these metals are produced by supernovae of a population of intra-cluster stars [2], different interaction processes between the cluster galaxies and the ICM contributing to the ICM's metallicity have been suggested: ram-pressure stripping, galactic winds, starbursts due to merging galaxies [3] and jets from active galaxies. With our simulations we are able to determine the dynamical state of the model clusters at each timestep and artificial X-ray maps can be compared to observations for better interpretation of the data. In this contribution we present techniques to study the gas dynamics and we investigate the time evolution and different efficiencies of ram-pressure stripping and galactic winds in merging galaxy clusters.

2 The Simulations

For the numerical simulations we use different modules to model the different components of galaxy clusters. Large-scale structure formation is treated with an N-body tree code. An additional semi-analytical model for galaxy formation gives the orbits and properties of the galaxies [4]. For the simulations of the ICM an Eulerian hydrodynamic code with shock capturing scheme PPM, radiative cooling and multiple grid refinement [5] is used. At every timestep the mass loss of each galaxy is calculated and added to the hydrodynamic simulation at the position of the galaxy.

3 Tracing Gas in Cluster Mergers

To study merging events one has to follow the gas which is initially situated within a subcluster. Eulerian hydrodynamic codes evolve local properties of a fluid (density, temperature, velocities), but they cannot trace certain fluid elements. Therefore we use tracer-arrays to mark a specific gas clump and advect this tracer with the flow.

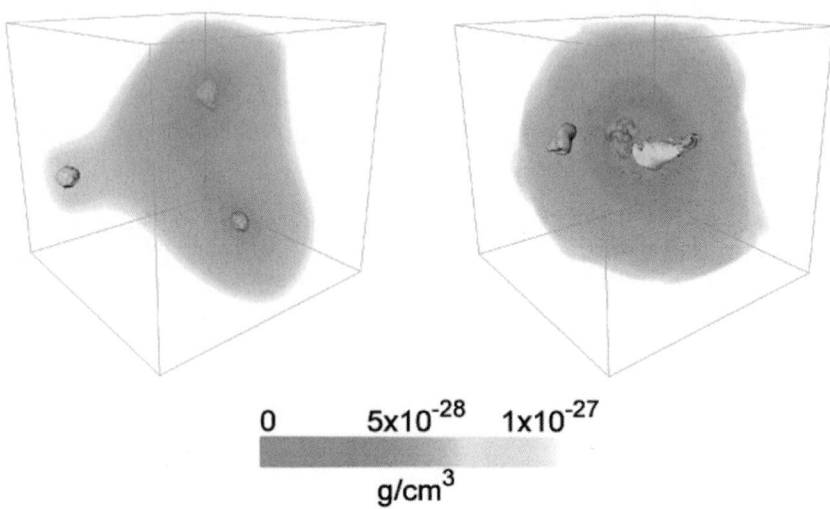

Fig. 1. Two snapshots of the density distribution of a merging galaxy cluster. The isosurfaces show the initial setup of the three gas tracers (left) and their location 3.6 Gyr later when the two main subclusters have already merged.

Figure 1 shows two snapshots of the gas density distribution of a merging galaxy cluster. The left image shows the initial setup of three tracer arrays marking the gas situated within the subclusters. The image on the right shows their location 3.6 Gyr later when the subclusters have already merged. The small subcluster to the left has lost its gas due to ram-pressure stripping before reaching the centre. Also the main subclusters have lost their gas in this way which is spread perpendicular to the initial collision axis.

4 Metal Enrichment of the Intra-Cluster Medium

Figure 2 shows the evolution of the metal mass loss due to ram-pressure stripping and galactic winds for a merging galaxy cluster. During the two merging events (at 3 and 5.5 Gyr) the mass loss due to ram-pressure stripping is enhanced because of the increased ICM pressure and velocity dispersion of the galaxies. In high-pressure regions galactic winds are suppressed, therefore the mass loss due to winds is slightly lowered during merging events.

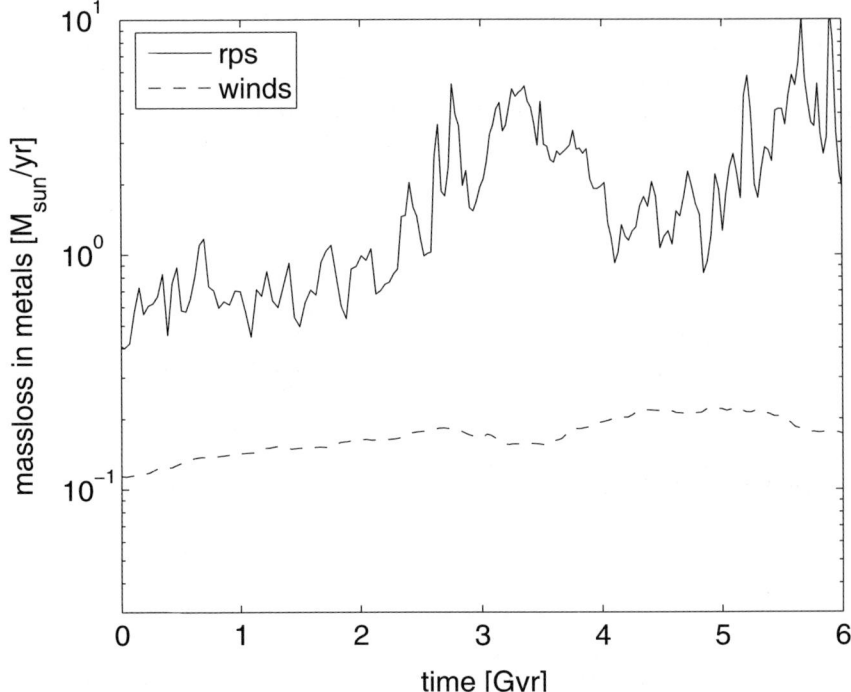

Fig. 2. The evolution of the metal mass loss from all galaxies due to ram-pressure stripping and galactic winds for a merging galaxy cluster.

For more details see Schindler et al. (2005) [6], Kapferer et al. (2006) [7] and Domainko et al. (2006) [8].

Acknowledgments. The work has been performed under the FWF-Project P15868, the Project HPC-EUROPA (RII3-CT- 2003-506079) and a Doktoratsstipendium of the Leopold-Franzens University of Innsbruck.

References

1. Durret, F., Lima Neto, G. B., & Forman, W.: A&A, **432**, 809 (2005)
2. Domainko, W., Gitti, print,M., Schindler, S., & Kapferer, W.: A&A, **425**, L21 (2004)
3. Kapferer, W., Knapp, A., Schindler, S., et al.: A&A, **438**, 87 (2005)
4. van Kampen, E., Jimenez, R., & Peacock, J. A.: MNRAS, **310**, 43 (1999)
5. Ruffert, M.: A&A, **265**, 82 (1992)
6. Schindler, S., et al.: A&A, **435**, L25 (2005)
7. Kapferer, W., Ferrari, C., Domainko, W., Mair, M. et al.: A&A, **447**, 827-842 (2006)
8. Domainko, W., Mair, M., Kapferer, W. et al.: A&A, **452**, 795-802 (2006)

Turbulence in Galaxy Clusters: Impact on the Abundance Profiles

P. Rebusco[1], E. Churazov[1,2], H. Böhringer[3], and W. Forman[4]

[1] Max-Planck-Institut für Astrophysik, Karl-Schwarzschild-Strasse 1, 85741 Garching, Germany, pao@mpa-garching.mpg.de
[2] Space Research Institute (IKI), Profsoyuznaya 84/32, Moscow 117810, Russia, churazov@mpa-garching.mpg.de
[3] MPI für Extraterrestrische Physik, P.O. Box 1603, 85740 Garching, Germany, hxb@mpe.mpg.de
[4] Harvard-Smithsonian Center for Astrophysics, 60 Garden St., Cambridge, USA, wrf@head.cfa.harvard.edu

1 Introduction

The abundance peaks in the clusters with cool core are likely associated with the brightest galaxy (BG) of the cluster/group. These peaks are probably produced by the stars of the BG after the cluster/group was assembled. Observations of the optical light and metal profiles indicate that abundance profiles are much shallower than the BG light profile (see Fig. 1) suggesting that there must be a mixing of the injected metals. Another feature of cool core sources is a short gas cooling time, which could lead at high rate to temperatures lower than 1 keV. However such temperatures are never observed. Heating due to viscous dissipation of the gas motion can compensate for such gas cooling. We estimate the parameters of stochastic gas motions in the core of cool core clusters/groups, under the assumption of diffusion approximation.

2 The Model

2.1 Iron Enrichment and Stochastic Diffusion

Consider the evolution of the iron abundance profile in the gas involved in stochastic motions with a characteristic spatial scale l and velocity scale v. Moreover assume that the gas density and temperature are constant with time. In the diffusion approximation the iron radial profile obeys:

$$\frac{\partial na}{\partial t} = \nabla \cdot (Dn\nabla a) + S, \tag{1}$$

where $n(r)$ is the gas density, $a(r,t)$ is the iron abundance and $S(r,t)$ is the iron injection from the BG alone. $D = const$ is the diffusion coefficient, of the order

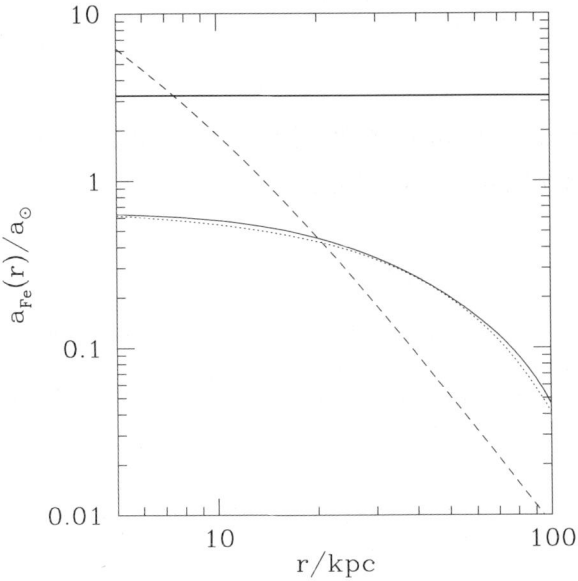

Fig. 1. Observed and expected iron abundance profiles for M87. The solid line shows the abundance profile adopted, from which a constant value of $\sim 0.2\ a_\odot$ is subtracted. For comparison we show the expected iron abundance profile (long dashed line) due to the ejection of metals from the BG, calculated assuming that the ejected metal distribution follows the optical light: it is much more peaked than the observed one. The horizontal line indicates the maximum value that the abundance can take, over which our approximation is not valid anymore. The short dashed line shows the profile derived with the additional effect of diffusion ($D \sim 8 \times 10^{28}\ cm^2/s$).

of $\frac{vl}{3}$, with v being the characteristic velocity and l the characteristic length scale. The total iron injection rate within a given radius r can be written as $s(<\ r,t) = \left(\frac{dM_{Fe}}{dt}\right)_* + \left(\frac{dM_{Fe}}{dt}\right)_{SN} \propto \left(\frac{L_B(<r)}{L_{B,\odot}}\right)$, where $\left(\frac{dM_{Fe}}{dt}\right)_*$ and $\left(\frac{dM_{Fe}}{dt}\right)_{SN}$ are the same as in [1]. The diffusion coefficients listed in Table 1 are obtained by integrating equation 1 starting from zero abundance at all radii and by comparing the observed and predicted abundance profiles (e.g. see Fig. 1).

2.2 Cooling and Heating

Let us assume that the dissipation of turbulent motions is the dominant source of heat and that the heating rate Γ_{diss} is equal to the gas cooling rate Γ_{cool}. For a given v and l the diffusion coefficient and the dissipation rate can be written as $D \sim 0.11vl$ and $\Gamma_{diss} \sim 0.4\rho v^3/l$ [2]. Using the value of D, determined in section 2.1 and setting $\Gamma_{diss} \approx \Gamma_{cool}$ at the fiducial radius of 20 kpc we can get constraints on v and l. These constraints are plotted in Fig.2. The intersection of the two bands gives the locus of the combinations of l and v such that on one hand D is approximately equal to the required value and on the other hand the dissipation rate is approximately equal to the cooling rate.

Table 1. Estimate of the turbulence parameters for each source in our sample: name (1), diffusion coefficient (2), lengthscale (3) and velocity (4) of the gas motion.

Name	$D\ [cm^2/s]$	$l\ [kpc]$	$v\ [km/s]$
NGC 5044	910^{28}	11	245
NGC 1550	1.210^{29}	16	225
M87	810^{28}	10	260
AWM4	4.510^{29}	36	376
Centaurus	1.510^{28}	2	260
AWM7	610^{27}	1	226
A1795	10^{29}	12	300
Perseus	210^{29}	21	300

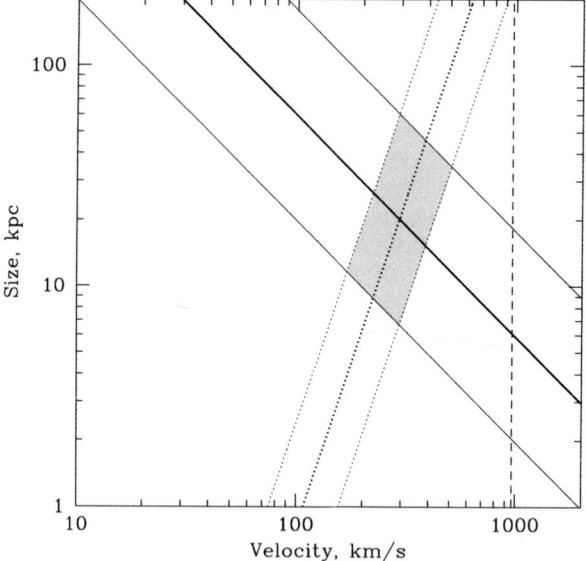

Fig. 2. Perseus cluster: range of v and l which provide the required diffusion and dissipation rates. Along the thick solid line the diffusion coefficient is equal to 2×10^{29} cm^2 s^{-1}. Along the thick dotted line Γ_{diss} is equal to the gas cooling rate. The thin solid and the dotted lines correspond to a variation of the coefficients by factors of 0.3 and 3. The vertical dashed line shows the sound velocity in the gas.

3 Results

In Table 1 the estimated values of D, v and l are listed. l are of the order of 10 kpc and v of the order of few hundred km/s: there is no obvious trend with the temperature of the source.

References

1. H.Böhringer, et al.: A&A, **416**, 21 (2004)
2. T.Dennis, B. Chandran: ApJ, **622**, 205 (2005)
3. P. Rebusco, et. al.: MNRAS, **359**, 1041 (2005)

Studying the Nature of Dark Energy with Galaxy Clusters

T.H. Reiprich[1], D.S. Hudson[1], T. Erben[1], and C.L. Sarazin[2]

[1] Argelander-Institut für Astronomie,[†] Auf dem Hügel 71, 53121 Bonn, Germany,
    thomas@reiprich.net
[2] Department of Astronomy, University of Virginia, P.O. Box 3818, Charlottesville, VA
    22903-0818, USA

Summary. We report on the status of our effort to constrain the nature of dark energy through the evolution of the cluster mass function. *Chandra* temperature profiles for 31 clusters from a local cluster sample are shown. The X-ray appearance of the proto supermassive binary black hole at the center of the cluster Abell 400 is described. Preliminary weak lensing results obtained with Megacam@MMT for a redshift $z = 0.5$ cluster from a distant cluster sample are given.

1 Introduction

Understanding the nature of dark energy is one of the major goals of contemporary cosmological and particle physics. The fate of the universe seems to be entirely determined by dark energy and a deeper understanding of its properties may shed light on the unification of general relativity and quantum theory.

At the moment, astronomical measurements seem most likely to provide further information about dark energy. Measurements of the evolution of the galaxy cluster abundance are among the most promising tools; they have the potential to yield tight constraints on the equation of state of dark energy. Here, we report on the status of our project to measure the evolution of the cluster mass function with very high quality observations of moderately sized local and distant cluster samples.

2 Local Cluster Sample

X-ray selection is well suited for the construction of complete cluster samples useful for cosmological tests, because the X-ray luminosity of clusters is tightly correlated with their total gravitational mass [1, 2]. The HIghest X-ray FLUx Galaxy Cluster Sample (*HIFLUGCS*) has been selected from the *ROSAT* All-Sky Survey and contains the X-ray brightest galaxy clusters in the sky, excluding ±20 deg around the Galactic plane as

---

[†] AIfA was founded by merging of the Institut für Astrophysik und Extraterrestrische Forschung, the Sternwarte, and the Radioastronomisches Institut der Universität Bonn.

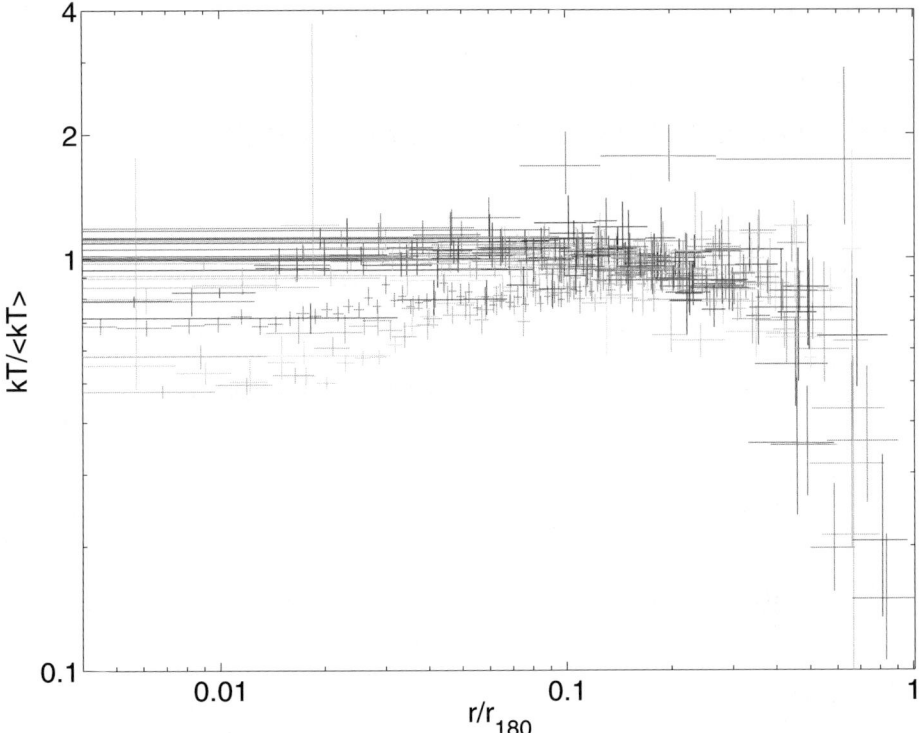

Fig. 1. Normalized intracluster gas temperature profiles for 31 clusters measured with *Chandra*.

well as the regions of the Virgo cluster and the Magellanic clouds [1]. All clusters have been reobserved with *Chandra* and almost all with *XMM-Newton*. We are currently analyzing these data to determine the most precise local X-ray galaxy cluster mass function.

Currently, cosmological constraints from X-ray galaxy clusters are limited by systematic effects, especially those affecting the total mass determination. The high quality *Chandra* and *XMM-Newton* observations now allow us to study the gas and temperature structure of clusters in much greater detail than previously possible.

Fig. 1 shows the normalized temperature profiles of about half of the *HIFLUGCS* clusters (Hudson et al., in prep.). A log scale is used to emphasize details of the innermost regions, showing *Chandra*'s unrivaled spatial resolution. This plot gives one of the most detailed views into the statistics of central temperature structures obtained to date (see also [3, 4, 5]). As expected, there is no universal temperature profile in the inner part – some profiles drop towards the center, others stay flat. Usually, the former are identified as relaxed clusters and the latter as merging clusters. More interesting is the fact that even the relaxed clusters do not appear to show a universal profile in the inner parts – quite a large spread becomes obvious in this log scale plot. This seems contrary to the universal profile for relaxed clusters that [3] found, although they used only 6 clusters. Scatter was also seen by [4] when they analyzed 13 clusters and, in

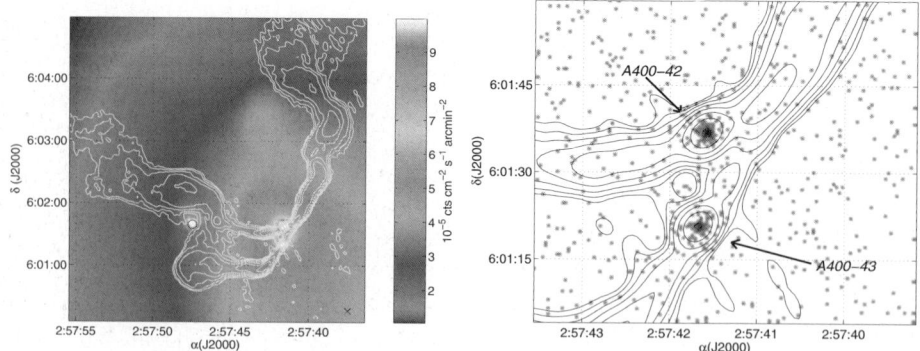

Fig. 2. *Left:* Adaptively smoothed *Chandra* image of the central region of Abell 400 [6]. Overlaid are the 4.5 GHz VLA radio contours. *Right:* Zooming further into the very center of the unsmoothed *Chandra* image.

fact, considering the complicated physics in cluster centers, non-universality might be expected. Note that these central regions account for only a small fraction of the total cluster mass.

The outer radial region where much of the cluster mass resides ($> r_{180}/2$) is rather compressed in this plot. There seems to be a clear indication for a temperature drop towards larger radii as observed by a number of previous works but the uncertainties become large, too. We aim to tighten these constraints by cross-correlating the *Chandra* data with our *XMM-Newton* results for the same clusters, taking advantage of *XMM-Newton*'s higher throughput and larger field-of-view.

Analyzing the data for all clusters in this sample, it is no surprise that exciting details about individual clusters are discovered. As an example, we show in Figs. 2 the very center of the galaxy cluster A400 [6], where the well-known radio source 3C75 – exhibiting a pair of double radio jets – resides. Both AGNs, separated by 15 arcsec corresponding to a projected separation of 7 kpc, are detected separately for the first time in X-rays. Detailed analysis of the X-ray data reveals further evidence that the two AGNs are physically close to each other and form a bound system – a proto supermassive binary black hole moving through the intracluster medium at the supersonic speed of about 1200 km/s.

3 Distant Cluster Sample

The 400 Square Degree (400d) *ROSAT* Survey (Burenin et al., in prep.) is the continuation of the 160d Survey [7, 8]. It contains clusters serendipitously detected in basically all useful *ROSAT* PSPC pointed observations. The covered search volume is larger than the volume of the entire local ($z < 0.1$) Universe. A complete subsample of the 41 most luminous, most distant clusters has been observed with *Chandra* in a large program.

Since the fraction of galaxy clusters undergoing a major merger is expected to increase with redshift [9], it would be ideal to have additional cluster mass estimates –

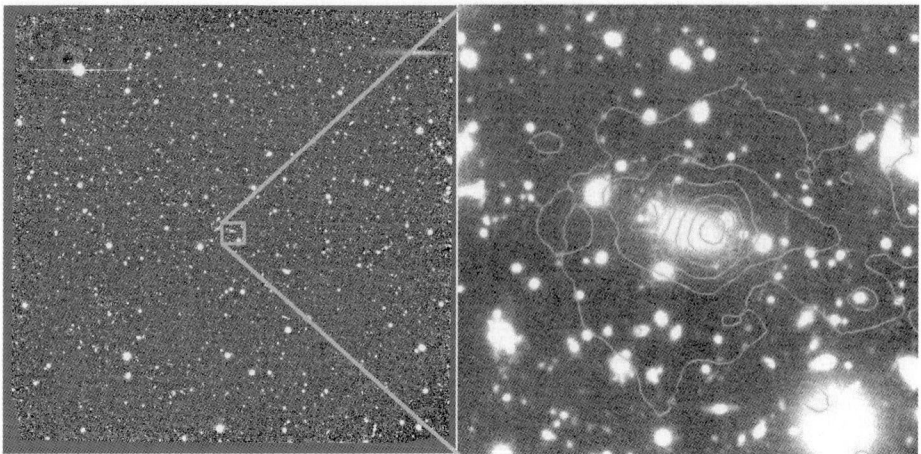

Fig. 3. *Left:* Coadded Megacam r'-band image of a $z = 0.5$ cluster (box size ∼30 arcmin). *Right:* Zoom into the very center. *Chandra* X-ray surface brightness contours are overlaid. Note the arc to the right of the bright central galaxy.

independent of their dynamical state – for these distant clusters. Therefore, we have engaged in a complete weak lensing follow-up of all 41 clusters. Observations of a redshift $z = 0.5$ cluster are shown as an example in Figs. 3 (Erben et al., in prep.). The images were taken at the 6.5m MMT telescope with the 36-CCD Megacam camera. The images of the 25 dithered observations were reduced and combined with the Ga-BoDS pipeline [10], adapted to Megacam. Zooming into the very center, a giant arc is detected. A preliminary weak lensing analysis shows that the cluster is clearly detected (Fig. 4), demonstrating for the first time that Megacam is ideally suited to perform weak lensing measurements for distant clusters.

Acknowledgments. This work was supported in part by NASA through *Chandra* Award GO4-5132X and *XMM-Newton* Grant NNG05GO50G as well as the DFG through Emmy Noether Research Grant RE 1462.

The MMT observations were supported in part by the F. H. Levinson Fund of the Peninsula Community Foundation.

References

1. T. H. Reiprich and H. Böhringer: ApJ **567**, 716-740 (2002)
2. B. J. Maughan, et al.: MNRAS **365**, 509-529 (2006)
3. S. W. Allen, R. W. Schmidt, and A. C. Fabian: MNRAS **328**, L37-L41 (2001)
4. A. Vikhlinin, et al.: ApJ **628**, 655-672 (2005)
5. M. Bonamente, et al.: ApJ **astro-ph/0512349** (2005)

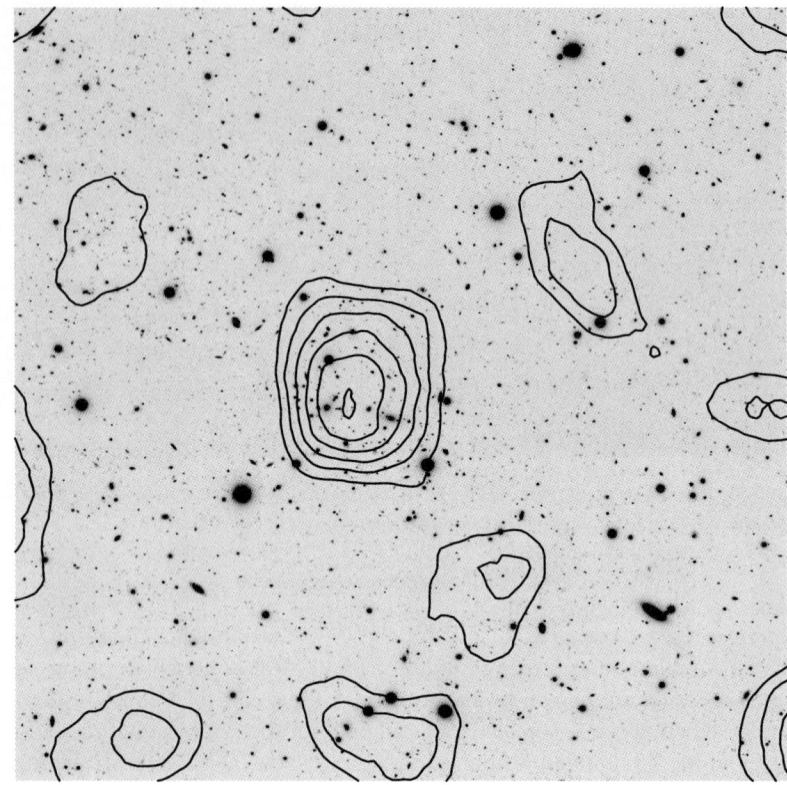

Fig. 4. Preliminary weak lensing contours (box size ∼10 arcmin). Contours indicate signal-to-noise ratio of 1.0, 1.5, .., 3.5; the cluster is clearly detected.

6. D. S. Hudson, et al.: A&A **453**, 2, 433-446 (2006)
7. A. Vikhlinin, et al.: ApJ **502**, 558-581 (1998)
8. C. R. Mullis, et al.: ApJ **594**, 154-171 (2003)
9. J. D. Cohn and M. White: Astroparticle Physics **24**, 316-333 (2005)
10. T. Erben, et al.: Astron. Nachr. **326**, 432-464 (2005)

Hydrodynamical Simulations of Cluster Formation with Central AGN Heating

D. Sijacki and V. Springel

Max-Planck-Institut für Astrophysik, Karl-Schwarzschild-Straße 1, 85740 Garching, Germany,
deboras@mpa-garching.mpg.de

Summary. We analyse the effects of central AGN heating on the formation of galaxy clusters by means of hydrodynamical simulations. Besides self-gravity of dark matter and baryons, our approach includes radiative cooling and heating processes of the gas component and a multiphase model for a self-consistent treatment of star formation and SNe feedback [1]. Additionally, we incorporate a periodic feedback mechanism in the form of hot buoyant bubbles, injected into the ICM during the active phases of accreting central AGN. We find that AGN heating can substantially affect the properties of the stellar and gaseous components, in particular reducing the mass deposition rate onto the central cD galaxy, thereby offering an energetically plausible solution to the cooling flow problem.

1 Method

Using the parallel TreeSPH-code GADGET-2 [2], we implemented a heating mechanism provided by an AGN, hosted in the central cluster galaxy. During the phases in which the AGN is "switched on", a certain amount of thermal energy is introduced into the ICM, generating centrally concentrated hot bubbles in regular time intervals. We parametrized our scheme in terms of the AGN duty cycle, the amount of feedback energy, the radius and the distance of the buoyant bubbles from the cluster centre. First, we studied bubble heating in isolated groups and clusters spanning a wide range of mass, and in a second step, we considered full cosmological simulations that follow the assembly of galaxy clusters self-consistently.

2 AGN Heating in Isolated Halos

In the left panel of Fig. 1, we analyse the bubble morphologies in terms of the projected mass-weighted temperature map of the centre of a $10^{15} M_\odot$ isolated galaxy cluster. The bubbles show well defined mushroom shaped structures and compress the surrounding gas above, which then forms characteristic cold rims. In the right panel of Fig. 1, we illustrate the unsharp masked X–ray emissivity map of the same cluster, where injection of bubbles into the ICM generates sound waves, which provide a source of ICM heating

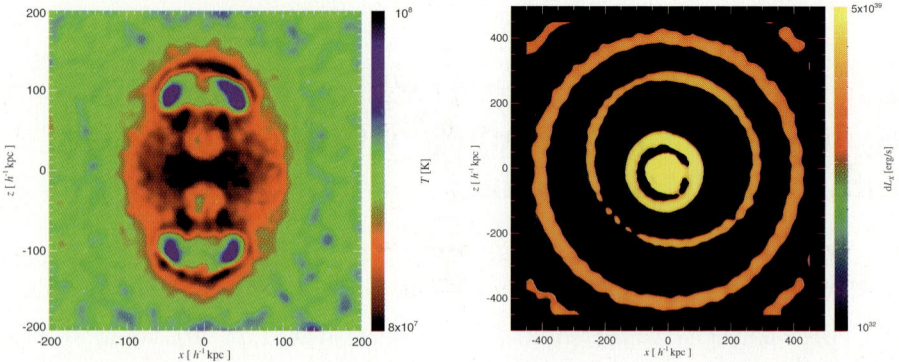

Fig. 1. The projected mass-weighted temperature map of a massive isolated cluster is shown in the left panel, while in the right panel the unsharp masked image of X-ray emissivity is represented.

on larger scales. The gas entropy profile and the mean mass inflow rate as a function of time are shown in Fig. 2. The intracluster gas is heated efficiently in the inner regions due to the AGN activity, and thus the entropy reaches a central floor, while the mass inflow rate stabilizes at a low level of $150 M_\odot/\mathrm{yr}$.

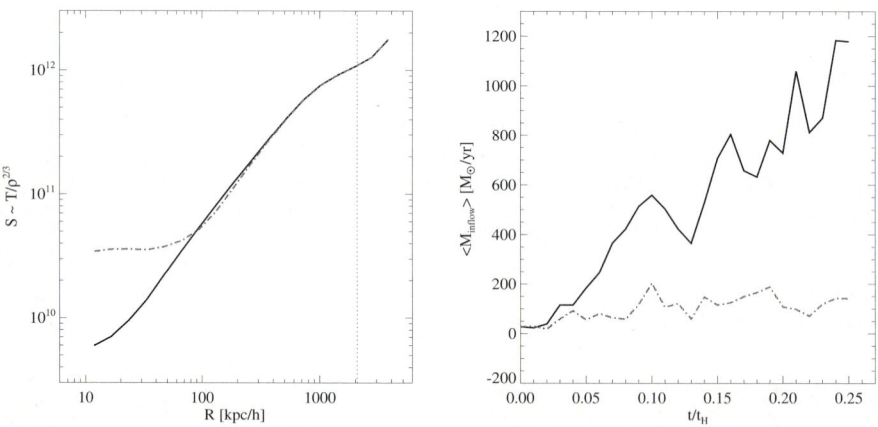

Fig. 2. The gas entropy profile (left panel) and the mean mass inflow rate as a function of time (right panel). Blue continuous line: cooling and star formation only; Red dot-dashed line: with additional AGN heating.

3 AGN Feedback in Cosmological Simulations

In cosmological simulations of cluster formation, AGN feedback provides a significant impact as well, similar to the case for isolated halos. The estimated cooling time in a $10^{15} M_\odot$ galaxy cluster at $z = 0$ is illustrated in the left panel of Fig. 3 (for a detailed description of the simulation characteristics and the models adopted see [3]). Bubble heating prevents the cooling time to drop substantially below the Hubble time, and thus the amount of central gas colder than 1keV is significantly reduced. Moreover, at lower redshifts the SFR of the central cluster galaxy decreases, and it appears older and redder, as shown in the right panel of Fig. 3. The white histogram is for the run without AGN heating, while the remaining two histograms show how SFR is suppressed at intermediate and low redshifts for the two different AGN feedback scenarios assumed.

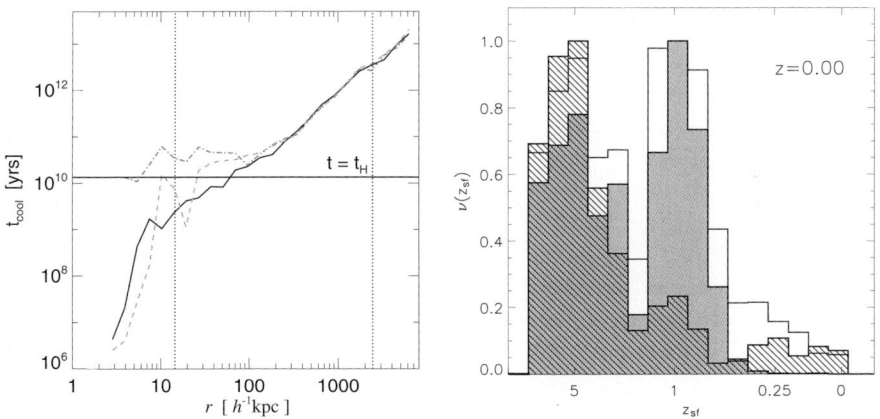

Fig. 3. The gas cooling time as a function of radius (left panel) and the histogram of stellar formation redshifts of the central cluster galaxy at z=0 (right panel). Blue continuous line: without AGN; Dashed and dot-dashed lines: different AGN feedback models.

4 Conclusions

The central heating mechanism provided by AGN activity, in the form of hot buoyant bubbles, appears to be a very important ingredient of galaxy cluster formation. AGN feedback affects both the gas and the stellar properties in the innermost regions, reducing the amount of cold IC gas and quenching recent star formation episodes, thus alleviating the cooling flow problem. Furthermore, the spreading of metals into the ICM is enhanced by AGN–driven bubbles, decreasing the radial metallicity gradient and making the distribution less lumpy.

References

1. V. Springel, L. Hernquist: MNRAS **339**, 289 (2003)
2. V. Springel: MNRAS **364**, 1105 (2005)
3. D. Sijacki, V. Springel: MNRAS **366**, 397-416 (2006)

Metal Enrichment Processes in the Intra-Cluster Medium

S. Schindler[1], W. Kapferer[1], W. Domainko[1], M. Mair[1], T. Kronberger[1,2], E. v. Kampen[1], S. Kimeswenger[1], M. Ruffert[3], and D. Breitschwerdt[4]

[1] Institut für Astro- und Teilchenphysik, Universität Innsbruck, Austria
[2] Institut für Astrophysik, Universität Göttingen, Germany
[3] School of Mathematics, University of Edinburgh, UK
[4] Institut für Astronomie, Universität Wien, Austria

1 Introduction

The components of clusters of galaxies – galaxies and intra-cluster medium (ICM) – interact with each other in many ways. There is more and more observational evidence that various types of processes are at work, which remove interstellar medium (ISM) from the galaxies. This metal enriched ISM mixes with the ICM, so that the currently observed ICM is a mixture of both. Hence the metals are a good tracer for the present and past interaction processes between galaxies and ICM. The new X-ray observations by CHANDRA and XMM make it possible to measure not only profiles but 2D distributions of metals: metallicity maps (e.g. Schmidt et al. 2002; Furusho et al. 2003; Sanders et al. 2004; Fukazawa et al. 2004; Hayakawa et al. 2004) showing that the metallicity has clearly a non-uniform, non-spherical distribution. These new results have stimulated a lot of discussion on the origin of metals. Given that most metals must come from cluster galaxies, the high ICM metallicity of around 0.3 solar or larger (Peterson et al. 2003; Tamura et al. 2004) and the small contribution of the galaxy masses to the baryonic matter in a cluster ($\approx 20\%$) the amount of material that has to be transported from the ISM to the ICM must be considerable.

Already many years ago supernova-driven galactic winds were suggested as a possible ISM transfer mechanism (De Young 1978). Recent observations and simulations indicate that other processes can contribute considerably, as well (e.g. Hayakawa et al. 2004). One process gaining increasing attention is ram-pressure stripping (Gunn & Gott 1972). In the Virgo cluster alone at least 7 ram-pressure affected spiral galaxies have been found (e.g. Cayatte et al. 1990; Vollmer et al. 2004). Also other processes like galaxy-galaxy interactions (Kapferer et al. 2005), jets from AGNs and intra-cluster supernovae (Domainko et al. 2004) can contribute to the metal enrichment.

Currently, several approaches are being made to explain the overall enrichment of the ICM by taking into account galactic winds. De Lucia et al. (2004) and Nagashima et al. (2005) use a combination of semi-analytic techniques and N-body simulations to calculate the overall ICM metallicity, but they do not predict the distribution of metals in a cluster. They find that mainly the massive galaxies contribute to the enrichment and that there is a mild metal evolution since $z = 1$. Tornatore et al. (2004) do simulations with smoothed particle hydrodynamics that include detailed yields from type Ia and II supernovae, but do not distinguish between the different transport processes and between gas in galaxies and in clusters.

Other groups have calculated the effect of ram-pressure stripping on single galaxies (e.g. Abadi et al. 1999; Quilis et al. 2000; Toniazzo & Schindler 2001; Roediger & Hensler 2005), but not the effect on the ICM. We take a complementary approach by investigating the transport of the metals from the galaxies into the intra-cluster medium and study in detail the efficiency of the various transport processes. In the simulations presented here, we have taken into account different transport processes – ram-pressure stripping, quiescent galactic winds and starbursts.

2 Numerical Method

We use different numerical techniques to calculate the various cluster components. The numerical method is described in Schindler et al. (2005), Kapferer et al. (2006) and Domainko et al. (2006).

3 Results

A comparison of ram-pressure stripping and quiescent winds shows that the two mechanisms result in a very different spatial distribution (see Fig. 1 and Schindler et al. 2005). Ram-pressure stripping is very efficient in the centre, because there the ICM density is very high ($\sim 6 \times 10^{-27}$ g/cm^3 within a radius of 0.1 Mpc) and leads to a centrally concentrated metallicity distribution. On the other hand the high central ICM density yields a high pressure which suppresses galactic winds resulting in a very extended distribution. Apart from the differences in the centre both processes yield pronounced inhomogeneities in the abundances. Single galaxies leave traces of metals in the shape of stripes behind them. Both the centrally concentrated metallicities and the general inhomogeneities are in good agreement with observations. The increase of metallicity (X-ray emission weighted) since redshift 1 averaged over the central cluster region within a radius of 400 kpc is 0.09 and 4×10^{-8} in solar units for the ram-pressure and the winds model, respectively.

The situation looks very different in a less massive cluster. In such a cluster the winds are less suppressed. Therefore the winds contribute also to the central metallicity. Inhomogeneities in the abundances do not disperse immediately (compare insets) even though the ICM in this cluster is stirred by a merger. The stripes are gradually spread out and at the end in a roughly homogeneous region of high metallicity is present at the centre. A given inhomogeneity at $t = 1.7$ Gyrs will be spread out on average over a 30 times larger volume at the end of the simulation ($t = 8$ Gyrs), depending strongly on the local dynamics of the surrounding ICM, e.g. shocks. For more figures and films see http://astro.uibk.ac.at/astroneu/hydroskiteam/index.htm.

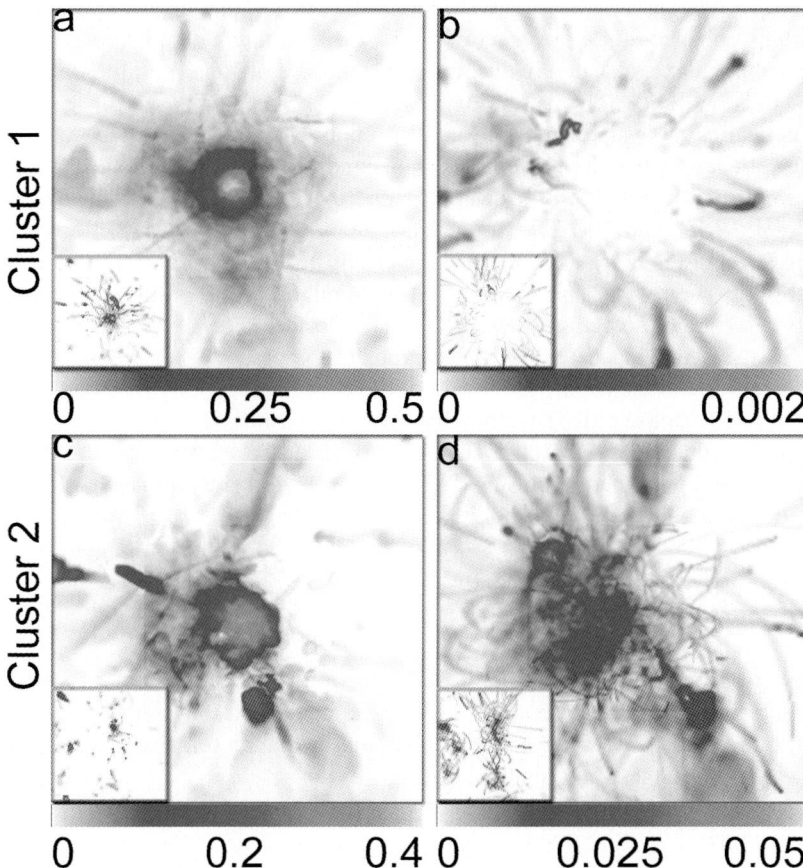

Fig. 1. X-ray weighted metal maps of two simulations at present epoch and insets at an epoch 6.7 Gyrs ago (a and b, a massive galaxy cluster; c and d, a merging cluster). Panels a and c show the metal maps for ram-pressure stripping and panels b and d those for galactic winds (non-starbursts). The quantities are given in solar abundances. Ram-pressure stripping always yields a centrally concentrated metallicity distribution. The distribution of metals transported by winds depends strongly on the cluster mass: in a massive cluster (b) the winds are suppressed in the centre resulting in a low abundance there. The comparison with the insets shows how the inhomogeneities spread/evolve (from Schindler et al. 2005)

Non-starburst driven winds are less efficient in transporting gas than ram-pressure stripping in the time interval between redshift 1 and 0 investigated here. During the non-merger phases ram-pressure stripping is about a factor of 5 more efficient than non-starburst driven winds.

The mass loss due to ram-pressure stripping shows large fluctuations compared to galactic winds, because non-starburst driven winds have an almost constant mass loss over a long period of up to 5 Gyrs, whereas ram-pressure stripping removes huge amounts of gas from a single galaxy on much shorter time scales. During the two merger events (at 3 Gyrs and 5.5 Gyrs) the mass loss due to ram pressure is much higher because of the enhanced ICM pressure. In contrast non-starburst driven galactic winds are suppressed in galaxies located in the high ICM pressure region. Therefore the mass loss due to winds is slightly lowered during mergers.

In a detailed analysis of ram-pressure stripping (Domainko et al. 2006) we found that the ram-pressure stripping efficiency depends strongly on the cluster mass. On average ram-pressure stripping can contribute 10-15% to the cluster metals within a radius of 1.3 Mpc and 100% to the cluster metals within a radius of 100 kpc. Ram-pressure stripping is more efficient during periods of merging. It is acting already at a redshift of 1. The stripes that we find in our simulations are similar to observational findings: see the plume behind NGC4388 in HI (Oosterloo & van Gorkom 2005).

An analysis of galactic winds and starbursts (Kapferer et al. 2006) showed that these processes can contribute about 5% to the cluster metallicity between redshifts 4 and 1, and 15% to the cluster metallicity between redshifts 1 and 0. The metal maps yield information on the dynamical state of the cluster: they can be used to distinguish pre-merger from post-merger clusters. We also predict how metal maps will look when observed with the future X-ray satellite XEUS (see Fig. 2).

4 Outlook

We are currently working on an inclusion of the effects of AGN into our simulations. AGN can influence the energetics as well as the metallicity of a cluster due to winds as the example RBS797 shows, where we found evidence for three periods of AGN activity in VLA images (Gitti et al. 2006).

References

1. Abadi, M.G., Moore, B., Bower, R.G.: MNRAS **308**, 947 (1999)
2. Cayatte V., et al.: AJ **100**, 604 (1990)
3. De Lucia G., Kauffmann G., White S.D.M.: MNRAS **349**, 1101 (2004)
4. De Young, D.S.: ApJ **223**, 47 (1978)
5. Domainko W., et al.: A&A **425**, L21 (2004)
6. Domainko, W., et al.: A&A **452**, 795-802 (2006)
7. Fukazawa, Y., Kawano, N., & Kawashima, K.: ApJ **606**, L109 (2004)
8. Furusho, T., Yamasaki, N.Y., & Ohashi, T.: ApJ **596**, 181 (2003)
9. Gitti, M., Feretti, L., Schindler, S.,: A&A **448**, 853-860 (2006)
10. Gunn J.E., Gott J.R.III: ApJ **176**, 1 (1972)
11. Hayakawa A., et al.: PASJ **56**, 743 (2004)
12. Kapferer, W., et al.: A&A **438**, 87 (2005)
13. Kapferer W., et al.: A&A **447**, 827 (2006)
14. Nagashima M., et al.: MNRAS **358**, 1247 (2005)
15. Oosterloo T., van Gorkom J.: A&A **437**, L19O (2005)
16. Peterson J.R., et al.: ApJ **590**, 207 (2003)
17. Quilis, V., Moore, B., Bower, R.: Science **288**, 1617 (2000)
18. Roediger E., Hensler G.: A&A **433**, 875 (2005)

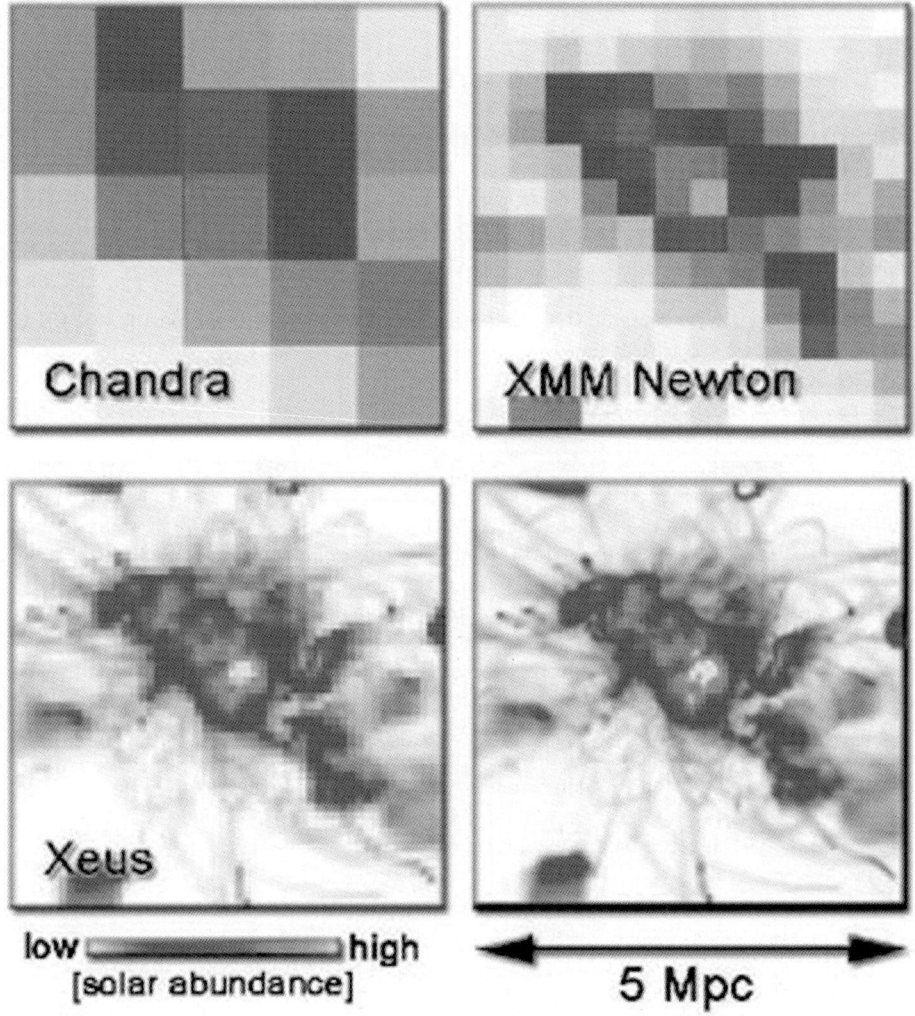

Fig. 2. Simulated metallicity maps of a cluster observed with different instruments. In the lower right the original metallicity map is shown (from Kapferer et al. 2006).

19. Sanders J.S., et al.: MNRAS **349**, 952 (2004)
20. Schindler S., et al.: A&A **435**, L25 (2005)
21. Schmidt R.W., Fabian A.C., Sanders J., 2002 MNRAS 337, 71
22. Tamura T., et al.: A&A **420**, 135 (2004)
23. Toniazzo T., Schindler S.: MNRAS **325**, 509 (2001)
24. Tornatore L., et al.: MNRAS **349**, L19 (2004)
25. Vollmer B., et al.: AJ **127**, 3375 (2004)

Ultraviolet-Bright, High-Redshift ULIRGS

G.M. Williger[1], J. Colbert[2], H.I. Teplitz[2], P.J. Francis[3], P. Palunas[4], and B.E. Woodgate[5]

[1] Depts. of Physics & Astronomy, JHU, Baltimore MD & U. Louisville, Louisville KY, USA, williger@pha.jhu.edu
[2] Spitzer Science Ctr, Pasadena CA, USA
[3] Austral. Nat'l U., Canberra, Australia
[4] U. Texas, Austin TX, USA
[5] NASA/GSFC, Greenbelt MD, USA

Summary. We present *Spitzer* Space Telescope observations of the z=2.38 Lyα-emitter over-density associated with galaxy cluster J2143-4423, the largest known structure (110 Mpc) above $z = 2$. We imaged 22 of the 37 known Lyα-emitters within the filament-like structure, using the MIPS 24μm band. We detected 6 of the Lyα-emitters, including 3 of the 4 clouds of extended (>50kpc) Lyα emission, also known as Lyα Blobs. Conversion from rest-wavelength 7μm to total far-infrared luminosity using locally derived correlations suggests all the detected sources are in the class of ULIRGs, with some reaching Hyper-LIRG energies. Lyα blobs frequently show evidence for interaction, either in *HST* imaging, or the proximity of multiple MIPS sources within the Lyα cloud. This connection suggests that interaction or even mergers may be related to the production of Lyα blobs. A connection to mergers does not in itself help explain the origin of the Lyα blobs, as most of the suggested mechanisms for creating Lyα blobs (starbursts, AGN, cooling flows) could also be associated with galaxy interactions.

1 Introduction

The 110 Mpc filament mapped out by 37 Lyα-emitting objects around the $z = 2.38$ galaxy cluster J2143-4423 is the largest known structure above $z = 2$ [24], comparable in size to some of the largest structures seen in the local Universe i.e. the Great Wall [17]. Initially identified from narrow-band imaging tuned to Lyα at z=2.38, it has since been spectroscopically confirmed [14, 15]. In addition to its compact Lyα-emitters, this high-redshift "Filament" is also home to four extended Lyα-emitting clouds, called Lyα blobs.

The Lyα blob is a relatively new class of objects found among high-redshift galaxy over-densities [27, 19, 24]. While similar in extent ($\sim$100 kpc) and Lyα flux ($\sim 10^{44}$ ergs s^{-1}) to high-redshift radio galaxies, blobs are radio quiet and are therefore unlikely to arise from interaction with jets. Current surveys have reported the discovery of roughly 10 of these giant Lyα blobs, but they are not isolated high redshift oddities. Matsuda et al. [21] demonstrated that the blobs are part of a continuous size distribution of resolved (>16 arcsec^2) Lyα emitters, with more than 40 presently known.

Their energy source is a mystery, as the measured UV flux from nearby galaxies cannot produce the observed Lyα fluxes. Lyα blobs may be powered by SNe-driven superwinds [23], driving gas into the surrounding ambient medium and producing shocks. Another model (i.e. [1]) considers obscured AGN, with the exciting UV illumination escaping along different lines of sight. Cooling flows have also been suggested [11, 13].

There is growing evidence that Lyα blobs mark regions of extreme infrared luminosity. Submillimeter flux has been detected in two of the giant Lyα blobs, SMM J221726+0013 [3] and SMM J17142+5016 [26]. The submm source SMM J02399-0136 is also likely surrounded by a Lyα blob halo, as its Lyα emission covered most of a 15 arcsec slit [18]. Geach et al. [16] recently detected submm flux from four of the smaller (< 55 square arcsec) and less luminous ($< 2 \times 10^{43}$ L$_\odot$) Lyα blobs: LAB5, LAB10, LAB14 & LAB18 [21]. Also, there is a single Lyα blob (SST24 J1434110+331733) in the NOAO Deep Wide-Field Survey with strong 24μm flux (0.86 mJy) [9].

Here we discuss *Spitzer* 24μm observations of these z=2.38 Lyα-emitters, both compact sources and blobs. We estimate the total far-IR luminosity for all detections and discuss connections between mergers and Lyα blobs. We assume an Ω_M=0.3, Ω_Λ=0.7 universe with H$_o$=70 km s^{-1} Mpc^{-1}.

2 Observations and Analysis

Our data were obtained using Multiband Imaging Photometer for *Spitzer* (MIPS) [25] in 24μm photometry mode. See [6] for details. The 5-σ detection limit, as measured within a 7.5$''$ aperture, is 58 μJy over most of the image, but can be as high as 180 μJy at the edges. Six out of 22 ($\sim$30%) of the Lyα-emitters within the filament are associated with 24μm sources (Figure 1), including three of the four resolved ($>$50 kpc) Lyα blobs. This blob association rate is much higher than for unresolved Lyα-emitters (3 of 18). We do not detect 16 of the 22 Lyα-emitters with MIPS. Six sources lie in noisy edge regions, where the 5σ detection limits are 100-180 μJy. The remaining Lyα-emitters are more centrally located, with detection limits of $<$60 μJy.

Two of the Lyα blobs, B6 and B7, appear to have additional MIPS sources associated with them. The B6 Lyα blob extends over 25$''$, or over 200 kpc at z=2.38, and contains more than one knot or concentration. The brightest Lyα knot is the central one and it is associated with a bright MIPS source, but the southern knot (not to be confused with the large diffuse area to the southeast) also has a MIPS source of almost identical brightness. There is also an area of more diffuse Lyα emission to the north which also appears to be associated with a bright MIPS source. Each source is separated by approximately 9.5$''$ from the central source, creating a potential triple system.

3 Ultraviolet-Bright ULIRGs

The detected 24μm flux densities range from 0.1 to 0.6 mJy, which at z=2.38 corresponds to roughly 2×10^{11} to 10^{12} L$_\odot$ in rest frame 7μm $\nu F \nu$. To convert from the mid-infrared to total bolometric luminosity (L$_{bol}$), we use the relationship from [5],

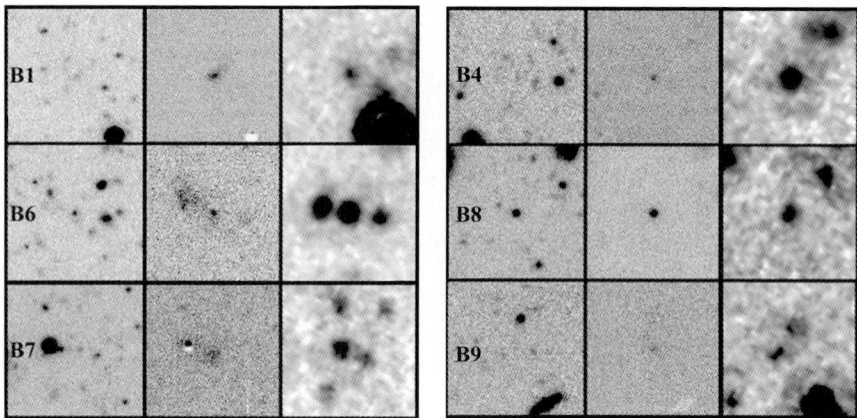

Fig. 1. Left) Images of the $40'' \times 40''$ field around the three MIPS-detected Lyα blobs: B1, B6, & B7. From left to right the images are CTIO B-band, CTIO continuum-subtracted Lyα, and MIPS 24μm. East is upwards, north to the right. One arcsecond corresponds to 8kpc at z=2.38. Right) Images of the $40'' \times 40''$ field around the three MIPS-detected non-extended Lyα-emitters. Format same as left.

hereafter CE01, calibrated by examining galaxies measured with both ISO and IRAS: $L_{IR} = 4.37^{+2.35}_{-2.13} \times 10^{-6} \times L^{1.62}_{6.7\mu m}$.

In an effort to avoid over-estimation of L_{IR}, the mid-infrared conversion used throughout this paper assumes the low end of the 1σ envelope, producing total luminosities roughly a factor of 2 lower than a direct application of the formula. Even this conservative mid-IR conversion puts all the sources in the class of ultra luminous infrared galaxies (ULIRGs; $> 10^{12} L_\odot$), with many achieving Hyper-LIRG ($> 10^{13} L_\odot$) status. It should be noted, however, that the CE01 relation is based on nearby starburst-dominated galaxies and it is possible it may not hold at high redshift or for the most extreme starbursts, if their spectral energy distributions are significantly different.

If we combine the 24μm detections from this study and from [9] with the submm detections of Lyα blobs [3, 26, 16], there are now at least 10 known high redshift ULIRGs surrounded by Lyα halos. We plot Lyα luminosity versus L_{bol} for all the identified Lyα blobs with infrared or submm detections in Figure 2.

Applying a similar analysis, there is a weak Lyα /IR relation for submm-detected Lyα blobs with a typical $L_{Ly\alpha}/L_{bol}$ ratio just under 0.1% [16]. Including our 24μm-detected Lyα blobs, which are brighter in Lyα than most of the objects from [16], we continue to see the same weak trend, with a $L_{Ly\alpha}/L_{bol}$ efficiency of 0.05-0.2%. Neither the detection of X-rays (SMM J17142+5016 & LAB18) nor CIV emission (B1, SMM J17142+5016, & SST24 J1434110+331733), both strong indicators of AGN activity, has any clear effect on this trend. While this relation is still tentative, it suggests a direct causal connection between the ULIRG infrared luminosity and the Lyα blobs.

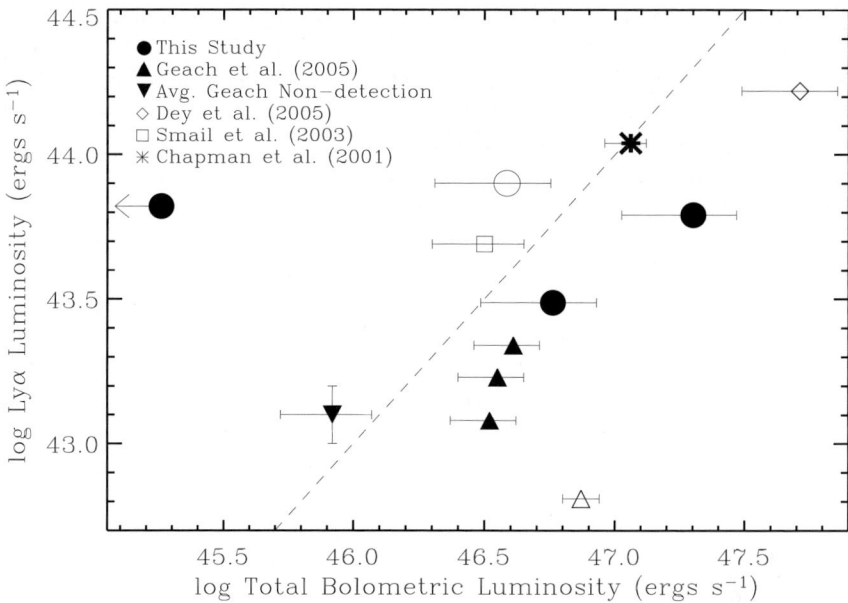

Fig. 2. Log $L_{Ly\alpha}$ vs. log L_{bol} for high-z Lyα blob ULIRGs. The 24μm-detected blobs are the circles (this study) and the diamond [9], with error bars based on the estimated reliability of the Chary & Elbaz [5] relation. The submm detections are the triangles [16], squares [26], and asterisks [3]. The upside-down triangle is the average of all non-detected Geach et al. blobs. Suspected AGN (with X-rays or CIV emission) are hollow symbols. The dashed line marks $L_{Ly\alpha}/L_{bol} = 0.001$.

The high rate of MIPS detection for the Lyα blobs (3 out of 4) demonstrates that these ultraviolet sources are locations of tremendous infrared energy. Two of the Lyα blobs (B6 & B7) are associated with multiple MIPS sources, suggesting possible interaction or even merger. These separations between possible ULIRG component galaxies are large (60-70 kpc) compared to the typical low-redshift ULIRG, which have a median separation ~ 2 kpc [22]. However, interacting ULIRGs with wide separations on the scale of 50 kpc do exist [20], and seem to occur when at least one the interacting galaxies is highly gas-rich [10].

In the local universe, the most energetic mergers are associated with ULIRGs, with roughly 90% of ULIRGs clearly interacting [2, 12]. The majority of local ULIRG mergers are between two low mass (0.3-0.5 L*) galaxies [7], but major mergers of high-mass galaxies, like those predicted to build giant ellipticals, must be occurring at higher redshift i.e. [8]. If merger-induced star formation is the source of the majority of the infrared flux in these objects, it implies star formation rates of 1000s of $M_\odot$ yr^{-1}, like those suggested for some sub-mm sources [4, 16]. Such massive SFRs would be capable of generating the supernova kinetic energy needed to drive a superwind and power the Lyα blobs.

Mergers could also help drive gas inwards into a supermassive black hole, making AGN another viable infrared energy source, with escaping ultraviolet radiation driving the Lyα blobs. Dey et al. [9] found that their 24μm-detected blob is better fit by an AGN spectral energy distribution than that of a star forming galaxy, but longer rest-wavelengths or a mid-IR spectrum are required for a definitive determination.

Cooling flows also remain a possibility, as it might be common to find multiple galaxies merging at the center of such a large inward flow of gas. It does become a less favoured model, however, as the energy from the cooling gas in no longer required to power the Lyα blobs. The ULIRGs would appear to have more than enough energy in their budget to do so on their own.

References

1. Basu-Zych, A. & Scharf, C.: ApJL **615**, L85 (2004)
2. Bushouse, H. A., et al.: ApJS **138**, 1 (2002)
3. Chapman, S. C., et al.: ApJ **548**, L17 (2001)
4. Chapman, S. C., et al.: ApJ **599**, 92 (2003)
5. Chary, R., & Elbaz, D.: ApJ **556**, 562 (2001)
6. Colbert, et al.: ApJL **637**, L89-L92 (2006)
7. Colina, L., et al.: ApJ **563**, 546 (2001)
8. Conselice, C. J.: ApJ **620**, 564-583 (2005)
9. Dey, A. et al.: ApJ **629**, 654-666 (2005)
10. Dinh-V-Trung et al.: ApJ **556**, 141 (2001)
11. Fardal, M. A. et al.: ApJ **562**, 605 (2001)
12. Farrah, D., et al.: MNRAS **326**, 1333 (2001)
13. Francis, P. J. et al.: ApJ **554**, 1001 (2001)
14. Francis, P. J. et al.: ApJ **614**, 75 (2004)
15. Francis, P. J. & Williger, G. M.: ApJL **602**, L77 (2004)
16. Geach, J. E. et al.: MNRAS **363**, 1398-1408 (2005)
17. Geller, M. J., & Huchra, J. P.: Science **246**, 897 (1989)
18. Ivison, R. J. et al.: MNRAS **298**, 583 (1998)
19. Keel, W. C. et al.: AJ **118**, 2547 (1999)
20. Kim, D. -C., Veilleux, S., & Sanders, D. B.: ApJS **143**, 277 (2002)
21. Matsuda, Y. et al.: AJ **128**, 569 (2004)
22. Murphy, T., et al.: AJ **111**, 1025 (1996)
23. Ohyama, Y. & Taniguchi, Y.: AJ **127**, 1313 (2004)
24. Palunas, P. et al.: ApJ **602**, 545 (2004)
25. Rieke, G. H. et al.: ApJS **154**, 25 (2004)
26. Smail, I. et al.: ApJ **583**, 551 (2003)
27. Steidel, C. C. et al.: ApJ **532**, 170 (2000)

Part VI

Gamma Ray Bursts

The Correlation between νF_ν Peak Energy and radiated Energy in Gamma–Ray Bursts

L. Amati

INAF - IASF Bologna, via P. Gobetti 101, 40129 Bologna, Italy
amati@iasfbo.inaf.it

Summary. I review the observational status and the main implications of the correlation between GRBs peak energy $E_{p,i}$, the photon energy at which the intrinsic (i.e. corrected for cosmological redshift) νF_ν spectrum peaks, and isotropic equivalent radiated energy E_{iso}. This correlation, discovered by Amati et al. (2002) based on BeppoSAX measurements and confirmed and extended to X–ray rich GRBs and X–ray Flashes (XRF) by HETE–2 measurements, can be used to constrain the parameters of the models for the prompt emission of GRBs, is a challenging test for jet and GRB/XRF unification models, provides hints on the nature of sub-classes of GRBs (e.g. short, sub–energetic), is commonly adopted as an input or a test for GRB synthesis models and can be also used to build up GRB redshift estimators.

1 Introduction

Cosmic Gamma–Ray Bursts (GRBs) are sudden and unpredictable bursts of hard X-/ soft gamma–rays photons, lasting from a fraction to hundreds of seconds, showing a hard X–ray flux that is normally much higher than that from all other celestial sources in the same energy band and coming from random and isotropically distributed directions in the sky with a rate of ~ 0.8 events/day (as measured by low Earth orbit satellites). In the last 10 years, the observational scenario has been substantially enriched by the discovery of fading X–ray, optical, radio counterparts (*afterglow emission*) and host galaxies, leading to the estimates of redshifts (mostly in the range 0.1–6.3) and thus of radiated energy (as huge as $\sim 10^{54}$ erg under the assumption of isotropic emission) for more than 70 events. Despite this enormous progress and the consolidation of the standard basic picture, in which the burst of X/gamma radiation ("prompt emission") is produced by the interaction of colliding shells within a possibly collimated plasma of photons, electrons and a small percentage of baryons expanding ultra–relativistically ("internal shocks" scenario), and the multi–wavelength afterglow emission is originated by the interaction of this "fireball" (or "jet") with the circum–burst environment ("external shock"), the study of the Gamma–Ray Burst (GRB) phenomenon is still characterized by various unsettled issues. Among these, the physics of prompt emission, jet structure and evolution, the impact of viewing angle, the different emission mechanisms at work in short GRBs($\lesssim 2$ s, likely originated by NS–NS or NS–BH mergers) and long GRBs ($\gtrsim 2$ s, likely originated by the asymmetric core–collapse of a fastly rotating massive star), the nature of sub–classes of events like X–ray Flashes (XRFs, events characterized) and sub–energetic GRBs (GRB980425), the connection between GRBs and supernovae. Under these respects, one of the most intriguing and useful observational evidences is the correlation (a.k.a. "Amati relation") between the frequency at which the νF_ν spectrum of GRBs peaks and their total radiated energy under isotropic assumption. In this paper, I briefly review the observations and main implications of this correlation.

2 The $E_{p,i} - E_{iso}$ correlation: observations

The prompt emission spectra of GRBs are non-thermal and show in many cases substantial evolution. Most average and time resolved spectra can be modeled with the Band function [1], a smoothly broken power–law introduced to describe the BATSE (25-2000 keV) data, whose parameters are the low energy spectral index, α, the high energy spectral index, β, the break energy, E_0, and the overall normalization. The photon energy at which the $\nu F\nu$ spectrum peaks is given by $E_p = E_0 \times (2 + \alpha)$ and is called the *peak energy*. Since the BeppoSAX breakthrough discoveries in 1997, more than 70 redshift estimates have become available, all concerning long duration GRBs. As a consequence, for these events it is possible to compute the intrinsic peak energy, $E_{p,i} = E_p \times (1 + z)$, and the total radiated energy in a given (cosmological rest–frame) energy band by exploiting the distance estimate and the measured average spectrum and fluence, following e.g. the methods described in [2, 3, 4]. In 2002, Amati et al. [2] presented the results of the analysis of the average WFC (2–28 keV) and GRBM (40–700 keV) spectra of 12 BeppoSAX GRBs with known redshift (9 firm measurements and 3 possible values). By fitting the redshift–corrected spectra with the Band function, they were able to estimate the intrinsic (i.e. in the source cosmological rest frame) values of the spectral parameters. Also, by integrating the best fit model of the intrinsic time integrated spectrum and adopting a standard cosmology, they computed the total radiated energy in the 1–10000 keV band assuming isotropic emission, E_{iso}, and performed correlations studies between intrinsic spectral parameters, E_{iso} and the redshift z. The more relevant outcome of this work was the evidence of a strong correlation between $E_{p,i}$ and E_{iso} : the correlation coefficient between $\log(E_{p,i})$ and $\log(E_{iso})$ was found to be 0.949 for the 9 GRBs with firm redshift estimates, corresponding to a chance probability of 0.005%. The slope of the power–law best describing the trend of $E_{p,i}$ as a function of E_{iso} was ~ 0.5. This work was extended by [3] by including in the sample 10 more events with known redshift for which new spectral data (BeppoSAX events) or published best fit spectral parameters (BATSE and HETE–2 events) were available. The $E_{p,i} - E_{iso}$ correlation was confirmed and its significance increased, giving a correlation coefficient similar to that derived by [2] but with a much higher number of events. Based on HETE–2 measurements, [5] not only confirmed the $E_{p,i} - E_{iso}$ correlation but remarkably extended it to XRFs, showing that it holds over three orders of magnitude in $E_{p,i}$ and five orders of magnitude in E_{iso}. The addition of new data, as more redshift estimates became available, confirmed the correlation and increased its significance, as found e.g. by [4] (29 events, chance probability of 7.6×10^{-7}). The most updated analysis of the $E_{p,i} - E_{iso}$ correlation is that reported by [6], who, based on a sample of 43 GRBs and XRFs with firm estimates of both z and E_p (which includes 11 Swift events) and by fitting with a method which accounts for extra–Poissonian dispersion, finds a power–law index and normalization of ~ 0.5 and ~ 100, respectively, and an extra–Poissonian logarithmic dispersion of ~ 0.15. This work includes also a discussion of the possible systematics affecting the estimates of E_p, E_{iso} and the extra–Poissonian dispersion. It is important to note that the peculiar sub–energetic event GRB980425 / SN1998bw, which is the prototype event for the GRB–SN connection and is a peculiar event also because of its very low redshift (0.0085), shows $E_{p,i}$ and E_{iso} values inconsistent with the correlation, as possibly the other sub–energetic event GRB031203 (but in this case the $E_{p,i}$ value is uncertain). This holds also for the few short GRBs for which redshift estimates were performed very recently, showing high $E_{p,i}$ values with respect to their (typically low) E_{iso}. Thus, there is evidence that the $E_{p,i} - E_{iso}$ correlation is a property of 'normal' long GRBs and XRFs, but not of short GRBs and of at least a fraction of sub–energetic GRBs. The present observational status ot the $E_{p,i} - E_{iso}$ correlation is shown graphically in Figure 1.

3 Main implications of the $E_{p,i} - E_{iso}$ correlation

Since its discovery in 2002 [2] and in particular its confirmation and extension to XRFs [3, 5, 7], the origin of the $E_{p,i} - E_{iso}$ correlation and its implications for GRB models have been investigated by several works. The impact of this robust observational evidence on prompt emission models concerns mainly the physics, the geometry (i.e. shape and properties of jets), viewing angle effects and GRB/XRF unification. Indeed, the existence of the $E_{p,i} - E_{iso}$ correlation and its properties are also often used as an ingredient or a test output for GRB synthesis models.

The physics of the prompt emission of GRBs is still far to be settled and a variety of scenarios, within the standard fireball picture, have been proposed, based on different emission mechanisms (e.g. SSM internal shocks, Inverse Compton dominated internal shocks, SSM external shocks, photospheric emission dominated models) and different kinds of fireball (e.g. kinetic energy dominated or Poynting flux dominated), see e.g. [8] for a review. In general, both $E_{p,i}$ and E_{iso} are linked to the fireball bulk Lorentz factor, Γ, in a way that varies in each scenario, and the existence and properties of the $E_{p,i} - E_{iso}$ correlation allow to constrain the range of values of the parameters, see, e.g., [8, 9]. For instance, as shown, e.g., by [8, 10, 11], for a power–law electron distribution generated in an internal shock within a fireball with bulk Lorentz factor Γ, it is possible to derive the relation $E_{p,i} \propto \Gamma^{-2} L^{1/2} t_\nu^{-1}$, where t_ν is the typical variability time scale. Clearly, in order to produce the observed $E_{p,i} - E_{iso}$ correlation the above formula would require that Γ and t_ν are approximately the same for all GRBs, an assumption which is difficult to justify. Things get even more complicated if one takes into account that the models generally assume $L \propto \Gamma^\beta$, with the value of β varying in each scenario and is typically ~ 2–3 [8, 9, 10]. An interesting possibility, which is currently the subject of many theoretical works, is that a substantial contribution to prompt radiation of GRBs comes from direct or Comptonized thermal emission from the photosphere of the fireball [8, 10, 11]. In this scenarios, $E_{p,i}$ is mainly determined by

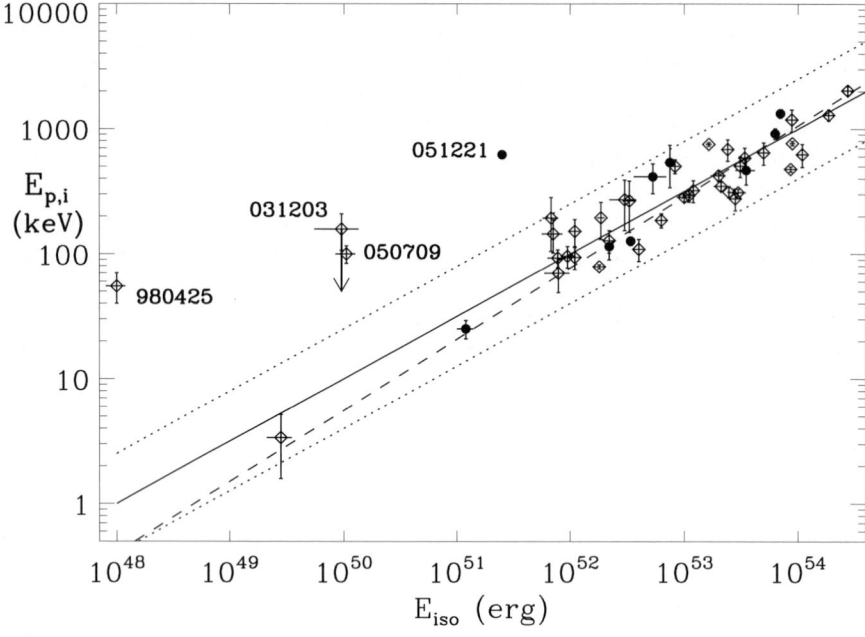

Fig. 1. $E_{p,i}$ and E_{iso} values for 43 GRBs/XRFs with firm redshift and $E_{p,i}$ estimates. Filled circles correspond to Swift GRBs. The continuous line is the best fit power–law $E_{p,i} = 99 \times E_{iso}^{0.49}$ obtained by accounting for sample variance; the dotted lines delimitate the region corresponding to a vertical logarithmic deviation of 0.4. The dashed line is the best fit power–law $E_{p,i} = 77 \times E_{iso}^{0.57}$ obtained by fitting the data without accounting for sample variance. Also shown are the $E_{p,i} - E_{iso}$ points of sub–energetic (980425, 031203) and short (050709, 051221) GRBs, which were not included in the fits. Data and models are taken from the preliminary version of [6].

the peak temperature T_{pk} of black–body distributed photons and thus naturally linked to the luminosity or radiated energy.

The validity of the $E_{p,i} - E_{iso}$ correlation from the most energetic GRBs to XRFs confirms that these two phenomena have the same origin and is a very challenging observable for GRB jet models. Indeed, these models have to explain not only how E_{iso} and $E_{p,i}$ are linked to the jet opening angle, θ_{jet}, and/or to the viewing angle with respect to the jet axis, θ_v, but also how E_{iso} can span over several orders of magnitudes. In the most simple scenario, the uniform jet model [12, 13], jet opening angles are variable and the observer measures the same value of E_{iso} independently of θ_v. In the other popular scenario, the universal structured jet model (e.g. Rossi, Lazzati & Rees), E_{iso} depends on θ_v. As argued by [13], the structured universal jet model, in order to explain the validity of the $E_{p,i} - E_{iso}$ correlation from XRFs to energetic GRBs, predicts a number of detected XRFs several orders of magnitude higher than the observed one ($\sim$1/3 than that of GRBs). In their view, the uniform jet model can overcome these problems by assuming a distribution of jet opening angles $N(\theta_{jet}) \propto \theta_{jet}^{-2}$. This implies that the great majority of GRBs have opening angles smaller than $\sim 1°$ and that the true rate of GRBs is several orders of magnitude higher than observed and comparable to that of SN Ic. On the other hand, [15] show that the requirement that most GRBs have jet opening angles less than 1 degree, needed in the uniform jet scenario in order to explain the $E_{p,i} - E_{iso}$ correlation, as discussed above, implies values of the fireball kinetic energy and/or of the interstellar medium density much higher than those inferred from the afterglow decay light curves. Recently, modifications of both the uniform and universal structured jet model that can reproduce the $E_{p,i} - E_{iso}$ correlation and predict the observed ratio between the number of XRFs and that of GRBs have been proposed, like Fisher–shape jets and quasi–universal Gaussian structured jet. Of particular interest are also the off–axis scenarios, in which the jet is typically assumed to be uniform but, due to relativistic beaming and Doppler effects, for $\theta_v > \theta_{jet}$ the measured emissivity does not sharply go to zero and the event is detected by the observer with E_{iso} and $E_{p,i}$ dropping rapidly as θ_v increases [16, 17, 18]. In these models, XRFs are those events seen off–axis and the XRFs rate with respect to GRBs and the $E_{p,i} - E_{iso}$ correlation can be correctly predicted.

As mentioned above, an important observational evidence is that the peculiar sub–energetic and very close ($z=0.0085$) event GRB980425 / SN1998bw (and possibly GRB031203) and short GRBs do not follow the $E_{p,i} - E_{iso}$ correlation. This shows the importance of the $E_{p,i} - E_{iso}$ plane for the identification of sub–classes of GRBs and gives hints on their different nature, or viewing condition, with respect to 'normal' GRB / XRFs. Concerning sub–energetic GRBs, the most common explanations assume that the peculiarity of these events is due to particular and uncommon viewing angles, as proposed e.g. by [19] for GRB980425 and [20] for GRB031203. An alternative explanation has been suggested by [21] in the framework of the CB model [22]. In this scenario, the νF_ν spectra of GRBs are characterized by two peaks, one at sub–MeV energies, the normal peak following the $E_{p,i} - E_{iso}$ correlation, and one in the GeV–TeV range. When a GRB is seen very off–axis, the same relativistic Doppler and beaming effects discussed above would shift the high energy peak to low energies confusing the normal low energy GRB peak. For short GRBs, [23] found that their emission properties are similar to those of the first ~ 1–2 s of long GRBs. This could indicate that the central engine is the same for the two classes, but it works for a longer time in long GRBs. This would explain the low radiated energy by short GRBs and, given that the emission would stop before the typical hard to soft evolution observed for long GRBs, also their high $E_{p,i}$ (with respect to their E_{iso}). The merger scenarios naturally explain the short life of the central engine, and thus the low radiated energy and the quitting of the emission before hard to soft spectral evolution.

Finally, the robustness of the $E_{p,i} - E_{iso}$ correlation and its validity from the closest to the farthest GRBs (excluding the peculiar cases discussed above) stimulated its use for the estimate of pseudo–redshifts for large samples of GRBs and, in turn, for the reconstruction of the GRB luminosity function. The more precise redshift estimates can be obtained with the redshift indicator developed by [24] or its refined version proposed very recently by [25], which provides redshift estimates accurate to a factor of ~ 2 and is currently used to estimate pseudo–redshifts of HETE–2 GRBs.

References

1. D. Band, J. Matteson, L. Ford et al: ApJ **413**, 281 (1993)
2. L. Amati, F. Frontera, M. Tavani: A&A **390**, 81 (2002)
3. L. Amati: ChJAA **3 Suppl.**, 455 (2003)
4. G. Ghirlanda, G. Ghisellini, D. Lazzati: ApJ **616**, 331 (2004)
5. D.Q. Lamb, T.Q. Donaghy, C. Graziani: NewAR **48**, 459 (2004)
6. L. Amati: MNARS, **372**, 6 (2006)
7. T. Sakamoto, D.Q. Lamb, C. Graziani et al: ApJ **602**, 875 (2004)
8. B. Zhang, P. Mészáros: ApJ **581**, 1236 (2002)
9. B.E. Schaefer: ApJ **583**, L71 (2003)
10. F. Ryde: ApJ **625**, L95 (2005)
11. M.J. Rees, P. Meśzaŕos: ApJ **628**, 847 (2005)
12. D.A. Frail, S.R. kulkarni, R. Sari et al: ApJ **562**, L55 (2001)
13. D.Q. Lamb, T.Q. Donaghy, C. Graziani: ApJ **620**, 355 (2005)
14. E. Rossi, D. Lazzati, M.J. Rees: MNRAS **332**, 94 (2002)
15. B. Zhang, X. Dai, N.M. Lloyd–Ronning, P. Meszáros: ApJ **601**, L119 (2004)
16. R. Yamazaki, K. Ioka, T. Nakamura: ApJ **593**, 941 (2003)
17. J. Granot, A. Panaitescu, P. Kumar, S.E. Woosley: ApJ **570**, L61 (2002)
18. D. Eichler, A. Levinson: ApJ **614**, L13 (2004)
19. R. Yamazaki, D. Yonetoku, T. Nakamura: ApJ **594**, L79 (2003)
20. E. Ramirez–Ruiz, J. Granot, C. Kouveliotou et al: ApJ **625**, L91 (2005)
21. S. Dado, A. Dar: ApJ **627**, L109 (2005)
22. A. Dar, A. De Rujula: Phys. Rep. **405**, 203 (2004)
23. G. Ghirlanda, G. Ghisellini, A. Celotti: A&A **422**, L55 (2004)
24. J–L. Atteia: A&A **407**, L1 (2003)
25. A. Pelangeon, The Hete-2 Science Team: AIPC **836**, 149 (2006)

Particle Acceleration and Radiative Losses at Relativistic Shocks

P. Dempsey and P. Duffy

UCD School of Mathematical Sciences, University College Dublin, Belfield, Dublin 4, Ireland,
(paul.dempsey,peter.duffy)@ucd.ie

Summary. A semi-analytic approach to the relativistic transport equation with isotropic diffusion and consistent radiative losses is presented. It is based on the eigenvalue method first introduced in Kirk & Schneider [5] and Heavens & Drury [3]. We demonstrate the pitch-angle dependence of the cut-off in relativistic shocks.

1 Background and Methods

In the late 80's Kirk & Schneider [5] and Heavens & Drury [3] introduced an eigenfunction approach to the relativistic transport equation for first order Fermi acceleration. The method was initially restricted to low Lorentz factors, $\Gamma < 5$, but was amended slightly in [4] to allow for much larger Γ-factors. We use a similar method to examine acceleration in the presence of synchrotron losses. The equation of interest is:

$$\Gamma\left(u+\mu\right)\frac{\partial f}{\partial z} = \frac{\partial}{\partial \mu}\left(D_{\mu\mu}\frac{\partial f}{\partial \mu}\right) + \lambda\frac{1}{p^2}\frac{\partial(p^4 f)}{\partial p} \tag{1}$$

where $\lambda = 4\sigma_T U_B/3$ for synchrotron losses in a tangled magnetic field. This holds separately upstream and downstream and we impose a boundary condition at the shock to get the full solution. When considering the losses we let $W = p^4 f$ and assume $W(p \to \infty) = 0$. Considering only momentum independent diffusion, we can now take the Laplace transform with respect to $1/p$ to give:

$$\Gamma\left(u+\mu\right)\frac{\partial R}{\partial z} = \frac{\partial}{\partial \mu}\left(D_{\mu\mu}\frac{\partial R}{\partial \mu}\right) - \lambda k R \tag{2}$$

where k is the Laplace variable and R is the Laplace transform of W. Now we write

$$R = \sum_i a_i(k)k^{-s+3} \exp\left(\frac{\Lambda_i(k)z}{\Gamma}\right) Q_i(\mu, k) \tag{3}$$

where Q_i is an eigenfunction with associated eigenvalue Λ_i which is a solution to

$$\frac{\partial}{\partial\mu}\left(D_{\mu\mu}\frac{\partial Q_i(k,\mu)}{\partial\mu}\right) = (\Lambda_i(k)(u+\mu)+\lambda k)Q_i(k,\mu) \tag{4}$$

$Q_i(k=0,\mu)$, $\Lambda_i(k=0)$ are solved as in [4] and the Q_i's are normalised such that

$$\int_{-1}^{1}(u+\mu)\,Q_iQ_j\,d\mu = \delta_{i,j}\,\mathrm{sign}\left(\frac{1}{2}-j\right) \equiv \eta_{i,j} \tag{5}$$

We solve for $k>0$ using the coupled differential equations

$$\frac{\partial Q_i}{\partial x} = \lambda\sum_{j\neq i}\frac{1}{\Lambda_j-\Lambda_i}\left(\int_{-1}^{1}g(\mu)Q_iQ_j\,d\mu\right)Q_j\,\eta_{j,j} \tag{6}$$

$$\frac{d\Lambda_i}{dx} = -\lambda\left(\int_{-1}^{1}g(\mu)Q_iQ_i\,d\mu\right)\eta_{i,i} \tag{7}$$

2 Results

In these proceedings we only show results for Fermi acceleration with synchrotron losses in a shock velocity profile using the Jutter-Synge equation of state in the hydrodynamic limit. If we relate the upstream and downstream k's by $k_{up}p_d = k_dp_{up}$ we can derive a matching condition [2].

$$\Gamma_{rel}^3(1+u_{rel}\mu_{up})^3\sum_{i>0}a_i^{up}(k_{up})k_{up}^{-s+3}Q_i^{up}(\mu_{up},k_{up})$$

$$= \sum_{i\leq0}a_i^d(k_d)k_d^{-s+3}Q_i^d(\mu_d,k_d) \tag{8}$$

To arrive at the solution we expand $a_i(k) = \sum_{j\geq0}a_{i,j}k^j$ and solve at $k=0$ for the derivatives of Eqn.(8). This will give s and all the $a_{i,j}$. The k^{-s+3} was factored out so that when we have no losses we recover the correct result. Having demonstrated that this method is successful its application for acceleration in GRBs & AGNs will be discussed in a forthcoming paper.

2.1 Validity of Matching Condition

The matching condition described in equation (8) is valid only if the eigenvalues $\Lambda_i(k)$ don't switch sign. Since these are continuous they can only switch sign if for some k^* we have $\Lambda_i(k^*)=0$. However for isotropic diffusion this implies

$$\frac{\partial}{\partial\mu}\left(D(1-\mu^2)\frac{\partial Q_i(k^*,\mu)}{\partial\mu}\right) = \lambda k^*Q_i(k^*,\mu) \tag{9}$$

This only has a continuous solution if $\lambda k^*/D = -l(l+1)$ for some natural number l (Legendre polynomials), which can never be true for $k^*>0$.

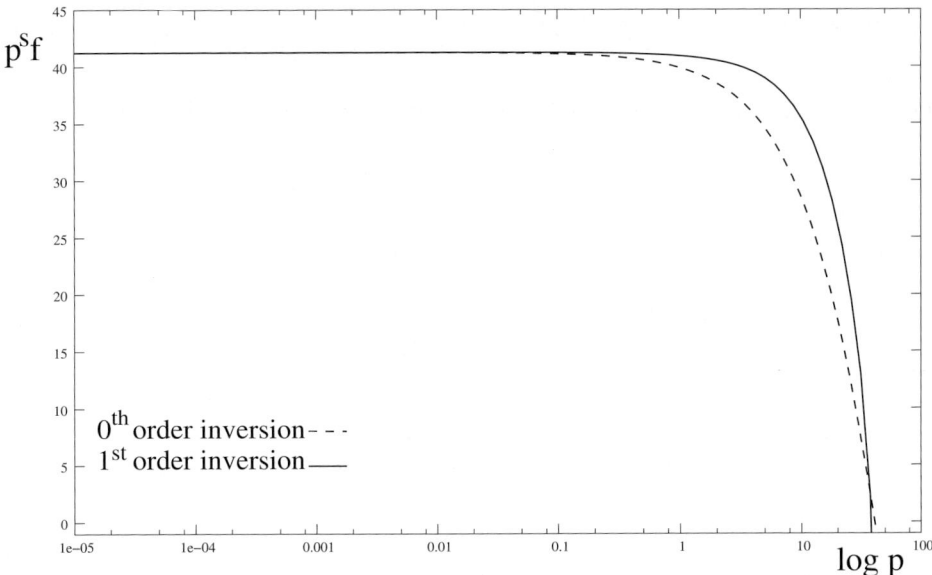

Fig. 1. The 0th and 1st order Post Widder inversion are plotted for $p^s f$ versus $\log p$ using 2 eigenfunctions. This plot used the parameters $u_{up} = .95$, $D_{\mu\mu} = (1 - \mu^2)$ and $\lambda = .025$ at $\mu = -1$.

2.2 Laplace Inversion

Since expressions can be derived for all derivatives of $Q_i(k)$ it proves useful to do the Laplace inversion using a Post-Widder Method [1]. By this method the zeroth order approximation to the solution is

$$f_0(p, \mu, 0) = p^{-s} \sum_i (a_i(x)Q_i(x)) \bigg|_{x=p} \tag{10}$$

while the first order approximation is

$$f_1(p, \mu, 0) = p^{-s} \left(\sum_i a_i(x)Q_i(x) + \frac{x}{3-s} \sum_i \frac{\partial}{\partial x}(a_i(x)Q_i(x)) \right) \bigg|_{x=2p} \tag{11}$$

Acknowledgments. Paul Dempsey would like to thank the Irish Research Council for Science, Engineering and Technology for their financial support. He would also like to thank Cosmogrid for access to their computational facilities.

References

1. Abate J. & Whitt W.: *INFORMS Journal on Computing,***Vol 7**, No. 1, 36 (1995)

2. Dempsey P. & Duffy P., 2006, *in preparation*.
3. Heavens A.F. & Drury L.O'C.: *Mon. Not. R. astr. Soc.*, **235**, 997 (1988)
4. Kirk J.G., Guthmann A. W., Gallant Y.A. & Achterberg A.: *ApJ*, **542**, 235 (2000)
5. Kirk J.G. & Schneider P.: *ApJ*, **323**, L87 (1987)

The Swift Gamma-Ray Burst Mission: First Results

N. Gehrels[1] on behalf of the Swift Team

NASA/Goddard Space Flight Center, Code 661, Greenbelt, Maryland, 20771, USA
gehrels@milkyway.gsfc.nasa.gov

Summary. Since its launch on 20 November 2004, the Swift mission is detecting 100 new gamma-ray bursts (GRBs) each year, and immediately (within tens of seconds) starting simultaneous X-ray and UV/optical observations of the afterglow. It has already collected am impressive database of bursts, including prompt emission to higher sensitivity than BATSE, uniform monitoring of afterglows, and rapid follow-up by other observatories notified through the GCN.

1 INTRODUCTION

Despite impressive advances over the roughly three decades since GRBs were first discovered [1], the study of bursts remains highly dependent on the capabilities of the observatories which carried out the measurements. The era of the Compton Gamma Ray Observatory (CGRO) led to the discovery of more than 2600 bursts in just 9 years. Analyses of these data led to the conclusion that GRBs are isotropic on the sky and occur at a frequency of roughly two per day all sky [2].

The BeppoSAX mission made the critical discovery of X-ray afterglows [3]. With the accompanying discoveries by ground-based telescopes of optical [4] and radio [5] afterglows, GRBs could start to be studied within the astrophysical context of identifiable objects in a range of wavelength regimes. Successful prediction of the light curves of these afterglows across the electromagnetic spectrum has given confidence that GRBs are the signal from extremely powerful explosions at cosmological distances, which have been produced by extremely relativistic expansion [6].

The Swift mission selected by NASA in 1997 combines the sensitivity to discover new GRBs with the ability to point high sensitivity X-ray and optical telescopes at the location of the new GRB as soon as possible. From this capability Swift has the goal to answer the following questions:

1. What causes GRBs?
2. What physics can be learned about black hole formation and ultra-relativistic outflows?
3. What is the nature of subclasses of GRBs?
4. What can GRBs tell us about the early Universe?

The general operations of the Swift observatory are as follows. The wide-field Burst Alert Telescope (BAT) detects the bursts in the 15-350 keV band and determines the position to a few arcminute accuracy. The position is provided to the spacecraft, built by Spectrum Astro General Dynamics, which repoints to it in less than 2 minutes. Two narrow-field instruments then observe the afterglow: the X-Ray Telescope (XRT) and UV-Optical Telescope (UVOT). Alert data from all three instruments is sent to the ground via NASA's TDRSS relay satellite. The full data set is stored and dumped to the Italian Space Agency's Malindi Ground Station.

Swift was built by an international team from the US, UK, and Italy. After five years of development it was launched form Kennedy Space Center on 20 November 2004. The spacecraft and instruments were carefully brought into operational status over an eight-week period, followed by a period of calibration and operation verification, which ended with the start of normal operations on 5 April 2004. A complete description of the Swift mission can be found in [7].

As of 30 November 2005, the Swift achievements include: discovery of 92 new GRBs by the Swift Burst Alert Telescope (BAT) instrument (with a typical error region of less than 2 arcmin radius); observation of 72 X-ray afterglows by the Swift X-Ray Telescope (XRT) instrument (with a typical error region of less than 3 arcsec radius); and observations of 20 afterglows by the Swift UV-Optical Telescope (UVOT) instrument (with a typical error region of less than 1 arcsec). More than half of the afterglow observations start within two minutes of the BAT GRB trigger with a record of only 54 seconds. Afterglow data have been obtained from non-Swift discovered bursts with typical response times of 3 hours.

2 SWIFT HIGHLIGHTS

2.1 BAT Detected GRBs

The BAT [8] on Swift has detected 77 GRBs between when it was turned on in mid-December 2004 and 30 November 2005. Thus in 345 days of operation, the BAT has detected GRBs at a rate of about 97 bursts per year. This value is close to the rate of 100 bursts estimated prior to launch.

Spectral analysis of the BAT bursts shows them to be consistent with the population of GRBs seen by the Compton Gamma-Ray Observatory BATSE experiment, both in the ratio of the fluxes in the 25-50 keV and 50-100 keV energy bands, and in flux and duration distributions.

2.2 XRT Detected GRBs

The XRT [9] is performing the first rapid-response observations of the X-ray afterglow of GRBs. In the first 80 cases, all but 5 of the BAT GRB triggers resulted in detection of an X-ray counterpart for the BAT source. In 3 cases the XRT observations started while the BAT was still detecting hard X-ray prompt emission from the GRB.

The Swift afterglow observations are rapid, with more than half of the observations started in less than 300 seconds after the burst. When XRT arrives this quickly it is very common to see a fast X-ray decline within the first 300 seconds.

In addition to the BAT detected events, Swift can also observe GRBs discovered by other satellites. Swift has discovered X-ray afterglow emission in 4 cases for HETE-2 and 5 cases INTEGRAL. In a particularly impressive case, Swift was able to respond to the ground control commands and start observations of the HETE-2 GRB050408 within 40 minutes of the GRB.

2.3 UVOT Detected GRBs

The UVOT [10] is co-aligned with the XRT and so observes the GRB afterglows just as promptly as the XRT. Despite these prompt observations the UVOT has detected far fewer UV/optical counterparts than the XRT.

Of the first 68 GRBs observed by the UVOT, only 20 had detected emissions. The UVOT has generated important upper limits for these early times, which are lower than those for bursts studied by previous missions. Combining UVOT plus ground-based observations 40 optical afterglows have been detected.

Reasons for this low number include the possibility that the Swift bursts are more distant than previous bursts; that a substantial number of GRBs have intrinsic dust extinction which suppresses the optical/UV emission compared to the I and R bands typically reported for earlier afterglows; or that some afterglows come from high magnetic field regions in the outflow which suppresses the optical and UV emission. These possibilities are discussed in [11].

Although not every GRB produces UV or optical flux, which can be detected by the UVOT, several bursts have produced interesting early time light curves, including GRB050318 [12] and GRB050319 [13].

2.4 XRT Early Light Curve Behavior

Swift has opened up a new regime for GRB afterglow studies. Never before has it been possible to study the X-ray behavior on timescales of minutes after the GRB happens. Swift has frequently started observations within a few minutes of the detection of GRBs by the BAT (with a record of only 52 seconds).

These extremely prompt observations have given rise to new findings. In roughly 40% of the cases, the X-ray afterglow can be characterized by a three-part light curve (see Figure 1 and 2). First comes an extremely rapid decay of a very bright source. At these early times the decay can be fit by a power law of index in the range of 2.5 or greater. After a few minutes the decay rate flattens, and we can fit it with an index approximately equal to 1 (plus or minus 0.5). Finally after a delay ranging from hours to days, the decay rate steepens again, sometime resulting in a behavior interpreted as a jet break [14] and [15]. Tagliaferri et al. [16] and Barthelmy et al. [17] each consider two early XRT afterglows. They show that the X-ray emission during the prompt phase (estimated from extrapolation of the BAT spectrum) connects to the bright early XRT afterglow (see Figure 3). This suggests that the bright early afterglow is an extension of the prompt phase.

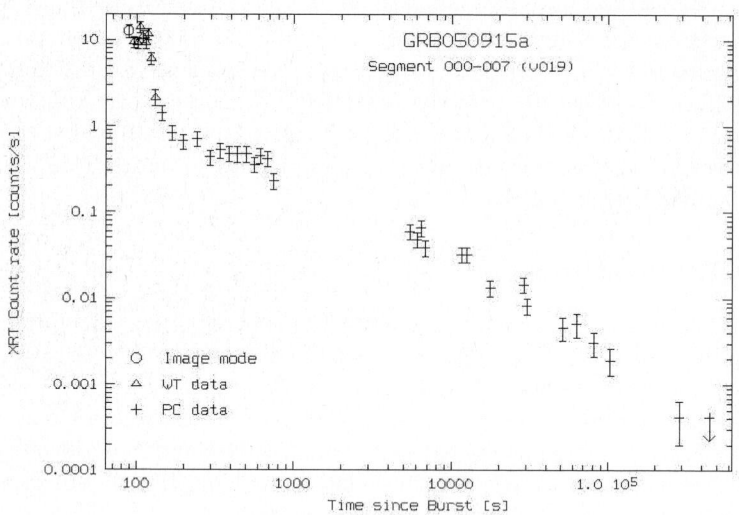

Fig. 1. Typical X-ray afterglow light curve.

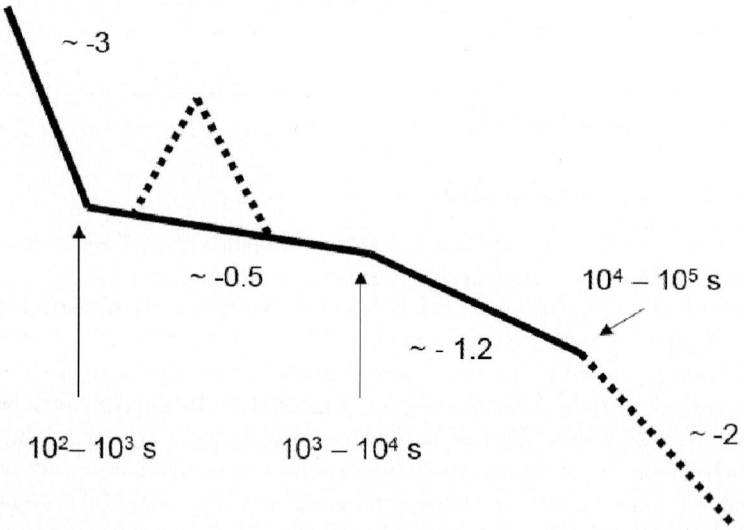

Fig. 2. Generic lightcurve observed by XRT. Numbers shown next to the segment are the power-law index of the time decay. From Zhang et al. (2006).

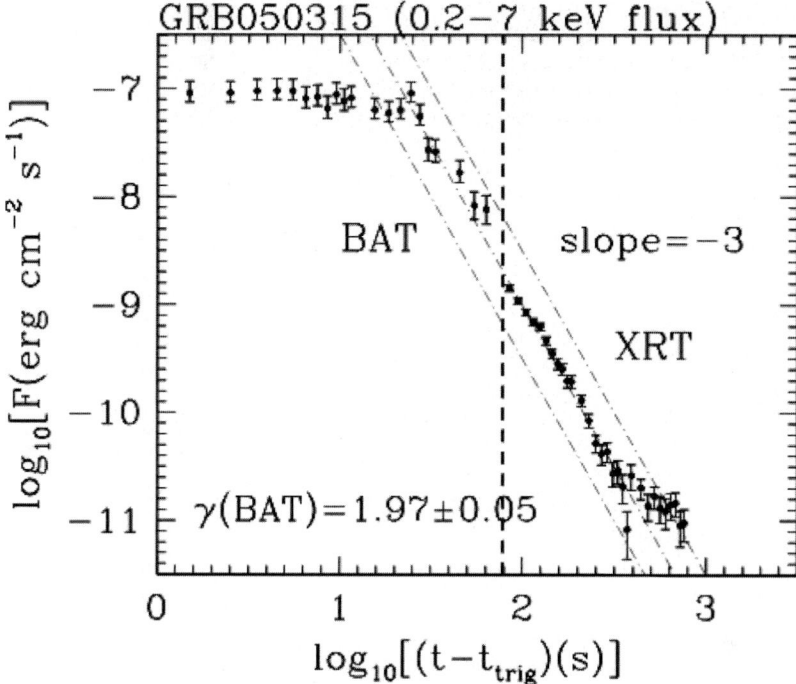

Fig. 3. The BAT spectrum is extrapolated to the 0.2 7 keV band. The early XRT lightcurve connects smoothly to the prompt emission. From Barthelmy et al. [17]

Swift also detects strong X-ray flares in afterglows at early times. In one case (GRB050502b) the X-ray flux increased by a factor of roughly 100. The dramatic flaring events seem to be superposed on a background, which follows the multipart behavior mentioned above. Burrows et al. [18] discuss the flaring behavior seen in GRB050502b and GRB050408. Flares are seen in 20%-30% of the observed afterglow.

2.5 Short GRBs

As of late November 2005, the BAT has detected 7 GRBs in the short-hard class. The first one of them (GRB 050202) had no prompt slew and no counterparts. From the next 2 events we have learned a great deal. GRB 050509b [19] had an X-ray afterglow that gave an error circle with a bright elliptical galaxy (cD galaxy in a cluster) in it (Figure 4). GRB 050724 [20] had an XRT afterglow, plus Chandra, optical and radio detections. The sub-arcsecond positions located it once again in the outer regions of a bright elliptical. The fact that these ellipticals have very low star formation rates argues strongly against a collapsar origin like that for long bursts. Also, the redshifts for the two are in the z = 0.2 to 0.3 range, a factor of 3 closer than typical long GRBs. The evidence to date is consistent with an origin of short burst in merging binary neutron stars.

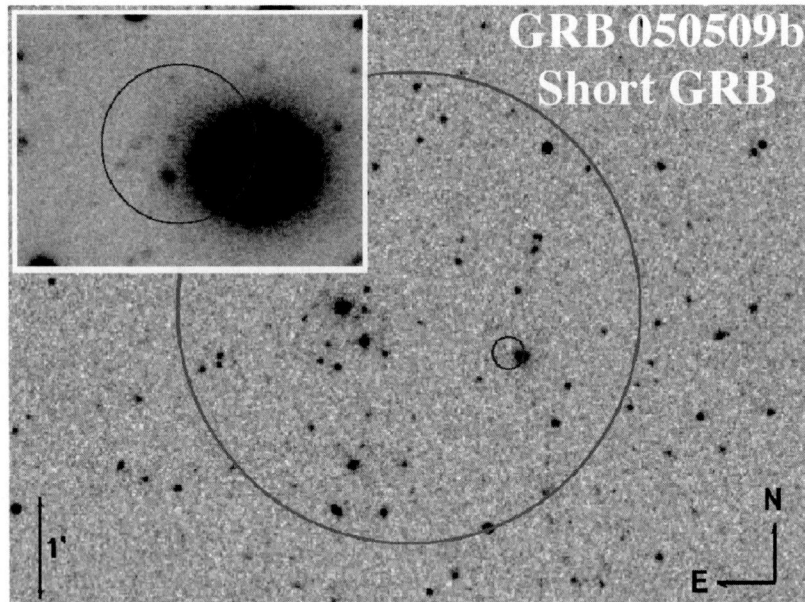

Fig. 4. Localization of short GRB 050509b. Large circle is BAT position; small circle is XRT position. The inset shows a bright elliptical galaxy in the XRT circle. From Gehrels et al. [19]

The picture for the last 4 short GRBs seen by Swift is still not well understood. GRB050813 appears to have a faint host at z=1.8. GRBs 050906 and 051105A did not have XRT detections despite rapid slews. GRB 050925 was in the galactic plane and had a soft spectrum; it may be a new galactic SGR.

2.6 GRB Redshifts

As of late November 2005, redshifts have been determined for 23 Swift GRBs. The average redshift (excluding short GRBs) is z=2.5. This is significantly higher than the pre-Swift average of z=1.2. The sensitivity of the Swift instruments is leading to a sampling of more distance GRBs.

On September 4, 2005, Swift detected a long, smooth GRB [21]. The redshift was found to be the very large value of z =6.29, one of the highest redshift objects ever seen. The light curve for this GRB is shown in Figure 5.

2.7 Giant Flare from SGR 1806-2

On 27 December 2004, the Solar System was struck by the brightest gamma-ray transient ever observed. Every orbiting gamma-ray observatory detected the flash produced by the soft gamma-ray repeater SGR 1806-20. Although Swift was not pointed toward the target, the flux was so high that the BAT detector was swamped by more than a billion gamma-rays per cm^2 passing through the structure of the spacecraft.

Swift/BAT Light curves

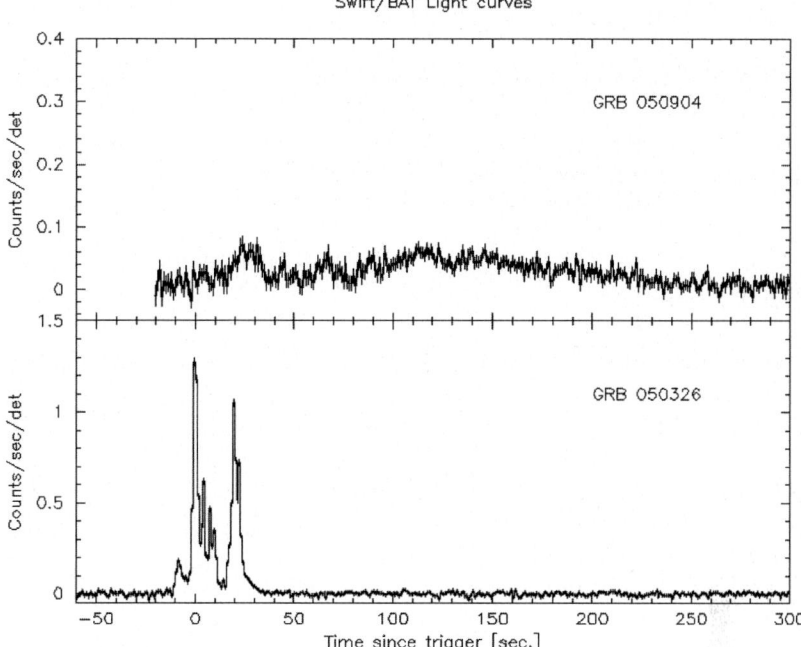

Fig. 5. Lightcurve of high-redshift GRB 050904 compared to a typical GRB. The long smooth nature of the lightcurve is due to cosmologic time dilation as the photons propagated to us from z=6.29.

Palmer et al. [22] present the Swift data on this dramatic event. Although the emitting system is located more than 10 kpc from the Earth, the received energy flux was brighter than the full Moon for the 0.2 seconds. This giant flare was more than 100 times more luminous than the two previous flares seen in 1998 from SGR1900+14 and in 1979 from SGR0526-66.

Such events may be the cause of at least some short GRBs, in that the rapid, extremely bright flash of gamma rays had a similar duration and energy profile to a short GRB. Such an event in an external galaxy would be detectable out to 60 Mpc.

2.8 UV/Optical & X-ray Observations of SN2005am

Type Ia supernovae are critical to our understanding of the fundamental fabric of our Universe. They are the standard candles used to measure distances over the range in which cosmological effects become significant. Observations of nearby supernovae in the ultraviolet can be quite important for understanding and calibrating their light curves and luminosities.

Missions with UV capability such as the International Ultraviolet Explorer (IUE) and the Hubble Space Telescope began these studies, but they are limited in the intrinsically slower operational response time than offered by Swift. Thus Swift has been an ideal observatory for early observations of nearby bright supernovae, of which SN2005am is a prime example.

Brown et al. [23] present ultraviolet and optical light curves for SN2005am, starting four days prior to maximum light, and extending to 69 days after peak. In addition, when the target was bright enough, Swift was able to carry out moderate-resolution grism UV/optical measurements. These data for SN2005am are the best sampled in time, and cover the widest range of any Type Ia supernova follow-up to date.

3 CONCLUSIONS

The Swift observatory is performing excellent scientific observations at high efficiency and with important progress toward its mission objectives.

The BAT is working flawlessly. The positional agreement with the XRT and ground-based detections suggests that the typical on-board positional accuracy for GRBs is roughly 65 arcsec, exceeding the pre-launch predictions.

The UVOT has demonstrated excellent UV and optical performance on GRBs and other sources. Source positions are accurate to 0.3 arcsec.

The XRT has demonstrated excellent X-ray sensitivity and rapid responsiveness. The average accuracy for the XRT positions confirmed with XMM or ground-based optical detection is 2.6 arcsec. XRT is observing afterglows at a level of 100 to 1000 times fainter than Beppo-SAX. This rapid acquisition with sensitive X-ray detection is revealing new lightcurve behaviors.

As Swift observations become more numerous we expect to build up a substantial database of prompt gamma-ray emission and early (to late) X-ray and UV/optical light curves. From these, new insights into GRB formation and GRB environments will be gleaned.

References

1. R.W. Klebesadel, I.B. Strong & R.A. Olson: ApJL, **182**, L85 (1973)
2. M.S. Briggs: ApJ, **459**, 40 (1996)
3. E. Costa, et al.: Nature, **387**, 783 (1997)
4. J. van Paradijs, et al.: Nature, **386**, 686 (1997)
5. D.A. Frail, et al.: Nature, **389**, 261, (1997)
6. R.A.M.J. Wijers, M.J. Rees & P. Meszaros: MNRAS, **288**, L51 (1997)
7. N. Gehrels, et al.: ApJ, **661**, 1005 (2004)
8. S. Barthemly, et al.: Sp Sci Rev,**120**, 143 (2005a)
9. D.N. Burrows, et al.: Sp. Sci. Rev., **120**, 165 (2005a)
10. P. Roming, et al.: Sp. Sci. Rev., **120**, 95 (2005a)
11. P. Roming, et al.: ApJ, submitted, (astro-ph 0511751) (2005b)
12. M. Still, et al.: ApJ, **635**, 1187 (2005)
13. K. Mason, et al.: ApJ, **639**, 311-315 (2006)
14. J. Nousek, et al.: ApJ, **642**, 389-400 (2006)
15. B. Zhang, et al.: ApJ, **642**, 354-370 (2006)
16. G. Tagliaferri, et al.: Nature, **436**, 985 (2005)

17. S. Barthemly, et al.: ApJ, **635**, L133-L136 (2005b)
18. D.N. Burrows, et al.: Sci, **309**, 1833 (2005b)
19. N. Gehrels, et al.: Nature, **437**, 851 (2005)
20. S. Barthemly, et al.: Nature, **438**, 994 (2005c)
21. G Cusumano, et al.: Nature, **440**, 7081, 164 (2006)
22. D. Palmer, et al.: Nature, **434**, 1107 (2005)
23. P.J. Brown: ApJ, **635**, 1192 (2005)

The Afterglow of the Gamma-Ray Burst 050502a: First Case of an Early (<1 hr) Multi–Colour Detection

C. Guidorzi[1], A. Monfardini[1,2], A. Gomboc[1,3],, C.G. Mundell[1], I.A. Steele[1], D. Carter[1], M.F. Bode[1], R.J. Smith[1], C.J. Mottram[1], M.J. Burgdorf[1], N.R. Tanvir[4], N. Masetti[5], and E. Pian[6]

[1] Astrophysics Research Institute, Liverpool John Moores University, Twelve Quays House, Egerton Wharf, Birkenhead, CH41 1LD, U.K.,
(crg,am,cgm,ias,dxc,mfb,rjs,cjm,mjb)@astro.livjm.ac.uk
[2] ITC–IRST and INFN, Trento, via Sommarive, 18 38050 Povo (TN) Italy
[3] FMF, University in Ljubljana, Jadranska 19, 1000 Ljubljana, Slovenia,
andreja.gomboc@fmf.uni-lj.si
[4] Centre for Astrophysics Research, University of Hertfordshire, Hatfield AL10 9AB, UK,
nrt@star.herts.ac.uk
[5] INAF - Istituto di Astrofisica Spaziale e Fisica Cosmica, Sezione di Bologna, Via Gobetti 101, I-40129 Bologna, Italy, masetti@bo.iasf.cnr.it
[6] INAF - Osservatorio Astronomico di Trieste, Via G.B. Tiepolo 11, 34131 Trieste, Italy, pian@ts.astro.it

1 Introduction

Gamma-Ray Bursts (GRBs) are detected as sudden flashes of gamma-rays seen at random locations, outshining the overall gamma-ray sky. The typical average rate is a few/day (1–3 for an open-sky detector, depending on the sensitivity) and the duration ranges from a fraction of seconds (typically few tenths for the so-called "short bursts") up to tens (seldom up to hundreds or, rarely, a thousand) of seconds. Originally discovered in 1967, only since 1997 it has been possible thanks to the Italian-Dutch *BeppoSAX* spacecraft to detect a long-lived fading counterpart at other wavelengths, called the *afterglow* [1]. In most cases the discovery of the counterpart in the X-rays and then in the optical bands allowed the measurement of redshift directly from the optical transient spectrum or, indirectly, through the detection of the host galaxy. Recently the *Swift* mission [2] greatly improved the rate of GRBs with measured redshift thanks to its capability of distributing fast and precise positions (arcmin-sized a few seconds and arcsec-sized a few minutes after the burst onset time) to ground facilities. The $\sim$70 GRBs with measured redshift so far detected (January 2006) turned out to lie at cosmological distances: the median redshift, $z_{\mathrm{med}} = 1.2$ (see Fig. 1).

It is now believed that long-duration GRBs are connected with the collapse of massive stars (e.g. [3]): at least three relatively nearby GRBs (980425, 030329 and 031203) have unambiguously been associated with simultaneous SNe. According to the fireball model, the gamma rays are thought to be produced by the mutual interaction of rela-

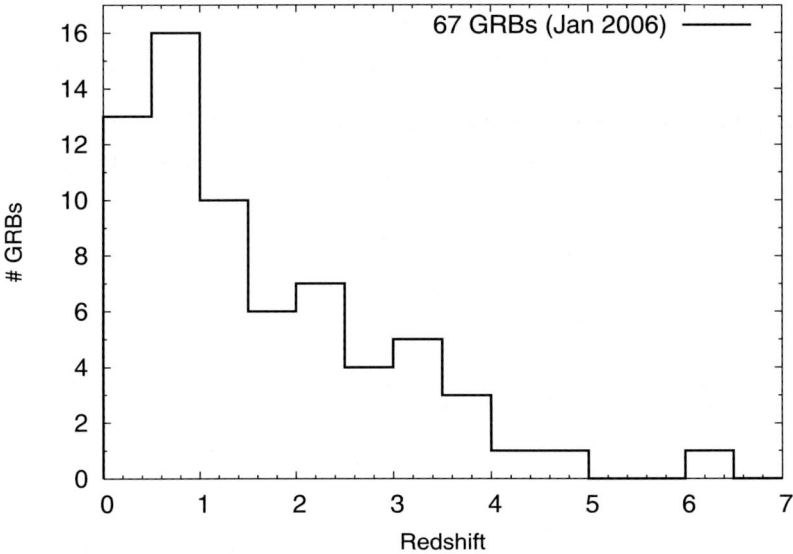

Fig. 1. Redshift distribution of 67 GRBs with measured redshift (January 2006).

tivistic expanding shells (Γ up to $\sim$100) (*internal shocks*), while the afterglow emission seen later at longer wavelengths is the result of the interaction between the expanding shells and the surrounding medium (*external shocks*).

Usually GRB afterglows decay following power laws, although there are a number of cases showing deviations, i.e. breaks in the power-law index, erratic variability, bumps, from X-rays to optical and IR wavelengths. Early observations of a GRB afterglow are very useful in probing the circumburst medium, thus providing strong clues to the nature of the progenitor.

2 GRB 050502a: first early multi-colour light curve

GRB 050502a was a 20-s long burst detected by *INTEGRAL* on May 2, 2005, at 02:13:57 UT [4] and followed up robotically by the 2-m Liverpool Telescope (LT) located in La Palma 3 minutes after the burst onset time. The LT automatic real-time GRB pipeline [5] discovered a new fading source initially of magnitude $r' = 15.7$ [6]. The LT kept observing it in four different filters ($BVr'i'$) for an hour. These observations represent the first case of a multi-colour light curve in the first hour of the burst, thus providing unprecedented early spectral information on the evolution of a GRB optical afterglow. Late follow-up observations were performed using the 2-m Faulkes Telescope North (FTN) located in Hawaii, which is, in addition to the LT, a member of the *RoboNet-1.0* consortium [7]. The redshift was determined spectroscopically: $z = 3.793$ [8].

Figure 2 shows the multi-colour light curve acquired with LT and FTN. A power-law index of $\alpha = 1.2 \pm 0.1$ ($F(t) \sim t^{-\alpha}$) fits each single colour light curve. However, evidence for a simultaneous bump is present in all of the four bands at $t \sim 0.02$ days. Guidorzi et al. [9] interpreted this feature as the signature of a uniform circumburst medium with clumps in density. The bump would be the result of the impact of the expanding shells on clumps: this gives rise to a sudden flux increase relaxing asymptotically. However, other interpretations of the bump such as the result of later shells catching up with the front shock are not ruled out, although less favoured.

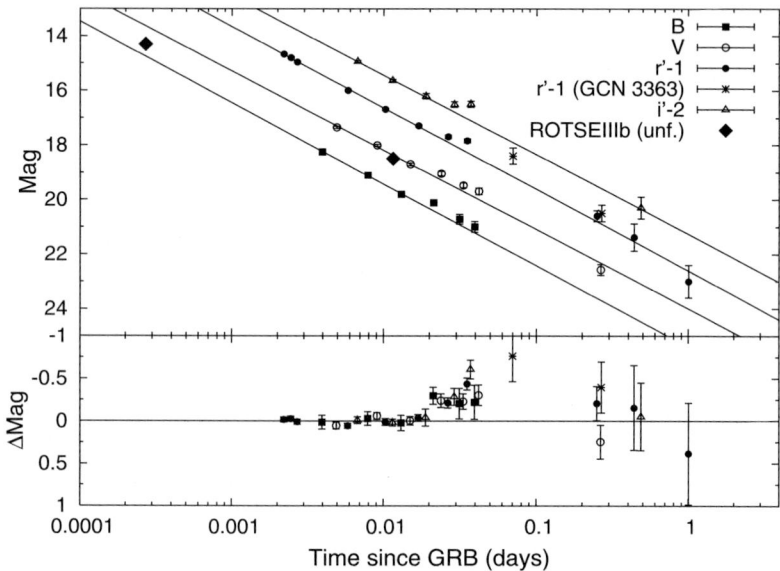

Fig. 2. *Top Panel*: Multi–colour light curve of GRB 050502a measured with the Liverpool and Faulkes North Telescopes. Also shown are the best-fit power laws: all of them are consistent with a power–law index of 1.2 ± 0.1 (see text). Two ROTSE–IIIb unfiltered points (Yost et al. [10]) and two r' points derived from Mirabal et al. ([11]) are plotted as well. *Bottom Panel*: residuals with respect to the best-fitting power laws. (Fig. taken from [9]).

References

1. Costa E. et al.: Nature, **387**, 783 (1997)
2. Gehrels N. et al.: ApJ, **611**, 1005 (2004)
3. Paczyński B.: ApJ, **494**, L45 (1998)
4. Götz D., Mereghetti S., Mowlavi N., Shaw S., Beck M. and Borkowski J.: GCN Circ., **3323** (2005)
5. Guidorzi C. et al.: PASP, **118**, 840, 288-296 (2006)
6. Gomboc A., Steele I.A., Monfardini A., Mottram C.J., Guidorzi C., Bode M.F. and Mundell C.G.: GCN Circ., **3325** (2005)
7. Gomboc A. et al.: Nuovo Cimento C, Vol. **028**, Issue 04-05, 723 (2005)
8. Prochaska J.X., Ellison S., Foley R.J., Bloom J.S. and Chen H.-W.: GCN Circ., **3332** (2005)

9. Guidorzi C. et al.: ApJ, **630**, L121 (2005)
10. Yost S. A., Swam H., Schaefer B. A., and Alatalo K.: GCN Circ., **3322** (2005)
11. Mirabal N., Boettcher M., Shields J., Joshi M., and Halpern J. P.: GCN Circ., **3363** (2005).

Kinetic Plasma Simulations of GRB Fireball Collisions: Synchrotron Features

C.H. Jaroschek[1], H. Lesch[1], and R.A. Treumann[2]

[1] Institut für Astronomie und Astrophysik der LMU, Munich, Germany
(cjarosch,lesch)@usm.uni-muenchen.de
[2] MPI für extraterrestrische Physik, Garching, Germany art@mpe.mpg.de

GRB polarimetry: Constraints on fireball models

Time-resolved polarimetry of GRB prompt emission and afterglow is an excellent tool to constrain the physical nature of the GRB emission mechanism. High linear polarizations of $\Pi > 50\%$ for prompt emission [1, 2] in combination with the required high radiative efficiencies strongly favour a synchrotron emission scenario. Associated spectra show power-law tails at high frequencies with Band indices of $\alpha \simeq -1.2$ and $\beta \simeq -2.5$, respectively. These features support the GRB fireball model, i.e. a highly variable outflow of non-thermal pair plasma collimated into fireball shells and moving at relativistic speeds. Colliding fireball shells feature the Coupled Two-Stream Weibel Instability (CTW) as the pervasive kinetic plasma mode constituted (CT) by the electrostatic Two-Stream (TSI) and the electromagnetic Weibel Instability (WBI).

$$CTW = TSI \times CT + WBI \tag{1}$$

Particle-In-Cell simulations show the efficient generation of magnetic fields $B_\perp$ (Fig. 1) perpendicular to the plasma bulk motion by the WBI. Near-equipartition Weibel magnetic fields in fireball collisions are the prerequisite of synchrotron emission scenarios.

Turbulent magnetic fields and synchrotron emission

Self-consistent kinetic plasma simulations of Weibel turbulence during fireball collisions [3, 4] identify the connection between equipartition ratio and collision energy, which is rooted in the typical coherence length of magnetic fields. Intriguingly, Weibel turbulence evolves in an 'Inverse Cascade' proceeding towards smooth power-law spectra on kinetic time scales (Fig. 2). The power-law index appears as a characteristic of the collision energy γ_0 assuming s=-1 for full 3D turbulence. This result suggests analogies to the 'viscosity damped regime' observed in simulations of 3D MHD turbulence [5]. Synchrotron emission proceeds on typical cooling times τ_{sc} far beyond kinetic time scales of the CTW instability and diffusive particle scattering in

$$\tau_{sc}\,\omega_{p0} = 3.5 \cdot 10^{17}\,\gamma_0^{-1}(\gamma_0 - 1)^{-1}\,(n/1\,\mathrm{cm}^3)^{-1/2}\,\epsilon_B^{-1/2}\,(m/m_e)^{-3/2} \qquad (2)$$

momentum space. Consequently self-similar solutions of diffusive particle scattering as obtained in [6] appear applicable. Then the turbulent Weibel fields directly correspond to power-law shapes in the synchrotron spectra. Simulations indicate the validity of

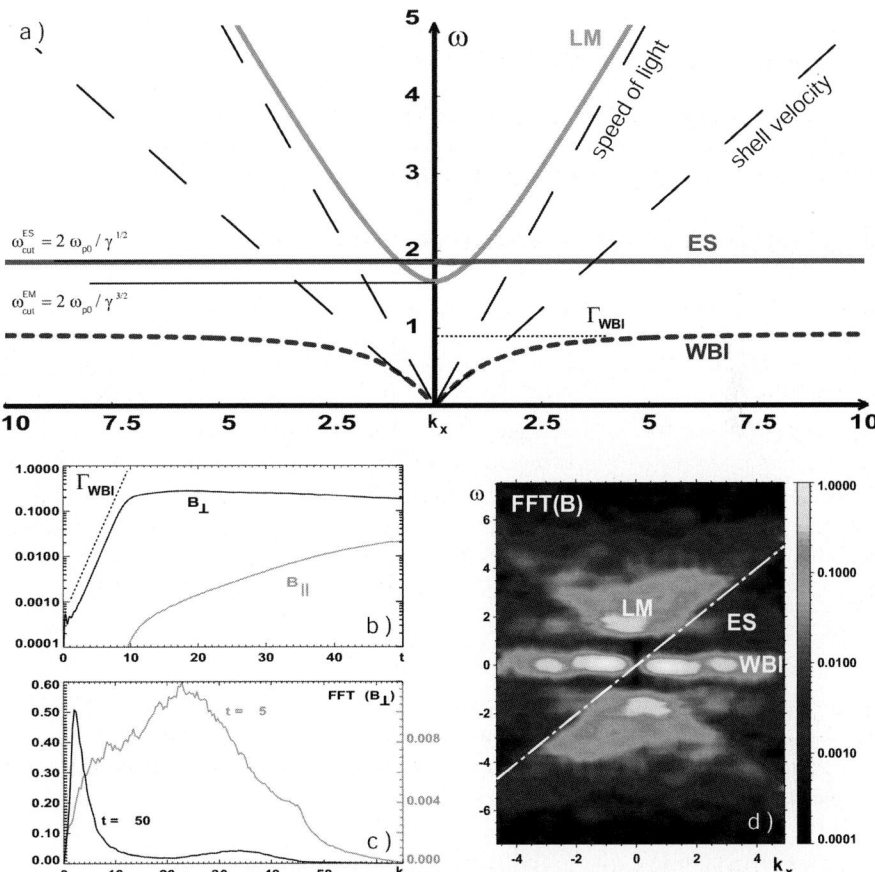

Fig. 1. Dispersion characteristics of the Weibel Instability (WBI): Linear theory (a) identifies WBI as purely growing mode. Non-linear evolution (b) and typical coherence lengths in Fourier space (c,d) are accessible in self-consistent Particle-In-Cell simulations.

the synchrotron 'jitter' mechanism [7] and the associated high degrees of polarization. Turbulent field spectra are linked to typical coherence lengths, which are also the key to understand the complementary data from afterglow observations indicating low, but sustained polarizations on the few %-level [8].

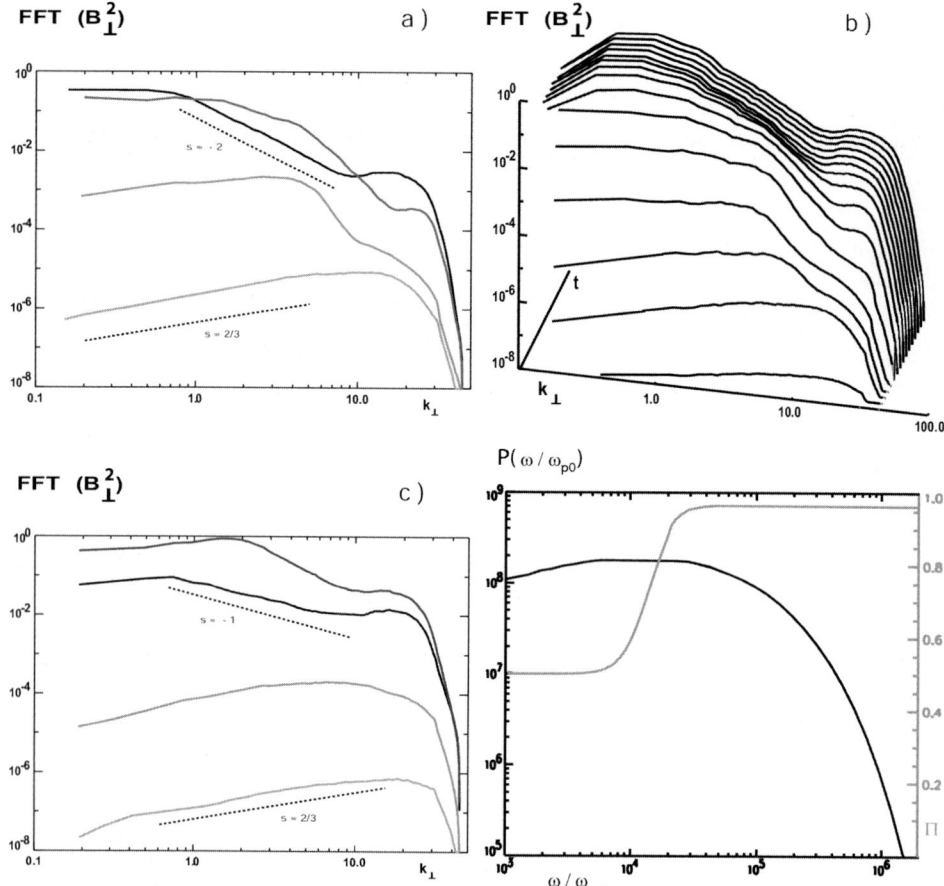

Fig. 2. Time evolution of Weibel magnetic fields observes an 'Inverse Cascade' (a-c) - bright grey lines correspond to the linear regime, darkening towards non-linear saturation. Power-law indices are characteristic for the fireball collision energy / typical coherence length of the magnetic field. Associated synchrotron spectra exhibit extreme degrees of polarization Π (grey line in the lower part of b).

References

1. W. Coburn & S.E. Boggs: Nature **423**, 415 (2003)
2. D.R. Willis et al: A&A **439**, 245 (2005)
3. J.T. Frederiksen et al: ApJ **608**, L13 (2004)
4. C.H. Jaroschek, H. Lesch & R.A. Treumann: ApJ **618**, 822 (2005)
5. J. Cho & A. Lazarian: MNRAS **345**, 325 (2003)
6. C. Lacombe: A&A **54**, 1 (1977)
7. M.V. Medvedev: ApJ **540**, 704 (2000)
8. J. Greiner et al: Nature **426**, 157 (2003)

Relativistic Jet Propagation in the Progenitor of GRBs

A. Mizuta[1], T. Yamasaki[1], S. Nagataki[1], and S. Mineshige[2]

[1] Max-Planck-Institute für Astrophysik, Karl-Schwarzschild-Str. 1, 85741 Garching, Germany,
`mizuta@mpa-garching.mpg.de`
[2] Yukawa Institute for Theoretical Physics, Kyoto University Oiwake-cho, Kitashirakawa,
Sakyo-ku, Kyoto 606-8502, Japan

1 Introduction

The gamma-ray bursts (GRBs) are one of the most energetic phenomena in the universe. It is realized that the GRB occurs at cosmological distances and the explosion is the relativistic collimated jet [1]. But the central engine of the GRBs is not fully understood yet.

It is proposed that the central engine of the long duration GRBs is a collapsar which is the death of massive stars at the last stage of the stellar evolution [2, 3]. The model predicts the formation of outflows from the center of the progenitor as follows. When the iron core of the progenitor collapses, a black hole or proto-neutron star and an accretion disk which surrounds the central object are formed. Some fraction of the collapsing gas can produce an outflow, although the formation mechanism of this outflow is not fully understood yet. Some observations directly support this hypothesis. GRB030329 was associated with SN2003dh (type Ic SN) [4, 5, 6, 7]. The host galaxy of the long duration bursts whose duration is longer than a few seconds is a star-forming galaxy [8, 9, 10].

The jet propagation in the collapsar was studied by numerical relativistic hydrodynamic simulations [11, 12, 13]. Those calculations showed that the jet emerged around the core can keep its collimated structure and successfully break out from the progenitor. But there still remains a question: which kind of outflow can keep the collimated structure and break out from the progenitor to be observed as a GRB. We performed numerical relativistic hydrodynamic simulations on the jet propagation to answer this problem.

2 Model and Numerical Method

We use the progenitor model developed by Hashimoto [14]. The progenitor has about 40 solar masses in the main sequence and 16 solar masses at the pre-supernova stage. The radius of the progenitor at the pre-supernova stage is 3.7×10^{10}cm. The outflow is injected at a distance of 2×10^8 cm from the core, assuming outflow formation

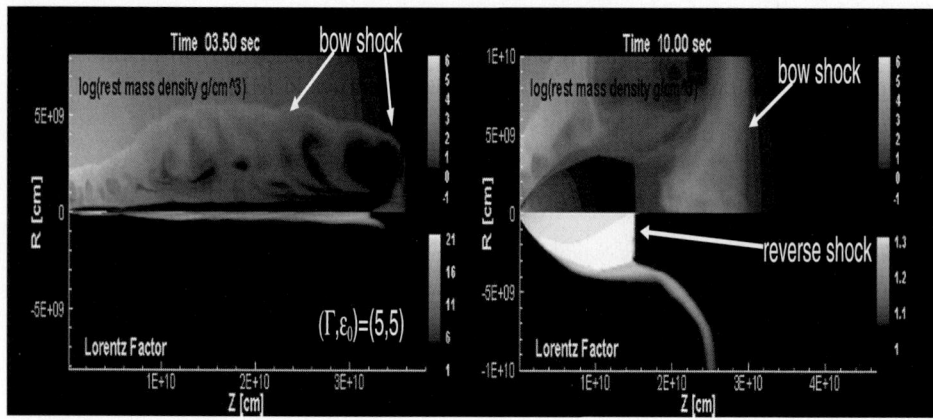

Fig. 1. Rest mass desnity(upper) and Lorentz factor(lower) contours of two models (Left:$(\Gamma_0, \epsilon_0) = (5,5)$, Right:$(\Gamma_0, \epsilon_0) = (1.15, 0.5)$)

around the center of the progenitor. We follow the outflow propagation to a distance of 6.5×10^{10} cm.

The power and radius of the injected outflow from the computational boundary which is at the distance of 2×10^8 cm from the center of the progenitor is assumed to be 10^{51} ergs s^{-1} and 7×10^7 cm, respectively. Two more parameters are necessary to define the injected outflow. We choose a Lorentz factor Γ_0 and specify the internal anergy ϵ_0 as the parameters. The pressure of the progenitor is ignored, since the pressure driven by the bow shock is much higher than that of the progenitor gas. The gravitational potential is also ignored, because the timescale of crossing the outflow is shorter than that of free fall.

Assuming axisymmetric geometry, two-dimensional relativistic hydrodynamic equations are solved by our relativistic hydrodynamic code [15]. The ideal EOS is assumed with constant adiabatic index (4/3).

3 Results and Discussion

Figure 1 shows the rest mass density (upper) and the Lorentz factor (lower) contours of two models. The left shows the result at $t = 3.5$s of the model ($\epsilon_0/c^2 = 5$, $\Gamma_0 = 5$). The right presents the result at $t = 10$s of the model ($\epsilon_0/c^2 = 0.1$, $\Gamma_0 = 1.15$). The left panel shows the succsessful eruption of the collimated jet and the high Lorentz factor along the axis. The maximum Lorentz factor in this model is 31. This value is good agreement with a simple formula, which is derived from energy conservation, $\Gamma_{\max} \sim \Gamma_0(1 + \epsilon_0/c^2)$, where c is speed of light. The bow shock and reverse shock do not separate. On the other hand the latter model shows an expanding outflow in which the bow shock and reverse shock separate as time goes on. We have performed a parametric study in the range of $1.05 \leq \Gamma_0 \leq 5$ and $0.1 \leq \epsilon/c^2 \leq 5$ [17]. As the velocity of the injected outflow decreases, the transition from a collimated flow to an expanding outflow is observed. The maximum Lorentz factor is in good agreement with the simple formula.

The collimated and high Lorentz factor jet could be observed as a GRB, although the Lorentz factor achieved in this study is still a little smaller than that of GRBs. The outflow, in which the maximum Lorentz factor achieved is not so large, could be observed as an X-ray flash which is simlar phenomenon with GRB but the energy peak is in the X-ray range [16]. The expanding outflow corresponds to an asymmetric supernova explosion which does not have an accompanying GRB.

4 Summary

The outflow propagation in the progenitor in the context of the collapse model is studied. The properties of the outflow dramatically change as the specific internal energy and/or the Lorentz factor of the injected outflow changes. For the high velocity ($> 0.7c$) injected outflows, the outflow keeps its collimated structure and successfully erupts from the progenitor. On the contrary the slow velocity ($\lesssim 0.7c$) injected outflows become expanding ones. The forward and reverse shock separate as time goes on. The achieved maximum Lorentz factor depends on the specific internal energy and the Lorentz factor and follows a simple formula derived from the energy conservation relation.

The different features could be related with different observed features, such as GRBs, X-ray flashes, and supernovae.

References

1. T. Piran: Phys. Rep., **314**, 575 (1999)
2. S. E. Woosley: ApJ, **405**, 273 (1993)
3. A. I. MacFadyen, and S. E. Woosley: ApJ, **524**, 262 (1999)
4. K. Z. Stanek et al.: ApJ, **591**, L17 (2003)
5. M. Uemura, et al.: Nature, **423**, 843 (2003)
6. J. Hjorth, et al.: Nature, **423**, 847 (2003)
7. P. A. Price, et al.: Nature, **423**, 844 (2003)
8. J. S. Bloom, S. R. Kulkarni, and S. G. Djorgovski, AJ, **123**, 1111 (2002)
9. E. Le Floc'h, et al.: A&A, **400**, 499 (2003)
10. N. R. Tanvir, et al.: MNRAS, **352**, 1073 (2004)
11. M. A. Aloy, et al. ApJ, **531**, L119 (2000)
12. W. Zhang, S. E. Woosley, and A. I. MacFadyen: ApJ, **586**, 356 (2003)
13. W. Zhang, S. E. Woosley, and A. Heger: ApJ, **608**, 365 (2004)
14. M. Hashimoto: Progress of Theoretical Physics, **94**, 663 (1995)
15. A. Mizuta, S. Yamada, and H. Takabe: ApJ, **606**, 804 (2004)
16. J. Heise, et al.: Gamma-ray Bursts in the Afterglow Era, Edited by Enrico Costa, Filippo Frontera, and Jens Hjorth. Berlin Heidelberg: Springer 16 (2001)
17. A. Mizuta, et al: ApJ ApJ, **651**, 960 (2006)

Gravitational Collapse and Neutrino Emission of Population III Massive Stars

K. Nakazato[1], K. Sumiyoshi[2,3], and S. Yamada[1,4]

[1] Department of Physics, Waseda University, 3-4-1 Okubo, Shinjuku, Tokyo 169-8555, Japan,
`nakazato@heap.phys.waseda.ac.jp`
[2] Numazu College of Technology, Ooka 3600, Numazu, Shizuoka 410-8501, Japan
[3] Division of Theoretical Astronomy, National Astronomical Observatory of Japan, 2-21-1 Osawa, Mitaka, Tokyo 181-8588, Japan
[4] Advanced Research Institute for Science & Engineering, Waseda University, 3-4-1 Okubo, Shinjuku, Tokyo 169-8555, Japan

Summary. We compute the collapse of Population III (Pop III) massive stars with 300 - 13500 $M_\odot$ for 18 models. In this study, we solve the general relativistic hydrodynamics and the neutrino transfer equations simultaneously to follow the evolution of space time under spherical symmetry. As a result, it is shown that the neutrino transfer plays a crucial role in the dynamics of the gravitational collapse and the emitted neutrino spectrum does not become harder for more massive stars. We also evaluate the flux of the relic neutrino background from Pop III massive stars and discuss the possibility to study the Pop III star formation history.

1 Introduction

Population III (Pop III) stars are the first stars formed in the universe. Recent theoretical studies suggest that Pop III stars may have a large population of very massive objects from hundreds to thousands of solar masses and lose little of their mass during the quasi-static evolution because of zero-metallicity. If an initial stellar mass is larger than $\sim 100M_\odot$, the star becomes unstable against gravitational collapse due to the pair-instability in the helium burning stage. As a result, if an initial stellar mass is smaller than $\sim 260M_\odot$, the collapse is bounced by the oxygen burning and makes a pair-instability supernova explosion (the so-called pair-instability supernova). On the other hand, more massive stars can not halt the collapse and form a black hole directory emitting a large amount of neutrinos. In this study, we investigate these black hole forming models. For more detailed discussion, we refer to our other article [1] and see also the previous studies [2][3].

These black holes will eventually have the mass of their progenitor. It is noted that the mass range is consistent with that of intermediate mass black holes, whose existence is recently suggested by observations. And more, some gamma ray bursts are proposed to be associated with Pop III massive stars.

2 Models and Methods

Collapses caused by the pair-instability occur in the helium burning stage and start the oxygen burning. Consequently, stars with over $\sim 260 M_\odot$ form the iron cores which are unstable to collapse due to the photodisintegration. These iron cores are isentropic and the entropy per baryon is determined by the initial core mass [4]. We make the iron cores which are equilibrium configurations as initial models solving the Oppenheimer-Volkoff equation.

The general relativistic hydrodynamics and the neutrino transports are solved simultaneously for the Misner-Sharp metric by the Boltzmann solver [5][6]. Our models are spherically symmetric and have 127 radial grid points. For the neutrino distribution, the energy space and the angular space are discretized to 12 and 4 grid points respectively. We compute 4 species of neutrino (ν_e, $\bar{\nu}_e$, $\nu_\mu = \nu_\tau$ and $\bar{\nu}_\mu = \bar{\nu}_\tau$).

3 Gravitational Collapse of Pop III Massive Stars

At first, we find that the neutrino cooling crucially affects the dynamics the collapse. For example, when we compare the results with and without neutrino transport for the model with $375 M_\odot$, the baryon mass coordinate where the apparent horizon forms for the first time is $4.08 M_\odot$ for the model with neutrino transport, whereas that of the model without neutrino transport is $13.2 M_\odot$ (Fig. 1). The inner core also becomes small owing to neutrino cooling.

When we compare different models, the qualitative features are common to all the models. As the initial mass M_i gets larger, the initial value of entropy per baryon in the iron core, s_{iron}, becomes greater. Our results show that the final value of entropy per baryon at the center of core, s_{core}, does not become larger proportionately. For massive models, in fact, s_{core} is saturated. The reason is that the neutrino cooling is more efficient for more massive stars, since electrons are non-degenerate and a large amount of electron-positron pairs exist. These results are summarized in Table 1.

As for the initial mass dependence, the total neutrino energy becomes larger as the initial mass is greater, while the average energy of neutrinos does not so large. We can see these features from the neutrino spectra in Fig. 2. This is firstly because the neutrino cooling becomes more efficient and the temperature does not rise very much as the stellar mass is increased. The second reason is that the thicker outer core of more massive models prevents high energy neutrinos from escaping.

4 Relic Neutrino from Pop III Massive Stars

We estimate the flux of the relic neutrino from Pop III massive stars computed above. Here we use four assumptions. (1) ΛCDM cosmology model. (2) The Pop III star formation efficiency is 10% . (3) The initial mass function (IMF) is the top heavy type [7] such as,

$$\frac{dn}{dm} \propto m^{-\beta-1} \quad (m > 100 M_\odot, 1 < \beta \le 3). \tag{1}$$

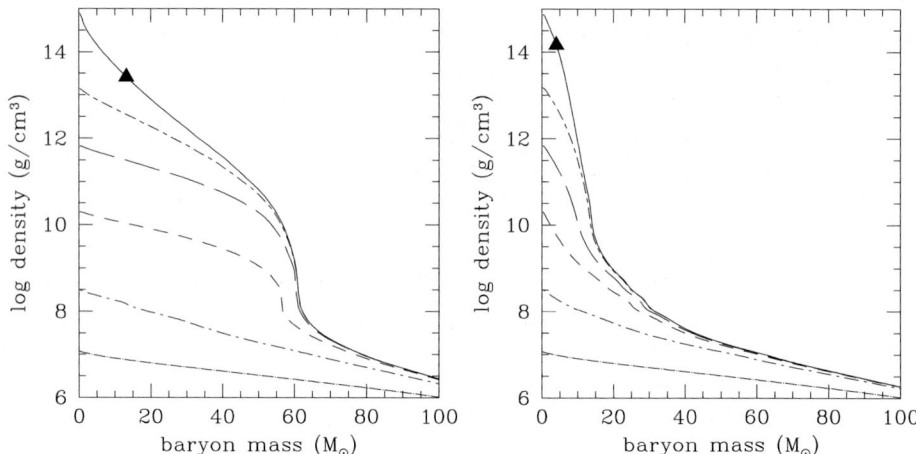

Fig. 1. Comparison of the results without (left) and with (right) neutrinos for the same model with $M_i = 375 M_\odot$. In the each panel, the triangles show the locations of the apparent horizon. In the left panel, the lines correspond, from bottom to top, to $t = -12.0\,\mathrm{s}$, $t = -343\,\mathrm{ms}$, $t = -42.3\,\mathrm{ms}$, $t = -4.92\,\mathrm{ms}$, $t = -0.699\,\mathrm{ms}$ and $t = 0\,\mathrm{ms}$. In the right panel, the lines correspond, from bottom to top, to $t = -8.88\,\mathrm{s}$, $t = -239\,\mathrm{ms}$, $t = -41.6\,\mathrm{ms}$, $t = -12.3\,\mathrm{ms}$, $t = -1.52\,\mathrm{ms}$ and $t = 0\,\mathrm{ms}$. Here we measure the time from the point at which the apparent horizon is formed.

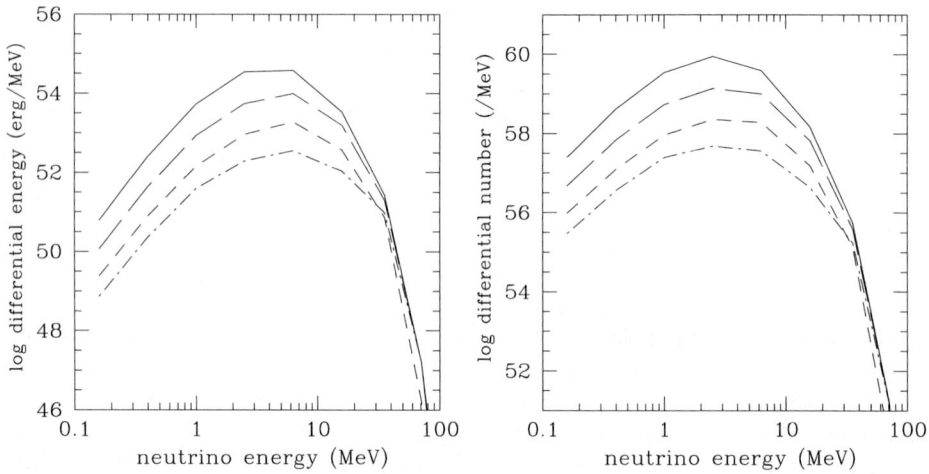

Fig. 2. Neutrino energy (left) and number (right) spectra. The dot-dashed line, short dashed line, long dashed line and solid line represent, respectively, the models with $M_i = 375 M_\odot$, $M_i = 1145 M_\odot$, $M_i = 3500 M_\odot$ and $M_i = 10500 M_\odot$.

(4) As for the star formation rate (SFR), we employ the following 3 models. (A) SFR follows the cosmic reionization history suggested by the WMAP observation. (B) SFR is given by the theoretical investigation by [8]. (C) SFR is proportional to the GRB rate [9].

Table 1. Initial Mass Dependence of Core Entropy, Core Mass and Location of Apparent Horizon.

M_i	300	375	470	585	730	915	1145	1430	1800
s_{iron}	15.98	17.50	19.17	20.96	22.97	25.25	27.77	30.54	33.74
s_{core}	7.33	7.37	7.70	7.88	8.07	8.28	8.50	8.77	8.90
M_{core}	13.6	14.1	15.8	16.8	18.2	19.5	21.0	23.4	24.7
M_{AH}	4.17	4.08	4.53	4.84	5.18	5.39	5.82	6.33	6.46

M_i	2250	2800	3500	4350	5500	6800	8500	10500	13500
s_{iron}	37.19	40.96	45.24	49.89	55.50	61.16	67.78	74.75	84.06
s_{core}	9.24	9.64	10.10	10.35	10.56	10.41	10.58	10.48	11.20
M_{core}	26.5	30.0	33.5	36.0	39.0	37.6	40.1	39.5	44.5
M_{AH}	6.96	7.60	8.42	8.44	9.11	9.10	9.40	9.26	9.91

M_i denotes the initial stellar mass in the unit of the solar mass ($M_\odot$). s_{iron} and s_{core} are the entropy par baryon of the initial model and the inner core at the center at $t = 0$, respectively, and it is in the unit of the Boltzmann constant (k_B). M_{core} is the mass of the inner core at $t = 0$ and M_{AH} is the location of the apparent horizon at $t = 0$. They are in the unit of the solar mass ($M_\odot$).

It is difficult for the currently operating detectors to detect this flux because the cosmological redshift reduces the peak energy to lower values. Moreover, this flux is overwhelmed by solar ν_e and $\bar{\nu}_e$ from nuclear reactors. However, because the solar and reactor neutrinos are not isotropic, removing them is possible at least in principle. For $\bar{\nu}_e$, in particular, Pop III massive stars are the largest cosmological sources. Thus, if ever observed, the spectrum will enable us to estimate the formation history of Pop III stars since the peak energy is mainly determined by the Pop III star formation model and not sensitive to the IMF (Fig. 3).

5 Conclusions

In this study, we have numerically studied the spherical gravitational collapse of the Pop III massive stars, taking into account the reactions and transports of neutrinos in detail. Neutrinos affect the dynamics of collapse crucially, determining the inner core mass, the location and formation time of the apparent horizon. Neutrino cooling is very efficient and the final value of core entropy is not sensitive to the initial value or the initial stellar mass. For very massive stars, even the outer core becomes opaque to neutrinos. As a result, the neutrino spectra do not become harder as the initial mass increases. Therefore, the peak energy of relic neutrinos is mainly determined by the redshift of the Pop III star formation and not sensitive to the IMF. At present, the detection of these relic neutrinos is difficult because the neutrino energy is too low and the other background neutrinos should be discriminated.

Acknowledgments. In this work, numerical computations were partially performed on Fujitsu VPP5000 at the Astronomical Data Analysis Center, ADAC, of the National

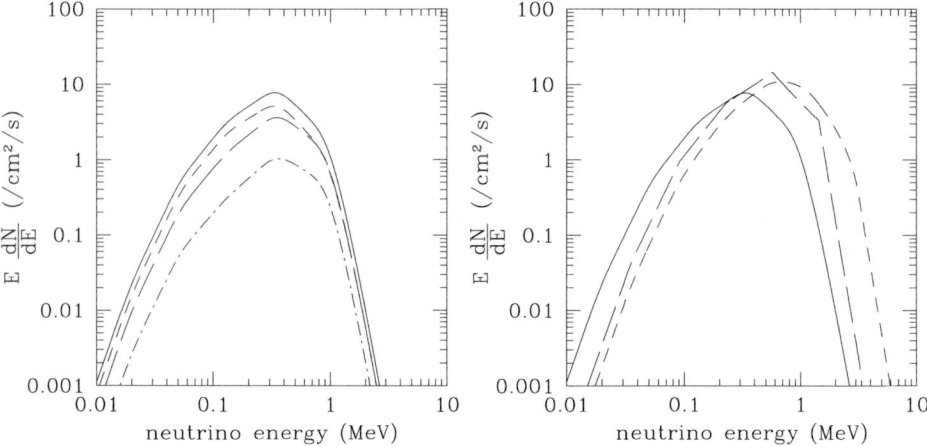

Fig. 3. Relic $\bar{\nu}_e$ number fluxes from Pop III massive stars for various values of β and different star formation histories. The left panel shows the result of model A and the short dashed line, solid line, long dashed line and dot-dashed line correspond to $\beta = 1.1$, $\beta = 1.35$, $\beta = 2$ and $\beta = 3$, respectively. The right panel shows the result for $\beta = 1.35$ and the solid line, long dashed line, and short dashed line represent models A, B and C, respectively.

Astronomical Observatory of Japan (VPP5000 System Projects wkn10b), and on the supercomputers in RIKEN and KEK (KEK Supercomputer Project No. 108). This work was partially supported by Grants-in-Aid for the Scientific Research from the Ministry of Education, Science and Culture of Japan through No.14079202, No.15740160, No.17540267, and The 21st century COE Program "Holistic Research and Education Center for Physics of Self-organization Systems".

References

1. K. Nakazato, K. Sumiyoshi, & S. Yamada: ApJ **645**, 519 (2005)
2. C.L. Fryer, S.E. Woosley, & A. Heger: ApJ **550**, 372 (2001).
3. F. Iocco, G. Mangano, G. Miele, G.G. Raffelt, & P.D. Serpico: Astropart. Phys. **23**, 303 (2005).
4. J.R. Bond, W.D. Arnett, & B.J. Carr: ApJ **280**, 825 (1984).
5. S. Yamada: ApJ **475**, 720 (1997).
6. K. Sumiyoshi, S. Yamada, H. Suzuki, H. Shen, S. Chiba, & H. Toki: ApJ **629**, 922 (2005).
7. F. Nakamura, & M. Umemura: ApJ **548**, 19 (2001).
8. E. Scannapieco, R. Schneider, & A. Ferrara: ApJ **589**, 35 (1984).
9. T. Murakami, D. Yonetoku, M. Umemura, T. Matsubayashi, & R. Yamazaki: ApJ **625**, L13 (2005).

Theoretical Interpretation of GRB 031203 and URCA-3

R. Ruffini[1,2], M.G. Bernardini[1,2], C.L. Bianco[1,2], P. Chardonnet[1,3], F. Fraschetti[1], and S.-S. Xue[1,2]

[1] ICRANet and ICRA, P.le della Repubblica 10, I–65100 Pescara, Italy, ruffini@icra.it
[2] Dip. Fisica, Univ. "La Sapienza", P.le A. Moro 5, I–00185 Roma, Italy
[3] Univ. de Savoie, LAPTH–LAPP, BP 110, F–74941 Annecy, France

1 Luminosity and Spectral Properties.

GRB 031203 was observed by IBIS, on board of the INTEGRAL satellite in the $20 - 200$ keV band [1], as well as by XMM [2] and Chandra [3] in the $2 - 10$ keV band, and by VLT [3] in the radio band. It appears as a typical long burst [4], with a simple profile and a duration of ≈ 40 s. The burst fluence in the $20-200$ keV band is $(2.0\pm0.4)\times10^{-6}$ erg/cm^2 [4], and the measured redshift is $z = 0.106$ [5]. We analyze in the following the gamma-ray signal received by INTEGRAL.

Such observations find a direct explanation in our theoretical model [6]. We determine the values of its two free parameters: the total energy stored in the Dyadosphere E_{dya} and the mass of the baryons left by the collapse $M_B c^2 = B E_{dya}$. We follow the expansion of the pulse, composed by the $e^- - e^+$ plasma initially created by the vacuum polarization process in the Dyadosphere. The plasma self–propels outward and engulfs the baryonic remnant left over by the collapse of the progenitor star. As such a pulse reaches transparency, the Proper–GRB is emitted. The remaining accelerated baryons interacting with the ISM produce the afterglow emission. The ISM is described by the two additional parameters of the theory: the average particle number density n_{ISM} and the ratio $\mathcal{R}$ between the effective emitting area and the total area of the pulse [7], which takes into account the ISM filamentary structure [8]. The best fit of the observational data (see Fig.2.Left) leads to a total energy of the Dyadosphere $E_{dya} = 1.85 \times 10^{50}$ erg and the amount of baryonic matter in the remnant is $B = 7.4 \times 10^{-3}$. The ISM parameters are: $< n_{ism} >= 0.3$ particle/cm^3 and $< \mathcal{R} >= 7.81 \times 10^{-9}$ [9].

The luminosity in selected energy bands is evaluated integrating over the EQui-Temporal Surfaces (EQTS) [10] the energy density released in the interaction of the accelerated baryons with the ISM measured in the co–moving frame, duly boosted in the observer frame. The radiation viewed in the co–moving frame of the accelerated baryonic matter is assumed to have a thermal spectrum [7]. In addition to the luminosity in fixed energy bands we can derive also the instantaneous photon number spectrum $N(E)$. Although the spectrum in the co–moving frame of the expanding pulse is thermal, the shape of the final spectrum in the laboratory frame is non thermal. In fact each single instantaneous spectrum is the result of an integration of hundreds of thermal

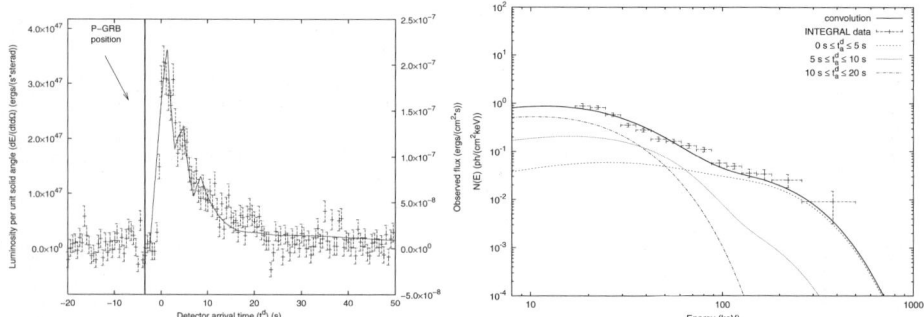

Fig. 1. Left: Theoretically simulated light curve of the GRB 031203 prompt emission in the 20 − 200 keV energy band (solid line) is compared with the observed data [4]. **Right:** Three theoretically predicted time–integrated photon number spectra $N(E)$ are presented here (dashed, dotted, dashed–dotted lines). The theoretically predicted time–integrated photon number spectrum $N(E)$ corresponding to the first 20 s of the prompt emission (solid line) is compared with the data observed by INTEGRAL [4].

spectra over the corresponding EQTS [10]. This calculation produces a non thermal instantaneous spectrum in the observer frame. We integrated the photon number spectrum $N(E)$ over the whole duration of the prompt event: in this way we obtain a typical non–thermal power–law spectrum which results to be in good agreement with the IN-TEGRAL data and gives a clear evidence of the possibility that the observed GRBs spectra originate from thermal emission [9] (see Fig.1.Right).

2 The GRB 031203/Sn2003lw/URCA-3 Connection.

In the early days of neutron star physics it was clearly shown by Gamow and Shoenberg [11] that the URCA processes are at the very heart of the Supernova explosions. The neutrino–antineutrino emission described in the URCA process is the essential cooling mechanism necessary for the occurrence of the process of gravitational collapse of the imploding core. Since then, it has become clear that the newly formed neutron star can be still significantly hot and in its early stages will be associated to three major radiating processes [12]: a) the thermal radiation from the surface, b) the radiation due to neutrino, kaon, pion cooling, and c) the possible influence in both these processes of the superfluid nature of the supra-nuclear density neutron gas.

We already proposed that the Supernova explosion can be the result of an induced gravitational collapse [13], and the X–ray emission observed is the sign of the cooling of the young neutron star born from the Supernova explosion (see GRB 980425/SN1998bw/URCA-1 [14] and GRB 030329/SN2003dh/URCA-2 [15]). This possibility was explored also for the system GRB 031203/SN2003lw. Also in this case the observations in other wavelengths could be related to the Supernova event, which we called URCA-3 (see Fig.2).

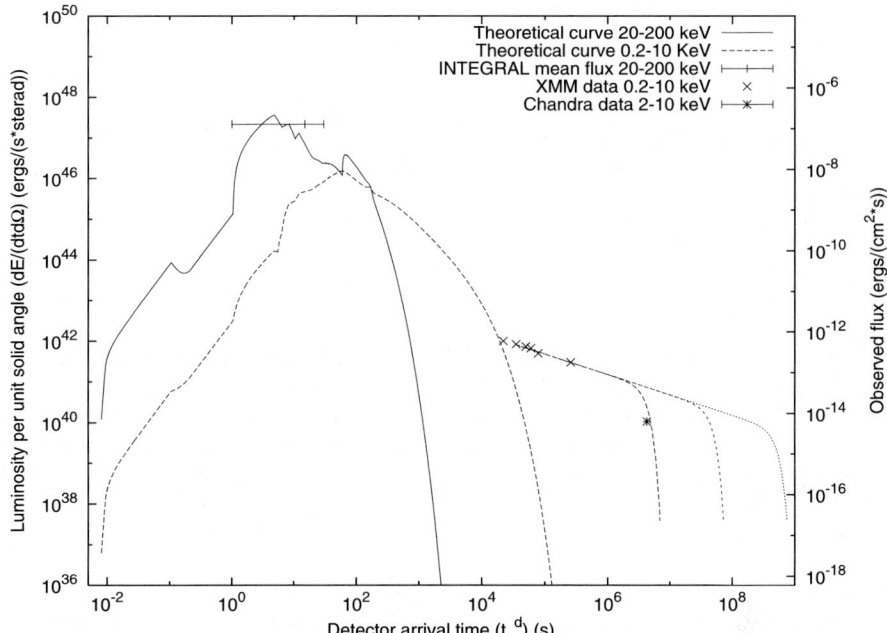

Fig. 2. Theoretically simulated light curves of GRB 031203 in $20 - 200$ KeV (solid line) and $2 - 10$ KeV (dashed line) energy bands are shown together with qualitative representative curves for the cooling processes.

References

1. S. Mereghetti, D. Göts: GCN 2460 (2003)
2. D. Watson et al: ApJ **605**, L101 (2004)
3. A.M. Soderberg et al: Nature **430**, 648 (2004)
4. S.Y. Sazonov, A.A. Lutovinov, R.A. Sunyaev: Nature **430**, 646 (2004)
5. J.X. Prochaska et al: ApJ **611**, 200 (2004)
6. R. Ruffini, C.L. Bianco, P. Chardonnet et al: New Perspectives in Physics and Astrophysics from the Theoretical Understanding of GRBs. In: *Cosmology and Gravitation: X Brazilian School of Cosmology and Gravitation*, 668, ed by M. Novello, S.E. Perez Bergliaffa (AIP, Melville New York 2003) pp 16–107
7. R. Ruffini, C.L. Bianco, P. Chardonnet et al: IJMP D **13**, 843 (2004)
8. R. Ruffini, C.L. Bianco, P. Chardonnet et al: IJMP D **14**, 97 (2005)
9. M.G. Bernardini, C.L. Bianco, P. Chardonnet et al: ApJ **634**, 29 (2005)
10. C.L. Bianco, R. Ruffini: ApJ **413**, 281 (2004)
11. G. Gamow, M. Schoenberg: Phys. Rev. **59**, 539 (1941)
12. S. Tsuruta: Phys. Rep. **56**, 237 (1979)
13. R. Ruffini, C.L. Bianco, P. Chardonnet et al: ApJ **555**, 117 (2001)
14. R. Ruffini, C.L. Bianco, P. Chardonnet et al: Adv. Sp. Res. **34**, 2715 (2004)
15. R. Ruffini, M.G. Bernardini, C.L. Bianco et al: Black Hole Physics and Astrophysics: The GRB–Supernova Connection and URCA-1 – URCA-2. In: *X Marcel Grossmann Meeting*, ed by M. Novello, S.E. Perez Bergliaffa (World Scientific, Singapore) in press, astro-ph/0503475

Baryonic Loading and e^+e^- Rate Equation in GRB Sources

R. Ruffini[1,2], C.L. Bianco[1,2], G. Vereshchagin[1,2], and S.-S. Xue[1,2]

[1] ICRANet and ICRA, P.le della Repubblica 10, I–65100 Pescara, Italy, ruffini@icra.it
[2] Dip. Fisica, Univ. "La Sapienza", P.le A. Moro 5, I–00185 Roma, Italy

Summary. The expansion of the electron-positron plasma in the GRB phenomenon is compared and contrasted in the treatments of Mészáros, Laguna and Rees, of Shemi, Piran and Narayan, and of Ruffini et al.. The role of the correct numerical integration of the hydrodynamical equations, as well as of the rate equation for the electron-positron plasma loaded with a baryonic mass, are outlined and confronted for crucial differences.

1 Introduction

One of the earliest contributions in the theoretical understanding of GRBs is found in the work of [1]. Some crucial ideas which later became very important in this field were presented there (e.g. the relevance of an electron-positron plasma, the possibility of baryonic loading of such plasma, and the evolution from an optically thick to an optically thin phase). The treatment, based on qualitative considerations, provided little information about the dynamics. The electron-positron plasma was there purported to evolve toward a proliferation of photons and pairs, leading to a degradation of the mean energy of particles in the plasma and a consequent cooling. Such a plasma would become transparent on a very short time scale and no dynamical phases are envisaged making this model inappropriate for GRBs. The physical reasons of the non-validity of these crucial conclusions are discussed elsewhere [2].

It soon became clear, however, following the work of [3] and [4] that the major characteristic of a sudden energy release process in the electron-positron plasma leads to a very rapid self acceleration of a shell of material, reaching ultra-relativistic regimes with Lorentz gamma factors in the range $10^2 < \gamma < 10^3$. The major results were obtained in [5, 6, 7, 8, 9].

In our model the analysis of the dynamical expansion of the electron-positron plasma is not just a topic of academic interest, it is indeed crucial to the description of the entire GRB phenomena. The initial process is the vacuum-polarization around the black hole [10], followed by the dynamical expansion of the pair plasma, leading to an ultrarelativistic accelerated motion [8, 9]. Such acceleration leads to a shell of baryons with a Lorentz gamma factor $\gamma \sim 100 - 300$ which, by interacting with the interstellar matter (ISM), gives origin to the afterglow. A first crucial signature in our model is carried by the radiation emitted at the moment of transparency, what we have called the Proper-GRB (P-GRB). The current models in the literature identify two different components in the long GRBs, the "prompt radiation" and the "afterglow", the first one being emitted by an "inner engine" (see e.g. [11] and references therein). In our model, instead, the "prompt radiation" is an essential part of the "afterglow" and the entire long GRB phenomena is uniquely due to external shock processes.

For the above reasons we return, in this communication, to a comparison and contrast between our results and the ones in the current literature. Now we can address with clarity the analogies and the differences in the treatments and pin down the source of such differences as due to inaccurate theoretical work or / and equally inaccurate numerical computations.

2 Contrasts in the Dynamical Description of the Expanding Plasma

Our results are represented in Fig. 1-Left, showing the energy release at transparency carried by photons and in the form of kinetic energy of the pulse, depending on the loading parameter $B = Mc^2/E_0$, where M is the total mass of the plasma, c is the speed of light, E_0 is initial energy of the system. In Fig. 1-Right we show the corresponding results by [5] which are markedly different from ours, and do not fulfill the basic requirement of the conservation of energy. In Fig. 2-Left we show the results of [12] about the gamma factor at transparency which overlap with ours. The major difference between our treatment and the one by [6] and collaborators stems from different accuracy in the description of the electron-positron pairs equations which we have explicitly integrated in full details with the dynamical equations of the pulse. In Fig. 2-Right we show some analogies and differences between the two treatments. In particular, the energy release in the form of photons at transparency, which is crucial for the detection of the P-GRB, is qualitatively correct in Ref. [12] but underestimated in view of the simplified analytical approach adopted, which does not take into proper account the explicit integration of the electron-positron rate equation. The moment when the pulse reaches transparency and the corresponding radius are also different in simple analytical models and our detailed numerical computations, which leads to different predictions for the energetics.

In addition to the above differences, the treatment in [9] clearly predicts an instability in the expanding plasma for a value $B > 10^{-2}$.

3 Conclusions

The analysis of the electron positron plasma expansion, far from being purely academical, is essential in describing the entire GRB phenomenon in our model. We are currently working in proving the uniqueness of our model. It is clear that any minimal deviation in the integration of the hydrodynamical equations, or any inadequacy on the integration of the rate equation, may lead to the impossibility of reaching the correct theoretical model of GRBs by lacking the necessary accuracy in the description of the fundamental process determining the energetics of GRBs. Indeed the agreement

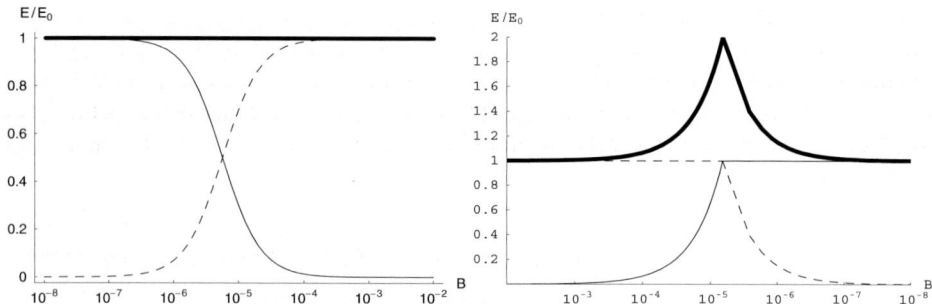

Fig. 1. Left: Relative energy release in the form of photons emitted at transparency point E_γ (solid line) and kinetic energy of the plasma E_k (dashed line) of the baryons in terms of the initial energy E_0 of the electron-positron plasma as computed in our quasi-analytic model [8, 9]. The quantities are given as a function of the baryon loading parameter B. The bold line denotes the total energy of the system in terms of initial energy E_0 which is, as it should, constant and equal to 1. **Right:** The same quantities are computed in the Mészáros, Laguna and Rees model [5]. Note that in Ref. [5] the parameterization is done as a function of the dimensionless parameter $\eta = 1/B$. The bold line denotes the total energy of the system in terms of initial energy E_0, which is not conserved in such a treatment [5].

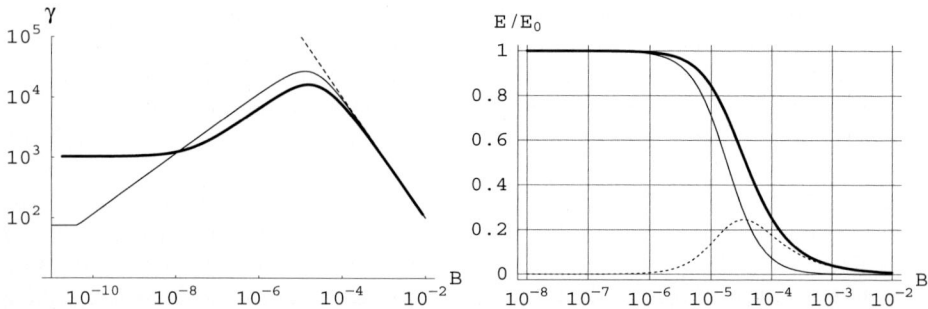

Fig. 2. Left: Relativistic gamma factor when transparency is reached as a function of the baryon loading parameter B. The bold line denotes the numerical results obtained in Ref. [8, 9], taking into due account the electron-positron rate equation. The plain line corresponds to the analytical estimate from Shemi and Piran model [12], neglecting the rate equation. The dashed line denotes the asymptotic value of the Lorentz gamma factor at transparency $\gamma_{asym} = B^{-1}$ (see Ref. [9] for details). **Right:** Relative energy release in the form of photons emitted at transparency point E_γ of the GRB in terms of initial total energy E_0 depending on the baryon loading parameter B. The bold line represents our numerical results, already given in Fig. 1-Left. The plain line shows the results for the analytic model of Shemi and Piran [12]. The dashed line shows the difference between our numerical analysis, taking into proper account the rate equation [8, 9] and the approximate analytical [12] results.

of our theoretical model [13, 14, 15, 16, 17, 18, 19] with the observations has been successfully tested in four different sources for the intensities of both the P-GRB and the afterglow in selected energy bands. These sources are: GRB 991216 [20], GRB 980425 [21], GRB 030329 [22], GRB 031203 [23] (see Fig. 3). It is particularly interesting that, in all the above sources, the value of the B parameter is smaller than 10^{-2} as clearly predicted in Ref. [9].

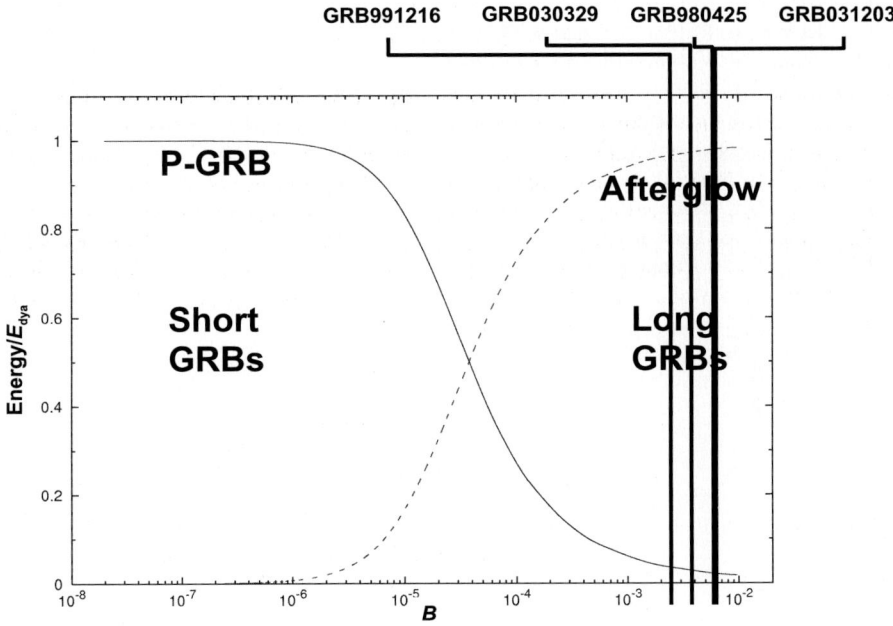

Fig. 3. The relative intensity of the P-GRB and the afterglow are given for four selected sources as a function of the baryon loading parameter B.

References

1. G. Cavallo, M.J. Rees: MNRAS **183**, 359 (1978)
2. R. Ruffini, G.V. Vereshchagin, A. Aksenov: in preparation
3. P. Vitello, M. Salvati: Phys. Fluids **19**, 1523 (1976)
4. J. Goodman: ApJ **308**, L47 (1986)
5. P. Mészáros, P. Laguna, M.J. Rees: ApJ **415**, 181 (1993)
6. T. Piran, A. Shemi, R. Narayan: MNRAS **263**, 861 (1993)
7. G.S. Bisnovatyi-Kogan, M.V.A. Murzina: Phys. Rev. D **52**, 4380 (1995)
8. R. Ruffini, J.D. Salmonson, J.R. Wilson, S.-S. Xue: A&A **350**, 334 (1999)
9. R. Ruffini, J.D. Salmonson, J.R. Wilson, S.-S. Xue: A&A **359**, 855 (2000)
10. G. Preparata, R. Ruffini, S.-S. Xue: A&A **338**, L87 (1998)

11. P. Mészáros: Theories of early afterglow. In *Proceedings of the 16th Annual October Astrophysics Conference in Maryland*, ed by S. Holt, N. Gehrels, J. Nousek. Melville New York, AIP Conf.Proc. **836**, 234-243 (2006)
12. A. Shemi, T. Piran: ApJ **365**, L55 (1990)
13. R. Ruffini, C.L. Bianco, P. Chardonnet et al: ApJ **555**, L107 (2001)
14. R. Ruffini, C.L. Bianco, P. Chardonnet et al: ApJ **555**, L113 (2001)
15. R. Ruffini, C.L. Bianco, P. Chardonnet et al: ApJ **555**, L117 (2001)
16. R. Ruffini, C.L. Bianco, P. Chardonnet et al: ApJ **581**, L19 (2002)
17. C.L. Bianco, R. Ruffini: ApJ **605**, L1 (2004)
18. C.L. Bianco, R. Ruffini: ApJ **620**, L23 (2005)
19. C.L. Bianco, R. Ruffini: ApJ **633**, L13 (2005)
20. R. Ruffini, C.L. Bianco, P. Chardonnet et al: New Perspectives in Physics and Astrophysics from the Theoretical Understanding of GRBs. In: *Cosmology and Gravitation: X Brazilian School of Cosmology and Gravitation*, 668, ed by M. Novello, S.E. Perez Bergliaffa (AIP, Melville New York 2003) pp 16–107.
21. R. Ruffini, C.L. Bianco, P. Chardonnet et al: Adv. Sp. Res. **34**, 2715 (2004)
22. R. Ruffini, M.G. Bernardini, C.L. Bianco et al: Black Hole Physics and Astrophysics: The GRB–Supernova Connection and URCA-1 – URCA-2. In: *X Marcel Grossmann Meeting*, ed by M. Novello, S.E. Perez Bergliaffa (World Scientific, Singapore) in press, astro-ph/0503475
23. M.G. Bernardini, C.L. Bianco, P. Chardonnet et al: ApJ **634**, L29 (2005)

Gamma-ray Bursts from X-ray Binaries

H.C. Spruit

Max Planck Institute for Astrophysics, Box 1317, DE-85748 Garching, Germany,
henk@mpa-garching.mpg.de

Summary. In this contribution I review the mechanism proposed earlier for producing a gamma-ray burst from the rapidly spinning neutron star in an X-ray binary (Spruit 1999), with a discussion of some more recent developments and outstanding issues.

1 Introduction

As observations of low-mass X-ray binaries and their descendants the millisecond pulsars show, neutron stars accreting from a companion star are spun up to periods as short as 2ms or less. At this period, the rotational energy of the star, $\sim 5\,10^{51}$ erg, is sufficient to power a good sized GRB, as was realized early on (e.g. Usov 1992, see also Kluzniak and Ruderman 1997, Dai and Lu 1998).

2 Magnetization of LMXBs

Spinning in a near vacuum, the neutron star in an LMXB would automatically satisfy the requirement of low baryon loading necessary to obtain the high Lorentz factors of GRB outflows. The extraction of rotational energy to power the burst in such a low density environment is most conveniently done by endowing the star with a magnetic field, sufficiently strong that the ordinary pulsar radiation mechanism will spin the star down on the required GRB time scale. Since long-duration GRBs are now known to be associated with supernovae, it is natural to look for the short bursts, with typical durations of 0.3s, as the class of objects to which the idea (like other off-mainstream ideas) might be applicable.

The magnetic field needed to spin a neutron star of 2ms down in 0.3 is $B_s = 10^{17}$ G, where B_s is the (dipole) field strength at the surface. This presents a problem that at first sight is rather unlikely to have a plausible solution: a field of this strength has to appear at the surface of the star on a time scale of the order of the burst duration, preferably less. For this reason, early models involving rapidly spinning neutron stars assumed, for example, the accretion-induced collapse of a white dwarf (Kluzniak and Ruderman 1998); a recent new suggestion is the formation of a heavy neutron star in the merger of a double neutron star binary (Dai et al. 2006). These environments are naturally rather baryon-loaded, hence require additional mechanisms to avoid baryon pollution of the candidate GRB.

To see how an LMXB neutron star may achieve this unlikely feat, note first that it is not necessary that the entire magnetic field of the star be generated within 0.3s. The field could be buried in the interior, and emerge to the surface only when it reaches a suffcient strength, much like active regions do on the Sun. Magnetic fields are naturally buoyant, but the fluid interior of a neutron star is stably stratified, and a minimal field strength is needed for the field to overcome this stratification. This critical field strength turns out to be just the 10^{17} G required for the GRB-from-XRB idea. The idea is thus to produce a strong field in the interior on a leisurely time scale, and allow it to become unstable when reaching the critical field strength. The eruption time scale would be the Alfvén time scale, of the order of 1ms B_{17}^{-1}, conveniently short compared with the burst duration.

3 Generation of Differential Rotation

A magnetic field buried almost entirely in the interior is produced when an initially weak poloidal field (a uniform field, say), is wound up by differential rotation. The resulting toroidal field (i.e. azimuthal with respect to the rotation axis) would be confined entirely to the interior, with only the unchanging poloidal field sticking through the surface. If the energy in differential rotation is large enough, at least equal to the energy of a 10^{17}G magnetic field, it can in principle wind up an arbitrarily weak initial field to 10^{17}G, given enough time. For an initial field of 10^7 G, for example, this would take a few months. The problem of producing a GRB from an XRB is thus reduced to finding a reason why the star would be rotating differentially, by some 10%. The process producing this differential rotation must operate on a time scale of months or less.

A spin period of 2ms happens to be close to the period at which the star becomes unstable to the emission of gravitational waves, by the Chandrasekhar-Friedman-Schutz mechanism. As shown by Andersson (1997) the most unstable mode of oscillation is an 'r-mode' (more properly called by its older geophysical name of inertial oscillation, cf. Rieutord 2001). In a neutron star, slowly spun up by accretion on a time scale of some 10^7 yr, the instability will develop very slowly when the spin reaches this critical value. Rather than saturating at some low amplitude, as one might have guessed, the instability does something more interesting, as was realized by Levin (1998) and Spruit (1999). The viscosity of a neutron star fluid has a negative temperature dependence, just like water or engine oil. Dissipation of oscillation energy heats this fluid, reducing its viscosity. This reduces the damping of the oscillation, and since at marginal stability viscous damping was just balanced by the intrinsic growth of the instability, the result of heating is an increasing *growth rate*. The consequence is that the instability grows faster than exponential, diverging on a finite time scale. The last few e-foldings can in principle take place on a very short time scale, days or hours.

The coupling of the oscillation to the gravitational wave is most effective in the outer parts of the star, hence the angular momentum carried by the wave is provided mostly by these outer parts. While some angular momentum is thus lost, the important consequence of the instability for the present purpose is that it generates differential rotation, and it turns out that some 30% is easily reached if the instability as described is the only process to consider.

Once a significant amount of angular momentum is lost the instability stops, and one is left with a differentially rotating star, which can wind an initial field of 10^7 G into a 10^{17} G field on a time scale of months. A prediction is thus the generation of a gravitational wave burst significantly *preceding* the GRB.

4 Open Issues

The gravitational wave instability thus is the key to producing, eventually, a magnetic field of some 10^{17} G that erupts to the surface on the required short time scale. The magnetic energy dissipated in the atmosphere during the eruption is itself a plausible source of high-energy radiation, but much more important is that once the surface is magnetized, this field will extract essentially all the remaining rotational energy of the star on a time scale of a second or so in the form of a large-amplitude electromagnetic wave, as in the case of pulsar spindown.

The strength of the scenario is that it combines a number of known and well-understood physical processes in neutron stars, while the critical rotation rate (GW instability) and critical field strength (for buoyant rise) are just in the right range to produce a GRB. Some problems remain, however. One is the question whether the GW instability really can reach the large amplitudes needed for angular momentum loss on a short time scale. Other damping mechanisms than just viscous damping, for example some form of turbulence, may limit the instability to low amplitudes, as argued by Arras et al. (2003 and refs therein). At these lower amplitudes, differential rotation will still be generated, but perhaps not sufficiently fast to wind up the field quickly enough. The result would be a lower fineal field strength, given by the balance between internal magnetic torque and the spindown torque due to the gravitational wave.

A second problem is the wound-up magnetic field itself. To reach the high field strengths that are possible on purely energetic grounds, the field has to be wound up in an orderly manner, without wasting differential rotation energy. A highly wound-up, nearly azimuthal magnetic field, however, is typically unstable (Tayler 1973), producing disorder and friction. In a normal stellar interior, a Tayler-type instability is likely to produce a small scale magnetic field, which produces a high effective 'viscosity'. The conditions for instability of this type are likely to be different in a neutron star interior, however, but the proper stability analysis has yet to be done.

References

1. N. Andersson: Astrophys. J. **502**, 708 (1997)
2. Z.G. Dai, X.Y. Wang, X.F. Wu, B. Zhang: Science, **311**, 1127 (2006)
3. Z. G. Dai, T. Lu, T.: Physical Review Letters, **81**, 4301 (1998)
4. W. Kluźniak, M. Ruderman: Astrophys. J. **505**, L113 (1998)
5. M. Rieutord: Astrophys. J. **550**, 443 (2001)

6. H.C. Spruit: Astron. Astrophys **341**, L1 (1999)
7. H.C. Spruit: Astron. Astrophys **381**, 923 (2002)
8. R.J. Tayler: MNRAS **161**, 365 (1973)
9. V. V. Usov; Nature **357**, 472 (1992)
10. P. Arras, et al.:ApJ **591**, 1129 (2003)

Part VII

X-ray Binaries and Jets

Accretion and Relativistic Jets in Galactic Microquasars

T. Belloni

INAF – Osservatorio Astronomico di Brera, Via E. Bianchi 46, I-23807 Merate, Italy
belloni@merate.mi.astro.it

Summary. Our knowledge of the phenomenology of accretion onto black holes has increased considerably thanks to ten years of observations with the RXTE satellite. In particular, it has been possible to schematize the outburst evolution of transient systems on the basis of their spectral and timing properties, and link them to the ejection of relativistic jets as observed in the radio. Here I concentrate on some aspects of this scheme, concentrating on the timing properties and on their link with jet ejection.

1 Introduction

The RossiXTE mission is changing our view of the high-energy emission of Black Hole Transients (BHT). The picture that is emerging is complex and difficult to interpret, yet it is a considerable step forward in our knowledge. The classification of states and state-transition derived from a few objects is applicable to most systems, indicating that the common properties are more than what previously known (see [1, 2, 3] but also [4, 5]). At the same time, a clear connection between X-ray and radio properties has been found (see [6, 7]).

In this paper, I concentrate on a few aspects of this picture, highlighting the connection between Quasi-Periodic Oszillation (QPO) and noise components, source states and radio/jet properties.

2 Source States in a Nutshell

For a more complete definition of the four bright source states see e.g. [1, 2, 8]. The evolution of the 2002/2003 outburst of GX 339-4 as observed by RXTE/PCA is shown in Fig. 1. The regions of the Hardness-Intensity Diagram (HID) corresponding to the four states are marked. Another important diagram is shown in Fig. 2: on the X-axis is the same hardness than in Fig. 1, while on the Y-axis is the integrated fractional RMS variability. In the following, I concentrate on some aspects of these states:

- Two states, the Low/Hard (LS) and Hard-Intermediate (HIMS), found at the right of the thick vertical line in Fig. 1 have a number of aspects in common. A strong hard component is visible in their energy spectrum, usually attributed to (thermal) Comptonization. In the LS, this can be approximated with a power law with spectral index 1.6-1.8 and a high-energy cutoff around 100 keV; in the HIMS, the slope is higher (2.0-2.4) and the high-energy cutoff moves to lower energies (see [9]). The power spectrum is dominated by a band-limited component, with the presence of a type-C QPO in the HIMS (see [10]). The fractional RMS decreases smoothly from 50% at the beginning of the LS to 10% at the end of the HIMS (see Fig. 2). These states correspond to detectable radio emission from the nuclear source, associated with compact jet ejection. Just before the transition to the Soft-Intermediate (SIMS), it was suggested that the jet velocity increases rapidly, giving origin to the fast relativistic jets [7].

• The other two states, the High/Soft (HS) and SIMS, are markedly different, The energy spectrum is dominated by a thermal disk component, with only a weak steep hard component visible without a high-energy cutoff (see [11]). The power spectrum does not show a strong band-limited noise, but rather a weaker power-law component. In the SIMS, strong type-A/B QPOs are observed, while in the HS occasionally weak QPOs are observed, which cannot be easily identified with the ABC types (see [10, 8]). As one can see from Fig. 2, the integrated fractional RMS is below 10% and goes down as low as 1% in GX 339-4. Moreover, the HS points (triangles and dots) follow a continuation of the RMS-hardness correlation of the LS/HIMS, while the SIMS points deviate from it, being at lower RMS values (boxed area in Fig. 2). Also notice that when the source returned to the HIMS at the end of the outburst (black circles), only one point can be attributed to a SIMS, while the others followed the RMS-hardness correlation all the way to the right. These states, to the left of the 'jet line', correspond to non-detectable emission from the central source (see [12, 7]).

Therefore, the four states can be grouped into two 'classes': the hard and radio-loud ones (LS and HIMS, although radio emission is observed in the HIMS only at high luminosities, see [13]) and the soft and radio-quiet ones (HS and SIMS). The evolution from hard to soft causes a crossing of the jet line: the velocity of the outflow increases rapidly when approaching the line, creating fast propagating shocks in the jet [7]. The evolution from soft to hard simply leads to the (slow) formation of a new outflow, only mildly relativistic, and cannot give rise to fast relativistic ejections.

The focus here is obviously on the transitions between states, which mark very clearly the division between them. The transition between LS and the HIMS is *clearly* marked by changes in the infrared/X-ray correlation and can be placed very precisely in the evolution of the outburst [14, 8]. The transition between HIMS and SIMS is marked by abrupt changes in the timing properties, with the appearing of type-A/B QPOs and the dropping of the noise level. From SIMS to HS, a recovery of the noise level and disappearing of type-A/B QPOs is observed.

3 Time Variability as a Tracer

No light curve is shown in the figures here. It is not necessary to examine the time evolution, as the evolution of the system can be synthesized with only the two plots in Figs.

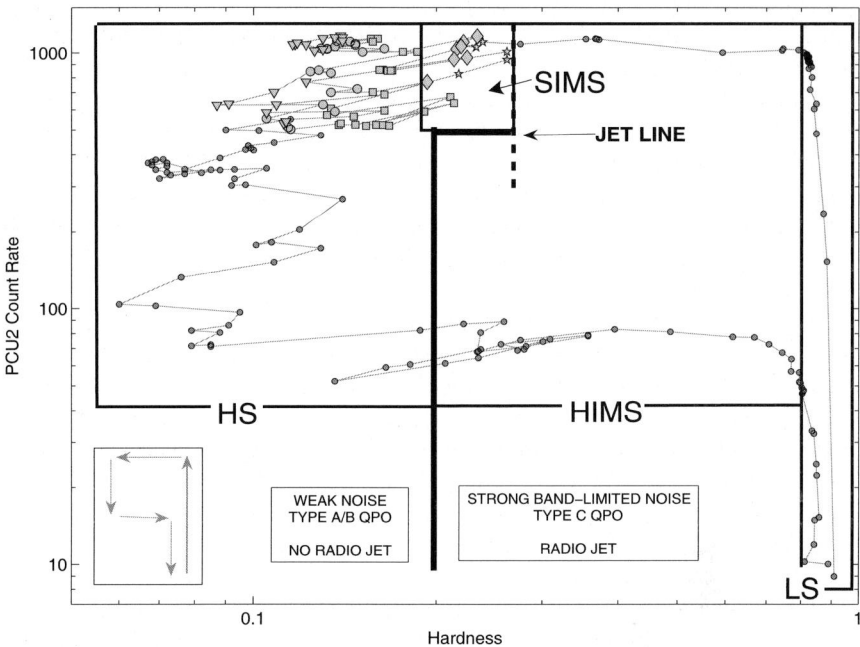

Fig. 1. Hardness-Intensity Diagram of the 2002-2003 outburst of GX 339-4 as observed by the RXTE/PCA, adapted from [2]. The lines mark the four source states described in the text. The different symbols are described in [8]. The dashed line shows the position of the 'jet line'. The thin line indicates the transition line that marks the presence/absence of strong band-limited noise in the power spectra. The inset on the lower left shows the general time evolution of the outburst along the q-shaped pattern.

1 and 2. Fig. 2 is particularly important, as it shows at the same time the color (spectral) and RMS (timing) evolution. It is evident from Fig. 2 that the timing properties are tracers of the evolution of the accretion flow and the jet properties. The crossing of the jet line can be identified by the drop in integrated RMS (although in some cases this effect is rather small and a plot of the full diagram is necessary to mark it) and by the appearance of type-A/B QPOs, easily identifiable in the power spectra. Type-C QPOs and band-limited noise components provide characteristic frequencies that trace the evolution of the accretion during the early and final part of the outburst; in the hard states, the strong correlations between hard X-ray emission and radio flux led to models for the jet production and X-ray emission which need to take these timing tracers into account (see e.g. [15]). Type-A/B QPOs are much less studied and their origin is unclear. However, the transient presence of type-B QPOs could be important for the study of the physical conditions of the accretion flow as the source crosses the jet line. The small range of long-term variability of their centroid frequency, its fast short-term variability

(see [16, 8]) and the absence of band-limited noise components are key ingredients for their understanding (see [10]).

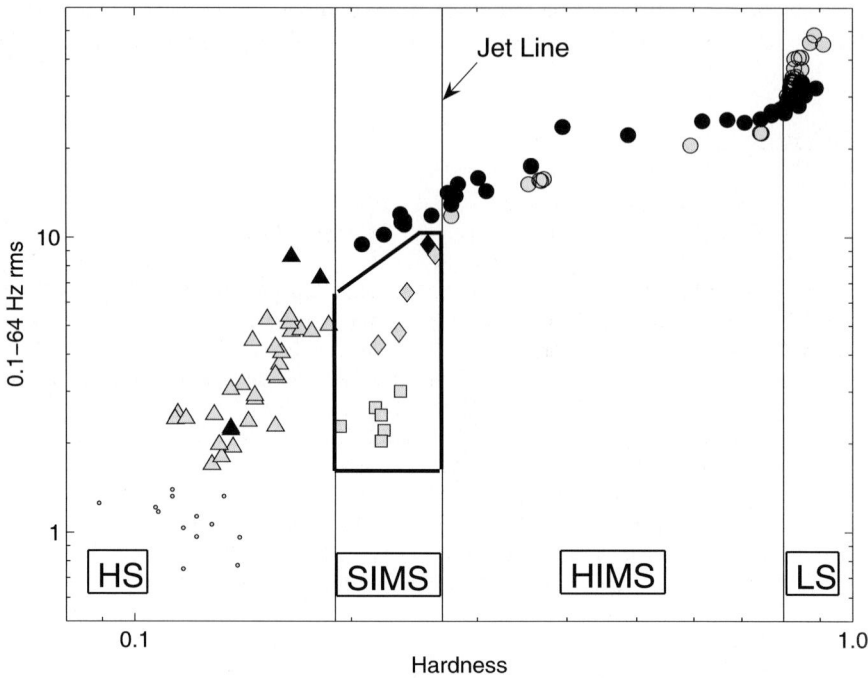

Fig. 2. RMS-hardness diagram corresponding to the observations in Fig. 1. The hardness intervals corresponding to the four states are marked. Gray points correspond to the first part of the outburst (followed right to left), black points to the second part (left to right). Different symbols indicate different power spectral shapes. Circles: strong band-limited noise (and type-C QPO); diamonds: type-B QPO; squares: type-A QPO; triangles: unidentified QPO; dots: only weak signal. The region marked with a thick line shows the SIMS points, characterized by a drop in RMS.

4 Conclusions: Noisy Accretion and Ejection

The general picture that is emerging is becoming more and more clear and it can be applied to a large sample of systems [3]. In particular, the details of the changes in the X-ray emission when crossing the jet line appear crucial for the modeling of the accretion/ejection connection. Although the timing properties change on a very short time scale, at low (<20 keV) energies spectral changes are minimal, as can be seen by the small difference in X-ray hardness (see e.g. [16]). Recently, evidence for a possible fast change in the spectral properties at energies >20 keV was gathered [9], although more high-energy data on this elusive transition should be obtained.

Although timing analysis of the fast variability of BHTs can give us direct measurements of important parameters of the accretion flow, up to now we do not have unique models that permit this. Recent results show that a clear association can be made between type-C QPO, strong band-limited noise and the presence of a relativistic jet. In the framework of unifying models, these results could play an important role.

References

1. J. Homan, T. Belloni: AP&SS **300**, 107 (2005)
2. T. Belloni: Adv. Sp. Res. in press (astro-ph/0507556) (2006)
3. J. Homan, T. Belloni: in preparation (2005)
4. J.A. Tomsick, S. Corbel, A. Goldwurm, P. Kaaret: ApJ **630**, 413 (2005)
5. J.A. Tomsick: Adv. Sp. Res. in press (astro-ph/0509110) (2006)
6. R.P. Fender: Jets from X-ray binaries. In: *Compact Stellar X-ray Binaries*, ed. by W.H.G. Lewin, M. van der Klis (Cambridge Univ. Press, Cambridge 2006), csxs.book 381F (2006)
7. R.P. Fender, T. Belloni, E. Gallo: MNRAS **355**, 1105 (2004)
8. T. Belloni, J. Homan, P. Casella, et al.: A&A **440**, 207 (2005)
9. T. Belloni, I. Parolin, M. Del Santo et al.: MNRAS **367**, 1113 (2006)
10. P. Casella, T. Belloni, L. Stella: ApJ **629**, 403 (2005)
11. J.E. Grove, W.N. Johnson, R.A. Kroeger, et al.: ApJ **500**, 899 (1998)
12. S. Corbel, R.P. Fender, J.A. Tomsick, et al.: ApJ **617**, 1272 (2004)
13. E. Kalemci, J.A. Tomsick, M.M. Buxton, et al.: ApJ **622**, 508 (2005)
14. J. Homan, M. Buxton, S. Markoff, et al.: ApJ **624**, 295 (2005)
15. D.L. Meier: Adv. Sp. Res. in press (astro-ph/0504511) (2006)
16. E. Nespoli, T. Belloni, J. Homan, et al.: A&A **412**, 235 (2003)

Spectral and Variability Properties of LS 5039 from Radio to very High-Energy Gamma-Rays

V. Bosch-Ramon[1], J.M. Paredes[1], and G.E. Romero[2,3]

[1] Departament d'Astronomia i Meteorologia, Universitat de Barcelona, Av. Diagonal 647, E-08028 Barcelona, Catalonia, Spain, vbosch@am.ub.es, jmparedes@ub.edu

[2] Instituto Argentino de Radioastronomía, C.C.5, (1894) Villa Elisa, Buenos Aires, Argentina, romero@iar.unlp.edu.ar

[3] Facultad de Ciencias Astronómicas y Geofísicas, UNLP, Paseo del Bosque, 1900 La Plata, Argentina

1 Introduction

Microquasars are X-ray binaries with relativistic jets. The microquasar LS 5039 turned out to be the first high-energy gamma-ray microquasar candidate due to its likely association with the EGRET source 3EG J1824−1514 [9]. Further theoretical studies supported this association [2], which could be extended to other EGRET sources ([7, 11, 3]). Very recently, [1] have communicated the detection of LS 5039 at TeV energies. This fact confirms the EGRET source association and leaves no doubt about the gamma-ray emitting nature of this object. The aim of the present work is to show that, applying a cold-matter dominated jet model to LS 5039, we can reproduce many of the spectral and variability features observed in this source. Jet physics is explored, and some physical quantities are estimated as a by-product of the performed modeling. Although at the moment only LS 5039 has been detected on the entire electromagnetic spectrum, it does not seem unlikely that other microquasars will show similar spectral properties. Therefore, an in-depth study of the first gamma-ray microquasar, on theoretical grounds supported by observations, can render a useful knowledge applicable elsewhere.

2 A Cold Matter Dominated Jet Model applied to LS 5039

The jet is modeled as dynamically dominated by cold protons and radiatively dominated by relativistic leptons. It considers accretion according to the orbital parameters and companion star properties, allowing for a consistent orbital variability treatment. The magnetic field energy density and the non-thermal particle maximum energy values along the jet depend on the cold matter energy density and the particle acceleration/energy loss balance, respectively, and the amount of relativistic particles within the jet is restricted by the efficiency of the shock to transfer energy to them. The model takes into account the external and internal photon and matter fields, as well as the jet magnetic field, all of them interacting with the jet relativistic particles and producing emission from radio to very high energies. For further details concerning the model, see [4].

The scenario for LS 5039 is that of a high-mass microquasar. It is an eccentric system (e=0.35) with an O7 star, likely harboring a black-hole ([6]). Concerning the jet geometry, we have adopted typical values present in the literature. Regarding other jet properties, for the jet to accretion rate ratio, we have adopted 0.1 and, for the magnetic field, we have fixed it to 0.1 the equipartition value. To explain observations, the shock energy dissipation efficiency the acceleration efficiency have been adjusted to reasonable suitable values ($\sim$10% of kinetic energy radiated and efficient acceleration with dE/dt=0.1qBc), similar to those obtained for other microquasars, SNRs, GRBs and AGNs. The contributions from the disk and the corona have been taken faint according to observational data.

In Fig. 1, we show photon-photon opacities due to the external photon fields for different distances from the compact object. Although this calculation does not take into account the angular dependence of photon-photon absorption, it shows clearly that absorption will affect the higher energy band of the spectrum. The computed SED for LS 5039 is presented in Fig. 2. In general, there is a good agreement between the model and the data, although the model predicts a spectrum in radio that is slightly too hard (see [8]). Also, in the TeV band the fluxes are slightly underpredicted. Both facts can be interpreted as hints of more intense high energy processes (particle acceleration and emission) taking place outside the binary system, since in our model radiation is emitted mainly within the 100 GeV-photon absorption region within the binary system. It is worth noting that the total jet kinetic power required for producing a SED like the one plotted in Fig. 2 is pretty reasonable from the energetic point of view, $\sim 2 \times 10^{36}$ erg s^{-1}. In Fig. 3, the lightcurve of one orbital period at different energy bands is presented. Variability has been modeled by orbital motion of the compact object through a slow asymmetric equatorial wind from a fast rotating stellar companion ([6]). The slow equatorial asymmetric wind can reproduce a peak at X-rays at phase 0.2, as well as the major peak at phase 0.8 ([5]). Moreover, at TeV energies there is a peak also around phase 0.8, which resembles what might be marginally present in HESS data ([6, 1]). For more details concerning the application of the model to LS 5039, see ref. [10].

References

1. Aharonian, F. A. et al.: Science **309**, 746 (2005)
2. Bosch-Ramon, V. & Paredes, J. M.: A&A **417**, 1075 (2004)
3. Bosch-Ramon, V., Romero, G. E., Paredes, J. M.: A&A **429**, 267 (2005)
4. Bosch-Ramon, V., Romero, G. E., Paredes, J. M.: A&A **447**, 263 (2006)
5. Bosch-Ramon, V., Paredes, J. M., Ribó, M. et al.: ApJ**628**, 388 (2005)
6. Casares, J., Ribó, M., Ribas, I. et al.: MNRAS **364**, 899 (2005)
7. Kaufman Bernadó, M., Romero G. E., Mirabel, F. A&A **385**, L10 (2002)
8. Martí, J., Paredes, J. M., Ribó et al.: A&A **338**, 71 (1998)

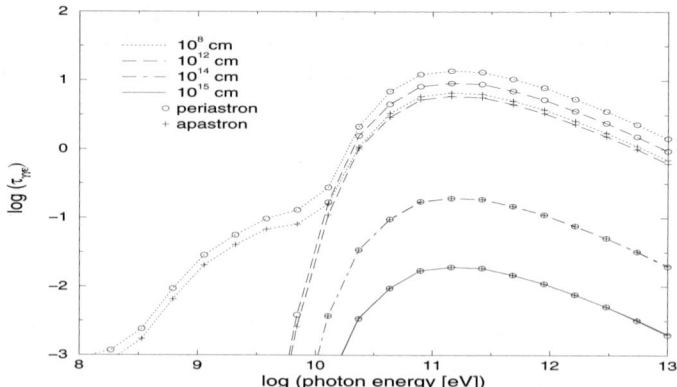

Fig. 1. Photon-photon opacity at different jet distances.

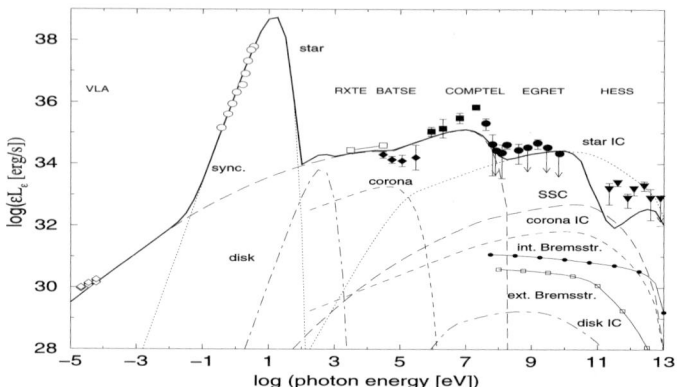

Fig. 2. Computed spectral energy distribution of LS 5039.

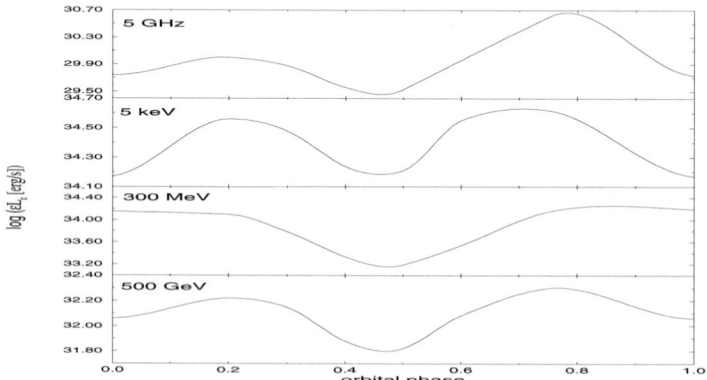

Fig. 3. Computed lightcurves of radio, X-ray and gamma-ray radiation.

9. Paredes, J. M., Martí, J., Ribó, M. et al.:Science **288**, 2340 (2000)
10. Paredes, J. M., Bosch-Ramon, V., Romero, G. E. A&A **451**, 259 (2006)
11. Romero, G. E., Torres, D. F., Kaufman Bernadó, M. et al.: A&A **410**, 1 (2003)

30 Years Blandford-Znajek Process – Are Black Hole Jets Driven by the Ergosphere ?

M. Camenzind

Landessternwarte Königstuhl, ZAH Heidelberg, D–69117 Heidelberg, Germany
`M.Camenzind@lsw.uni-heidelberg.de`

Summary. The collapse of massive stars produces the most bizarre objects in the Universe. In the last ten years, astronomers have been successfully searching for real black holes in the Universe. There is now clear evidence for about 20 stellar mass black holes and a hundred thousand of supermassive black holes in galactic centers. All these black holes are expected to be rapidly rotating objects, and the energy stored in the rotation can be extracted by means of pulsar–like processes when strong magnetic fields are advected by accretion. In the last few years, progress in the simulation of such processes has enabled us to test the fundamental idea of the Blandford-Znajek mechanism and its relation to the launch and collimation of jets.

1 The Two Hairs of Black Holes

Unlike neutron stars, the gravitational field of astrophysical black holes is determined by only two parameters. The mass M is the essential parameter, which is then naturally used to classify astrophysical black holes [5]:

- $M < 2 M_\odot$: **Primordial Black Holes**. These black holes cannot be generated in a normal stellar collapse. They have been postulated to be born in the early Universe, e.g. during the hadronic phase transitions. There is, however, no clear evidence for the existence of such black holes.
- $2.2 M_\odot < M < 100 M_\odot$: **Stellar Black Holes**. They are formed in the stellar collapse of stars initially more massive than $\simeq 25 M_\odot$. In binary systems, a lot of mass can be lost during the evolution so that the final mass of the resulting black hole need not exceed 3 solar masses.
- $100 M_\odot < M < 10^5 M_\odot$: **Intermediate Mass Black Holes**. A few candidates have been found in the form of ultra–luminous X–ray sources, but they wait for confirmation. Middleweight black holes in this mass range have been hard to pin down.
- $10^5 M_\odot < M < 10^{10} M_\odot$: **Supermassive Black Holes** are found in centers of nearby galaxies and in all type of active galactic nuclei [6].

Masses of black holes in the center of galaxies can now be measured very accurately for various types of galaxies and even for high redshift quasars. The masses of black

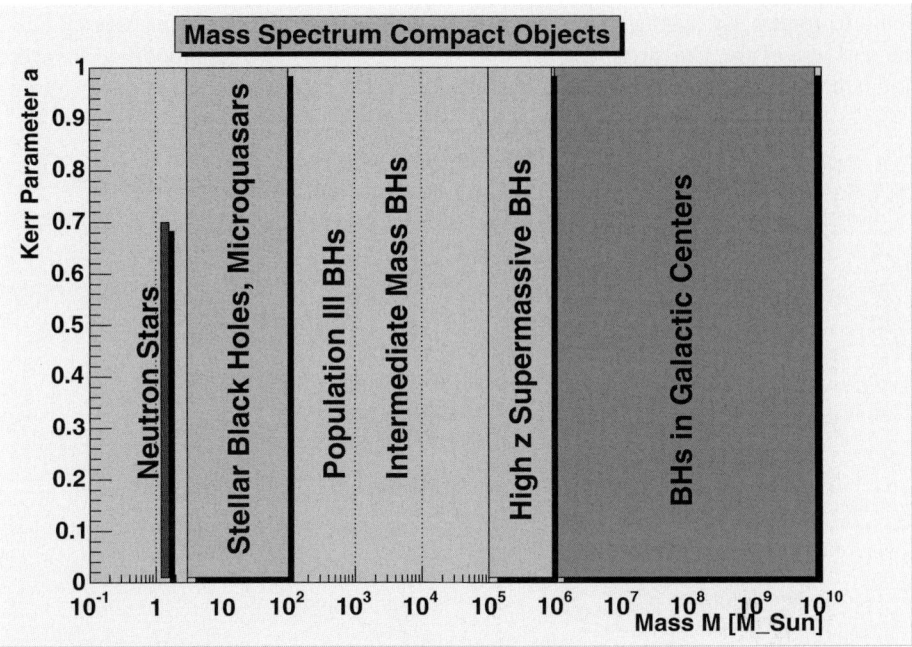

Fig. 1. Two–Hair plane for cosmic black holes [6]. In the last 15 years, masses have been determined very accurately for neutron stars, stellar and supermassive black holes. The spin of these black holes is still largely undetermined. Population III stars at redshifts $z \simeq 15 - 25$ will leave behind black holes in the intermediate mass range, which can grow by merging and accretion towards supermassive black holes. Alternatively, these black holes could grow by accretion from high redshift black holes with masses of a few hundred thousand solar masses formed in supermassive cloud collapse at redshift $z \simeq 12$ [19].

holes in the center of nearby galaxies correlate with the properties of the underlying bulge of the galaxy. There is a tight correlation between the mass of the black hole and the mass of the bulge [13], and there is the famous Magorrian relation which correlates the black hole mass with the velocity dispersion of the stars in the bulge, $M_H \propto \sigma_*^4$.

The mass M and the spin parameter a of a black hole form the two–hair plane (M, a) (Fig. 1). In this plane, the neutron stars cluster around a very narrow peak at $1.4 \, M_\odot$ with a maximum spin parameter a/M not exceeding $\simeq 0.7$. Stellar black holes populate the region between 3 and $100 \, M_\odot$. We expect that population III stars in the early Universe leave black holes even in the range of a few hundred solar masses. These black holes could grow by merging processes and accretion towards masses of even beyond $10^5 \, M_\odot$. For sure, all galaxies seem to harbor black holes with masses in the range from 10^6 to $10^{10} \, M_\odot$. Black holes are now well established cosmic objects, though there are still no reliable estimates for their spin parameter.

We expect however that black holes are in general spinning objects. The spin of a black hole would be caused by the angular momentum of the star or cloud that formed it, particularly if that progenitor is a spinning neutron star. Alternatively, a black hole could form directly during a supernova or the collapse of some supermassive clouds in the dark era of the Universe. The original spin is certainly strongly modified by accretion. Chandra and XMM-Newton observations of iron atoms in the hot gas orbiting 3

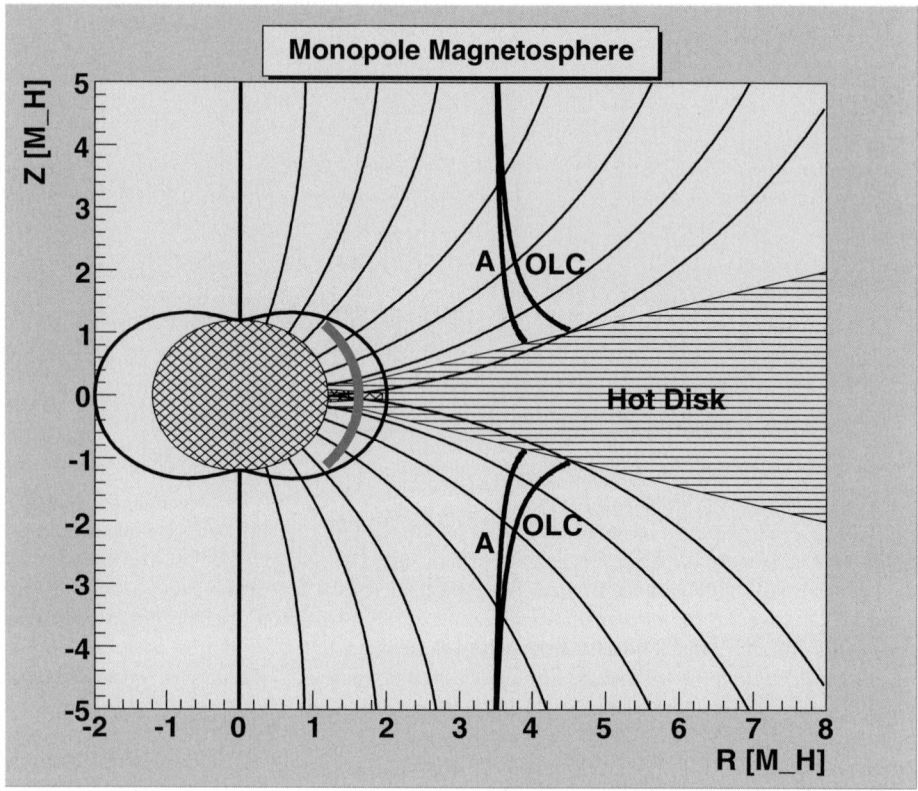

Fig. 2. The Black Hole magnetosphere in the strong field limit. Magnetic fields are advected from the parsec–scale in an elliptical galaxy and embed the horizon into a spherical magnetic structure. Plasma from the inner part of the hot disk (thick semi–circle near the horizon) is driven away and crosses the Alfvén surface (A) and the outer light cylinder (OLC). Collimation of these plasma carrying field lines is expected to occur near 100 Schwarzschild radii.

stellar black holes and of a handful active galactic nuclei have allowed to investigate the gravitational effects and spin of these black holes. Broad, asymmetric Fe Kα emission lines are found in the spectra of both supermassive black holes and stellar–mass Galactic black holes. Gravity of a black hole shifts X-rays from iron atoms to lower energies, producing a strongly skewed X-ray signal.

2 Black Hole Magnetospheres

Historically, black hole magnetospheres have been discussed in the force–free limit [2, 18, 1, 10], in analogy to the discussion of pulsar magnetospheres [8, 16]. In this approximation, the magnetic field structure, represented by the axisymmetric magnetic flux function $\Psi(\varpi, z)$ is a solution of the general relativistic Grad–Shafranov equation [3, 1]

$$\nabla \cdot \left[\frac{\alpha D}{\varpi^2} \nabla \Psi \right] + \frac{\Omega_F - \omega}{\alpha} \, \Omega_F{}' |\nabla \Psi|^2 + \frac{16\pi^2 \, II'}{\alpha \varpi^2} = 0 \,. \tag{1}$$

α is the redshift factor of the Kerr geometry in BL coordinates, ϖ the corresponding cylindrical radius, ω the frame–dragging potential, and D the light cylinder function. $D = 0$ gives the position of the inner light cylinder surface (ILC) and the classical outer one (OLC). Field lines rotate differentially with angular frequency $\Omega_F(\Psi)$ and the poloidal current function $I(\Psi)$ is arbitrary in the force–free limit. Except for the Blandford–Znajek solution of slowly rotating holes [2], no analytic solutions are known. Fig. 2 shows the global structure of a magnetosphere around rotating black holes. In fact it turns out to be much easier to find solutions with a time–dependent relaxation method [15] than to solve eq (1).

3 Time–Dependent GRMHD and Jet Outflows

In the last years, new methods have been developed to treat Maxwell's MHD equations in a time–dependent way [5]. As in the case of pure plasma dynamics [11], Maxwell's equations are formulated in conservative form [12, 15]

$$\partial_t(\sqrt{\gamma}\,\mathbf{U}) + \partial_i(\sqrt{\gamma}\,\mathcal{F}^i) = \sqrt{\gamma}\,\mathcal{S}(g) \,, \tag{2}$$

where $\mathbf{U}$ represents the state vector of the hyperbolic system and $\mathcal{F}^i$ the corresponding fluxes. Komissarov [15] has recently shown how to formulate Maxwell's equations for magnetospheres on Kerr spaces in a coordinate system which is regular at least at the outer horizon, $r = r_+$ (Kerr–Schild e.g.).

As an example, Komissarov [15] has investigated the relaxation of the Wald solution, augmented by some plasma (electron–positron pairs e.g.). Though the initial state is not rotating, all field lines connected to the ergosphere are driven to rotation by the frame–dragging effect, confirming the BZ mechanism. Field lines outside the ergosphere are still not rotating. In the time–dependent problem, the horizon is no special boundary surface, no currents are flowing along the horizon, as required in the membrane formalism [20]. Instead of this, a current sheet is formed within the ergosphere along the equatorial plane, where the force–free condition $\mathbf{B}^2 - \mathbf{E}^2 > 0$ breaks down.

Plasma outflows driven away along field lines connecting the ergosphere are then collimated on the scales of $\simeq 100$ Schwarzschild radii [3], [6]. This is consistent with recent observations on M87 [9]. The jet in M87 is straight and well collimated on the scales from 2 mas to about 200 mas (Fig. 3). Some flaring is triggered at a distance of

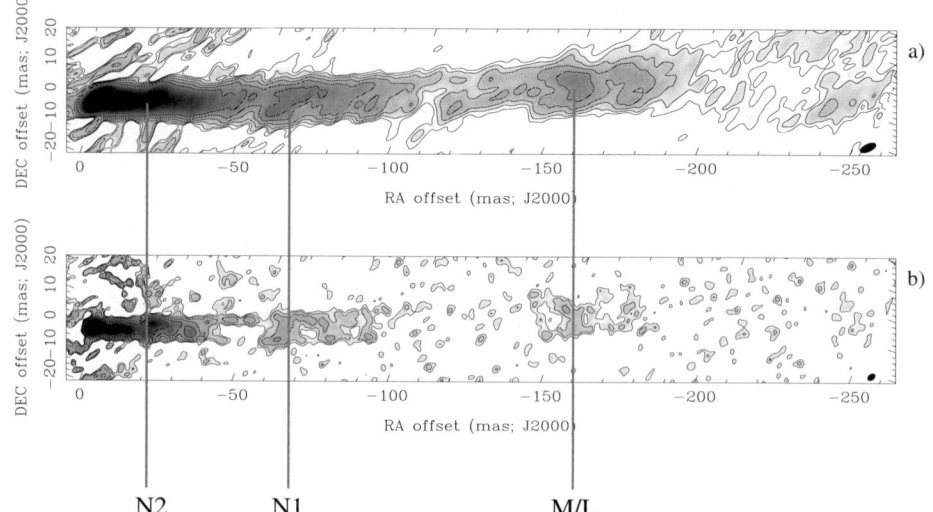

N2   N1   M/L

Fig. 3. M87 Jet at 1.6 GHz (a) and 4.8 GHz (b) as observed by the VLBA [9]. The 1.6 GHz resolution is 5.9×2.6 and the 4.8 GHz resolution is 2.9×1.9. The images have been rotated by $-20.62°$ so that the jet runs left to right. 1 mas corresponds to 200 Schwarzschild radii.

about 100 jet radii, and the jet recollimates at the position of the HST-1 knot, where superluminal motion is observed (about 800 mas from the core). In this region, Poynting–flux is converted to kinetic energy. The successful simulation of this process is still open.

Acknowledgments. Research on supermassive black holes has been supported by the DFG Sonderforschungsbereich 439 in Heidelberg.

References

1. V.S. Beskin, I.V. Kuznetsova, Il Nuovo Cimento B, **115**, 795 (2000)
2. R.D. Blandford, R.L. Znajek, MNRAS **179**, 433 (1977)
3. M. Camenzind, in: *Relativistic Astrophysics*, eds. H. Riffert et al., Vieweg (1998)
4. M. Camenzind, Mem. Soc. Astron. It. **76**, 98 (2005)
5. M. Camenzind, in *Cosmic Magnetic Fields*, ed. R. Wielebinski and R. Beck, Lecture Notes Phys. **664**, 255 (2005)
6. M. Camenzind, Ann. Physik **15**, 60 (2005)
7. M. Camenzind, *Astrophysics of Compact Objects – White Dwarfs, Neutron Stars and Black Holes*, Springer–Verlag (Berlin 2006), in press
8. I. Contopoulos, D. Kazanas, C. Fendt, ApJ **511**, 351 (1999)
9. R. Dodson et al., PASJ **58**, 243 (2006)
10. C. Fendt, E. Memola, A&A **365**, 631 (2001)
11. J.A. Font, Living Rev. Rel. **6**, 4 (2003)
12. C.F. Gammie, J.C. McKinney, G. Toth, ApJ **589**, 444 (2003)

13. N. Haering, H. W. Rix, ApJL **604**, L89 (2004)
14. S. Komissarov, MNRAS **303**, 343 (1999)
15. S. Komissarov, MNRAS **350**, 427/1431 (2004)
16. S. Komissarov, MNRAS **367**, 19 (2006)
17. A. Müller, M. Camenzind, A&A **413**, 861 (2004)
18. I. Okamoto, MNRAS **254**, 192 (1992)
19. M. Spaans, J. Silk, MNRAS (2006); astro–ph/0601714
20. K.S. Thorne, R.H. Price, D.A. Macdonald, *Black Holes: the Membrane Paradigm*, Yale University Press (New Haven 1986)

Radiative Acceleration and Collimation of Jets from TCAF Discs

I. Chattopadhyay

Dept. of Astronomy and Space Sc., Chungnam National University, Daejeon 305-764, South Korea. indra@canopus.cnu.ac.kr

1 Introduction

Matter spiraling onto black holes are known to suffer centrifugal pressure supported shocks [1]. In the post-shock region entropy goes up, density goes up, and the hot and dense post-shock matter puffs up in the form of a torus, and is abbreviated as CENBOL (CENtrifugal pressure supported BOundary Layer). Hot and puffed up CENBOL intercepts 'soft photons' from cooler Keplerian disc, and inverse-Comptonize to produce power-law tail of 'hard photons' [2]. However, if the Keplerian accretion rate increases, excess soft photons will cool down the hot CENBOL electrons. In the process the puffed up CENBOL slumps, cannot intercept significant soft photons. On the top of that the electrons in CENBOL being cold, cannot produce the high energy photons, thus generating the so-called soft state. Such a disc model is called Two Component Accretion Flow (hereafter TCAF) model.

Excess thermal gradient force in CENBOL can also drive matter along the axis of symmetry to produce jets [3]. However only thermal gradient force may not be sufficient to produce relativistic terminal speeds observed in some black hole candidates [4]. Chattopadhyay and Chakrabarti [5] showed, that normal plasma cannot be accelerated by disc radiation to over 90% velocity of light (c), though pair dominated jets may achieve relativistic terminal speed v_∞ [6]. The question which can be raised in this context is, what should be the spectral state of the disc in which such relativistic outflows are collimated? We aim to do so in this paper. In the next section, we discuss model assumptions, equations of motion and the result. In the last section we draw conclusion.

2 Model Assumptions and Results

In this paper we do not consider the generation mechanism of jets. We assume the post-shock CENBOL produces significant pairs, and at the base they are rotating with the local rotational velocity of disc. We ignore the role of magnetic field and assume the fluid to be radiation pressure dominated. In Fig. 1, we present a cartoon diagram of the TCAF disc with jets. The two components (SKH and KD) are shown.

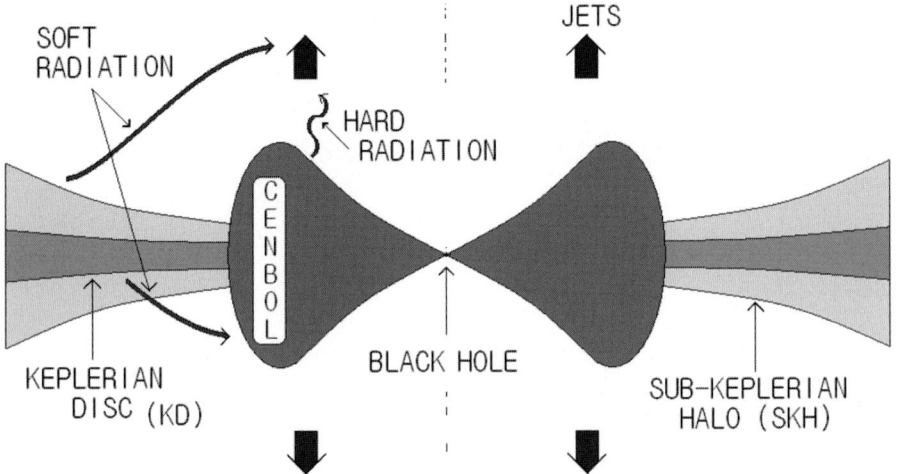

Fig. 1. Schematic diagram of TCAF disc model.

The source of radiation are two, the Keplerian disc (KD) which supplies soft black-body type radiation, and the sub-Keplerian halo (SKH), which supplies hard but power-law photons. Photons coming out of the disc will be Doppler shifted due to the disc motion. Though the radiation is locally isotropic (at the disc surface) but the geometry of the disc and the disc motion will severely influence the radiation field above the disc and would be different from purely thick disc [7] and thin disc [8].

The full set of relativistic equations of hydrodynamics is provided in earlier works [9, 10], we do not present them here for brevity of space. The radiation field for TCAF disc model is characterized by a very strong contribution from the CENBOL region and a much weaker contribution from KD. Therefore KD radiation does not account for the acceleration of jets. Since the net radiation force (including the drag terms) due to CEN-BOL peaks much closer to the black hole, it accelerates jets to very high speeds much closer ($\sim 100 r_g$; r_g = Schwarzschild radius) to the black hole. On the other hand drag terms due to KD radiation are stronger than the accelerating terms at such ($\sim 100 r_g$) distances above the disc plane. If CENBOL luminosity (ℓ_c) is low ($\lesssim 0.1$), then increasing KD luminosity (ℓ_k) can start accelerating the flow $\sim 1000 r_g$, however if $\ell_c \gtrsim 0.2 L_{\mathrm{Edd}}$ ($L_{\mathrm{Edd}} \equiv$ Eddington luminosity), then increasing ℓ_k will actually decrease the v_∞ of the jet. Such acceleration/deceleration phenomena for TCAF discs has been discussed elsewhere [6, 10] in detail.

Though in general, contribution due to CENBOL in various moments dominates that due to KD, radial flux and therefore the corresponding radial force from KD radiations are comparable to that due to CENBOL. Moreover the radial flux due to KD is directed towards the axis of symmetry above the CENBOL region. Thus instead of an extreme hard state ($\ell_c \gg \ell_k$), if we have an inter-mediate hard state, i.e. $\ell_c \sim \ell_k$, then radial force of KD will push the matter towards the axis of symmetry and will enhance collimation. Figure (2a), shows that increasing ℓ_k has a very marginal effect in deter-

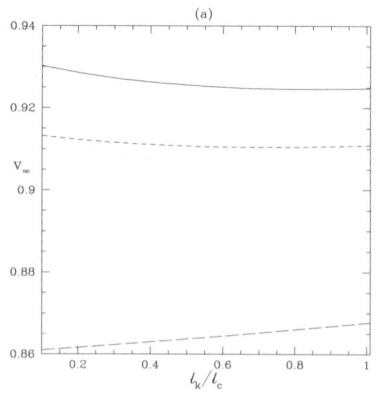 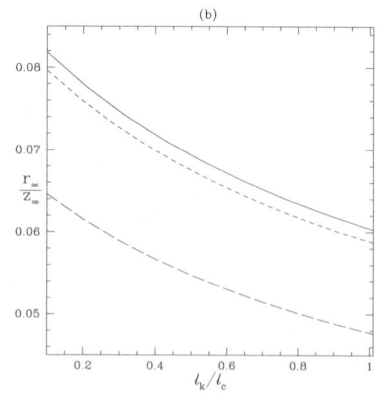

Fig. 2. (a) v_∞, (b) r_∞/z_∞ with ℓ_k/ℓ_c. $\ell_c = 0.5$ (solid), $\ell_c = 0.3$ (dashed) and $\ell_c = 0.1$ (long-dashed); $x_s = 20 r_g$.

mining v_∞, but Fig. (2b) shows that for the same parameters the collimation parameter r_∞/z_∞ (r & z at $v = v_\infty$), decreases rapidly with increasing ℓ_k.

3 Conclusion

We found that the radiation field produced by a TCAF disc can both accelerate and colli-mate pair dominated jets, photons from the post-shock region are responsible for accel-eration and photons from the Keplerian disc in collimating the flow. Therefore relativis-tic and collimated jets are possible not in typical low/hard states but high/intermediate-hard states.

Acknowledgments. The author acknowledges KOSEF grant R01-2004-000-10005-0.

References

1. S. K. Chakrabarti: Astrophys. J. **347**, 365 (1989)
2. S. K. Chakrabarti, L. Titarchuk: Astrophys. J. **455**, 623 (1995)
3. S. Das, I. Chattopadhyay, A. Nandi, S. K. Chakrabarti: Astrophys. Astron. **379**, 683 (2001)
4. I. F. Mirabel, L. F. Rodriguez: Nature, **371**, 46 (1994)
5. I. Chattopadhyay, S. K. Chakrabarti: MNRAS, **333**, 454 (2002)
6. I. Chattopadhyay, S. Das, S. K. Chakrabarti: MNRAS, **348**, 846 (2004)
7. K. Watarai, J. Fukue: PASJ, **51**, 725 (1999)
8. Y. Tajima, J. Fukue: PASJ, **48**, 529 (1996)
9. D. Mihalas, B. W. Mihalas: *Foundations of Radiation Hydrodynamics*, Oxford University Press, Oxford (1984)
10. I. Chattopadhyay: MNRAS, **356**, 145 (2005)

Relativistic Jets in Active Galactic Nuclei: Importance of Magnetic Fields

I. Duţan[1] and P.L. Biermann[1,2]

[1] Max-Planck-Institut für Radioastronomie, Auf dem Hügel 69, D-53121 Bonn, Germany,
idutan@mpifr-bonn.mpg.de
[2] University of Bonn, Dept of Astronomy & Physics, Auf dem Hügel 71, D-53121 Bonn,
Germany, plbiermann@mpifr-bonn.mpg.de

Summary. We present a theoretical model for powering relativistic jets in Active Galactic Nuclei (AGN) from the accretion flow power, which can be increased by the power coming from the spindown of the black hole (BH). The BH rotational energy is transferred to the accretion disk through closed magnetic field lines that connect them. Adding these two energy sources, the accreting mass flow energy and the rotational energy, we derive the jets-launching power, as well as the efficiency of the jets drive for both the Eddington accretion rate and low accretion rates. In the case of low accretion rates, the efficiency of the jets drive can reach unity, that means almost all of the power used to drive the jets comes from the BH rotational energy.

1 Power of the Jets Launching

For a steady, axisymmetric, and thin accretion disk around a Kerr BH, the general relativistic equations of angular momentum conservation and energy conservation have been investigated by Page & Thorne [5]. By our assumptions, the energy transported by the accreting mass of the disk, and the energy transported by the magnetic torques in the disk are carried away from the disk surfaces by the jets. From the energy conservation law, we can obtain the power of the jets extracted from the accretion disk and from the black hole rotation:

$$P_{jets} = (1 - q_m) \, \dot{M}_D c^2 \left[E^\dagger (r_{sl}) - E^\dagger (r_{ms}) \right] + 4\pi \int_{r_{ms}}^{r_{sl}} r H \Omega_D dr \qquad (1)$$

where $q_m = \dot{M}_{jet}/\dot{M}_D \simeq 0.05$ is the ratio of the mass flow rate into the jet to the mass accretion rate into the hole [3], $E^\dagger$ is the specific energy of the particles that orbit in the disk, H is the flux of the angular momentum transferred from the black hole to the disk by poloidal magnetic fields [4], and Ω_D is the angular velocity of particles in the disk. Here, the first term describes the rest energy of the accreting material onto the black hole, and the second term describes the energy transfer from the rotating black hole to the disk, which we then assume can go into the jets. The energy and angular momentum are transported from the BH to the accretion disk by poloidal magnetic field lines that connect them (variant of Blandford-Znajek mechanism [2]). Tapping the energy of the rotation of the black hole implies additional energy for the jets. The poloidal magnetic field threading the event horizon of the black hole B_H^p and the poloidal magnetic field threading the inner edge of the disk $B_D^p(r_{ms})$ are related by

$$B_H^p = \zeta B_D^p(r_{ms}), \text{ where } \zeta \geq 1 \tag{2}$$

Since the maximum of the power of the jets is obtained when $\zeta = 1$, in the following calculations we take this value of ζ.

For a black hole of mass $10^9 M_\odot$, the strength of the poloidal magnetic field near the BH cannot exceed $10^4 Gauss$ (Znajek [6]). Following Blandford [1], we consider the poloidal magnetic field strength, which threads the accretion disk ($r_* = r/r_g$), to be:

$$B_D^p \propto r_*^{-n}, \text{ where } 0 < n < 3. \tag{3}$$

Considering a continuum of magnetic flux between the BH and the accretion disk and the scaling index $n = 5/4$, we derive the launching power of the jets as a function of the Kerr BH spin parameter a_* for different values of accretion rate $\dot{m}$, which is depicted in Fig. 1 in the case of rapidly spinning black holes $0.950000 \leq a_* \leq 0.999999$. The energy source of the jets coming from the gravitational binding energy between the BH and the accretion disk is increased by the rotational energy of the BH.

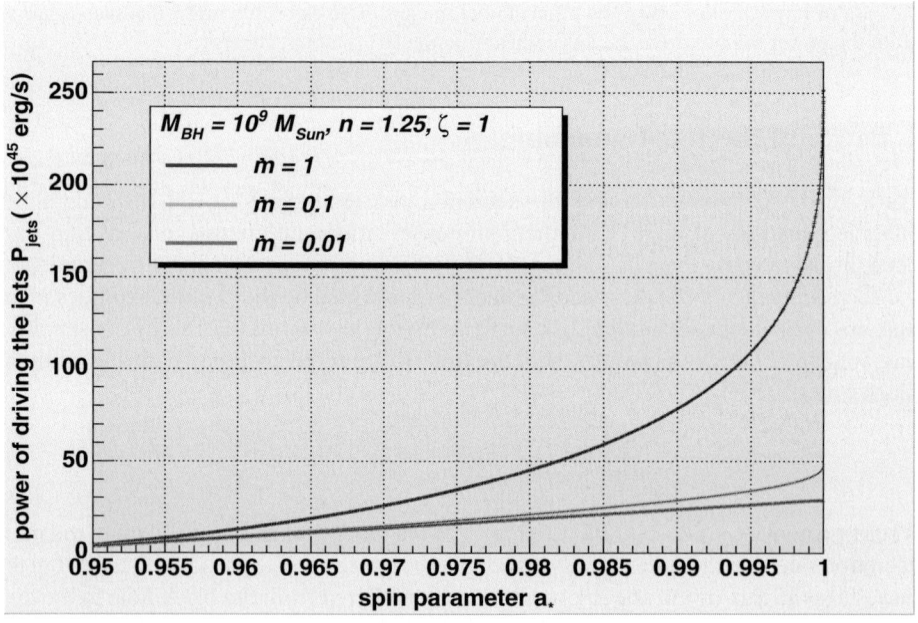

Fig. 1. The power of the jets launching as a function of spin parameter a_* of Kerr BH for different values of accretion rate $\dot{m}$.

2 Efficiency of the Jet Launching

We define the efficiency of the jet launching as the ratio of the total energy of the jets to the total contribution coming from the rest energy of the accreting mass flow and from the rotational energy of the black hole; thus

$$\eta = \frac{P_{jets}}{\dot{m} M_{Edd} c^2 + P_{jets}^{rot}} \tag{4}$$

At low accretion rates ($\dot{m} < 1$), the efficiency of the jet launching can reach values close to unity. In this case, even though the contribution of the rest energy of the accreting mass is very low, the jets can still be driven due to the energy transferred from the black hole to the disk and then into the jet.

3 Conclusions

- We derive the power of the jet launching (Eq. 1) and the efficiency of this process (Eq. 4), while the whole energy released at the innermost stable orbit is used to power the jets.
- The poloidal magnetic field connects the BH to the accretion disk; therefore, the BH exerts a magnetic torque on the disk. Energy and angular momentum are transferred from the BH to the accretion disk, increasing the energy available due to accreting gas.
- For $\zeta = 1$ (Eq. 2) and $n = 5/4$ (Eq. 3), respectively, the power of the jet launching ranges from $\sim 10^{45}$ to $\sim 10^{47} erg/s$, while $\dot{m}$ ranges from 0.01 to 1 (Fig. 1).
- In the case of a very low accretion rate, the efficiency of the jet launching (Eq. 4) can reach values close to unity, that is, almost 100 percent of the energy released from the inner edge of the accretion disk comes from the rotational energy of the BH.

Acknowledgments. I. D.'s studies are being funded by the International Max Planck Research School (IMPRS) for Radio and Infrared Astronomy at the Universities of Bonn and Köln, and the Max-Planck-Institut für Radioastronomie.

References

1. R. D. Blandford: MNRAS **176**, 465 (1976)
2. R. D. Blandford, R. L. Znajek: MNRAS **179**, 433 (1977)
3. H. Falcke, P. L. Biermann: A&A **342**, 49 (1999)
4. L.- X. Li: Ap. J. **567**, 463 (2002)
5. D. N. Page, K. S. Thorne: Ap. J. **191**, 499 (1974)
6. R. L. Znajek: MNRAS **185**, 833 (1978)

String Mechanism for Relativistic Jet Formation

S.A. Dyadechkin[1], V.S. Semenov[1], B. Punsly[2,3], and H.K. Biernat[4]

[1] Institute of Physics, State University St.Petersburg, Russia,
   (ego,sem)@geo.phys.spbu.ru
[2] Boeing Space and Intelligence Systems, Torrance, USA.
[3] International Center for Relativistic Astrophysics (ICRA), University of Rome La Sapienza,
   I-00185 Roma, Italy, brian.m.punsly@boeing.com
[4] Space Research Institute, Austrian Academy of Sciences, Graz, Austria,
   helfried.biernat@oeaw.ac.at

Summary. Here we present our latest studies of relativistic jet formation in the vicinity of a rotating black hole where the reconnection process has been taken into account. In order to simplify the problem, we use Lagrangian formalism and develop a method which enables us to consider a magnetized plasma as a set of magnetic flux tubes [5, 6]. Within the limits of the Lagrangian approach, we perform numerical simulations of the flux tube (nonlinear string) behavior which clearly demonstrates the process of relativistic jet formation in the form of outgoing torsional nonlinear waves. It turns out that the jet is produced deep inside the ergosphere where the flux tube takes away spinning energy from the black hole due to the nonlocal Penrose process [2]. This is similar to the Blandford-Znajek (BZ) mechanism to some extent [8], however, the string mechanism is essentially time dependent. It is shown that the leading part of the accreting tube gains negative energy and therefore has to stay in the ergosphere forever. Simultaneously, another part of the tube propagates along the spinning axis away from the hole with nearly the speed of light. As a result, the tube is continuously stretching and our mechanism is essentially time dependent. Obviously, such process cannot last infinitely long and we have to take into account the reconnection process. Due to reconnection, the topology of the flux tube is changed and it gives rise to a plasmoid creation which propagates along spin axis of the hole with relativistic speed carrying off the energy and angular momentum away from the black hole.

Relativistic jets are widespread in the Universe and reveal themselves in different objects as quasars, micro-quasars, active galactic nuclei etc. [1]. At present time, it is generally believed that the power source of jets are drawn from spinning energy of a black hole by means of an interaction of a rotating black hole with a magnetized plasma. However, the magnetohydrodynamic (MHD) interaction of a rotating black hole with accreting plasma is extremely complicated. Numerical simulations performed recently [3, 4] did not clarify the physical mechanism of relativistic jet creation completely. Two main questions are still unsolved: what is the mechanism of jet acceleration and collimation, is it located inside or outside the ergosphere?

Our simulation shows that the physical mechanism of the energy extraction from the rotating black hole and jet formation is rather simple. Due to the inhomogeneous rotation of the space around the black hole (Lense-Thirring effect), the flux tube be-

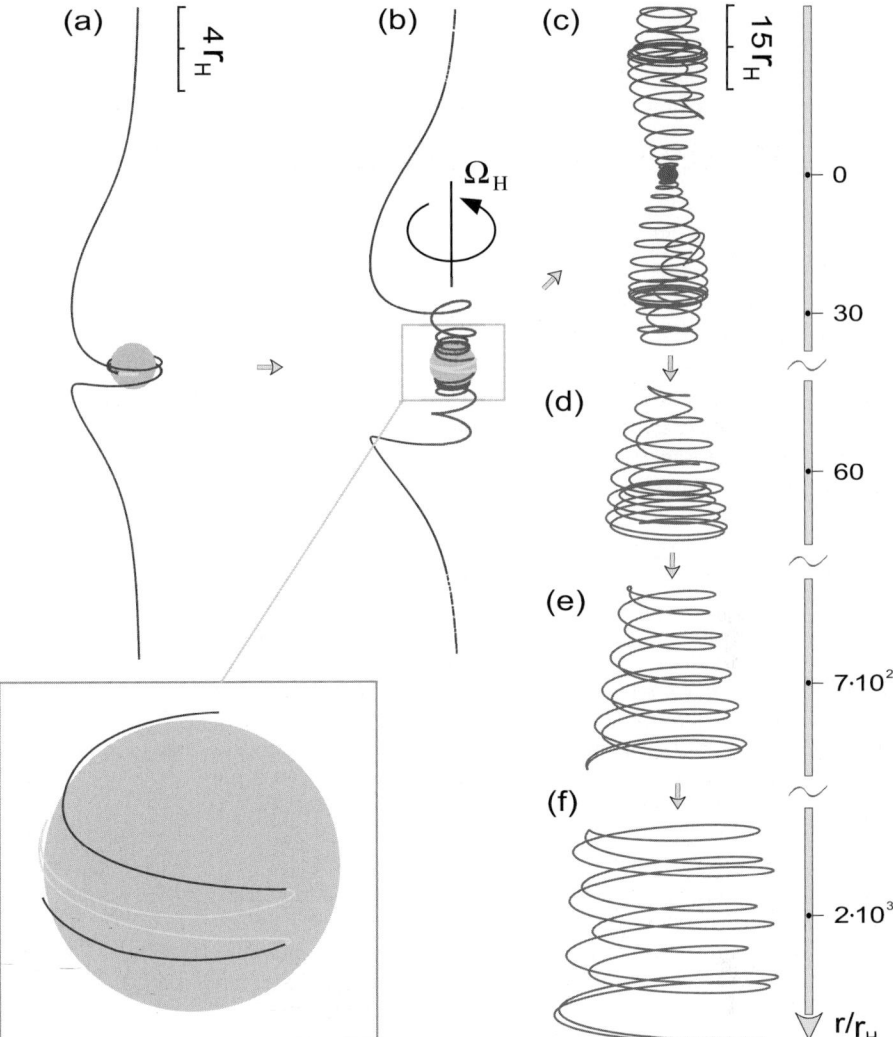

Fig. 1. Time evolution of the magnetic flux falling into a rotating black hole. This figure shows the different moments of the simulations, throughout the figure r_H is black hole horizon. (a) corresponds to the negative energy creation onset (part of the string with negative energy is labeled by yellow / light grey); (b) beginning of the spiral structure creation; (c) represents the initial stage of reconnection (red / dark grey marks the part of the string which is separated from the jet); (d) , (e) and (f) represent the development of the separated flux tube in the course of time. The result of reconnection is a plasmoid which carries energy and angular momentum away from the black hole (f).

comes stretched and twisted, leading to string braking (Fig. 1a) [7]. The deeper the tube falls into the hole, the more the tube is stretched, the stronger the magnetic field gets, the slower the central part of the tube rotates, and eventually the central part of the tube gains negative energy. Since the energy of the tube as a whole has to be conserved, the energy is redistributed along the flux tube which leads to an energy extracting from the hole by means of torsional Alfvén waves (Fig. 1b,c).

The continuous stretching of the flux tube in the course of the jet creation, evidently cannot last infinitely long, hence, we need to take into account another physical process, namely magnetic reconnection, which can change the magnetic flux topology and release magnetic tensions. We include this process in our simulation and present preliminary results. Reconnection unavoidably leads to flux tube breaking, and as a result, to the creation of a plasmoid (Fig. 1d), which propagates with relativistic speed along the rotation axis and carries energy and angular momentum far away from the black hole (Fig. 1c,d,e,f). Furthermore, the simulation demonstrates other interesting properties of the plasmoid behavior. First of all, a sufficiently strong jet collimation occurs (Fig. 1d,e,f), which obviously is attained by the equilibrium between the centrifugal forces of the plasma rotation and magnetic field stresses. Secondly, gradual simplification of the magnetic structure in the course of time, which implies that magnetic energy is transformed to plasma energy.

As evidenced by the foregoing, our mechanism is essentially time dependent and cannot be reduced to steady state as for instance the BZ mechanism [8]. Our mechanism can be applied to the explanation of different jet phenomena.

References

1. I.D. Novikov: *Black Holes and the Universe*, (Cambridge University Press 1990)
2. R. Penrose: Nuovo Cimento **1**, 252 (1969)
3. S.S. Komissarov: Mon. Not. R. Astron. Soc. **359**, 801 (2005)
4. S. Koide: ApJ. **606**, 45 (2004)
5. V.S. Semenov: Physica Scripta. **62**, 123 (2000)
6. V.S. Semenov, S.A. Dyadechkin, I.B. Ivanov and H.K. Biernat: Physica Scripta. **65**, 13 (2002)
7. V.S. Semenov, S.A. Dyadechkin and B. Punsly: Science. **305**, 978 (2004)
8. R.D. Blandford and R.L. Znajek: Mon. Not. R. Astron. Soc. **179**, 433 (1977)

Shock Location in Funnel Flows onto Magnetized Neutron Stars

S. Karino[1], M. Kino[1], and J.C. Miller[1,2]

[1] SISSA, via Beirut 2/4, Trieste, Italy, `karino@sissa.it`
[2] Department of Physics, University of Oxford, UK

In this study, we have considered funnel-flow accretion onto a magnetized neutron star, including taking into account the possible occurance of a standing shock. We have assumed that the funnel-flow goes strictly along the magnetic field lines and that the field is a dipole field aligned with rotation axis. The funnel-flow leaves from the inner edge of the disc and accelerates smoothly into the supersonic regime, passing through a critical point. Then, at some point after this, a standing shock occurs beyond which the flow then accretes sub-sonically onto the neutron star [1, 2]. We calculate the location of the standing shock, neglecting energy losses, and obtain a full solution for given boundary conditions at the stellar surface and at the inner edge of the disc. Such a simplified accretion flow can be described by, magnetic field geometry, mass conservation law under the MHD condition, and bernoulli equation [3, 4]

$$B_{\mathrm{p}}^2 = \frac{\mu^2}{r^6}\left(4 - \frac{3r}{R_{\mathrm{d}}}\right), \ 4\pi\rho\frac{v}{B_{\mathrm{p}}} = K,$$

$$E = \frac{1}{2}v^2 + \frac{a^2}{\gamma - 1} - \frac{Gm}{r} - \frac{1}{2}\Omega^2 r^2 \sin^2\theta, \tag{1}$$

where E is the specific energy (assumed to be constant), v is the poloidal velocity along the field line, and a is the sound speed, K is a constant.

Our procedure for calculating the entire flow solution, including the shock, starting from the inner edge of the disc (R_{d}) and ending at the surface of the NS (r_{NS}), is as follows: (i) Choose a position for the shock $r = r_{\mathrm{s}}$, with $r_{\mathrm{NS}} \leq r_{\mathrm{s}} \leq r_{\mathrm{c}}$. (ii) Solve for the flow in the region upstream of the shock, $r_{\mathrm{s}} \leq r \leq R_{\mathrm{d}}$, using above equation. This gives the value of $\rho_1 = \rho(r_{\mathrm{s}})$, just before the shock, from which can be calculated the corresponding values of p_1 and M_1. (iii) Using the Rankine-Hugoniot relations, we then calculate the values of density and pressure beyond the shock (ρ_2 and p_2) and hence the new value of the specific entropy s_2. (iv) Using these values, we then calculate the flow beyond the shock using Bernoulli equation, including obtaining the value of ρ at the stellar surface and hence also the values of the other fluid parameters there. (v) In practice, we would prefer to specify the value of the pressure at the stellar surface rather than the location of the shock and so we iterate the solution until the shock location corresponding to the desired surface pressure is found.

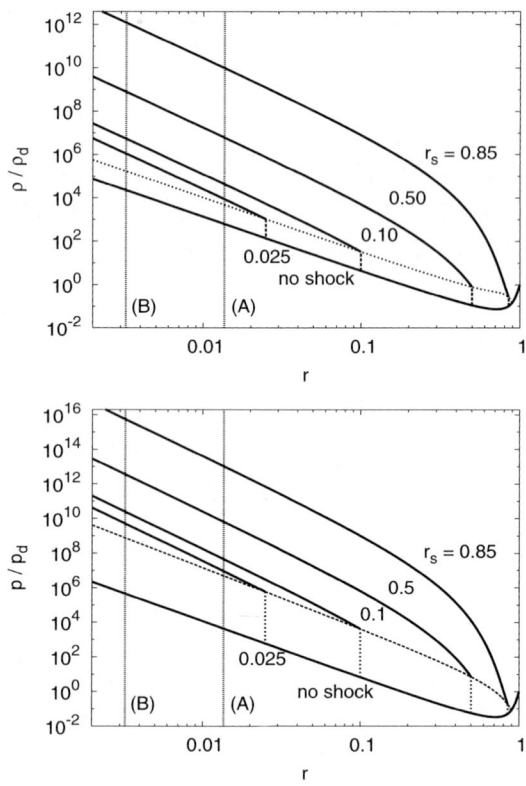

Fig. 1. Density (top) and Pressure (bottom) distributions for flows including shocks, taking $\gamma = 1.3$ and $s = 0.005$ (specific entropy). The upper solid curves are for flows with shock discontinuities at $r = 0.85, 0.5$ and 0.1 ; the dotted curve is the envelope of the density values on the downstream side of the shock; the lowest thick solid curve is for a flow with no shock. The vertical lines denote positions of the NS surface: (A) is the case for a disc model with $\alpha = 1.0$ and (B) denotes the the case of a free-fall accretion model.

Fig. 1 show the variation of density and pressure along the flowlines; the lowest solid curve corresponds to transonic flow without a shock, and the upper curves are for flows with having an adiabatic shock at different locations. From Fig. 1, we see that the resulting density and pressure at the stellar surface (which comes at the locations labelled (A) and (B) according to the different boundary conditions) depend sensitively on the location of the shock. If the pressure (or density) at the stellar surface is given as a boundary condition, the shock location can then be determined and the whole solution completed.

The results suggest that the shock position depends strongly on the boundary conditions. If the free-fall type of accretion occurs as discussed previously [2], the shock location must be rather far above the stellar surface.

We discuss first the case where the outer boundary of the funnel flow is matched onto the inner edge of a Shakura–Sunyaev accretion disc [5]. This then gives the value of R_d, using $B_p(R_d)^2/8\pi = \mu^2/8\pi R_d^6$ and taking $\dot{m}/\dot{m}_{crit} = 2/3$ as being a typical value for the accretion rate. For $\alpha = 1.0$, the set of parameter values at the inner edge of the disc is $R_d = 7.3 \times 10^7$cm, $\rho_d = 6.6 \times 10^{-6}$g cm^{-3}, $p_d = 2.6 \times 10^{11}$dyn cm^{-2}; For $\alpha = 0.1$, we get $R_d = 4.1 \times 10^7$cm $\rho_d = 1.2 \times 10^{-4}$g cm^{-3}, $p_d = 8.9 \times 10^{12}$dyn cm^{-2}. Also, we consider the situation when the outer accretion flow consists of matter coming essentially in free-fall from a companion star, until it is stopped at the Alfvén radius r_A given by $B_p^2/8\pi = \frac{1}{2}\rho(r_A)v^2(r_A)$. In this case, we have $R_d = r_A$, which gives $R_d = 3.1 \times 10^8$cm, $\rho_d = 7.6 \times 10^{-11}$g cm^{-3}, $p_d = 4.6 \times 10^7$dyn cm^{-2}.

The boundary condition on the stellar surface is imposed by the pressure where the balance condition between the magnetic pressure and the gas pressure maintains. This pressure will be up to 10^{23}dyn cm^{-2} for $B = 10^{12}$G.

In Table 1, we show the calculated shock positions for a relevant range of values of p_{NS}. If we take p_{NS} to be that given by the balance between magnetic pressure and gas pressure, the predicted shock position would be ~ 500 km away from the stellar surface. The shock location may affect the spectra, luminosities and light curves of X-ray pulsars.

Table 1. Shock positions for selected surface conditions in the cases of standard-disc accretion and free-fall accretion.

NS Surface Condition p_{NS}[dyn/cm^2]	Shock Position r_s[cm] (r_s/R_d)	Fluid Density at r_{NS} $\rho(r_{NS})$[g cm^{-3}]
$\alpha = 1.0$		
1.0×10^{19}	6.4×10^6 (0.09)	2.4×10^{-1}
1.0×10^{20}	2.0×10^7 (0.28)	2.4×10^0
1.0×10^{21}	3.4×10^7 (0.46)	2.4×10^1
1.0×10^{22}	4.4×10^7 (0.60)	2.4×10^2
1.0×10^{23}	5.2×10^7 (0.71)	2.4×10^3
$\alpha = 0.1$		
1.0×10^{19}	1.8×10^6 (0.04)	2.4×10^{-1}
1.0×10^{20}	8.2×10^6 (0.20)	2.4×10^0
1.0×10^{21}	1.6×10^7 (0.35)	2.4×10^1
1.0×10^{22}	2.2×10^7 (0.55)	2.4×10^2
1.0×10^{23}	2.7×10^7 (0.66)	2.4×10^6
free-fall		
1.0×10^{20}	1.4×10^8 (0.46)	3.6×10^0
1.0×10^{21}	1.9×10^8 (0.60)	3.6×10^1
1.0×10^{22}	2.2×10^8 (0.71)	3.6×10^2
1.0×10^{23}	2.5×10^8 (0.80)	3.6×10^3

References

1. H. Inoue: PASJ, **27**, 311 (1975)
2. M. M. Basko, R. A. Sunyaev: MNRAS, **175**, 395 (1976)
3. R. V. E. Lovelace, C. Mehanian, C. M. Mobarry, M. E. Sulkamen: ApJS, **62**, 1 (1986)
4. A. V. Koldoba, R. V. E. Lovelace, G. V. Ustyugova, M. M. Romanova: AJ, **123**, 2019 (2002) (KLUR)
5. N. I. Shakura, R. A. Sunyaev: A & A, **24**, 337 (1973)

General Relativistic Simulation of Jet Formation in Kerr Black Hole Magnetosphere

S. Koide

Faculty of Engineering, Toyama University, 3190 Gofuku, Toyama 930-8555, Japan,
koidesin@eng.toyama-u.ac.jp

Summary. We report a numerical result of jet formation driven by magnetic field due to a current loop around a rapidly rotating black hole. Beside the current loop, there are magnetic flux tubes that bridge the region between the ergosphere and the rotating disk, which we call 'magnetic bridges'. The numerical result shows that the magnetic bridges can not be stationary and expand explosively to form a jet when the black hole rotates rapidly.

1 Introduction

Relativistic jets were observed from quasars (QSOs) and active galactic nuclei (AGNs) [1, 2], and from binary systems called micro-quasars (μQSOs) in our galaxy [3, 4]. Recently, it is revealed that gamma-ray bursts (GRBs) also contain the relativistic jets [5].

It is believed that these relativistic jets are formed due to the violent phenomena around black holes [6]. However, the distinct mechanism of its formation, that is, the acceleration and collimation of the jet, is not revealed yet. Many models were proposed to fix the mechanism. Among them, the models of the magnetic mechanisms become most promising because they can explain the acceleration and the collimation of the jet all at once [7, 8, 9, 10].

To investigate the magnetic mechanism around the black hole, we have developed a numerical calculation code to solve the equations of the general relativistic magnetohydrodynamics (GRMHD), which is a simplest model of the plasma interacting with the magnetic field around the black hole. We have applied the code to the several cases with the uniform and radial magnetic field configurations [11, 12, 13, 14, 15, 16]. In this report, we consider more realistic situation with the magnetic field of the current loop and the accretion disk around the rapidly rotating black hole. In this situation, there are closed magnetic flux tubes bridging the region between the ergosphere and the accretion disk. We call them 'magnetic bridges'. We expect these magnetic bridges to cause the drastic phenomena around the rapidly rotating black hole as suggested by the results of previous non-relativistic calculations of the magnetic bridges between the star and the accretion disk [17, 18, 19]. In this report, we show that the magnetic bridges between the ergosphere and the disk expand explosively and form the jet.

2 Results

As the initial condition of the present calculation, we assume the hydrostatic equilibrium of the plasma around the rapidly rotating black hole (rotation parameter, $a = 0.99995$). We set the magnetic field caused by the current loop at the intersection of the surface of the ergosphere and the equatorial plane ($r = r_S$, $\theta = \pi/2$, r_S is the Schwarzschild radius of the black hole) initially as shown in Fig. 1. The accretion disk with the circular orbital velocity locates around the equatorial plane, $r \geq 2r_S$, $\theta = \pi/2$ (Fig. 2(a)). In this situation, there are closed magnetic flux tubes bridging the region between the ergosphere and the disk, called 'magnetic bridges'.

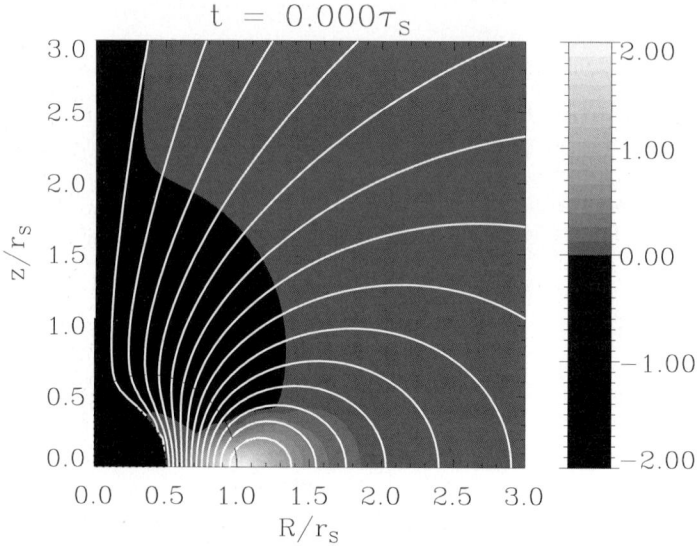

Fig. 1. Initial condition: Current density J_ϕ (gray-scale) and magnetic field configuration (solid white lines) of the magnetic field. The white solid lines are magnetic flux surfaces. The black quarter-circle at the origin indicates the event horizon of the rapidly rotating black hole ($a = 0.99995$). The black dashed line very near the horizon shows the inner boundary of the calculation region at $r = 1.006 r_H$. The black broken line shows the boundary of the ergosphere.

Figure 2 shows the time evolution of the magnetic field and the plasma around the rotating black hole. At $t = 20.03 \tau_S$, the radial outflow is formed (Fig. 2(b)). At $t = 40.02 \tau_S$, the outflow begins to be collimated (Fig. 2(c)), and at $t = 79.29 \tau_S$, the jet is formed (Fig. 2(d)). The maximum velocity of the jet is $v_{max} = 0.6c$ at $t = 79.29 \tau_S$. Around the disk surface, part of the magnetic surface is elongated horizontally. This separation of the outflow to the jet along the axis and the horizontal flow along the disk is also seen in the non-relativistic MHD simulation [20]. The magnetic island (plasmoid) is also seen in the region. However, as we use the ideal MHD condition (zero electric resistivity), the magnetic reconnection never happens. The formation of the magnetic island (plasmoid) is due to the numerical diffusion, while it suggests that such anti-parallel magnetic field configuration is realized easily.

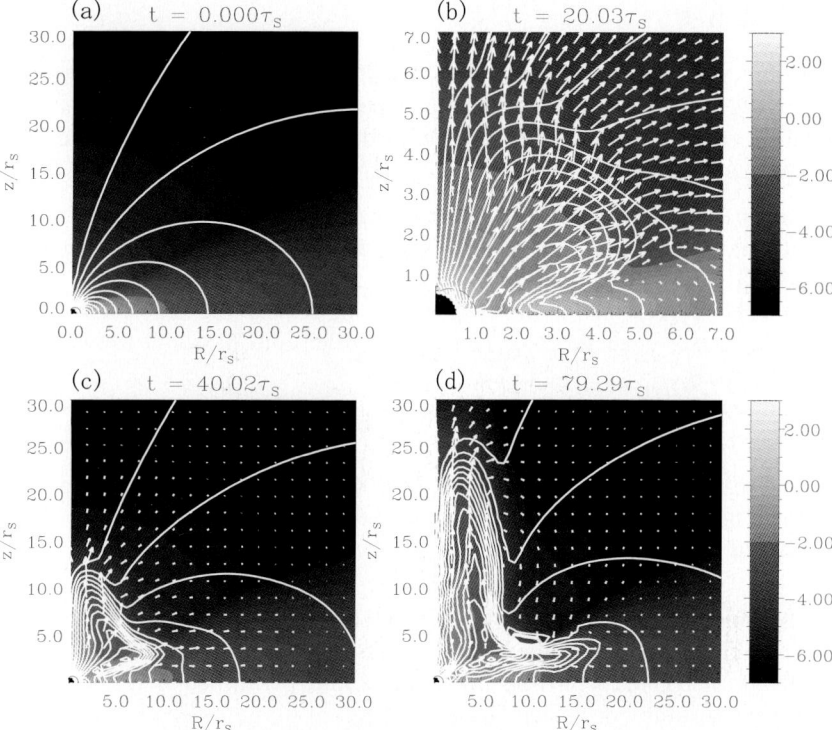

Fig. 2. Time evolution of a system of the current loop, the equilibrium coronal plasma, the coro-tating disk, and the rapidly rotating black hole ($a = 0.99995$). The gray-scale portion shows the logarithm of the mass density of the plasma. The arrows show the poloidal component of the plasma velocity, $\mathbf{v}^{P}$.

To identify the acceleration mechanism of the jet, we evaluate the power density of the Lorentz force and the gas pressure gradient to the jet plasma acceleration at $t = 20.03\tau_{S}$. It is shown that the jet is accelerated by the Lorentz force. The sign of the azimuthal component of the magnetic field shows that the twist of the magnetic filed lines is mainly caused by the frame-dragging effect in the jet region. Therefore, the jet is accelerated by the frame-dragging effect through the magnetic bridges. In this case, both the magnetic pressure and tension accelerate the jet.

To confirm the collimation mechanism of the jet, we consider the contribution of the Lorentz force density and the gas pressure gradient to the collimation of the plasma element. The force to the plasma element for the collimation of the outflow should act on the plasma element perpendicularly to the velocity vector and should be directed inward (toward axis) because the parallel component of the force just accelerates the plasma element. It shows that the jet is collimated by the Lorentz force, that is, the jet is collimated magnetically.

3 Conclusion

In this report, we presented the numerical results of the GRMHD simulations of the magnetic bridges between the ergosphere and the disk. The numerical results show the elemental processes of the jet formation caused by the magnetic bridges around the rapidly rotating black hole as follows.

1. The magnetic bridges between the ergosphere and the disk are twisted mainly by the plasma of the ergosphere, if the plasma density is not so low compared to that of the disk.
2. The magnetic pressure in the magnetic bridges near the ergosphere increases rapidly and the magnetic force (pressure and tension) blows of the plasma near the ergosphere.
3. The outflow driven by the magnetic force is pinched by the magnetic tension and is collimated to be a jet.

In conclusion, the magnetic bridges between the ergosphere and the disk around the rapidly rotating black hole can not be steady, and they will be swallowed by the black hole or expand explosively to form the jet.

Acknowledgments. The author thanks Kazunari Shibata and Takahiro Kudoh for their essential contribution to this study. He also thanks Mika Koide for her useful comments on this manuscript. The calculations in this study were performed by the supercomputers of National Institute for Fusion Science, Japan.

References

1. J. A. Biretta, W. B. Sparks, and F. Macchetto, ApJ **520**, 621 (1999).
2. T. J. Pearson and J. A. Zensus in *Superluminal Radio Sources*, edited by J. A. Zensus and T. J. Pearson (Cambridge Univ., London, 1987), p. 1.
3. I. F. Mirabel, and L. F. Rodriguez, Nature **374**, 141 (1994).
4. S. J. Tingay *et al.*, Nature **374**, 141 (1995).
5. S. R. Kulkarni, Nature **398**, 389 (1999).
6. D. L. Meier, S. Koide, and Y. Uchida, Science **291**, 84 (2001).
7. R. D. Blandford and D. Payne, MNRAS **199**, 883 (1982).
8. Y. Uchida and K. Shibata, Publ. Astron. Soc. Jpn. **37**, 515 (1985).
9. K. Shibata and Y. Uchida, Publ. Astron. Soc. Jpn. **38**, 631 (1986).
10. T. Kudoh, R. Matsumoto, and K. Shibata, ApJ **508**, 186 (1998).
11. S. Koide, K. Shibata, and T. Kudoh, ApJ **495**, L63 (1998).
12. S. Koide, K. Shibata, and T. Kudoh, ApJ **522**, 727 (1999).
13. S. Koide, et al.: ApJ **536**, 668 (2000).
14. S. Koide, et al.: Science **295**, 1688 (2002). Published online 24 January 2002; 10.1126/Science. 1068240.
15. S. Koide, Physical Review D **67**, 104010 (2003).
16. S. Koide, ApJ Lett. **606**, L45 (2004).
17. M. R. Hayashi, K. Shibata, and R. Matsumoto, ApJ Lett. **468**, L37 (1996).
18. T. Kudoh, R. Matsumoto, and K. Shibata, Proc. IAU Symposium No. **195**, p. 407 (2000).
19. Y. Kato, S. Mineshige, K. Shibata, ApJ **605**, 307 (2004).
20. T. Kudoh, R. Matsumoto, and K. Shibata, Pub. Astron. Soc. Japan **54**, 267 (2002).

Radio Jets as Decelerating Relativistic Flows

R.A. Laing[1], A.H. Bridle[2], and J.R. Canvin[3]

[1] ESO, Karl-Schwarzschild-Straße 2, 85748 München, Germany
   rlaing@eso.org
[2] National Radio Astronomy Observatory, 520 Edgemont Road, Charlottesville,
   VA 22903-2475, U.S.A.
   abridle@nrao.edu
[3] School of Physics, University of Sydney, A28, Sydney, NSW 2006, Australia
   jcanvin@physics.usyd.edu.au

The largest-scale manifestation of Special Relativistic aberration [3] in contemporary astrophysics is the appearance of initial asymmetries in kpc-scale radio-galaxy jets. It is well known that synchrotron radiation from intrinsically symmetrical, oppositely directed bulk-relativistic outflows will appear one-sided as a result of Doppler beaming. The jet/counter-jet ratio, $I_j/I_{cj} = [(1+\beta \cos \theta)/(1-\beta \cos \theta)]^{2+\alpha}$ for isotropic emission in the rest frame, where βc is the jet velocity, θ is its angle to the line of sight and α is the spectral index. Less well known is the way in which Special Relativity modifies the observed linear polarization of such jets. As aberration acts differently on radiation from the approaching and receding jets, their observed polarization images represent two-dimensional projections of the magnetic-field structure viewed from different directions θ'_j and θ'_{cj} in the rest frame of the flow: $\sin \theta'_j = \sin \theta [\Gamma(1 - \beta \cos \theta)]^{-1}$ and $\sin \theta'_{cj} = \sin \theta [\Gamma(1 + \beta \cos \theta)]^{-1}$, where $\Gamma = (1 - \beta^2)^{-1/2}$. This is the key [1] to breaking the degeneracy between β and θ and to estimating the physical parameters of the jets.

We model jets in low-luminosity, FR I extragalactic radio sources [4] as intrinsically symmetrical, axisymmetric, relativistic, stationary flows in which the magnetic fields are assumed to be disordered but anisotropic. We adopt simple parameterized functional forms for the geometry and the spatial variations of velocity (allowing both deceleration and transverse gradients), emissivity and field-component ratios. We then optimize the model parameters by fitting to deep VLA images in Stokes I, Q and U. The model brightness distributions are derived by integration along the line of sight, including the effects of anisotropy in the rest-frame emission, aberration and beaming. The model and observed images for four sources, described in refs. [1, 2, 5], are compared in Fig. 1. The asymmetry in total intensity close to the nucleus is characteristic of FR I radio jets. It is correlated with an asymmetry in linear polarization: the apparent magnetic field is *longitudinal* on-axis in the bright bases, but *transverse* at the corresponding locations in the counter-jets. We attribute the asymmetries in brightness and polarization to the effects of aberration and the decrease of both asymmetries with distance from the

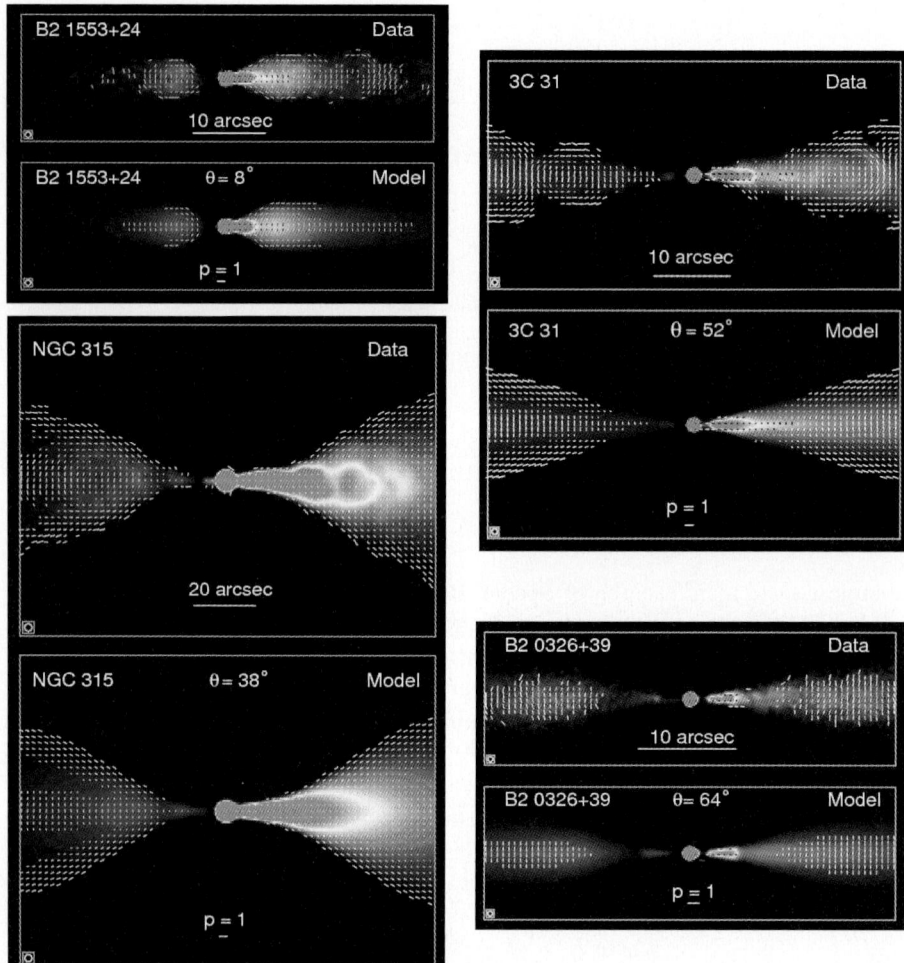

Fig. 1. The four sources we have observed and modelled. Vectors proportional to the degree of polarization, p, aligned with the apparent magnetic-field direction are superimposed on images of total intensity.

nucleus to deceleration of the jet by mass loading. We assume that the anisotropic field is disordered on small scales; our results would be unchanged if the toroidal component is vector-ordered but we can rule out a helical field [2]. Our principal conclusions are:

1. Our relativistic jet model provides an excellent description of the observed total intensity and linear polarization and we can use it to estimate the angle to the line of sight and the three-dimensional distributions of velocity, emissivity and magnetic-field structure.
2. The jets in all of the sources decelerate from $\beta \approx 0.8$ to $\beta \approx 0.1 - 0.4$ over short distances within the region of rapid expansion. Further out they recollimate and

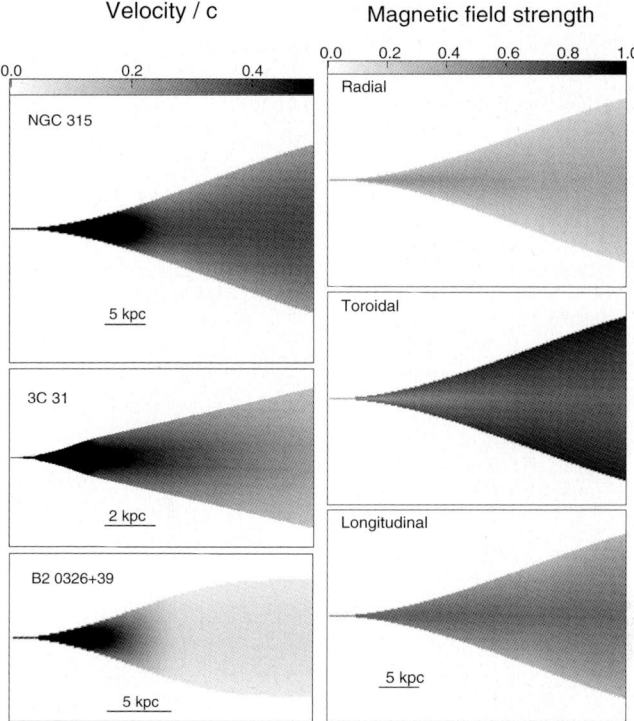

Fig. 2. Left: model velocity fields in the range $\beta = 0 - 0.5$ for three sources. Right: fractional field components for NGC 315.

subsequent deceleration is slower or completely absent. The ratio of edge to on-axis velocity is typically ≈ 0.7 (Fig. 2).

3. The jets are intrinsically centre-brightened.
4. The magnetic field (e.g. Fig. 2) evolves from mainly longitudinal close to the nucleus to mainly toroidal at large distances, qualitatively but not quantitatively as expected from flux freezing in an expanding flow; the behaviour of the (weaker) radial component differs between the sources.
5. Given a kinematic model derived from the radio data and knowing the external density and pressure from X-ray data, we can apply the laws of conservation of mass, momentum and energy for a bulk relativistic flow to estimate the variation of dynamical quantities along the jets [6].

References

1. Canvin J.R., Laing R.A.: MNRAS, **350**, 1342 (2004)
2. Canvin J.R., Laing R.A., Bridle A.H., Cotton W.D.: MNRAS, **363**, 1223 (2005)
3. Einstein, A.: Annalen der Physik, **17**, 140 (1905)

4. Fanaroff B.L., Riley J.M.: MNRAS, **167**, 31P (1974)
5. Laing R.A., Bridle A.H.: MNRAS, **336**, 328 (2002)
6. Laing R.A., Bridle A.H.: MNRAS, **336**, 1161 (2002)
7. Laing R.A., Bridle A.H.: MNRAS, **348**, 1459 (2004)

Extragalactic Relativistic Jets and Nuclear Regions in Galaxies

A.P. Lobanov and J.A. Zensus

Max-Planck-Institut für Radioastronomie, Auf dem Hügel 69, 53121 Bonn, Germany,
(alobanov,azensus)@mpifr-bonn.mpg.de

Summary. Past years have brought an increasingly wider recognition of the ubiquity of relativistic outflows (jets) in galactic nuclei, which has turned jets into an effective tool for investigating the physics of nuclear regions in galaxies. A brief summary is given here of recent results from studies of jets and nuclear regions in several active galaxies with prominent outflows.

1 Introduction

Substantial progress achieved during the past decade in studies of active galactic nuclei (see [25] for a review of recent results) has brought an increasingly wider recognition of the ubiquity of relativistic outflows (jets) in galactic nuclei [5, 43] turning them into an effective probe of nuclear regions in galaxies [21]. Emission properties, dynamics, and evolution of an extragalactic jet are intimately connected to the characteristics of the supermassive black hole, accretion disk and broad-line region in the nucleus of the host galaxy [25]. The jet continuum emission is dominated by non-thermal synchrotron and inverse-Compton radiation [40]. The synchrotron mechanism plays a more prominent role in the radio domain, and the properties of the emitting material can be assessed using the turnover point in the synchrotron spectrum [20], synchrotron self-absorption [19], and free-free absorption in the plasma [12, 42].

High-resolution radio observations access directly the regions where the jets are formed [11], and trace their evolution and interaction with the nuclear environment [30]. Evolution of compact radio emission from several hundreds of extragalactic relativistic jets is now systematically studied with dedicated monitoring programs and large surveys using very long baseline interferometry (such as the 15 GHz VLBA[1] survey [13] and MOJAVE [18]). These studies, combined with optical and X-ray studies, yield arguably the most detailed picture of the galactic nuclei. Presented below is a brief summary of recent results from studies outlining the relation between jets, supermassive black holes, accretion disks and broad-line regions in prominent active galactic nuclei (AGN).

---

[1] Very Long Baseline Array of National Radio Astronomy Observatory, USA

Table 1. Characteristic scales in the nuclear regions in active galaxies

	l [R_g]	l_8 [pc]	θ_Gpc [mas]	τ_c [yr]	τ_orb [yr]
Event horizon:	1–2	10^{-5}	5×10^{-6}	0.0001	0.001
Ergosphere:	1–2	10^{-5}	5×10^{-6}	0.0001	0.001
Accretion disk:	10^1–10^3	10^{-4}–10^{-2}	0.005	0.001–0.1	0.2–15
Corona:	10^2–10^3	10^{-3}–10^{-2}	5×10^{-3}	0.01–0.1	0.5–15
Broad line region:	10^2–10^5	10^{-3}–1	0.05	0.01–10	0.5–15000
Molecular torus:	$>10^5$	>1	>0.5	>10	>15000
Narrow line region:	$>10^6$	>10	>5	>100	>500000
Jet formation:	$>10^2$	$>10^{-3}$	$>5 \times 10^{-4}$	>0.01	>0.5
Jet visible in the radio:	$>10^3$	$>10^{-2}$	>0.005	>0.1	>15

Column designation: l – dimensionless scale in units of the gravitational radius, $G\,M/c^2$; l_8 – corresponding linear scale, for a black hole with a mass of $5 \times 10^8\,\mathrm{M}_\odot$; θ_Gpc – corresponding largest angular scale at 1 Gpc distance; τ_c – rest frame light crossing time; τ_orb – rest frame orbital period, for a circular Keplerian orbit. Table is reproduced from [25]

2 Anatomy of Jets

Jets in active galaxies are formed in the immediate vicinity of the central black hole, and they interact with every major constituent of AGN (see Table 1). The jets carry away a fraction of the angular momentum and energy stored in the accretion flow [9] and in the rotation of the central black hole [15, 16, 38]. At distances of $\sim 10^3\,R_\mathrm{g}$, the jets become visible in the radio regime, which makes high-resolution VLBI observations a tool of choice for probing directly the physical conditions in AGN on such small scales [11, 17]. Recent studies indicate that at 10^3–$10^5\,R_\mathrm{g}$ ($\lesssim 1\,\mathrm{pc}$) the jets are likely to be dominated by pure electromagnetic processes such as Poynting flux [39] or have both MHD and electrodynamic components [29]. The magnetic field is believed to play an important role in accelerating and collimating extragalactic jets on parsec scales [41]. Three distinct regions with different physical mechanisms dominating the observed properties of the jets can be identified: 1) ultracompact jets where collimation and acceleration of the flow occurs; 2) parsec-scale flows dominated by relativistic shocks and 3) large-scale jets where plasma instabilities are dominant.

2.1 Ultracompact Jets

Ultracompact jets observed on sub-parsec scales typically show strongly variable but weakly polarized emission (possibly due to limited resolution of the observations). In many cases, the emission is optically thick, indicating that opacity effects may play a prominent role [19]. Ultracompact outflows are probably dominated by electromagnetic processes [29, 39], and they become visible in the radio regime (identified as compact "cores" of jets) at the point where the jet becomes optically thin for radio emission [19, 23]. The ultracompact jets do not appear to have strong shocks [20], and their evolution and variability can be explained by smooth changes in particle density of the flowing plasma, associated with the nuclear flares in the central engine [23]. Compelling evidence exists for acceleration [3] and collimation [11], [17] of the flows on these scales.

2.2 Parsec-Scale Flows: Shocks and Instabilities

Parsec-scale outflows are characterized by pronounced curvature of trajectories of superluminal components [13, 23], rapid changes of velocity and flux density and predominantly transverse magnetic field [10]. Relativistic shocks are prominent on these scales, which is manifested by strong polarization [37] and rapid evolution of the turnover frequency of synchrotron emission [26]. Mapping the turnover frequency distribution provides also a sensitive diagnostic of plasma instabilities in relativistic flows [20]. Complex evolution of shocked regions is revealed in observations [7, 10, 23] and numerical simulations [1] of parsec-scale outflows. However, the shocks are shown to dissipate rapidly [23], and shock dominated regions are not likely to extend beyond $\sim 100\,\mathrm{pc}$. Starting from these scales, instabilities (most importantly, Kelvin-Helmholtz instability) determine at large the observed structure and dynamics of extragalactic jets [24, 27, 36]. Non-linear evolution of the instability [33, 34] and stratification of the flow [35] are important for reproducing the observed properties of jets. Similarly to stellar jets, rotation of the flow is expected to be important for extragalactic jets [6], but observational evidence remains very limited on this issue.

3 Jets and Nuclear Regions in AGN

A number of recent studies have used jets to probe physical conditions in the central regions of AGN. Opacity and absorption in the nuclear regions of AGN have been probed effectively using the non-thermal continuum emission as a background source [21]. The free-free absorption studies indicate the presence of dense, ionized circumnuclear material with $T_e \approx 10^4\,\mathrm{K}$ distributed within a fraction of parsec from the central nucleus [19, 42].

Absorption due to several atomic and molecular species (most notably due to HI, CO, OH, and HCO^+) has been detected in a number of extragalactic objects. OH absorption has been used to probe the conditions in warm neutral gas [8, 14], and CO and HI absorption were used to study the molecular tori [4, 32] at a linear resolution often smaller than a parsec [30]. These observations have revealed the presence of neutral gas in a molecular torus in NGC 4151 and in a rotating outflow surrounding the relativistic jet in 1946+708 [31].

Connection between accretion disks and relativistic outflows [9] has been explored, using correlations between variability of X-ray emission produced in the inner regions of accretion disks and ejections of relativistic plasma into the flow [28]. The jets can also play a role in the generation of broad emission lines in AGN. The beamed continuum emission from relativistic jet plasma can illuminate atomic material moving in a sub-relativistic outflow from the nucleus, producing broad line emission in a conically shaped region located at a significant distance above the accretion disk [2]. Magnetically confined outflows can also contain information about the dynamic evolution of the central engine, for instance that of a binary black hole system [22]. This approach can be used for explaining, within a single framework, the observed optical variability and kinematics and flux density changes of superluminal features embedded in radio jets.

4 Conclusion

Extragalactic jets are an excellent laboratory for studying physics of relativistic outflows and probing conditions in the central regions of active galaxies. Recent studies of extragalactic jets show that they are formed in the immediate vicinity of central black holes in galaxies and carry away a substantial fraction of the angular momentum and energy stored in the accretion flow and rotation of the black hole. The jets are most likely collimated and accelerated by electromagnetic fields. Relativistic shocks are present in the flows on small scales, but dissipate on scales of $\lesssim 100\,\mathrm{pc}$. Plasma instabilities dominate the flows on larger scales. Convincing observational evidence exists connecting ejections of material into the flow with instabilities in the inner accretion disks. In radio-loud objects, continuum emission from the jets may also drive broad emission lines generated in sub-relativistic outflows surrounding the jets. Magnetically confined outflows may preserve information about the dynamics state of the central region, allowing detailed investigations of jet precession and binary black hole evolution to be made. This makes studies of extragalactic jets a powerful tool for addressing the general questions of physics and evolution of nuclear activity in galaxies.

References

1. I. Agudo, J.L. Gómez, J.M. Martí, et al.: ApJ **549**, 183 (2001)
2. T.G. Arshakian, A.P. Lobanov, V.H. Chavushyan, et al.: A&A *subm.* (2006), astro-ph/0512393
3. U. Bach, M. Kadler, T.P. Krichbaum, et al.: Multi-Frequency and Multi-Epoch VLBI Study of Cygnus A. In: *Future Directions in High Resolution Astronomy: The 10th Anniversary of the VLBA*, ASP Conference Proceedings, Vol. 340, ed. by J. Romney, M. Reed (ASP, San Fransisco 2005) pp. 30–34
4. J.E. Conway: New Astron. Rev. **43**, 509 (1999)
5. H. Falcke: Rev. Mod. Astron. **14**, 15 (2001)
6. C. Fendt: A&A **323**, 999 (1997)
7. J.L. Gómez, A.P. Marscher, A. Alberdi, et al.: ApJ **561**, 161 (2001)
8. J.R. Goikoechea, J. Martín-Piintado, J. Chernicharo: ApJ **619**, 291 (2005)
9. M. Hujeirat, M. Livio, M. Camenzind, et al.: A&A **408**, 415 (2003)
10. S.G. Jorstad, A.P. Marscher, M.L. Lister, et al.: AJ **130**, 1418 (2005)
11. W. Junor, J.A. Biretta, M. Livio: Nature **401**, 891 (1999)
12. M. Kadler, E. Ros, A.P. Lobanov, et al.: A&A **426**, 481 (2004)
13. K.I. Kellermann, M.L. Lister, D.C. Homan, et al.: AJ, **609**, 539 (2004)
14. H.R. Klöckner, W.A. Baan: ApSS **295**, 277 (2005)
15. S. Koide, K. Shibata, T. Kudoh, et al.: Science **295**, 1688 (2002)
16. S.S. Komissarov: MNRAS **359**, 801 (2005)

17. T.P. Krichbaum, D.A. Graham, A. Kraus, et al.: Towards the Event Horizon – The Vicinity of AGN at Micro-Arcsecond Resolution. In: *Proceedings of the 7th Symposium of the European VLBI Network*, ed. by R. Bachiller, F. Colomer, J.-F. Desmurs, P. de Vicente (Observatorio Astronomico Nacional, Madrid 2004) pp. 15–18
18. M.L. Lister, D.C. Homan: AJ **130**, 1389 (2005)
19. A.P. Lobanov: A&A **390**, 79 (1998)
20. A.P. Lobanov: A&ASS **132**, 261 (1998)
21. A.P. Lobanov: MemSAItS **7**, 12 (2005)
22. A.P. Lobanov, J. Roland: A&A **431**, 831 (2005)
23. A.P. Lobanov, J.A. Zensus: ApJ **521**, 509 (1999)
24. A.P. Lobanov, J.A. Zensus: Science **284**, 291 (2001)
25. A.P. Lobanov, J.A. Zensus: Active Galactic Nuclei at the Crossroads of Astrophysics. In: *Exploring the Cosmic Frontier: Astrophysical Instruments for the 21st Century*, ESO Astrophysical Symp. Series, ed. by A.P. Lobanov, J.A. Zensus, C. Cesarsky, P.J. Diamond (Springer, Heidelberg 2006) pp. 147–162
26. A.P. Lobanov, E. Carrara, J.A. Zensus: Vistas in Astronomy **41**, 253 (1997)
27. A.P. Lobanov, P.E. Hardee, J.A. Eilek: New Astron. Rev. **47**, 629 (2003)
28. A.P. Marscher, S.G. Jorstad, J.L. Gómez, et al.: Nature **417**, 625 (2002)
29. D.L. Meier: New Astron. Rev. **47**, 667 (2003)
30. C.G. Mundell, J.M. Wrobel, A. Pedlar, et al.: ApJ **583**, 192 (2003)
31. A.B. Peck, G.B. Taylor: ApJ **554**, 147 (2001)
32. A. Pedlar: ApSS **295**, 161 (2004)
33. M. Perucho, M. Hanasz, J.M. Martí, et al.: A&A **427**, 415 (2004)
34. M. Perucho, J.M. Martí, M. Hanasz: A&A **427**, 431 (2004)
35. M. Perucho, J.M. Martí, M. Hanasz: A&A **443**, 863 (2005)
36. M. Perucho, A.P. Lobanov, J.M. Martí: *these proceedings* (2006)
37. E. Ros, J.A. Zensus, A.P. Lobanov: A&A **354**, 55 (2000)
38. V. Semenov, S. Dyadechkin, B. Punsly: Science **305**, 978 (2004)
39. M. Sikora, M.C. Begelman, G.M. Madejski, et al.: ApJ **625**, 72 (2005)
40. S.C. Unwin, A.E. Wehrle, A.P. Lobanov, et al.: ApJ **480**, 596 (1997)
41. N. Vlahakis, A. Konigl: ApJ **605**, 656 (2004)
42. R.C. Walker V. Dhawan, J.D. Romney, et al.: ApJ **530**, 233 (2000)
43. J.A. Zensus: Ann. Rev. Astron. Astrophys. **35**, 607 (1997)

Modeling the Relativistic Jets in SS 433 Using Chandra X-ray Spectroscopy

H.L. Marshall[1], C.R. Canizares[1], S. Heinz[1], T.C. Hillwig[2], A.J. Mioduszewski[3], and N.S. Schulz[1]

[1] MIT Kavli Institute, Cambridge, MA 02139 USA, hermanm@space.mit.edu
[2] Valparaiso Univ., Valparaiso, IN 46383 USA, todd.Hillwig@valpo.edu
[3] NRAO, Socorro, NM 87801 USA, amiodusz@nrao.edu

Summary. The unusual X-ray binary SS 433 has been observed with the *Chandra* X-ray observatory on four occasions at different orbital and precessional phases. These data have provided excellent views of the hottest parts of the oppositely directed jets. Emission line widths directly provide the opening angle of the jet. We can also determine the composition, density, temperature, and ionization state of the jet gas from the X-ray spectra of the twin jets. We report on new, very long observations using Chandra that have been obtained in August 2005 with simultaneous radio and optical coverage.

1 Introduction

SS 433 is a very unusual binary system that has been the subject of many astrophysical research studies. It is well known for its optical spectra, which show Doppler shifted emission lines that periodically vary. These periodic variations are well-modeled as originating in twin, oppositely directed jets with flow velocities of about $0.26c$, whose orientations sweep out cones with a half-angle of about $20°$ due to "slaving" of the accretion disk orientation to the companion star's precession. In addition, there is a periodic torque exerted by the companion that can cause the disk to nutate slightly. See reviews by Margon [1] or Fabrika [2] for observational details of SS 433 and for discussions of physical models.

1.1 Previous Observations

The X-ray spectrum of the jet shows emission lines [3] that are well modeled as thermal emission from an expanding, cooling plasma [4]. Previous observations using the *Chandra* High Energy Transmission Grating Spectrometer (HETGS, [5]) usually show abundant emission lines (Fig. 1). Models based on thin, collisionally ionized plasma at several temperatures are generally good fits to the data, as shown in Fig. 1, making it possible to test models of the jets' thermal evolution.

Table 1. Some *Chandra* Spectroscopic Observations of SS 433

Date	ϕ_{prec}	ϕ_{orb}	Ref.
1999 Sep 23	0.92	0.67	[4]
2000 Mar 16	0.58	0.97	[6]
2001 Nov 21	0.68	0.69	[6]
2001 May 12	0.60	0.29	[9]
2005 Aug 12	0.18	0.01	this work
2005 Aug 16	0.20	0.25	this work
2005 Aug 18	0.21	0.44	this work

1.2 Physically Modeling the Jets

The emission lines provide a wealth of information that can be used to constrain models of the jets [4], such as:

- Lines from many ionizations are apparent and are collisionally excited, so the plasma is thermal
- The emission measure drops with temperature T, consistent with an adiabatically expanding jet
- The continuum level relative to the emission lines gives high metal abundances
- Lines of different ionizations have the same Doppler shifts and widths, indicating that the flow is uniform and conical and that all ions are accelerated to the same terminal velocity
- Lines are mildly Doppler broadened as one would expect from a highly collimated flow
- The Si XIII triplet line strengths can be used to estimate the density in the flow at $\sim 10^{14}$ cm^{-3}

The emission lines belong mostly to one of two systems: either the red jet or the blue jet. The models for the two systems seem quite similar, so we assume that they are perfectly opposed jets at an angle α to the line of sight. In this case, the Doppler shifts of the blue and red jets are given by

$$z = \gamma(1 \pm \beta\mu) - 1 \qquad (1)$$

where $v_j = \beta c$ is the velocity of the jet flow, $\gamma = (1 - \beta^2)^{-1/2}$, and $\mu = \cos\alpha$. The blue (red) Doppler shift is obtained by using the $-$ (+) sign. With high accuracy redshifts, we can determine γ (and β) by adding the redshifts to cancel the $\beta\mu$ terms:

$$\gamma = \frac{z_b + z_r}{2} + 1 \qquad (2)$$

Thus, we may determine the velocity and direction of the jet from accurate determinations of the average Doppler shifts of the blue- and red-shifted emission lines.

By modeling the X-ray spectra during eclipse, it was possible to infer the size of the companion star [6], which is very difficult to determine by optical spectroscopy [7, 8].

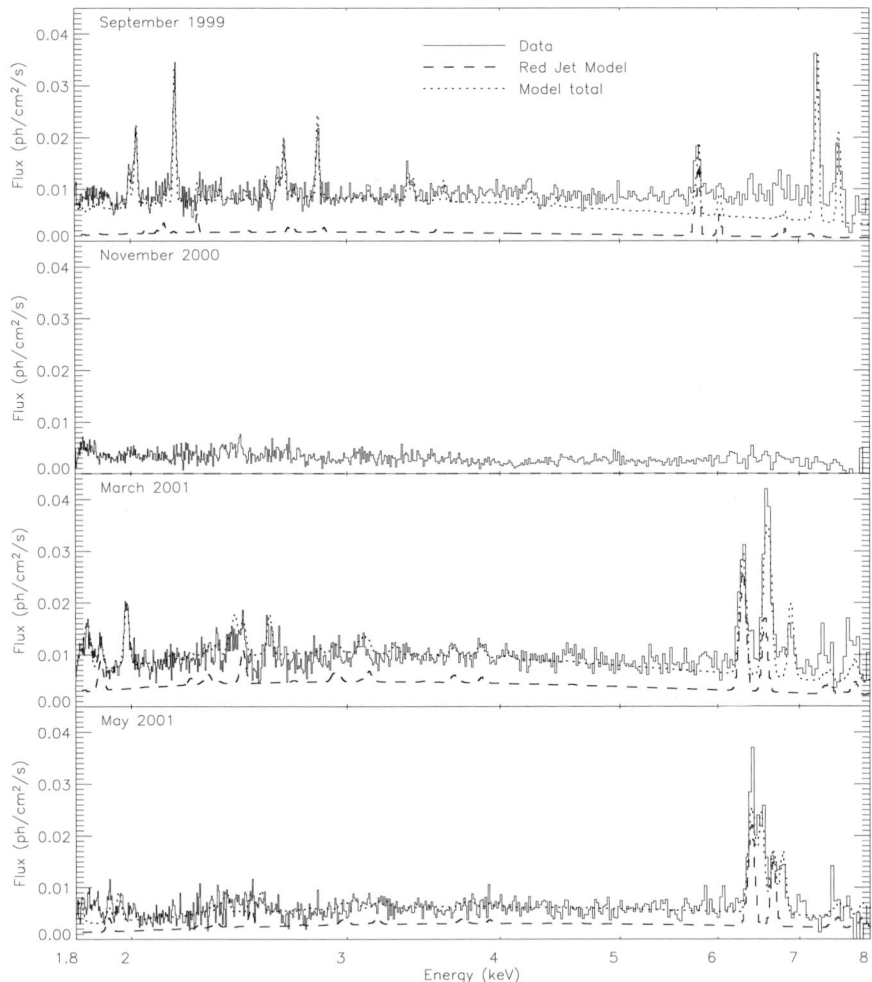

Fig. 1. Four X-ray spectra of SS 433 taken with the *Chandra* High Energy Transmission Grating Spectrometer. The models were taken from [4] and [6] for the 1999 and March 2001 data sets. The model for the May 2001 data set was based on the model for March 2001 except that the Doppler shifts were changed to 0.0460 and 0.0265 [9] and a factor of $0.85(E/1\text{keV})^{-0.5}$ was applied to reduce the continuum and the flux in the soft band. There are no clear emission lines in the November 2000 spectrum and the source was at least ×2 fainter than during other observations.

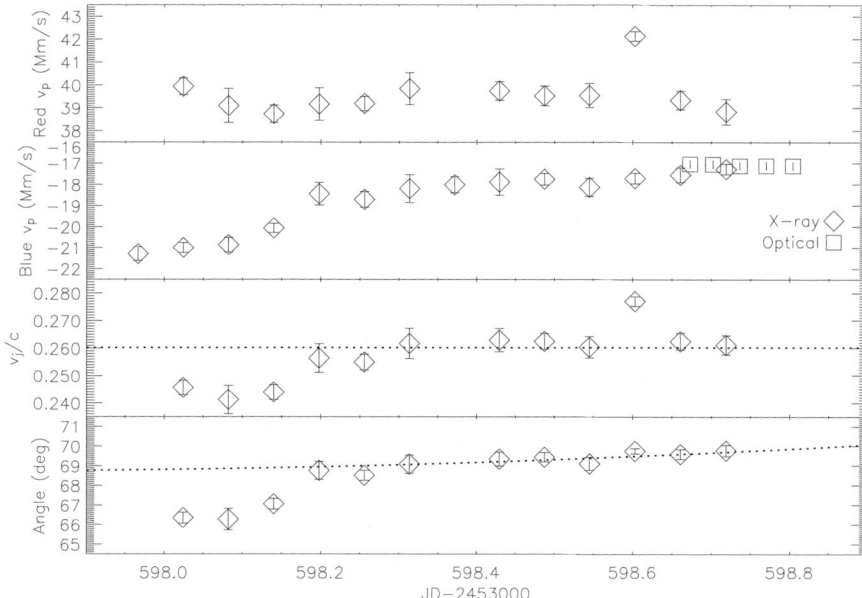

Fig. 2. Preliminary results from fits to new X-ray spectra. The observation taken on Aug. 16 was divided into 5 ks portions so that the velocities of the blue- and red-shifted emission lines could be measured as a function of time (top two panels). Optical emission line Doppler velocities are shown as open squares in the second panel and match the contemporaneous X-ray emission line Doppler velocities to within 200 km s^{-1}. The bottom two panels show the results from calculations of the jet velocity and angle to the line of sight using (2). This figure shows that it is unlikely that both of these two assumptions are true, given that only one of the two jets shows a significant trend in the first half of the observation while both of the derived parameters increase during this period.

2 Observations

New observations (see Table 1) were obtained in order to detect weak emission lines and to track variations in the line Doppler shifts. Simultaneous optical spectroscopy (by TCH) and VLBA (by AJM) were obtained in order to model the relationship between the various emission regions.

3 Preliminary Results

One of the new observations was taken on Aug. 16 (table 1). The 73 ks spectrum showed double-peaked Si XVI and Fe XXV lines, so the observation was divided into 5 ks portions so that the velocities of the blue- and red-shifted emission lines could be measured as a function of time (Fig. 2). Also shown in Fig. 2 are the results from calculations using (2), from which the jet velocity and angle to the line of sight can be determined under the assumption that the jets have the same speed and opposite directions. Given that only one of the two jets shows a significant trend in the first half of the observation while both of the derived parameters increase during this period, it seems unlikely that both of these two assumptions are true. One interpretation of these data is that the jets' terminal velocities are determined by environmental effects that perturb the direction or speed of the jet. This observation shows that X-ray emission lines from a long integration may appear broadened due to systematic short time-scale deviations from simple periodic motions. These deviations occur on a time scale that is very short compared to the orbital or precessional periods, which are included in the model shown in Fig. 2.

Another important result is that the optical and X-ray emission lines of the blue jet have the same Doppler shifts to within 200 km s^{-1}, which is of order the uncertainties in the optical and X-ray Doppler shifts. Thus, it appears that there is no significant deceleration of the jet between the optical and X-ray line emission regions. If they are not cospatial, then we may place a limit on the travel time delay between the X-ray and optical line emission regions using Fig. 2 of less than 0.4 days. Any longer than this and the optical lines would be related to X-ray lines over 3000 km s^{-1} different.

4 Further Work

Once we compensate for the time-variable Doppler shifts, we will be able to combine spectra obtained at different times in order to model the emission line spectrum in more detail by detecting fainter emission lines. We will also relate the Doppler shifts derived from the X-ray and optical spectra to the models of the VLBA radio images to test for knot deceleration.

References

1. B. Margon: ARA&A, **22**, 507 (1984)
2. S. Fabrika: Astrophysics and Space Physics Reviews **12**, 1 (2004)
3. T. Kotani, N. Kawai, M. Matsuoka, et al.: PASJ, **48**, 619 (1996)
4. H. L. Marshall, C. R. Canizares and N. S. Schulz: ApJ, **564**, 941 (2002)
5. C. R. Canizares, J. E. Davis, D. Dewey, et al.: PASP, **117**, 1144 (2005)
6. L. A. Lopez, H. L. Marshall, C. R. Canizares, et al.: ApJ, submitted (2006)
7. T. C. Hillwig, D. R. Gies, W. Huang, et al.: ApJ, **615**, 422 (2004)
8. A. D. Barnes, J. Casares, P. A. Charles, et al.: MNRAS, **365**, 296 (2006)
9. M. Namiki, N. Kawai, T. Kotani, et al.: PASJ, **55**, 281 (2003)

General Relativistic MHD Simulations of Relativistic Jets from a Rotating Black Hole Magnetosphere

Y. Mizuno[1], K.-I. Nishikawa[2], P. Hardee[3], and S. Koide[4]

[1] NRC Research Fellow, NASA Marshall Space Flight Center, NSSTC, 320 Sparkman Drive, XD12, Huntsville, AL 35805, USA, Yosuke.Mizuno@msfc.nasa.gov
[2] National Space Science and Technology Center, 320 Sparkman Drive, XD12, Huntsville, AL 35805, USA, Ken-Ichi.Nishikawa@nsstc.nasa.gov
[3] Department of Physics and Astronomy, The University of Alabama, Tuscaloosa, AL 35487, USA, hardee@athena.astr.ua.edu
[4] Department of Engineering, Toyama University, Toyama 930-8555, Japan, koidesin@eng.toyama-u.ac.jp

Summary. We have performed 3-dimensional general relativistic magnetohydrodynamic (GRMHD) simulations of jet formation from an accretion disk with/without initial perturbation around a rotating black hole. We input a sinusoidal perturbation ($m = 5$ mode) in the rotation velocity of the accretion disk. The simulation results show the formation of a relativistic jet from the accretion disk. Although the initial perturbation becomes weakened by the coupling among different modes, it survives and triggers lower modes. As a result, complex non-axisymmetric density structure develops in the disk and the jet. Newtonian MHD simulations of jet formation with a non-axisymmetric mode show the growth of the $m = 2$ mode but GRMHD simulations cannot see the clear growth of the $m = 2$ mode.

1 Simulations

Only a few 3D MHD simulations of jet formation have solved the accretion disk self-consistently [1, 2]. These simulations have shown turbulent and complex structures in the accretion disks caused by the non-axisymmetric mode of Magnetorotational instability (MRI). However the influence of non-axsymmetric modes on jet formation has not been studied extensionaly. Recently Kigure & Shibata (2005) [3] have performed 3-dimensional non-relativistic MHD simulations of jet formation by the interaction between an accretion disk and a large scale magnetic field including non-axisymmetric modes. They found that the jets have non-axisymmetric structure with $m = 2$ (m is the azimuthal wave number). The non-axisymmetric structure in the jet originates in the accretion disk, not in the jet itself, the Kelvin-Helmholtz instability is ruled out by the stability condition. However, we do not know whether jets formed in GRMHD simulations show the same behavior because the stability and growth rate of the instability is different from the non-relativistic cases. Thus, we investigate the connection between the non-axisymmetric structure in the jet and that in the disk by using GRMHD simulations.

In order to study the formation of relativistic jets from an accretion disk- black hole system we use a 3-dimensional general relativistic magnetohydrodynamic (GRMHD) code [4, 5]. We consider the following initial conditions for the simulations: a geometrically thin accretion disk rotates around the rotating black hole ($a = J/M_{BH}c = 0.95$), where the disk density is 100 times higher than the coronal density. The background corona is free-falling into the black hole (Bondi solution). The initial magnetic field is assumed to be uniform and parallel to the rotational axis. In one case we input a sinusoidal perturbation ($m = 5$ mode, 15% of Keplerian velocity) into the rotation of the disk ($\delta v_\phi = 0.15 v_K \sin 5\phi$).

2 Results

The matter in the disk loses its angular momentum through the magnetic field and falls towards the black hole. A centrifugal barrier decelerates the falling matter and creates a shock at around $r = 2r_S$. The matter near the shock region is accelerated by the $\mathbf{J} \times \mathbf{B}$ force and the gas pressure and forms relativistic outflows. These results are the same as found in previous GRMHD simualtions [6, 7]. As a relativistic outflow propagates outward, it shows a waving structure. The outflow structure is almost axisymmetric, even though this simulation runs in three dimensions. In the sinusoidal perturbation case the general properties are almost the same as that of the non-perturbed case (shock and jet formation). However, due to the initial perturbation the disk structure is slightly different from that of non-perturbation case, i.e., there is a thicker and lower density disk near the black hole.

Figure 1 shows a 2D snapshot (x-y plane) of the density in the accretion disk and the magnetic field structure in both cases. In the non-perturbation case the density shows clear symmetric structure. On the other hand, in the sinusoidal perturbation case, the density structure in the accretion disk shows complex non-axisymmetric structure. Part of the non-axisymmetric density structure reflects the initial sinusoidal mode ($m = 5$) and indicates interaction with other (lower?) modes. The initial perturbation in the accretion disk makes some of the matter fall into the black hole immediately and some of the matter rotates around the black hole for a long time. Although the initial perturbation is weakened by the coupling among different modes, it survives and triggers lower modes. As a result, complex non-axisymmetric density structure develops in the disk.

In the jet some non-axisymmetric structure is seen in both cases, but it is very faint. The non-axisymmetric structure in the sinusoidal perturbation case reflects the initial sinusoidal $m = 5$ perturbation to the disk and may mix with lower other modes, (This cannot be seen clearly in Figure 1).

3 Summary and Discussion

We have performed 3D GRMHD simulations of jet formation from an accretion disk with/without an initial perturbation (sinusoidal perturbation) around a rotating black

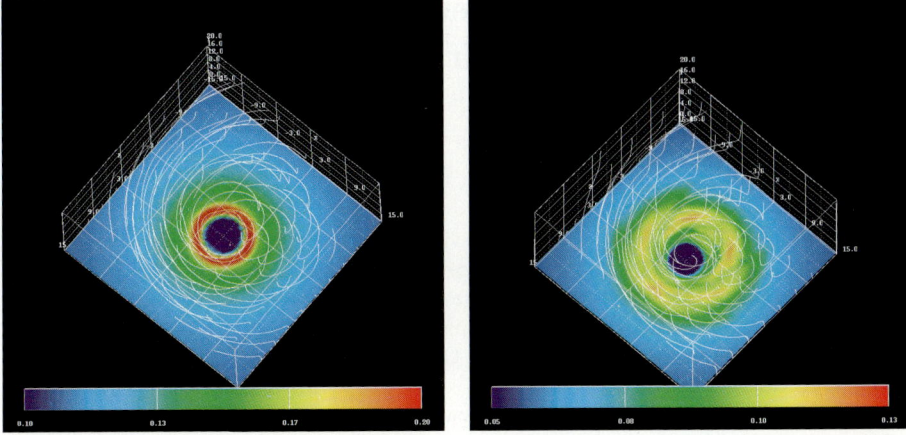

non-perturb case

(a) $time/\tau_s = 120.$
$z=1.0r_s$

sinusoidal perturb case

(b) $time/\tau_s = 91.3$

Fig. 1. 2D snapshot (x-y plane) of the density in the disk ($z = 1r_S$) for the non-perturbation case (a) and for the sinusoidal perturbation case (b).

hole. Although the initial perturbation is weakened by the coupling between different modes, it survives and triggers lower other modes. As a result, complex non-axisymmetric density structure develops in the disk. Previous Newtonian MHD simulations of jet formation with non-axisymmetric modes [3] show the growth of the $m = 2$ mode. However, the present GRMHD simulations do not clearly show growth of the $m = 2$ mode. This difference may be caused by a different magnetic field strength or a different structure of the accretion disk. The dependence of structure on the magnetic field strength will be investigated in future work.

References

1. J.F. Hawley & J.H. Krolik: ApJ **548**, 348 (2001)
2. M. Machida & R. Matsumoto: ApJ **585**, 429 (2003)
3. H. Kigure & K. Shibata: ApJ **634**, 879 (2005)
4. S. Koide: Phys. Rev. D **67**, 104010 (2003)
5. Y. Mizuno, S. Yamada, S. Koide & K. Shibata: ApJ **615**, 389 (2005)
6. S. Koide, D.L. Meier, K. Shibata & T. Kudoh: ApJ **536**, 668 (2000)
7. K.-I. Nishikawa, et al: ApJ **625**, 60 (2005)

Particle Acceleration, Magnetic Field Generation, and Emission in Relativistic Pair Jets with Weibel Instability

K.-I. Nishikawa[1], C. Hededal[2], P. Hardee[3], G.J. Fishman[4], C. Kouveliotou[4], and Y. Mizuno[1]

[1] NSSTC, 320 Sparkman Drive, XD12, Huntsville, AL 35805, USA,
   ken-ichi.nisikawa@nsstc.nasa.gov, yosuke.mizuno@msfc.nasa.gov
[2] Dark Cosmology Center, Niels Bohr Institute, Juliane Maries Vej 30, 2100 Copenhagen Ø, Denmark hededal@astro.ku.dk
[3] Department of Physics and Astronomy, The University of Alabama, Tuscaloosa, AL 35487, USA, hardee@athena.astr.ua.edu
[4] NASA-Marshall Space Flight Center, National Space Science and Technology Center, Huntsville, AL 35805, USA,
   (jerry.fishman,chryssa.kouveliotou)@nasa.gov

Summary. We have applied numerical simulations and modeling to the particle acceleration, magnetic field generation, and emission from relativistic shocks and plan to compare them with the observed gamma-ray burst emission. In collisionless shocks, plasma waves and their associated instabilities (e.g., the Weibel, Buneman and other two-stream instabilities) are responsible for particle (electron, positron, and ion) acceleration and magnetic field generation. A 3-D relativistic electromagnetic particle (REMP) code is used to study shock processes including spatial and temporal evolution of shocks in unmagnetized electron-positron plasmas with three different jet velocity distributions. The "jitter" radiation from the shocks is different from synchrotron radiation. The dynamics of shock microscopic process evolution may provide some insight into afterglows.

1 Simulations

It has been proposed that the needed particle acceleration occurs in shocks produced by differences in flow speed within the jet (e.g., [1]). Three simulations were performed using an $85 \times 85 \times 640$ grid with a total of 380 million particles (27 particles/cell/species for the ambient plasma) and an electron skin depth, $\lambda_{ce} = c/\omega_{pe} = 9.6\Delta$, where $\omega_{pe} = (4\pi e^2 n_e/m_e)^{1/2}$ is the electron plasma frequency and Δ is the grid size. In all simulations, jets are injected at $z = 25\Delta$ in the positive z direction. Radiating boundary conditions were used on the planes at $z = 0, z_{max}$. Periodic boundary conditions were used on all other boundaries. The ambient and jet electron-positron plasma has mass ratio $m_e/m_p \equiv m_{e-}/m_{e+} = 1$. The electron thermal velocity in the ambient plasma is $v_{th} = 0.1c$ where c is the speed of light. The electron number density of the jet is $0.741n_b$ where n_b is the ambient electron number density. The first case is as same as one of the simulations in a previous paper [4, 2] ($\gamma V_\parallel = 5$) (case A). The second case has a larger Lorentz factor ($\gamma V_\parallel = 15$) (case B). The third case is a special case and corresponds to cold jet electrons and positrons created by photon annihilation with $4 < \gamma V_\parallel < 100$ (e.g., [5]) (case C). For all cases the temperature of jet particles are very cold ($0.01c$ in the rest frame).

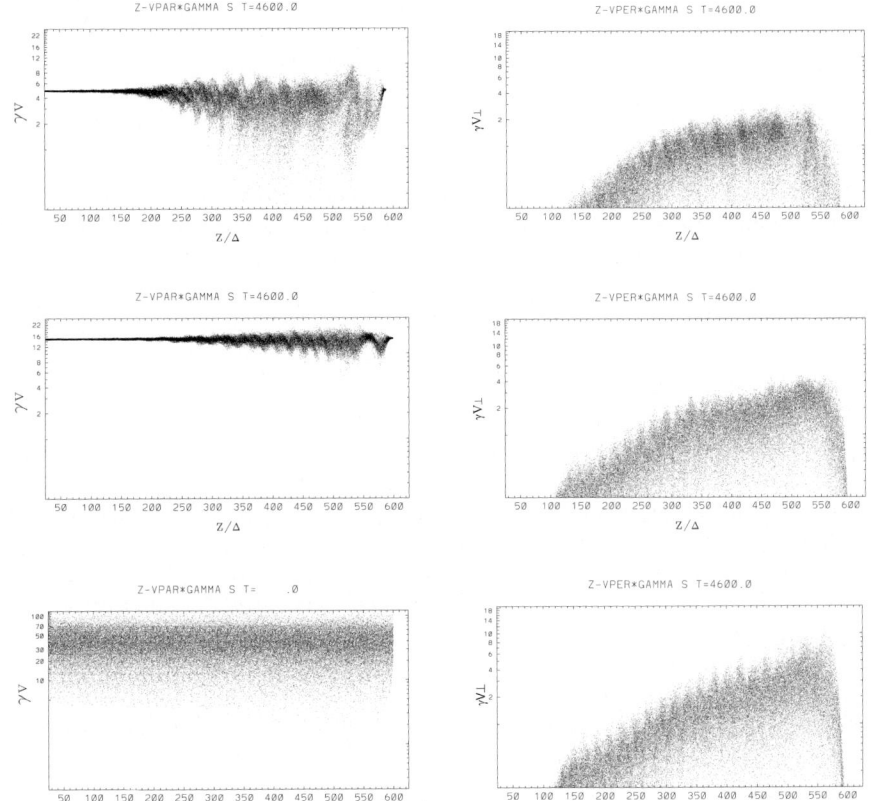

Fig. 1. Distributions of jet electrons in $Z/\Delta - \gamma V_\parallel$ (left column) and $Z/\Delta - \gamma V_\perp$ (right column) phase spaces for $\gamma v_\parallel = 5$ (top row) (case A) and 15 (middle row) (case B), and broad pair injection distribution $4 < \gamma v_\parallel < 100$ (bottom row) (case C) at $t = 59.8/\omega_{\mathrm{pe}}$. These jet electrons ($25 < Z/\Delta < 625$) are randomly selected for these plots.

As shown in Fig. 1, jet particles are accelerated and decelerated in the jet direction. The panels show the jet electron distributions in $z/\Delta - \gamma V_\parallel$ phase spaces. Two upper panels show that the cold jet electron injected distribution ($\gamma V_\parallel = 5$, and 15).

The acceleration of particles by electric fields toward or away from the current filaments and deflection of particles due to the Lorentz force increases as the magnetic field perturbation grows in amplitude. Basically we see that the transient force due to the radial electric field with circular magnetic field leads to acceleration in the perpendicular direction relative to the initial motion, and an average deceleration in the parallel direction but with an acceleration of some electrons in the parallel direc-

464   K.-I. Nishikawa et al.

tion to produce a high energy thermal distribution. The complicated filamented structures resulting from the Weibel instability have diameters on the order of the electron skin depth ($\lambda_{ce} = 9.6\Delta$) [6, 7]. This is in good agreement with the prediction of $\lambda \approx 2^{1/4} c \gamma_{th}^{1/2} / \omega_{pe} \approx 1.188 \lambda_{ce} = 10\Delta$ [7]. Here, $\gamma_{th} \sim 1$ is a thermal Lorentz factor.

2 Summary and Discussion

We have performed self-consistent, three-dimensional relativistic particle simulations of relativistic electron-positron jets propagating into unmagnetized electron-positron ambient plasmas for a longer time and a larger simulation system than our previous simulations [3, 4, 2] in order to investigate the nonlinear stage of the Weibel instability. The main acceleration of electrons takes place in the downstream region. Processes in the relativistic collisionless shock are dominated by structures produced by the Weibel instability. This instability is excited in the downstream region behind the jet head, where electron density perturbations lead to the formation of current filaments. On average the nonuniform electric field and magnetic field structures associated with these current filaments decelerate the jet electrons and positrons, while accelerating the ambient electrons and positrons, and accelerate the jet and ambient electrons and positrons in the transverse direction. The nonlinear region current channels generated by the Weibel instability dissipate. Dissipation levels depend on the initial jet electron parallel velocity distributions.

References

1. T. Piran, *Rev. Mod. Phys.* **76**, 1143–1210 (2005).
2. K.-I. Nishikawa, P. Hardee, C. B. Hededal, and G. J. Fishman: ApJ, **642**, 1267 (2006)
3. K.-I. Nishikawa, P. Hardee, G. Richardson, R. Preece, H. Sol, and G. J. Fishman: ApJ **595**, 555–563 (2003)
4. K.-I. Nishikawa, P. Hardee, G. Richardson, R. Preece, H. Sol, and G. J. Fishman: ApJ **622**, 927–937 (2005)
5. P. Meszaros, E. Ramirez-Ruiz, and M. J. Rees: ApJ **554** 660–666 (2001)
6. E. S. Weibel: Phys. Rev. Letters **2**, 83-84 (1959)
7. M. V. Medvedev and A. Loeb: ApJ **526**, 697–706 (1999)

Analytical and Numerical Studies of Fluid Instabilities in Relativistic Jets

M. Perucho[1], A.P. Lobanov[1], and J.M. Martí[2]

[1] Max-Planck-Institut für Radioastronomie. Auf dem Hügel 69, 53121 Bonn, Germany,
   (perucho,alobanov)@mpifr-bonn.mpg.de
[2] Departament d'Astronomia i Astrofísica. Universitat de València. C/ Dr. Moliner 50 46100 Burjassot, València, Spain, jose-maria.marti@uv.es

Summary. Relativistic outflows represent one of the best-suited tools to probe the physics of AGN. Numerical modelling of internal structure of the relativistic outflows on parsec scales provides important clues about the conditions and dynamics of the material in the immediate vicinity of the central black holes in AGN. We investigate possible causes of the structural patterns and regularities observed in the parsec-scale jet of the well known quasar 3C 273. We compare the model with the radio structure observed in 3C 273 on parsec scales using very long baseline interferometry (VLBI) and constrain the basic properties of the flow. Our results show that Kelvin-Helmholtz instabilities are the most plausible mechanism to generate the observed structures.

1 Introduction

3C 273 is the brightest quasar known. Its relative proximity ($z = 0.158$, [1]) has made it a paradigmatic object studied throughout the entire spectral range [2]. One of the most impressive features of 3C 273 is an apparently one-sided relativistic outflow emanating from the quasar nucleus [2] and extending up to a deprojected distance of about 60 kpc. The jet in 3C 273 has been observed on parsec scales using ground and space VLBI [3, 4, 5, 6]. Space VLBI observations made at 5 GHz with the VSOP[3] revealed a double helical patterns inside the parsec-scale jet in 3C 273 [5]. Linear perturbation analysis of Kelvin-Helmholtz (K-H) instability [7] applied to model these structures yielded an accurate estimate of the bulk parameters of the flow [5]. A different interpretation suggests that a helical magnetic field could generate such a structure [6]. We study the first possibility. A bulk Lorentz factor $W = 2.1$ obtained in [5] is below the values inferred from superluminal motions observed in the jet ($W = 5 - 10$, [4]), which suggests that the K-H instability develops in a slower underlying flow, while the fast components are shock waves inside the jet. Four K-H modes were found acting on parsec scales in the jet in 3C 273: the second helical body mode with a wavelength of 2 mas, the first elliptical and helical body modes, both with wavelengths close to 4 mas, and the elliptical surface mode, with a wavelength of 12 mas. Additionally a

---

[3] VLBI Space Observatory Program

structure with a wavelength of 18 mas was interpreted as an externally driven helical surface mode. Our previous works [8, 9] have shown that numerical simulations can be used effectively to connect the linear perturbation analysis with studies of non-linear regime of K-H instability. Our aim here is to use the jet parameters determined in [5] as initial conditions for numerical RHD simulations of a perturbed jet and compare the numerical results with the observed structures.

2 Numerical Simulations

2.1 Simulation 3C 273-A

We start with a steady jet with a Lorentz factor $W = 2.1$, a density contrast with the external medium $\eta = 0.023$, a sound speed $c_{s,j} = 0.53\,c$ in the jet and $c_{s,a} = 0.08\,c$ in the external medium, and the perfect gas equation of state (with adiabatic exponent $\gamma = 4/3$). Assuming an angle to the line of sight $\theta = 15°$, the observed jet extends up to $\approx 170\,\mathrm{pc}$. With the jet radius $R_j = 0.8\,\mathrm{pc}$ [5], the numerical grid covers $211\,R_j$ (axial) $\times 8\,R_j \times 8\,R_j$ (transversal), i.e., $169\,\mathrm{pc} \times 6.4\,\mathrm{pc} \times 6.4\,\mathrm{pc}$. The resolution is $16\,\mathrm{cells}/R_j$ in the transversal direction and $4\,\mathrm{cells}/R_j$ in the direction of the flow. A shear layer of $2\,R_j$ in width is included in the initial rest mass density and axial velocity profiles to keep numerical stability of the initial jet. Elliptical and helical modes are induced at the inlet.

Frequencies of the excited modes are computed from the observed wavelengths of modes, λ^{obs}, corrected for projection effects, relativistic bulk motion and wave pattern speed, $v_w = 0.23\,c$, according to $\omega = 2\,\pi\,v_w/\lambda^{theor}$, where

$$\lambda^{theor} = \frac{\lambda^{\mathrm{obs}}(1 - v_w/c\,\cos\theta)}{\sin\theta}, \tag{1}$$

Three wavelengths have been identified in the simulation: a helical $\lambda_1^{\mathrm{sim}} = 4\,R_j$ perturbation, a helical $\lambda_2^{\mathrm{sim}} = 25\,R_j$ perturbation and an elliptical $\lambda_3^{\mathrm{sim}} = 50\,R_j$ perturbation. The wave speeds $v_{w,2} \simeq 0.38\,c$ and $v_{w,3} \simeq 0.2\,c$ are estimated for the λ_2^{sim} and λ_3^{sim} modes, respectively. An upper limit for $v_{w,1}$ is given by the flow speed ($0.88\,c$). With these wave speeds, λ_1^{sim} would be observed as a 2.27 mas structure, λ_2^{sim} as a 3.37 mas structure and λ_3^{sim} as a 5.5 mas structure. These are similar to the shorter-wavelength structures found in the observations (2.27 mas vs 2 mas, 3.37 and 5.5 mas vs 4 mas). It remains to be studied why the simulations do not reproduce readily the longer modes. The longest mode, with the wave speed given in [5], corresponds to a simulated wavelength of $150\,R_j$, which is difficult to reproduce even in the grid of this simulation, in particular when shorter harmonics grow fast and disrupt the flow. The disruptive evolution may be caused by a number of factors, including the absence of a stabilizing magnetic field in the simulations.

2.2 Simulation 3C 273-B

In the second simulation, we try to study the effect of precession and injection of fast components on the observed structures in the jet. The precession frequency is given by the observed 15 yr periodicity of the position angle variations in the inner jet ([4]). The frequency of ejections of components is set by the reported 1 yr periodicity ([3]). The duration of each ejection is estimated to be 2 months, from the approximate inspiralling time from an orbit at $\sim 6\,R_G$ around a $5.5\,10^8\,M_\odot$. Velocity of the components is taken as constant, with the mean value of those given in [4], i.e., Lorentz factor $W \sim 5$. The components are treated as shells of diameter $0.5\,R_j$ ejected at the center of the transversal grid. The numerical grid for this simulation covers $30\,R_j$ (axial) $\times 6\,R_j \times 6\,R_j$ (transversal), i.e., 24 pc $\times$ 4.8 pc $\times$ 4.8 pc. The resolution of the grid is 16 cells/R_j in both transversal directions and 32 cells/R_j in the direction of the flow.

We find two structures in the simulation: a pinching perturbation with a wavelength of $0.4\,R_j$ caused by the injection of components, and a helical perturbation with a wavelength of $4.0\,R_j$ associated with the precession. The wave speed associated with these structures is $\leq 0.98\,c$. This upper limit for the wave speed results in observed wavelengths (from Eq. 1) of 0.6 and 6 mas, much smaller than the 2-4 mas and 18 mas expected from the observations. This results further supports the identification of the shorter modes with K-H instability modes. We also find that the 15 yr precession period cannot account for the observed longer, 18-mas wavelength. Thus, either the longer-wavelength structure is driven by a different mechanism, or the 15 yr periodicity in the source is not associated with the precession.

References

1. M.A. Strauss, J.P. Huchra, M. Davies, et al.: ApJSS **83**, 29 (1992)
2. T. J.-L. Courvoisier: A&ARv **9**, 1 (1998)
3. T.P. Krichbaum, A. Witzel, J.A. Zensus: From centimetre to millimetre wavelengths: A high angular resolution study of 3C273. In: *Proc. of the 5th EVN Symposium*, ed by J.E. Conway, A.J. Polatidis, R.S. Booth, Y.W. Pihlström (Onsala Space Observatory 2000) pp 25
4. Z. Abraham, E.A. Carrara, J.A. Zensus, S.C. Unwin: A&AS **115**, 543 (1996)
5. A.P. Lobanov, J.A. Zensus: Science **294**, 128 (2001)
6. K. Asada, M. Inoue, Y. Uchida, et al.: PASJ **54**, L39 (2002)
7. P.E. Hardee: ApJ **533**, 176 (2000)
8. M.P. Perucho, M. Hanasz, J.M. Martí, H. Sol: A&A **427**, 415 (2004)
9. M.P. Perucho, J.M. Martí, M. Hanasz: A&A **427**, 431 (2004)

Forced Oscillations in Relativistic Accretion Disks and QPOs

J. Pétri

Max-Planck-Institut für Kernphysik, Saupfercheckweg 1, 69117 Heidelberg, Germany,
j.petri@mpi-hd.mpg.de

1 Introduction

To date, quasi-periodic oscillations (QPOs) have been observed in about twenty Low Mass X-ray Binaries (LMXBs) containing an accreting neutron star. Among these systems, the high-frequency QPOs (kHz-QPOs) which mainly show up by pairs, denoted by frequencies ν_1 and $\nu_2 > \nu_1$, possess strong similarities in their frequencies, ranging from 300 Hz to about 1300 Hz, as well as in their shapes [1].

Recent observations of accretion disks orbiting around white dwarfs, neutron stars, or black holes have shown a strong correlation between the low and high frequency QPOs [2]. This relation holds over more than 6 orders of magnitude and strongly supports the idea that the QPO phenomenon is a universal physical process independent of the nature of the compact object.

We propose a new resonance in the star-disk system arising from the response of the accretion disk to either a non-axisymmetric rotating gravitational field, for a hydrodynamical disk, or to a rotating magnetic field, for an MHD disk as an explanation of this phenomenon.

2 Hydrodynamical Disk [3]

2.1 Linear Analysis

By perturbing the hydrodynamical equations governing the motion in the disk, and introducing the Lagrangian displacement, the weak motion in the radial and vertical directions can be cast into a partial differential equation for the radial and vertical displacement respectively. A careful analysis of this equation shows the emergence of three kinds of resonances corresponding to :

- a corotation resonance ;
- an inner and outer Lindblad resonance ;
- a parametric resonance when $m \left| \Omega_* - \Omega \right| = 2\, \kappa_{\mathrm{r/z}}/n$, where Ω_* is the stellar spin, $\kappa_{\mathrm{r/z}}$ the radial and vertical epicyclic frequencies, n is a natural integer and m the azimuthal mode of the gravitational perturbation.

The results for an accretion disk evolving in a Kerr spacetime are shown in Table 1. For a neutron star, we adopt the typical parameters, mass $M_* = 1.4\,M_\odot$, angular velocity $\nu_* = \Omega_*/2\pi = 300/600$ Hz, corresponding to slow and fast rotators respectively, moment of inertia $I_* = 10^{38}$ kg m^2, angular momentum $a_* = c\,I_*\,\Omega_*/G\,M_*^2$. For these typical parameters, $a_* = 0.1/0.2$. Therefore the vertical epicyclic frequency remains close to the orbital one $\kappa_z \approx \Omega$. As a consequence, for the vertical resonance, the Newtonian approximation remains valid within a few percent.

Azimuthal mode m	Orbital frequency $\nu(r, a_*)$ (Hz)			
	$\nu_* = 600$ Hz		$\nu_* = 300$ Hz	
	$n = 1$	$n = 2$	$n = 1$	$n = 2$
1	—- / 200	—- / 300	—/ 100	—/ 150
2	—- / 300	1198 / 400	—/ 150	599 / 200
3	1790 / 360	899 / 450	898 / 180	450 / 225

Table 1. Orbital frequencies at the parametric resonance locii.

2.2 Two-dimensional Simulations

From the analytical analysis of the linear response of a thin accretion disk in the 2D limit, we know that waves are launched at the aforementioned resonance locii. They propagate in some permitted regions inside the disk, according to the dispersion relation obtained by a WKB analysis. We confirm and extend these results by performing non linear hydrodynamical numerical simulations using a pseudo-spectral code solving Euler's equations in a 2D cylindrical coordinate frame. Simulations were performed for the Newtonian as well as for a pseudo-Newtonian potential. The simulations agree with the linear analysis.

The aforementioned treatment (linear analysis and numerical simulations) has been extended to a magnetised accretion disk. We recover exactly the same conclusions [4].

3 Slow vs Fast Rotator

Focusing on accreting neutron stars in LMXBs, observations reveal that they can be divided into two categories[5] : the slow rotators possessing a rotation rate $\nu_* \approx 300$ Hz, and for which the frequency difference between the two peaks is around $\Delta\nu \approx \nu_*$ and $\nu_2 \approx 3\nu_1/2$ and the fast rotators having $\nu_* \approx 600$ Hz, for which this difference is around $\Delta\nu \approx \nu_*/2$.

The model presented in this work can account for this segregation if the innermost stable circular orbit (ISCO) is taken into account. Indeed, for a typical neutron star, the orbital frequency at the ISCO is $\nu_{\text{isco}} = 1571$ Hz which is therefore the upper limit for any QPO frequency. Discarding the resonance frequencies in the relativistic disk which are higher than ν_{isco}, we conclude from Table 1 that :

- for slow rotators, the two highest frequencies are less than ν_{isco} and given by $\nu_1 = 599$ Hz and $\nu_2 = 898$ Hz, therefore $\Delta\nu = 299$ Hz which is very close to ν_* ;

- for fast rotators, the highest frequency is not observed because the resonance is located inside the radius of the ISCO. Therefore the two highest observable frequencies are $\nu_1 = 899$ Hz and $\nu_2 = 1198$ Hz, having a difference $\Delta\nu = 299$ Hz which is close to $\nu_*/2$.

These conclusions apply to hydrodynamical [6] as well as to magnetised [7] accretion disks.

4 Conclusion

The consequences of a weakly rotating asymmetric stellar gravitational or magnetic field on the evolution of a thin accretion disk are as follows. Corotation, driven and parametric resonances are excited at some preferred radii. The kHz-QPOs are interpreted as the orbital frequency of the disk at locations where the vertical response to the resonances are maximal. The 3:2 ratio is predicted for the strongest modes and a clear distinction exists between slow and fast rotators, a direct consequence of the presence of an ISCO. Nevertheless general relativistic effect are not required to excite these resonances. They behave identically in the Newtonian as well as in the Kerr field. Therefore the QPO phenomenology is unified in a same picture, irrespective of the nature of the compact object. Indeed the presence or the absence of a solid surface, a magnetic field or an event horizon plays no relevant role in the production of the X-ray variability.

Acknowledgments. This work was supported by a grant from the G.I.F., the German-Israeli Foundation for Scientific Research and Development.

References

1. van der Klis, M.: ARA&A, **38**, 717 (2000).
2. Mauche, C. W.: ApJ, **580**, 423 (2002).
3. Pétri, J.: Ap&SS **302**, 117 (2005).
4. Pétri, J.: A&A., **439**, 443 (2005).
5. van der Klis, M.: astro-ph/0410551 (2004).
6. Pétri, J.: A&A, **439**, L27 (2005).
7. Pétri, J.: A&A **443**, 777 (2005).

QPOs: Einstein's Gravity Non-Linear Resonances

P. Rebusco[1] and M.A. Abramowicz[2,3]

[1] Max-Planck-Institut für Astrophysik, Karl-Schwarzschild-Strasse 1, 85741 Garching, Germany, pao@mpa-garching.mpg.de
[2] Department of Physics, Göteborg University, S 412 96 Göteborg, Sweden, number44@fy.chalmers.se
[3] N. Copernicus Astronomical Center, Polish Academy of Sciences, Warsaw, Poland

Summary. There is strong evidence that the observed kHz Quasi Periodic Oscillations (QPOs) in the X-ray flux of neutron star and black hole sources in LMXRBs are linked to Einstein's General Relativity. Abramowicz & Kluźniak (2001) suggested a non-linear resonance model to explain the QPOs origin: here we summarize their idea and the development of a mathematical toy-model which begins to throw light on the nature of Einstein's gravity non-linear oscillations.

1 Introduction

QPOs are highly coherent Lorentzian peaks observed in the X-ray power spectra of compact objects in LMXRBs. They cover a wide range of frequencies (from mHz to kHz) and they show up alone, in pairs or in triplets. In the present review we focus on kHz QPOs which occur in pairs (the so called twin peak QPOs). Many models have been proposed in order to understand kHz QPOs: some of them involve orbital motions (e.g. [14, 8]), but there are also models that are based on accretion disk oscillations (e.g. [17, 7, 13, 9]). Here we present semi-analytical results, connected in particular with the Kluźniak&Abramowicz model.

2 QPOs and General Relativity

High frequency (kHz) QPOs lie in the range of orbital frequencies of circular geodesics just few Schwarzschild radii outside the central source. Moreover the frequencies scale with $1/M$, where M is the mass of the central object [10]. These two facts support the strong gravity orbital oscillations models.
Consider a test particle rotating around a compact source: the radial epicyclic frequency of planar motion and the vertical epyciclic frequency of nearly off-plane motion are respectively defined as

$$\omega_r^2 = \left(\frac{1}{2g_{rr}} \frac{\partial^2 U_{eff}}{\partial r^2} \right)_{\ell, r_0, \pi/2} \quad , \quad \omega_z^2 = \left(\frac{1}{2g_{\theta\theta}} \frac{\partial^2 U_{eff}}{\partial \theta^2} \right)_{\ell, r_0, \pi/2} \quad ,$$

where $U_{eff} = g^{tt} + lg^{t\phi} + l^2 g^{\phi\phi}$ is the effective potential and (r, θ, ϕ) are spherical coordinates. The eigenfrequencies depend only on the metric of the system, hence on strong gravity itself. In Newtonian gravity there is degeneracy between these eigenfrequencies and the Keplerian frequency (all three frequencies are equal) while in General Relativity this degeneracy is broken and as a consequence two or three different characteristic frequencies are present, opening the possibility of internal resonances . As a consequence, while Newtonian orbits are all close, in GR they do not close after one loop (this is for the same reason as the well-known advance of the perihelion of Mercury).

For a spherically symmetric gravitating fluid body (a better model of the accretion disk) these are the frequencies at which the center of mass (initially on a circular geodesic) oscillates. the example of the NS source Sco X-1 was studied by [3] and[12] (see Fig.1).

3 Kluźniak-Abramowicz Resonance Model

QPOs often occur in pairs, and the centroid frequencies of these pairs are in rational ratio (e.g. [15]): these features suggested that high frequency QPOs are a phenomenon due to non-linear resonance, and that there may be an analogy between the radial and vertical oscillations in a Shakura-Sunyaev disk and the motion of a pendulum with oscillating point of suspension ([4]). Since in GR $\omega_r < \omega_z$ the first allowed resonance would appear for $\omega_z : \omega_r = 3 : 2$ and it would be the strongest.

In all four microquasars which exhibit double peaks, the ratio of the two frequencies is $3 : 2$, as well as in many neutron star sources. Moreover combinations of frequencies and subharmonics have been detected: these are all signatures of non-linear resonance and they confirm the validity of the model.

3.1 Toy Model

A mathematical toy-model for the Kluźniak-Abramowicz resonance idea for QPOs was recently developed [1]. It describes the QPOs phenomenon in terms of two coupled non-linear forced oscillators,

$$\ddot{\delta r} + \omega_r^2 \delta r = F(\delta r, \delta z, \dot{\delta r}, \dot{\delta z}) + C \cos(\omega_0 t) + N_r(t), \qquad (1)$$

$$\ddot{\delta z} + \omega_z^2 \delta z = G(\delta r, \delta z, \dot{\delta r}, \dot{\delta z}) + D \cos(\omega_0 t) + N_z(t), \qquad (2)$$

where F and G are polynomes of second or higher degree (obtained in terms of expansion of the deviations from a Keplerian flow). The $\cos(\omega_0 t)$ terms represent an external forcing: they are mostly important in the case of NS, where ω_0 can be the NS spin frequency. N_r and N_z describe the stochastic noise due to the Magneto-Rotational Instability .

In absence of turbulence, the first finding, which can be derived by using any perturbative method, is that in the approximate solution there are terms with the denominators in the form $n\omega_r - m\omega_z$ (with n and m being integers). These n and m cannot take any value, but they depend on the symmetry of the metric and of the perturbation: for

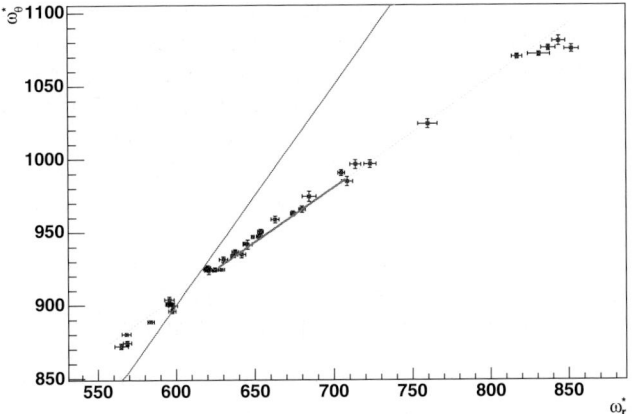

Fig. 1. The dotted line is the least-squares best-fit to the data points (the observed kHz QPOs frequencies in Sco X-1); the thin solid line corresponds to a slope of $3 : 2$ (for reference) . The thick solid segment is the analytic approximation, in which the frequencies are scaled for comparison with observations.

a plane symmetric configuration one can demonstrate (e.g. [12], [6]) that $m = 2\,p\,(p$ integer). Due to this, the regions where $n\omega_r = 2p\omega_z$ are candidate regions of internal resonance , and the strongest one would be for $\omega_z : \omega_r = 3 : 2$ (a similar reasoning can be done for the forcing term).

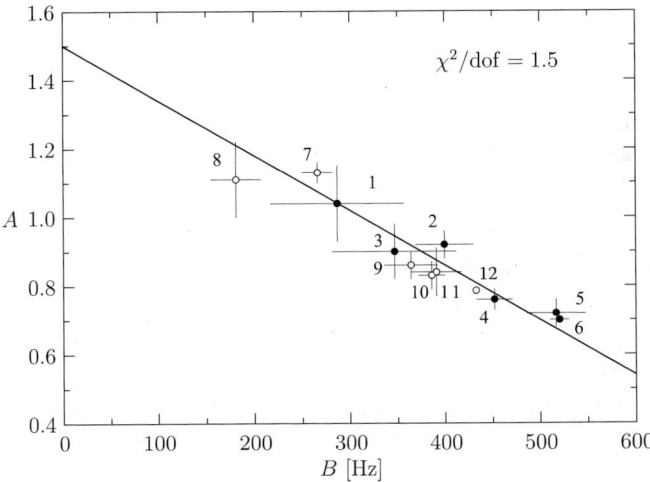

Fig. 2. The anticorrelation between shifts (B) and slopes (A). The points correspond to the individual Z and atoll sources listed in [2]. The best fit line goes through the point $(0, 1.5)$ (Courtesy of Gabriel Török).

3.2 Correlation and Anticorrelation

A well known property of weakly non-linear oscillators is the fact that the frequencies of oscillation (ω_r^* and ω_z^*) depend on the amplitudes, even if their value remains close to that of the eigenfrequencies (ω_r and ω_z). Perturbative methods help us again (e.g. [11]): indeed one can write

$$
\begin{aligned}
\omega_r^* &= \omega_r + \epsilon^2 \omega_r^{(2)} + \epsilon^3 \omega_r^{(3)} + O(\epsilon^4) \quad , \\
\omega_z^* &= \omega_z + \epsilon^2 \omega_z^{(2)} + \epsilon^3 \omega_z^{(3)} + O(\epsilon^4)
\end{aligned}
\tag{3}
$$

and find the frequencies corrections ($\omega_i^{(j)}$) by constraining the solution to be stationary at a given timescale (of the order of ϵ^{-j}). For small perturbations one would get that $\omega_z^* = A\omega_r^* + B$. In this way one can qualitatively explain the observed linear correlation between the twin frequencies in NS sources:

A direct consequence of the internal resonance is the fact that for different sources the coefficients A and B should be nearly linearly anticorrelated ([1]): $A = \omega_z/\omega_r - B/f_0(M)$, where $f_0(M)$ is a function which depends on the mass of the central object. Note that at higher orders the relation is expected to slightly deviate from linearity: anyway up to now this feature fits very well the available data (see Fig.2).

3.3 The Effect of Turbulence

Accretion disks are characterized by a huge number of degrees of freedom. The turbulent processes can be assumed to have a stochastic nature. In particular, we have investigated a simplified model for the Kluźniak-Abramowicz nonlinear theory and showed that a small noise in the vertical direction can trigger coupled epicyclic oscillations (see Fig. 3 and 4). On the other hand too much noise would disrupt the quasi-periodic motion [16]. This is similar to the stochastically excited p-modes in the Sun, and it may help in estimating the strength and the nature of the turbulence itself.

4 Conclusions

Non-linear parametric resonances occur everywhere in Nature: together with GR they could explain the mechanism at the basis of kHz QPOs. In this way the mass and the angular momentum (e.g.[5]) of the central compact object could be precisely measured, but most of all Einstein's strong gravity could be proved: a good motivation to keep on investigating on this puzzling phenomenon.

References

1. M.A. Abramowicz (editor), "Nordita Workdays on QPOs": Astron.Nachr., **326**, 9 (2005)
2. M.A. Abramowicz et al. in: S. Hledík, Z. Stuchlík (eds.), *Proceedings of RAGtime 6/7*, Silesian University Opava (2005)

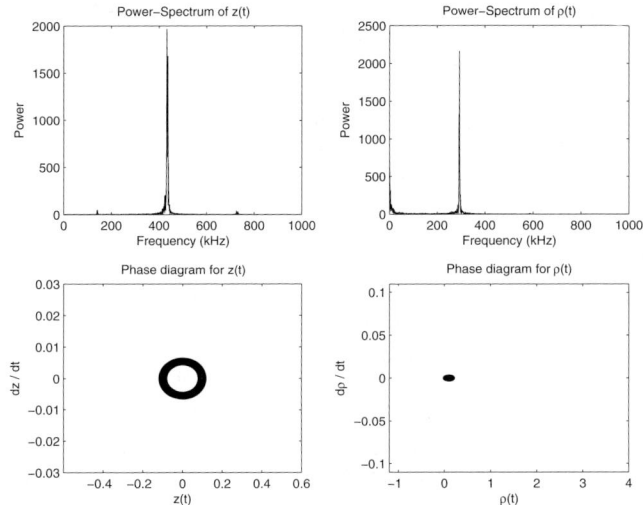

Fig. 3. Low turbulence: power spectra (upper part) and phase diagrams (lower part) for r and z. The displacements are in units of r_g, the frequencies are scaled to kHz (e.g. assuming a central mass M of $2M_\odot$). It does not differ much from the behavior in absence of turbulence.

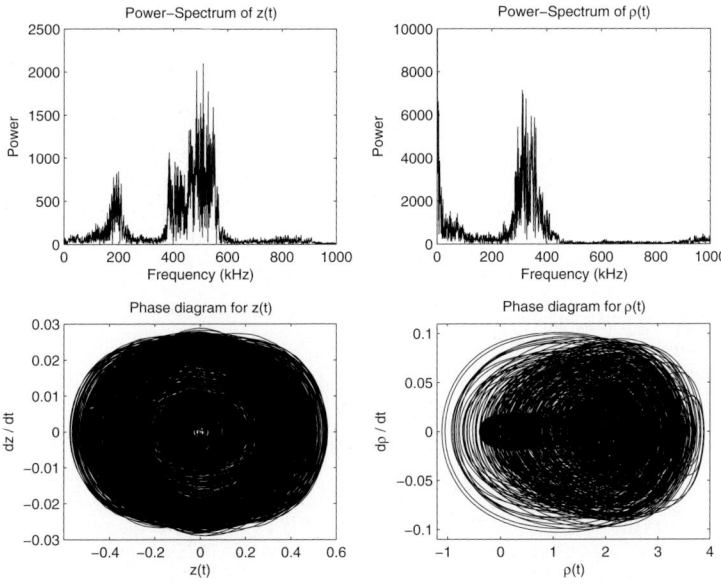

Fig. 4. The same as in the previous figure. In this case however the turbulence is strong enough to feed the resonance: as a consequence the amplitudes of oscillation are much greater.

3. M.A. Abramowicz, et al.: PASJ **55**, 467 (2003)
4. M.A. Abramowicz, W. Kluźniak: A&A, **374**, L19 (2001)
5. B. Aschenbach: A&A, **425**, 1075 (2004)
6. J. Horák in: S. Hledík, Z. Stuchlík (eds.), *Processes in the Vicinity of Black Holes and Neutron Stars*, Silesian University Opava (2004)
7. S. Kato: PASJ, **53**, L37 (2001)
8. F.K. Lamb, M.C. Miller: astro-ph/0308179 (2003)
9. L-X Li, R. Narayan: ApJ, **601**, L414 (2004)
10. J.E. Mc Clintock, R.A. Remillard in Compact Stellar X-ray Sources, ed. W.H.G. Lewin & M. van der Klis (Cambridge:Cam. Univ.Press), astro-ph/0306213 (2004)
11. D.A. Mook, A.H. Nayfeh: *Nonlinear oscillations*, John Wiley & Sons, New York (1976)
12. P. Rebusco: PASJ, **56**, 553 (2004)
13. L. Rezzolla, et al.: MNRAS, **344**, L37 (2003)
14. L. Stella, M. Vietri: ApJL, **492**, L59 (1998)
15. T.E. Strohmayer:ApJL, **552**, L49 (2001)
16. R. Vio, et al.: A&A **452**, 2, 383-386 (2006)
17. R.V. Wagoner, A.S. Silbergleit, M. Ortega-Rodríguez: ApJ, **559**, L25 (2001)

Cosmic-Ray Acceleration and Viscosity

F.M. Rieger and P. Duffy

UCD School of Mathematical Sciences, University College Dublin, Belfield, Dublin 4, Ireland,
frank.rieger@ucd.ie

Summary. Shear flows are ubiquitous phenomena in astrophysical environments. We present recent results concerning the acceleration of energetic particles in turbulent shear flows showing that cosmic ray particles can be accelerated efficiently in the relativistic jets of AGNs and GRBs. Efficient cosmic ray acceleration results in an induced flow viscosity that may explain part of the deceleration effects observed.

1 Introduction

Relativistic shear flows are a natural outcome of the density and velocity gradients in the extreme astrophysical environments of AGNs and GRBs, and in the case of AGN, for example, observationally well-established, cf. [4, 7]. The acceleration of energetic particles occurring in such flows can represent an efficient mechanism for converting a significant part of the bulk kinetic energy of the flow into nonthermal particles and radiation as has been successfully shown in several contributions, e.g. [6, 7, 10, 11]. Somewhat similar to the microscopic Fermi picture, shear acceleration is based on the fact that particles can gain energy by scattering off (small-scale) magnetic field inhomogeneities that have different local velocities due to being embedded in a collisionless shear flow, cf. [9], for a recent review: In a scattering event particles are assumed to be randomized in direction with their energies being conserved in the local (comoving) flow frame. As the momentum of a particle travelling across a velocity shear changes with respect to the local flow frame, scattering results in a net increase in momentum for an isotropic particle distribution. In contrast to 2nd order Fermi acceleration, however, not the random but the systematic velocity component of the scattering centres is assumed to play the important role.

2 Cosmic-Ray Acceleration in Shear Flows

The acceleration of energetic particles in non-relativistic gradual shear flows is known to lead to (local) steady-state power-law particle distributions with momentum indices that depend on the mean scattering time $\tau \propto p^\alpha$, i.e.,

$$f(p) \propto p^{-(3+\alpha)} \quad \text{for } \alpha > 0 \text{ and } p > p_0, \tag{1}$$

where p_0 is the injection momentum, e.g., [1, 8]. In general, efficient shear acceleration requires that energetic particles can sample sufficient shear within a mean scattering time, i.e., that diffusion across the shear is strong enough. In the simplest case of almost isotropic diffusion, a common approximation in shear acceleration theory, one physically expects that the stronger the shear and the larger the spatial diffusion coefficient $\kappa \propto p^\alpha$, the higher the possible impact of a scattering event and thus the more efficient the acceleration will be. This suggests that in the case of shear acceleration the typical acceleration timescale should be inversely related both to the velocity gradient as well as to the diffusion coefficient of the flow, in particular $t_{\text{acc}} \propto 1/\kappa$ in marked contrast to shock-type acceleration processes where $t_{\text{acc}} \propto \kappa$ [3]. Detailed calculations have shown indeed, cf. [7, 9], that the minimum (local) acceleration timescale for a longitudinal gradual shear flow $\boldsymbol{u} = u_z(r)\, \boldsymbol{e}_z$ decreasing linearly from relativistic $u_z(r_1) \sim c$ to non-relativistic speeds $u_z(r_2) \ll c$ over a distance $\Delta r = (r_2 - r_1)$, is of order

$$t_{\text{acc}}(\text{min}) \sim \frac{3\,(\Delta r)^2}{\gamma_b(r_1)^4\, \lambda'\, c}, \tag{2}$$

where γ_b is the position-dependent bulk Lorentz factor of the flow. Note that for the convenient choice of a gyro-dependent particle mean free path λ', i.e., $\alpha = 1$, this yields a timescale that scales with γ in the same way than the synchrotron loss timescale, so that losses are no longer able to constrain the mechanism once it has started to work efficiently. Application to collimated large-scale relativistic AGN jets (with typical parameters $B \sim 10^{-5}$ G, $\gamma_b(r_1) \sim 5$ and $\Delta r \sim 0.3$ kpc) shows that the so achievable energies for protons can exceed

$$E \sim 10^{19} \text{ eV} \tag{3}$$

and probably be as high as 10^{20} eV if non-gradual shear acceleration at the jet side boundary is fully taken into account, e.g. see [6].

For a radially expanding outflow on the other hand, the minimum (comoving) acceleration timescale becomes [10]

$$t_{\text{acc}}(\text{min}) \simeq \frac{5\, r^2}{2\, \gamma_{b0}^2\, \lambda'\, c}. \tag{4}$$

Application to the ultra-relativistic jets in GRBs with typical bulk Lorentz factors on the jet axis of order $\gamma_{b0} \sim 300$ shows that protons may reach UHE in excess of $\sim 10^{20}$ eV under a wide range of conditions and probably even more efficient than via internal shock scenarios, cf. [10].

Both results give strong support to the notion that relativistic AGN and GRB jets are indeed the sources of UHE cosmic rays, see [5] for consequences.

3 Cosmic-Ray Viscosity

Efficient shear acceleration of cosmic rays essentially draws on the kinetic energy of the background flow. The resultant dynamical effects on the flow can be modelled by means of an induced viscosity coefficient $\eta_s > 0$ describing the associated decrease in flow mechanical energy per unit time, e.g.,

$$\dot{E}_{\mathrm{kin}} = -\eta_s \int \left(\frac{\partial u_z}{\partial x}\right)^2 \mathrm{d}V \,, \tag{5}$$

in the case of a simple (non-relativistic) one-dimensional gradual shear flow $u_z(x)\,\boldsymbol{e}_z$. The viscosity coefficient may then be determined from the equation $\dot{E}_{\mathrm{kin}} = -\dot{E}_{\mathrm{cr}}$, where $E_{\mathrm{cr}} = \int \epsilon_{\mathrm{cr}} \mathrm{d}V$ is the energy gained by the cosmic ray particles, cf. also [2]. If particles are injected with momentum p_0 into the acceleration process at a density rate Q_0, the density $\dot{\epsilon}_{\mathrm{cr}}$ of power gained becomes

$$\dot{\epsilon}_{\mathrm{cr}} = 4\pi \int_0^\infty p^2\, E(p)\, \frac{\partial f}{\partial t}\, \mathrm{d}p - 4\,\pi\, Q_0\, E(p_0)\, p_0^2 \,, \tag{6}$$

where $E(p)$ denotes the relativistic kinetic energy. In the simplest steady-state case where particle escape above a momentum threshold p_{max} – corresponding to a mean free path larger than the width of the acceleration region – is assumed and where simple powerlaw-type cosmic ray distributions may be found (see above), one obtains $\epsilon_{\mathrm{cr}} \simeq 4\,\pi\, Q_0\, p_0^2\, p_{\mathrm{max}}\, c$, assuming $m_0\, c^2 \ll p_0 \ll p_{\mathrm{max}}$, which implies a viscosity coefficient (cf. Rieger & Duffy, in preparation)

$$\eta_s \simeq \frac{3\,\alpha}{15}\, \lambda(p_0)\, n_0\, p_{\mathrm{max}} \,, \tag{7}$$

where $\lambda(p_0)$ denotes the mean free path for a particle with momentum p_0, and n_0 is the cosmic ray particle number density in the acceleration region.

Efficient shear acceleration of cosmic ray particles thus provides an additional (non-negligible) force acting on the flow that may explain part of the deceleration observed in relativistic jets.

Acknowledgments. Inspiring discussions with John Kirk, Valenti Bosch-Ramon and Rob Laing are gratefully acknowledged.

References

1. E.G. Berezhko, G.F. Krymskii: Sov. Astr. Lett. **7**, 352 (1981)
2. J.A. Earl, J.R. Jokipii, G. Morfill: ApJ **331**, L91 (1988)
3. J.G. Kirk, R.O. Dendy: J.Phys.G: Nucl. Part.Phys. **27**, 1589 (2001)
4. R.A. Laing, A.H. Bridle: MNRAS **336**, 328 (2002)
5. K. Mannheim, R.J. Protheroe, J.P. Rachen: Phys. Rev. D **63**, 023003 (2001)
6. M. Ostrowski: A&A **335**, 134 (1998)
7. F.M. Rieger, P. Duffy: ApJ **617**, 155 (2004)
8. F.M. Rieger, P. Duffy: ChJAA **5S**, 195 (2005a)
9. F.M. Rieger, P. Duffy, in: Proc. of the Texas Symposium on Relativistic Astrophysics (Stanford 2004), eds. P. Chen et al., eConf:C041213, 2521 (2005b)
10. F.M. Rieger, P. Duffy: ApJ **632**, L21 (2005c)
11. L. Stawarz, M. Ostrowski: ApJ **578**, 763 (2002)

Gamma-Ray Emission from Microquasars: Leptonic vs. Hadronic Models

G.E. Romero[1], V. Bosch-Ramon[2], J.M. Paredes[2], and M. Orellana[1]

[1] Instituto Argentino de Radioastronomia, C.C.5, (1894) Villa Elisa, Buenos Aires, Argentina, romero@irma.iar.unlp.edu.ar

[2] Departament d'Astronomia i Meteorologia, Universitat de Barcelona, Av. Diagonal 647, 08028 Barcelona, Spain, vbosch@am.ub.es

Summary. In this work we discuss two different types of models for the high-energy gamma-ray production in microquasars. On the one hand, we introduce a new leptonic model where the emission arises from inverse Compton interactions with both internal (synchrotron) and external photon fields, and relativistic Bremsstrahlung from interactions with cold protons in the jet and the stellar wind material. On the other hand, we introduce a hadronic model where the gamma rays are the result of interactions between relativistic protons in the jet and cold protons from the anisotropic stellar wind. Spectral differences and similarities between both types of models are briefly discussed.

1 Introduction

Microquasars are characterized by the ejection of jets formed by relativistic plasma. These jets are detected at radio wavelengths as non-thermal synchrotron sources, indicating the presence of relativistic leptons. Synchrotron emission has been found in the extended jets of sources like XTE 1550-564 at X-rays, strongly suggesting the acceleration of electrons up to TeV energies (e.g. Corbel et al. 2002 [1]). The presence of such energetic particles in an environment with strong photon fields provided by the stellar companion, the accretion disk, and the hot inner corona around the compact object will result in the production of high-energy gamma rays [2, 3].

The matter content of the jets is not well established, but the large perturbations they produce in the interstellar medium suggest that they should have a large baryonic load (e.g. Heinz 2005 [4]). If particles are accelerated by shocks, we can expect that not only electrons, but also protons could reach relativistic energies. These protons might interact with the ambient medium inside the binary system, producing high-energy gamma rays through the decay of neutral pions created in pp collisions [5].

The recent detection of high-energy gamma rays from the microquasar LS 5039 by HESS telescopes [6] shows that relativistic particles reach multi-TeV energies in these objects. In what follows we outline both leptonic and hadronic models for gamma-ray production in microquasars and we compare the expected spectral energy distributions.

2 A Leptonic Model for Microquasars

We have developed a new model based on a freely expanding magnetized jet, whose internal energy is dominated by a cold proton plasma extracted from the accretion disk [7]. The cold proton plasma and its attached magnetic field, that is frozen to matter and similar to or below equipartition, provides a framework where internal shocks accelerate a fraction of the leptons up to very high energies. These accelerated leptons radiate by synchrotron, Bremsstrahlung and inverse Compton (IC) processes. In this model, the seed photons that interact with the jet leptons by IC scattering come from the star, the disk and the corona. A blackbody spectrum is assumed for the star and the disk, and a power-law plus and exponential cut-off spectrum for the corona. The synchrotron self-Compton radiation is also computed as well as the Bremsstrahlung and Compton self-Compton emission, where the radiation contribution of the latter two mechanisms is negligible.

The dissipated shock kinetic energy that goes to relativistic particles comes from the mean bulk motion kinetic energy, directly related to the kinetic energy carried by shocks in the plasma. The number of relativistic particles that can be produced along the jet is constrained by the limited capability of transferring energy from the shock itself to the particles, and by the fact that the relativistic particle pressure must be kept below the cold proton pressure. Their maximum energy is limited by the acceleration mechanism properties: strong shock, trans-relativistic velocities, almost parallel shock geometry, and diffusion coefficient.

We have applied the model to a high-mass microquasar with efficient acceleration, a compact object of 3 solar masses, macroscopic Lorentz factor around 1.5 (this factor actually evolves as the bulk energy is transferred to particles), and a jet viewing angle of 45 degress (see Ref. [7], Model A, for additional details). At high energies the spectral energy distribution (SED) is dominated by the inverse Compton upscattering of the stellar photons. There is significant absorption of gamma rays in the stellar photon field between 10 GeV and a few TeV, which results in a broad "valley" in the spectrum. The maximum luminosity is around a few times 10^{34} erg s^{-1} at ~ 10 GeV.

3 A Hadronic Model for Microquasars

In microquasars with high-mass stars, the stellar wind can provide a matter field for interactions with relativistic protons from the jet [5]. We have developed a model for a system similar to LSI +61 303, where the stellar companion is a Be star with a dense and slow equatorial wind and the compact object is in an eccentric orbit. The matter from the wind partially penetrates the jet from the side and pp interactions result in the production of neutral pions, which decay producing gamma-rays. Since the star is bright, at the periastron passage the opacity effects of the stellar photon field to the propagation of the high-energy gamma-rays can be important. At energies around 100 GeV the system is transparent only beyond distances $> 10^{12}$ cm, i.e. gamma-ray radiation is strongly absorbed inside the binary. The resulting gamma-ray SED from neutral pion decays mimics the proton spectrum at a few GeV, where absorption is unimportant, but then steepens showing a minimum around 100 GeV during the periastron. At high energies the spectrum hardens again, as the absorption decreases. For additional details we refer the reader to Ref. [8].

4 Discussion

Both leptonic and hadronic processes can produce significant gamma-ray emission in the inner part of microquasar jets when the companion star is of an early-type. In general, the intrinsic emission is higher in the case of IC scattering. The general shape of the SED is similar. As noted by Romero et al. [5] and Aharonian et al. [9], neutrino emission, associated with the charged pion decays in the hadronic scenario, can be used to differentiate the mechanism responsible for the radiation. Rapid variability at high energies could be used as well, since protons are expected to reach higher energies in the inner jet, where they do not suffer as strong losses as leptons. In particular, variability observations of objects with dense winds like LSI +61 303 which might be performed with the MAGIC telescope) might be very useful to shed light on the gamma-ray production in microquasars. In the case of systems with low-mass stars, gamma rays could be the result of interactions of relativistic particles (both leptons and hadrons) with the photon field of the putative corona around the compact object or produced by synchrotron self-Compton mechanism [10]. Such emission has yet to be found with Cherenkov telescopes.

References

1. Corbel, S., et al.: Science, **298**, 196 (2002)
2. Paredes, J.M., et al.: Science, **288**, 2340 (2000)
3. Bosch-Ramon, V., Romero, G. E., & Paredes, J.M.: A&A, **429**, 267 (2005)
4. Heinz, S.: ApJ **636**, 316-322 (2006)
5. Romero, G.E., et al.: A&A, **410**, L1 (2003)
6. Aharonian F., et al.: Science, **5717**, 1938 (2005)
7. Bosch-Ramon V., Romero G. E., & Paredes J. M.: A&A **447**, 1, 263-276 (2006)
8. Romero, G.E., Christiansen, H., & Orellana, M.: ApJ, **632**, 1093 (2005)
9. Aharonian F., et al.: J.Phys.Conf.Ser. **39**. 408-415 (2006)
10. Grenier, I.A., Kaufman Bernadó, M.M., & Romero, G.E.: Ap&SS, **297**, 109 (2005)

Magnetized Supernovae and Pulsar Recoils

H. Sawai, K. Kotake, and S. Yamada

Science & Engineering, Waseda University, 3-4-1 Okubo, Shinjuku, Tokyo 169-8555, Japan,
(hsawai,kkotake,shoichi)@heap.phys.waseda.ac.jp

1 Introduction

Observed pulsars generally have large recoil velocities, which are up to 500 km/s or even larger. Although the origin of such kicks has not been known yet, the most promising one is an asymmetric supernova (SN) explosion. We consider here a new mechanism for asymmetric SN which can produce large recoil velocities; that is magnetohydrodynamically-induced kicks. We consider initial offset dipole-like fields which are supported by a recent study which computed evolution of fossil magnetic field in A stars and white dwarfs [1]. Although our mechanism is, at present, applicable only to magnetars, it also may yield some implications for ordinary pulsar recoils.

2 Models

In our two-dimensional computations, we solve the ideal magnetohydrodynamic equations with the numerical code ZEUS-2D developed by Stone & Norman [2]. We use the central $1.4M_\odot$ core of $15M_\odot$ pre-SN stellar model [3]. We adopt a phenomenological parametric equation of state as in a paper of [4]. The parameters are chosen so that no explosion occurs with spherically symmetric models. Assuming the axial asymmetry, we solve only the half of the meridional plane. The numerical meshes cover the whole core. In order to set up ainitial dipole-like field, we put the uniform toroidal current in the region within the radius of about 1000 km. We initially set offset the dipole-like field by sliding the current region by 300 km from the center along the rotation axis. The initial differential rotation is given by a shell-type profile; that is $\Omega(r) = \Omega_0 r^2_0/(r_0^2+r^2)$, where Ω and r are the angular velocity and the distance from the center of the core, respectively. We calculate four models, model A1, A5, B1 and B5. In the name of models, A denotes the non-offset dipole-like field while B the offset one; 1 and 5 denote that the ratio of the magnetic energy to gravitational energy is 0.01 % and 0.05 %, respectively.

3 Results

As a result of the calculations, while models denoted by A form equatorially-symmetric jets, model B1 and B5 produce equatorially-asymmetric SN explosions, exploding more strongly towards the direction to which dipole-like field is shifted (see Fig. 1). The reason why such asymmetries occur is as follows. Since rotation is slowed down by magnetic field, the matter can collapse more deeply at the hemisphere with more intense magnetic field. (Note that magnetic field is weak enough not to affect dynamics during collapse.) This leads to stronger bounce at the hemisphere and results in asymmetric explosions.

As a consequence of the momentum conservation, the proto-neutron-star is accelerated toward the inverse direction and gains transient large recoil velocities[1] in each B model (see Fig. 1). As the proto-neutron-star moves, however, the pressure gradient decelerates it because the pressure around the head-on region is intensified. Although the acceleration by gravitational force is also large, the deceleration by pressure force is superior to it and soon the proto-neutron-star is "kicked back". This mechanism is repeated several times and produces oscillation of proto-neutron-star-velocity (see Fig. 1 again). Finally, the recoil velocity is damped on the time scale of a hundred milliseconds.

4 Discussion and Conclusion

Our results show that offset strong dipole-like fields generate large recoils of up to 500 km/s but that they are soon damped due to the "kick back" mechanism. The reason why such mechanism appears in our simulations is because the asymmetry already exists at the bounce-phase, and then only dense matter in a deep region (within radius of about 10 km) is kicked. The mass of the dense matter is prevented from moving away by the still dense surrounding matter. This implies that pulsar recoils originated from prompt explosion may be difficult.

In numerical SPH simulations by [5], such oscillations also appeared apart from origin of those are different from ours. Actually, these oscillating features can be seen only when full-domain computation is carried out instead of hollowing out a proto-neutron-star from the computational domain (e.g. [6]). If damping mechanisms similar to those also act for a "delayed kick", they surely diminish the recoil velocity. In case this is true, it is important to numerically simulate pulsar recoils without "hollowing".

5 Acknowledgments

Some of the numerical simulations were done on the supercomputer VPP700E/128 at RIKEN and VPP500/80 at KEK (KEK Supercomputer Projects No. 108). This work was partially supported by the Japan Society for Promotion of Science (JSPS) Research Fellowships, the Grants-in-Aid for the Scientific Research (14079202, 17540267) from

---

[1] We define the velocities of neutron stars by their center-of-mass velocities

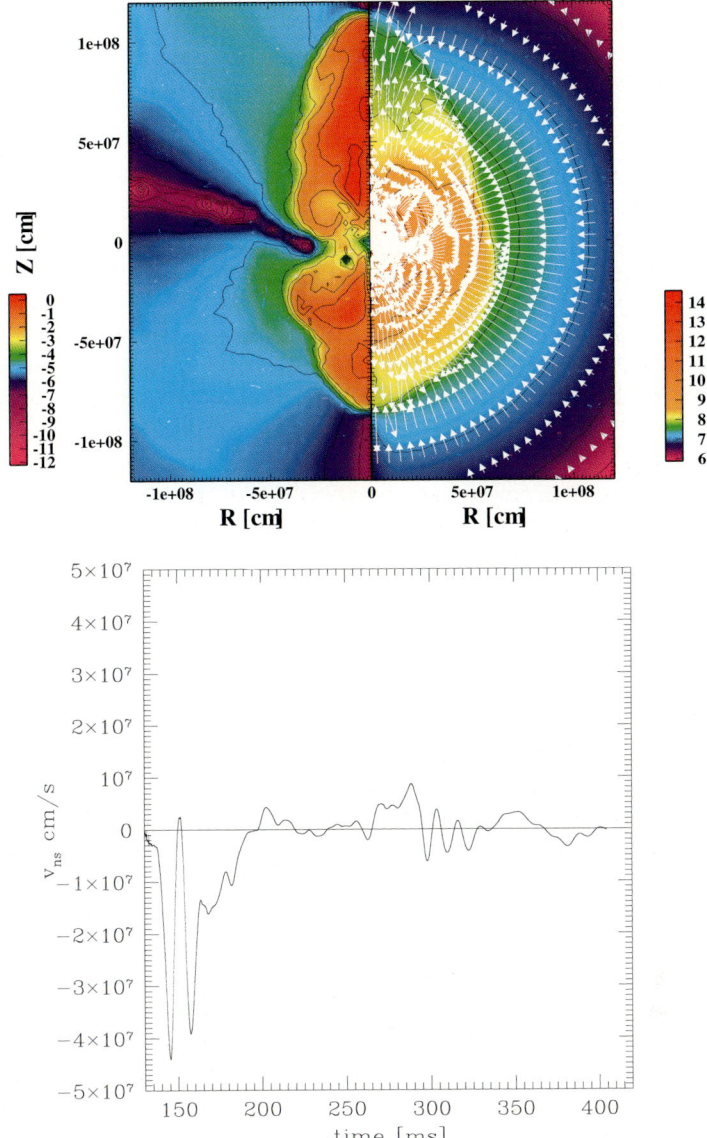

Fig. 1. The velocity field superposed on the density color contours (the right half of top panel) and the contours of the ratio of the magnetic pressure to matter pressure (the left half of top panel), at 195 ms after collapse for model B5. The time evolution of proto-neutron-star-velocity where proto-neutron-star is defined by the region in which the density is more than 10^{11} g/cm^3 (bottom panel).

Ministry of Education, Science and Culture of Japan, and by the Grants-in-Aid for the 21th century COE program "Holistic Research and Education Center for Physics of Self-organizing Systems".

References

1. J. Bralthwalte, and H. C. Spruit: Nature **431**, 819 (2004)
2. J. M. Stone, and M. L. Norman: ApJS **80**, 791 (1992)
3. S. E. Woosley: private communication (1995) .
4. H. Sawai, K. Kotake, and S. Yamada: ApJ **631**, 446 (2005)
5. C. L. Fryer: ApJ **601** L175 (2004)
6. L. Scheck, T. Plewa, K. Kifonidis, H.-Th. Janka, and E. Müller: Procs. 12th Workshop on Nuclear Astrophysics, Ringberg Castle (2004)

Jet Deceleration: the Case of PKS 1136-135

F. Tavecchio[1], L. Maraschi[1], and R.M. Sambruna[2]

[1] INAF-OAB, via Brera 28, 20121 Milano, Italy, tavecchio@merate.mi.astro.it
[2] NASA/GSFC, Code 661, Greenbelt, MD, 20771, USA

1 Introduction

Despite decades of intense efforts, the present knowledge of the physical processes acting in relativistic jets is still rather poor and basic questions are awaiting answers (e.g. Blandford 2001). Among these problems, one of the most fundamental concerns the speed of the flow and the processes leading to deceleration. The present evidence suggests that FRI jets decelerate, becoming trans-relativistic, quite early, within a few kiloparsecs (e.g. Bridle & Perley 1984), while the situation for FRII jets appears more ambiguous. The interpretation of multiwavelength observations of extended jets in QSOs points toward highly relativistic speeds ($\Gamma \sim 10$) even at very large scales (~ 100 kpc; Tavecchio et al. 2000, Celotti et al. 2001; see Stawarz et al. 2004 and Atoyan & Dermer 2004 for some criticisms to the model). In the same sources, recent multiwavelength observations suggest that also these jets undergo deceleration close to their terminal hot spots. This (model-dependent) conclusion is based on the observed increase of the radio to X-ray flux along the jet, which, interpreted in the framework of the synchrotron-IC/CMB emission model, uniquely implies deceleration.

The possibility of "observing" the gradual slowing-down of a jet could in principle provide precious information on the physical processes at work. This approach was successfully developed in great detail for a few FRI jets where the morphology could be well studied thanks to the large angular scale (e.g. Laing & Bridle 2002). Here we report on the analysis performed on the FRII jet of the quasar PKS 1136-135, for which the excellent data (Sambruna et al. 2006), allow us to apply a similar (but less detailed) approach. More details can be found in Tavecchio et al. (2006).

2 Modelling Deceleration

The profiles of the interesting physical quantities of the jet (the bulk Lorentz factor, Γ, the intensity of the magnetic field, B, and the density of the non-thermal electrons, K) can be derived by applying the IC/CMB emission model to the measured radio and X-ray fluxes at different emission knots along the jet. The derived values for different regions in the jet are shown in Fig. 1. The errorbars take into account as much as possible all the uncertainties (associated to the measurements and the modelling) affecting the derivation of the parameters. Clearly, the Lorentz factor appears to decrease along the jet, going from $\Gamma \sim 6$ at B to $\Gamma \sim 2.5$ at F. At the same time, the inferred magnetic field and the particle density increase, as expected in the case of deceleration (Georganopoulos & Kazanas 2004).

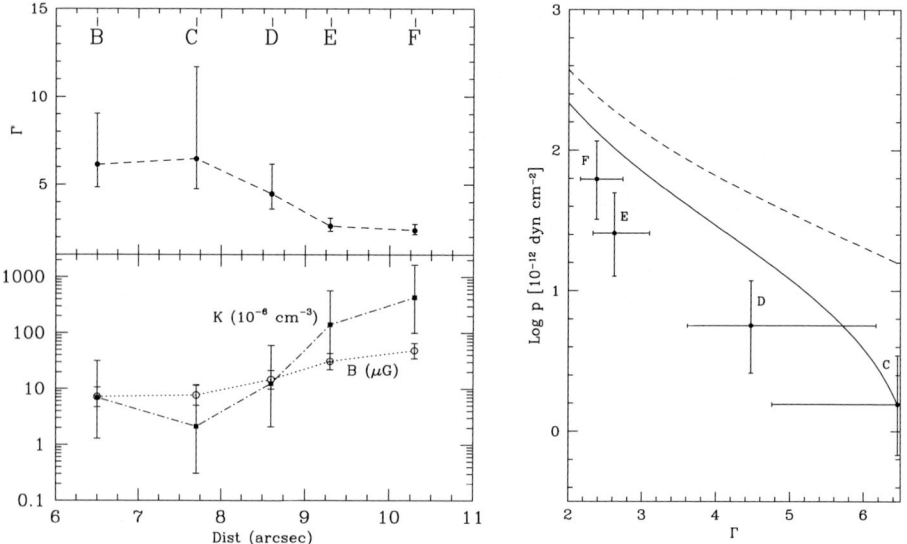

Fig. 1. *Left Panel:* profiles of the relevant quantities (Γ, top panel, B and K, lower panel) for regions B–F of the jet of PKS 1136-135 estimated from the radiative model. *Right Panel:* the pressure inside the jet as a function of the Lorentz factor of the jet, calculated with the momentum and energy flux conservation laws (Bicknell 1994), assuming the initial conditions inferred for knot C. The solid line refers to the case in which the initial pressure of protons is negligible, while the dashed line is calculated assuming an initial pressure in protons ten times that supported by the non-thermal component (relativistic electrons and magnetic field). Crosses indicate the value of the non-thermal pressure (provided by magnetic field and non-thermal electrons) calculated through the modelling of the observed emission (both plots are adapted from Tavecchio et al. 2006).

In the case in which the jet inertia is dominated by protons (as supported by several indications, e.g. Maraschi & Tavecchio 2003), we explored the possibility that entrainment of external gas is effective in decelerating the jet. Basically, deceleration through entrainment can be understood to happen through a continuous series of inelastic collisions between the moving plasma and the external gas at rest. As a result of the collision, part of the kinetic energy is dissipated and converted into internal energy of the jet, thus increasing the internal pressure. We applied the hydrodynamical treatment, based on the use of energy and momentum conservation, developed by Bicknell (1994) to describe the deceleration of the jet of 1136-135 in order to discuss the plausibility of the entrainment mechanism for this particular case. The predicted run of the pressure is reported in Fig. 1.

3 Discussion

The proposed scenario can explain in a plausible way the deceleration inferred for the jet of 1136-135 (and possibly in the other sources showing the same radio-to-X-rays increasing trend). A more detailed understanding of the processes at work and the comparison with the observed properties of jets necessarily involves several still poorly-known physical issues. An important feature of the entrainment-induced deceleration is that the jet starts to slow down significantly when the mass of collected gas is of the order of $1/\Gamma$ of the mass transported by the jet. In this framework, jets characterized by different mass fluxes will experience different behaviours. Large mass fluxes will assure that, under the same conditions of external gas density and entrainment rate, the jet will reach its hotspot almost unperturbed. On the other hand, jets with a small mass flux will be decelerated soon. It is tempting to further speculate along these lines, associating the FRII morphology to jets with large mass flux and FRI objects to jets characterized by small mass fluxes.

References

1. Atoyan, A. & Dermer, C.D: ApJ, **613**, 151 (2004)
2. Bicknell, G. V.: ApJ, **422**, 542 (1994)
3. Blandford, R. D.: ASP Conf. Ser. 250, 487 (2001)
4. Bridle, A. H., & Perley, R. A.: ARAA, **22**, 319 (1984)
5. Celotti, A., Ghisellini, G., & Chiaberge, M.: MNRAS, **321**, L1 (2001)
6. Georganopoulos, M., & Kazanas, D.: ApJ, **604**, L81 (2004)
7. Laing, R. A., & Bridle, A. H.: MNRAS, **336**, 1161 (2002)
8. Maraschi, L., & Tavecchio, F.: ApJ, **593**, 667 (2003)
9. Sambruna, R. M., et al.: ApJ **641**, 717-731 (2006)
10. Stawarz, L. et al: ApJ, **608**, 95 (2004)
11. Tavecchio, F. et al.: ApJ, **544**, L23 (2000)
12. Tavecchio, F. et al.: ApJ **641**, 732-739 (2006)

Some Conclusions on The Magnetic Fields of Neutron Stars in Atoll and Z Sources

C.M. Zhang

National Astronomical Observatories, Chinese Academy of Sciences, Beijing 100012, China,
zhangcm@bao.ac.cn

Summary. The main conclusions of the accretion induced neutron star magnetic field evolution model are described, and it is found that a bright Z source possesses a stronger field than that of on Atoll source and local strong field may exist in the recycled neutron stars or millisecond pulsars on account of the field line distribution.

1 Introduction

The magnetic fields (MFs) of neutron stars (NSs) in low mass X-ray binaries (LMXBs) are as weak as those of millisecond radio pulsars (MSPs)($<\sim 10^{10}$ G), much lower than those of NSs in high mass X-ray binaries (HMXBs) or normal pulsars ($\sim 10^{12}$ G)[1, 2]. The NSMF in LMXB is derived from the magnetosphere scale which is inferred by the detected star spin frequency, however NSMF in HMXB can be measured directly by the cyclotron lines, firstly discovered by J. Trümper in 1976 [3], which is shown to be the same magnitude order of the typical value of normal radio pulsar derived from the presumed magnetic dipole radiation model[1, 2]. The two typical X-ray sources in LMXBs are classified by G. Hasinger and M. van der Klis [4, 5], the luminous Z source of about Eddington luminosity L_{Edd} and less luminous Atoll source of $\sim 0.01 L_{Edd}$, however from the analysis of their spectra the magnetic field strength of Z source is stronger than that of Atoll[4], which implies a proportional correlation between the magnetic field strength and luminosity in LMXB. So, it raises the interesting question of which mechanism is responsible for the magnetic field phenomena in Z and Atoll sources and in HMXBs and LMXBs.

The generation and evolution of neutron star (NS) magnetic field (MF) have not yet been settled[1, 2, 6, 7]. Some models have been proposed, such as Ohmic decay and crustal plate tectonics models[8, 9, 10, 11] for the isolated NSs. However, the observations of NSs in the binary systems show that NSMF decay is inversely related to the mass accreted from the companions [12, 13], this scenario is supported by the recently discovered double pulsars J0737-3039AB [14, 15, 16], where the millisecond pulsar (MSP) possesses the low field and short spin period ($B = 7 \times 10^9$ G, P=22.7 ms) and the normal pulsar possesses high field and long spin period ($B = 6 \times 10^{12}$ G, P=2.77 s). This shows convincingly, that a MSP is recycled in the binary system where the accreted matter weakened its magnetic field and accelerated its spin [16]. Further statistics of the MSPs in binary systems by van den Heuvel and Bitzaraki [17] show that the NSMF evolution stops at some 'bottom' field strength $\sim 10^8$ G, which is expected to be the result of the NS magnetosphere reaching the surface of NS [17, 18, 19]. Moreover, the accretion induced NSMF decay ideas have been proposed by a couple of authors [6, 8, 20, 21, 22, 23, 24, 25, 26],

1.1 Main Conclusions of Model

The basic idea of model was first proposed by van den Heuvel [17] and refined later by Burderi and his collaborators [18, 19], based on which the accretion induced polar cap expansion model has been constructed by Zhang and its collaborators [23, 24]. The main conclusions of the model are described below. The accretion induced expansion of the magnetic poles cap drags the field lines aside, from the polar to equator, as a result the polar field strength is diluted and the equatorial field lines become dense but sunk into the inner crust. This process lasts until the NS magnetosphere radius matches the star radius, where the bottom magnetic field is inferred to be $\sim 10^8$ G. (i) The polar field decays with the accreted mass, which has been implied by the observational statistics [12, 13]; (ii) The bottom magnetic field strength of about 10^8 G can occur when NS magnetosphere radius approaches the star radius, and it proportionally depends on the luminosity or accretion rate as $\propto L^{1/2}$[27], which can account for the stronger field implied in Z source (Eddington luminosity $\sim L_{Edd}$) than that in Atoll source ($\sim 0.01 L_{Edd}$) hinted from their X-ray spectra by Hasinger and van der Klis [4]; (iii) The NS magnetosphere radius decreases with the accretion until it reaches the star radius, and its evolution is little influenced by the initial field and the accretion rate after accreting ~ 0.01 solar mass, which implies that the magnetosphere radii of NSs in LMXBs would be homogeneous if they accreted the comparable masses; (iv) A local strong field may exist in the equatorial zone because of the field lines dragged from the poles to equator, and the observation in the millisecond pulsar PSR 1821-14 (with its polar field $\sim 4.5 \times 10^9$ G) by Becker and his collaborators [28] seems to support this possibility, because from the X-ray spectral feature of PSR 1821-14 with the assumed electron cyclotron line a high field strength of $\sim 3 \times 10^{11}$ G is inferred; (v) The occurrence of the bottom field has nothing to do with the initial field, and this conclusion is consistent with the scenario of the field-period diagram of pulsars that most MSPs enter into the narrow distribution at the field ranges of 10^{8-9} G while the field distribution of the isolated or non-recycled pulsars spans 10^{11-14} G [29]. Moreover, the homogeneous distributions of kHz QPOs of Atoll and Z sources [30, 31] hints similar magnetosphere radii for both types of sources, and the same magnetosphere scale implies the magnetic field to be proportionally related to the accretion rate. In addition, the conception of bottom field will help us to understand the parallel line phenomena in the plot of kHz QPO frequency vs. the account rate across sources (Atoll to Z) [30], where the homogeneous kHz QPO distribution confronts the luminosity variations of about two magnitude orders across the sources, and the correlation of the kHz QPOs vs. the accretion rate after dividing by the field strength would present a unified picture. Furthermore, as a model prediction, the existence of the low magnetic field ($\sim 3 \times 10^7$ G) neutron stars or millisecond pulsars is possible if the long-term accretion luminosity is about $0.01 L_{Edd}$.

Acknowledgments. Author is very grateful for the helpful discussions with V. Burwitz, G. Hasinger, J. Trümper, W. Becker, U. Geppert, E. Phinney and R. Blandford when he attended the meeting and visited MPE, Garching. Many criticisms and comments from A. Melatos are highly appreciated.

References

1. D. Bhattacharya, E.P.J. van den Heuvel: Phys. Rep. **203**, 1 (1991)
2. E.S. Phinney, S.R. Kulkarni: Ann. Rev. A&A **32**, 591 (1994)
3. J.E. Truemper, W. Pietsch, C. Reppin et al: ApJ **219**, L105 (1978)
4. G. Hasinger, M. van der Klis: A&A, **225**, 79 (1989)
5. G. Hasinger: Reviews of Modern Astronomy, **3**, 60 (1990)
6. A. Melatos, E. S. Phinney: Pub. of Astron. Soc. of Australia **18**, 421 (2001)
7. R. Blandford, J. Applegate, L. Hernquist: MNRAS **204**, 1025 (1983)
8. U. Geppert, V. Urpin: MNRAS **271**, 490 (1994)
9. V. Urpin, U. Geppert: MNRAS **275**, 1117 (1995)
10. M. Ruderman: ApJ **366**, 261 (1991)
11. M. Ruderman, T. Zhu, K. Chen: ApJ **492**, 267 (1998)
12. R.E. Taam, E.P.J. van den Heuvel: ApJ **305**, 235 (1986)
13. N. Shibazaki, et al.: Nature, **342**, 656 (1989)
14. M. Burgay, N. D'Amico, A. Possenti, et al: Nature, **426**, 531 (2003)
15. A.G. Lyne, M. Burgay, M. Kramer, et al: Science, **303**, 1089 (2004)
16. E.P.J. van den Heuvel: Science, **303**, 1143 (2004)
17. E.P.J. van den Heuvel, O. Bitzaraki: A&A, **297**, L41 (1995)
18. L. Burderi, N. D'Amico: ApJ **490**, 343 (1997)
19. L. Burderi, A. King, G. Wynn: MNRAS **283**, L63 (1996)
20. D. Payne, A. Melatos: Mont. Not. Roy. Astron. Soc. **351**, 569 (2004)
21. R. Lovelace, M. et al.: ApJ **625**, 957 (2005)
22. S. Konar, A. Choudhury: MNRAS **348**, 661 (2004)
23. C.M. Zhang, Y. Kojima: MNRAS **366**, 137-143 (2006)
24. K.S. Cheng, C.M. Zhang: A&A **337**, 441 1998
25. E.F. Brown, L. Bildsten: ApJ **496**, 915 (1998)
26. G.W. Romani: Nature, **347**, 741 (1990)
27. N.E. White, W. Zhang: ApJ **490**, L87 (1997)
28. W. Becker, D. Swartz, G. Pavlov, et al: ApJ **594**, 798 (2003)
29. N. Vranesevic, et al: ApJ **617**, L139 (2004)
30. M. van der Klis: Ann. Rev. A&A **38**, 717 (2000)
31. C.M. Zhang, et al: MNRAS **366**, 1373-1377 (2006)

Part VIII

Conference Summary

Einstein's Legacy: a Summary

R. Blandford

KIPAC, Stanford University, USA

Albert Einstein 1905

I. **Special Relativity**
- *2 papers*
- *->GR->Cosmology, Holes, Lenses, Grav. Waves*
- *Fearless Deduction*

II. **Sizes of Atoms and Molecules**
- *3 papers*
- *-> Kinetic theory, Fluctuation-Dissipation*
- *Observing the Invisible*

III. **Photoelectric Effect**
- *1 paper*
- *->Photons, atomic physics, QM, NP, EP*
- *Connect Radical Idea to Experiment*

11 xi 05   Einstein's Legacy, Munich   1

I. Special Relativity

· *Postulates*
- *Principle of Relativity*
- *Speed of light independent of source*

· *Kinematics*
- *LT, Doppler, Aberration, Maxwell*

· *Dynamics*
- *E=mc², E/v invariant*

Fearless Deduction

11 xi 05   Einstein's Legacy, Munich   2

Superluminal Jets

· **Aberration, Doppler effect...**
· **AGN jets**   *Marshall, Zensus*
- *VLBI studies of jet structure, composition*
- *GLAST/HESS*
· **Galactic superluminals**
- *Jet disk connection*
- *Episodic vs steady jets*   *Mirabel, Belloni*
- *Extended sources*
- *TeV jet*
· **Emission from disk or jet in SgrA***

Falcke

11 xi 05   Einstein's Legacy, Munich   3

Gamma Ray Bursts

· **Application of SR principles to 19ᶜ physics**
- *Fluid dynamics*
- *Thermodynamics*
- *Kinetic theory*
- *Electromagnetism*
- *Optics*
· **Hyperluminal "jets" Γ ~ 300?**
· **Burst-afterglow observations of long bursts**
· **Flares, no sign of reverse shock**
- *Electromagnetic model?*

Gehrels, Amati Nakazato

11 xi 05   Einstein's Legacy, Munich   4

Short GRB Summary

Name	Redshift	Detections	Host	E_{iso}(15-350keV)
050509B	0.225	XT	Elliptical	1.1×10^{48} erg
050709	0.161	XT, OT	SF galaxy	6.0×10^{49}
050724	0.258	XT, OT, RT	Elliptical	3.0×10^{50}
050813	≥0.722	XT	? (cluster)	1.7×10^{50}
050925	short-soft	no afterglow		
051105A	short-hard	no afterglow		

Gehrels, Sprust, Phinney

Conclusions:
- Short GRBs have >10^3 lower E_{iso} than long GRBs
- Association with ellipticals argues against collapsar origin
- Observations support a NS-NS merger model
- Just the beginning of the study of short GRBs
- Advanced LIGO (~2013) sensitive to mergers at z~0.1

11 xi 05   Einstein's Legacy, Munich   5

General Relativity

· **General Relativity (Einstein 1915)**
- *Singular "simple" theory of classical gravity*
- *G=8πT*
- *Many, more elaborate alternatives*
  · *Scalar tensor, bimetric, extra dimensions, PPN...*
· **Experimental Program**
- *Classical tests*
  · Redshift, Mercury. Light deflection
- *Modern tests*
  · Shapiro delay, gravitational radiation, EP, inverse square law...

GR/AE vindicated at level from 10^{-2} to 10^{-4}!

11 xi 05   Einstein's Legacy, Munich   6

Double Pulsar

- Special relativity
- 3 classical tests
- Shapiro delay
- Grav. radiation
- 0.001 test
- Spin orbit EOS

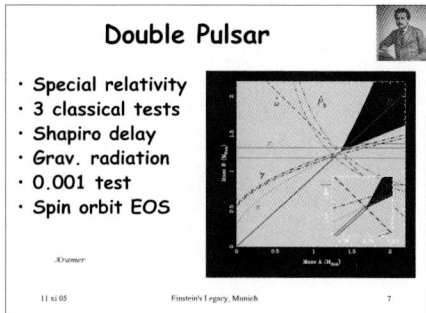

Kramer

11 xi 05   Einstein's Legacy, Munich   7

X-ray Spectroscopy

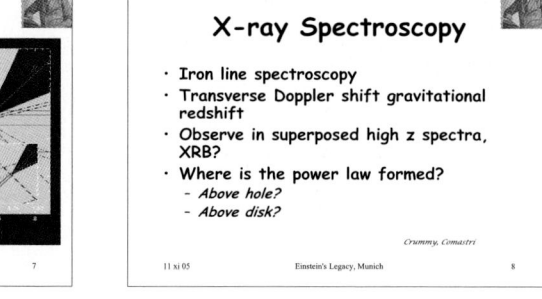

- Iron line spectroscopy
- Transverse Doppler shift gravitational redshift
- Observe in superposed high z spectra, XRB?
- Where is the power law formed?
  - *Above hole?*
  - *Above disk?*

Crummy, Comastri

11 xi 05   Einstein's Legacy, Munich   8

Cosmology

- Einstein 1916
  - *$G + \Lambda g = 8\pi T$ - Cosmological Constant*
    - Vacuum energy: $P = -\rho$.

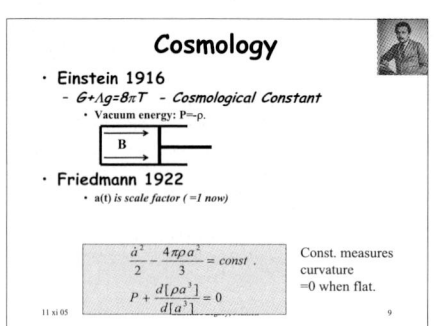

- Friedmann 1922
  - *a(t) is scale factor (=1 now)*

$$\frac{\dot{a}^2}{2} - \frac{4\pi\rho a^2}{3} = const.$$

$$P + \frac{d[\rho a^3]}{d[a^3]} = 0$$

Const. measures curvature =0 when flat.

11 xi 05   9

Λ taken seriously

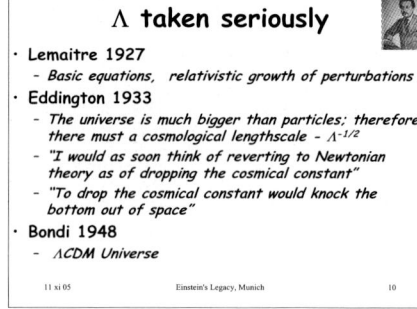

- Lemaitre 1927
  - *Basic equations,   relativistic growth of perturbations*
- Eddington 1933
  - *The universe is much bigger than particles; therefore there must a cosmological lengthscale - $\Lambda^{-1/2}$*
  - *"I would as soon think of reverting to Newtonian theory as of dropping the cosmical constant"*
  - *"To drop the cosmical constant would knock the bottom out of space"*
- Bondi 1948
  - *ΛCDM Universe*

11 xi 05   Einstein's Legacy, Munich   10

Simple World Models

- Λ only
  - *ρ const, a ~ e^t*
  - *De Sitter Universe*
- Matter only
  - *ρ ~ a^{-3} , a ~ $t^{2/3}$*
  - *Einstein - De Sitter Universe*
  - *Deceleration*
- Matter plus Λ
  - *Singular "simple" theory*
  - *a ~ $(\sinh t)^{2/3}$*
  - *ΛCDM universe*
  - *Deceleration -> acceleration*

$$j \equiv \frac{\ddot{a} a^2}{\dot{a}^3} = 1$$

Mattarese, Martini

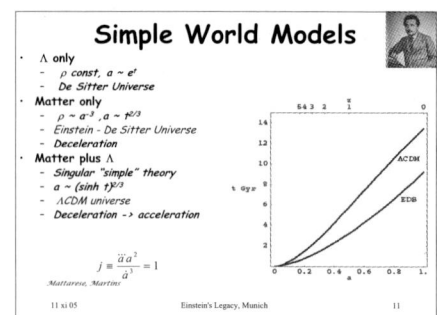

11 xi 05   Einstein's Legacy, Munich   11

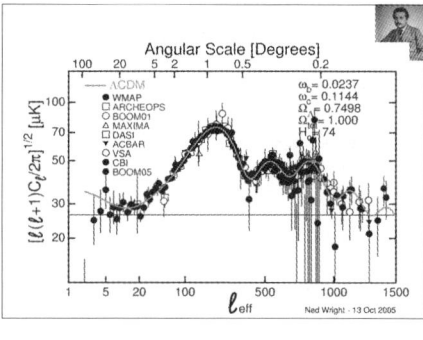

Lay of the Low *l* Land

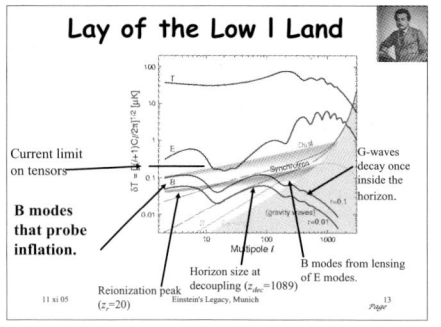

Current limit on tensors

B modes that probe inflation.

G-waves decay once inside the horizon.

Horizon size at decoupling (z_{dec}=1089)

Reionization peak (z_r=20)

B modes from lensing of E modes.

11 xi 05   Einstein's Legacy, Munich   13
Page

Observing Dark Energy

- ΛCDM is default model; a~$\sinh^{2/3}$t
- No preferred alternative
- Use d(0.00092), d'(1)=H_0^{-1}
- Kinematics
  - Distances
    - SNIa *Hillenbrand*
    - Baryon Oscillations
      - d back to recombination
  - Volumes
  - Δm(EDS;0.5))=0.6
  - Linear deviations for curvature etc

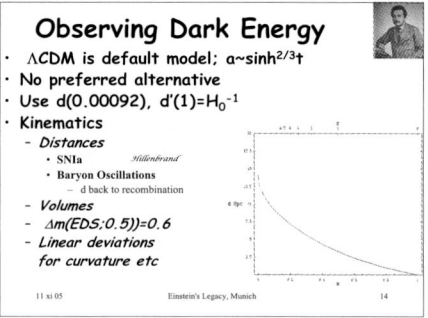

11 xi 05   Einstein's Legacy, Munich   14

Observing Dark Energy
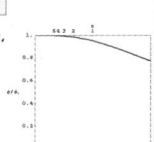

- Dynamics using H(a), q(a)
- Observe potential fluctuations ϕ at recombination
- Predicted linear growth from

$$\ddot{\phi} + 4H\dot{\phi} + H^2(1-2q)\phi = 4\pi\delta P$$
$$[\ddot{\phi} + \tfrac{5}{3}\coth t\dot{\phi} + \tfrac{4}{3}\phi = 0 \;\; for \;\; \Lambda CDM\,]$$

- Measure growth using weak lensing, cluster counts, ISW etc
- Trying to measure source term
- Will need major effort cf CMB

11 xi 05   Einstein's Legacy, Munich

Clusters of Galaxies

- Use rich clusters to probe dark energy

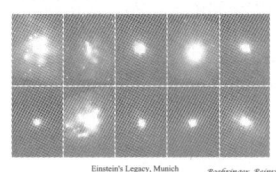

11 xi 05   Einstein's Legacy, Munich   *Boehringer, Reiprich*   16

Clusters are Complex

- Where does the cooling gas go?
- Is there enough binding energy in holes to heat gas for a Hubble time?
- Dissemination and dissipation
  - *Internal waves*
  - *Magnetic viscosity*
- Why are weak/strong lensing, difficult to reconcile?
- How much gas accretes onto black hole?

11 xi 05   *Churazov, Fabian, Forman, Schindler, Kausch*
Einstein's Legacy, Munich   17

Importance of Cosmic Rays

- Cluster gas has high entropy
- Acquired at high Mach number shocks
- Shocks accelerate ionic cosmic rays with partial pressure ~ 0.2-0.3 of total
- Cosmic ray pressure supports gas after it cools by ~1/3
- Inhibits compression and infall

11 xi 05   Einstein's Legacy, Munich   18

Non-radiative cooling

- Warm gas permeated by small quantity of cold filaments
- Size < electron mfp
- Cool off warm gas
- Radiate energy in UV
- Cold phase settles onto galaxy
  - *Star formation and black hole accretion*
- Feedback

11 xi 05   Einstein's Legacy, Munich   19

Gravitational Radiation

- Einstein's quadrupole formula correct
- Stellar source rate estimates higher
- Are short bursts coalescing NS?
- LIGO, GEO600 about to collect science data
  - *Stay tuned!*
- LISA
  - *Binary black holes, WD noise EMRB, recoil*
  - *EM channel*

Phinney, Kocsis, Sesana, Gopakamar

11 xi 05   Einstein's Legacy, Munich   20

Gravitational Lenses

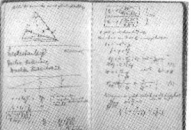

- Einstein's notebook
- 1911
- Microlensing formula missing factor 2
- M31 MACHOs
- Self-lensing
- Not dark matter

Seitz, Misztajn

11 xi 05   Einstein's Legacy, Munich   21

II. Existence of Atoms and Molecules

- Osmotic Pressure
- Viscosity
- Brownian Motion
- Markov Process
- Avogadro's Number

Observing the Invisible

11 xi 05   Einstein's Legacy, Munich   22

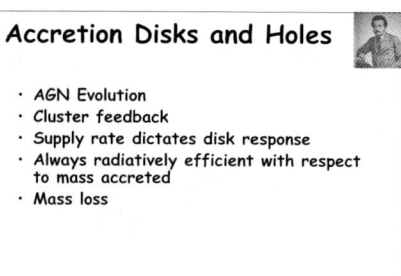

Accretion Disks and Holes

- AGN Evolution
- Cluster feedback
- Supply rate dictates disk response
- Always radiatively efficient with respect to mass accreted
- Mass loss

11 xi 05   Einstein's Legacy, Munich   23

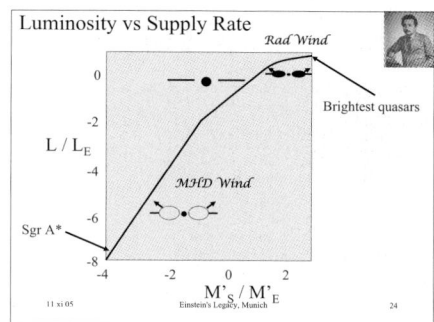

Luminosity vs Supply Rate

Rad Wind

Brightest quasars

L / L_E

MHD Wind

Sgr A*

M'_S / M'_E

11 xi 05   Einstein's Legacy, Munich   24

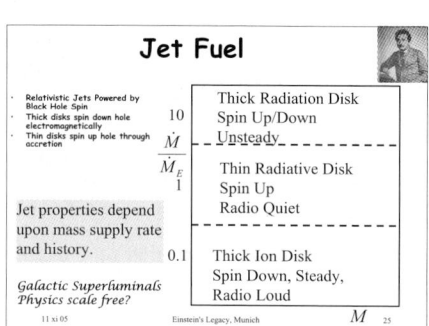

Jet Fuel

- Relativistic Jets Powered by Black Hole Spin
- Thick disks spin down hole electromagnetically
- Thin disks spin up hole through accretion

Jet properties depend upon mass supply rate and history.

Galactic Superluminals
Physics scale free?

10

$\dot{M}$

$\dot{M}_E$

1

0.1

Thick Radiation Disk
Spin Up/Down
Unsteady

Thin Radiative Disk
Spin Up
Radio Quiet

Thick Ion Disk
Spin Down, Steady,
Radio Loud

11 xi 05   Einstein's Legacy, Munich M 25

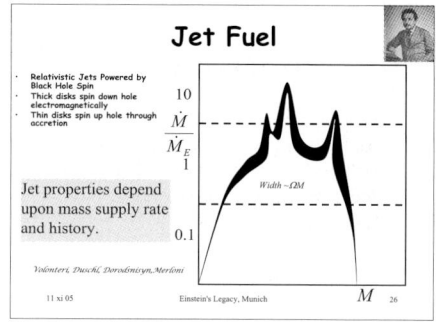

Jet Fuel

- Relativistic Jets Powered by Black Hole Spin
- Thick disks spin down hole electromagnetically
- Thin disks spin up hole through accretion

Jet properties depend upon mass supply rate and history.

Volonteri, Duschl, Dorodnitsyn, Merloni

10

$\dot{M}$

$\dot{M}_E$

1

0.1

Width ~ΩM

11 xi 05   Einstein's Legacy, Munich M 26

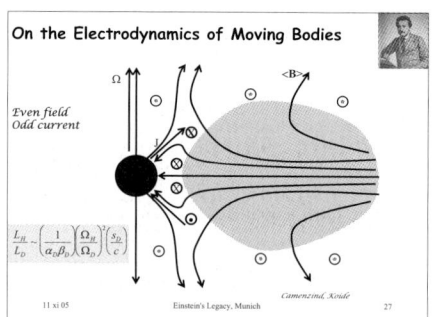

On the Electrodynamics of Moving Bodies

Ω

<B>

Even field
Odd current

$$\frac{L_H}{L_D} \sim \left(\frac{1}{a_D \beta_D}\right)\left(\frac{\Omega_H}{\Omega_D}\right)^2 \left(\frac{s_D}{c}\right)$$

Camenzind, Koide

11 xi 05   Einstein's Legacy, Munich   27

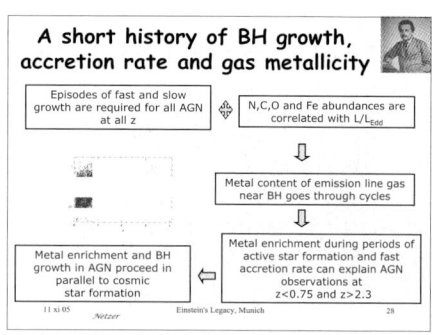

A short history of BH growth, accretion rate and gas metallicity

Episodes of fast and slow growth are required for all AGN at all z

N,C,O and Fe abundances are correlated with L/L_{Edd}

Metal content of emission line gas near BH goes through cycles

Metal enrichment and BH growth in AGN proceed in parallel to cosmic star formation

Metal enrichment during periods of active star formation and fast accretion rate can explain AGN observations at z<0.75 and z>2.3

11 xi 05   Netzer   Einstein's Legacy, Munich   28

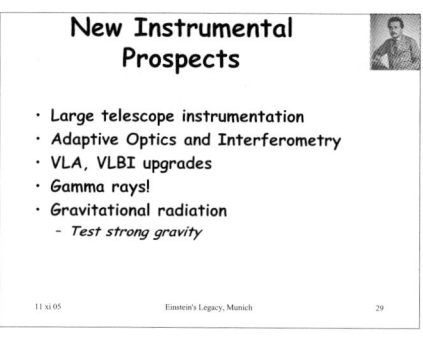

New Instrumental Prospects

- Large telescope instrumentation
- Adaptive Optics and Interferometry
- VLA, VLBI upgrades
- Gamma rays!
- Gravitational radiation
  - Test strong gravity

11 xi 05   Einstein's Legacy, Munich   29

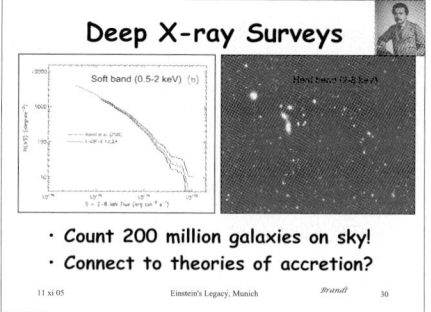

Deep X-ray Surveys

Soft band (0.5-2 keV)

- Count 200 million galaxies on sky!
- Connect to theories of accretion?

11 xi 05   Einstein's Legacy, Munich   Brandt   30

340 GHz Sgr A* Polarization
(Marrone, Moran, Zhao, & Rao, ApJ, submitted)

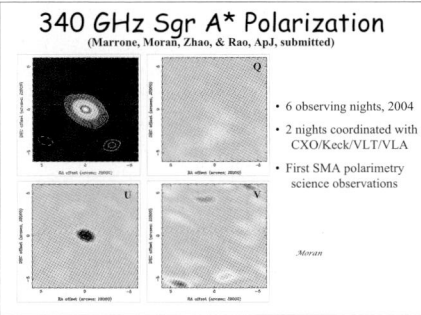

- 6 observing nights, 2004
- 2 nights coordinated with CXO/Keck/VLT/VLA
- First SMA polarimetry science observations

Moran

Power of Suzaku   preliminary
Suzaku Team

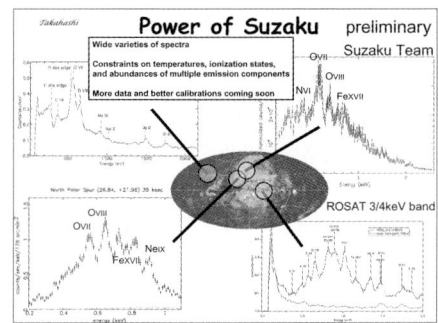

Takahashi

Wide varieties of spectra

Constraints on temperatures, ionization states, and abundances of multiple emission components

More data and better calibrations coming soon

ROSAT 3/4keV band

III. Photons

- Planck Law fitted data
- Oscillators only exchange energy with radiation in quanta, E=hν
- Radiation comprises photons
- Explain photoelectric effect

Connect Radical Idea to Experiment

11 xi 05   Einstein's Legacy, Munich   33

Galactic Center Stars

- 3-4 million solar mass hole
  - *orbits*
- ~100 young stars (only) formed ~6Myr ago in two disk
- Remnants??
  - *Resonances?*
- How do they get there?
- Also M31

Genzel, Cuadra, Bender

11 xi 05   Einstein's Legacy, Munich   34

S2 Orbit (Genzel et al.)

11 xi 05   Einstein's Legacy, Munich   35

The paradox increases: young massive stars in the central light month

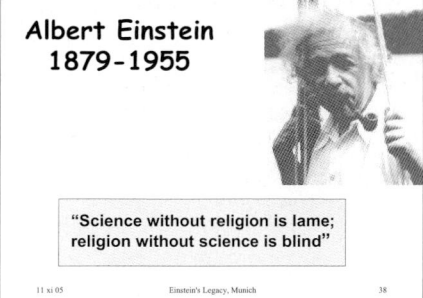

>90% of all K<16 stars in the central R=0.5" are young main sequence B stars

SINFONI 2005
70 mas resolution

○ late type stars
● early type stars

11 xi 05   Einstein's Legacy, Munich   36

Eisenhauer et al. 2005, ApJ 628, 246, Paumard et al. 2005, in prep.

Quasi-periodic Oscillations

- Rich phenomenology
- 3:2 resonance, split, const, ~ ν_R, $\nu_R/2$
- Like atomic spectroscopy a century ago
- General worry
  - *Mostly seen as HX phenomenon.*
  - *Disks are SX objects*
- QP magnetospheric cycling
  - *Modulate current*
  - *soft photons at footpoints, hot electrons in magnetosphere*
- Galactic Center IR observations

Rebusco, Eckart

11 xi 05   Einstein's Legacy, Munich   37

Albert Einstein
1879-1955

"Science without religion is lame; religion without science is blind"

11 xi 05   Einstein's Legacy, Munich   38

Author Index

ESO ASTROPHYSICS SYMPOSIA
European Southern Observatory

Series Editor: Bruno Leibundgut

Printing: Krips bv, Meppel, The Netherlands
Binding: Stürtz, Würzburg, Germany